T0192290

14 MeV Neutrons

14 MeV Neutrons

Physics and Applications

Vladivoj Valković

CRC Press is an imprint of the
Taylor & Francis Group, an **informa** business

CRC Press
Taylor & Francis Group
6000 Broken Sound Parkway NW, Suite 300
Boca Raton, FL 33487-2742

First issued in paperback 2021

ISBN 13: 978-0-367-78342-6 (pbk)
ISBN 13: 978-1-4822-3800-6 (hbk)

Visit the Taylor & Francis Web site at
http://www.taylorandfrancis.com

and the CRC Press Web site at
http://www.crcpress.com

Dedication

To my wife Georgia (Đurđa in Croatian) who accompanied me on this long and exciting journey.

Contents

Preface

I have been with and around 14 MeV neutrons for more than half a century, since 1961. During this period, I witnessed many exciting developments, both in fundamental nuclear physics and in applications of fast neutrons to many practical problems, the most challenging being inspection of large volumes in complex environments for the presence of threat materials.

Doing experiments with 14 MeV neutrons is often a difficult, time-consuming, and frustrating job. It is much more convenient, faster, and less painful to do experiments with charged particles; however, neutrons can offer a special insight into nucleus and other materials because of the absence of charge.

For obvious reasons, the book covers, in more detail, subjects that were of research interest to me, which were described in publications where I was author or coauthor. These specific topics include

- Studies of nuclear reactions with 14 MeV neutrons, including measurements of energy spectra, angular distributions, and deductions of reaction mechanism
- Studies of nuclear reactions with three particles in the final state induced by either neutrons or protons; identification of effects of final state interaction, quasi-free scattering, rescattering, and charge dependence of nuclear forces, on low and medium bombarding energies
- Development of charged particle and neutron detection methods, in particular position-sensitive detectors
- Development and application of analytical methods based on the detection of induced x-ray and gamma ray emission; use of focused beams for the measurement of concentration profiles and maps
- Study of the role of chemical elements in biological systems by using nuclear analytical techniques
- Study of environmental pollution by radionuclides and toxic elements, identification of pathways, and evaluation of impact
- Industrial applications of nuclear analytical methods, especially in the metallurgy and coal industry

- Quality assurance and quality control measures for nuclear analytical methods
- Development of nuclear and atomic physics–based technology for combating illicit trafficking and terrorism

The experiments reported here are mainly done at the Experimental Physics Department (which changed its name several times during the period covered here) of the Institute Ruder Boskovic, Zagreb, Croatia. The recent experiments (last ten years) I have done have been with a small group of collaborators: Dr. Davorin Sudac, Mr. Karlo Nad, and Dr. Jasmina Obhodas, who worked many hours in data collection and interpretation. In addition, all the graphic work presented in the book was created by Dr. Jasmina Obhodas.

Without their effort, this book would not have been possible, and I am very thankful to them all.

Vladivoj (Vlado) Valković
Zagreb, Croatia

Author

Dr. Vladivoj (Vlado) Valković, retired professor of physics, is a fellow of the American Physical Society and Institute of Physics (London). He earned his PhD in physics in 1964, with the thesis "Nuclear reactions with 14.4 MeV neutrons."

Dr. Valković authored 13 books and more than 380 scientific and technical papers. As of April 9, 2014, Google Scholar showed 472 entries for V. Valkovic and 3784 citations with an h-index of 28. He has published his works in many journals, including *Nuclear Physics, Physics Review, Nuclear Instruments and Methods, Physical Review Letters, Physics Letters, Nature, Contemporary Physics, Lettere al Nuovo Cimento, International Journal of Applied Radiation and Isotopes, Periodicum Biologorum, Environmental Science & Technology, Origins of Life, Fizika, Spectrochimica Acta, Fuel, Journal of Radioanalytical and Nuclear Chemistry, Science of the Total Environment, X-Ray Spectrometry, Journal of Biological Trace Element Research, Analytica Chimica Acta, Journal of Trace and Microprobe Techniques*, and others in the research areas of nuclear physics, instrumentation application of nuclear techniques to problems in biology, medicine, environmental research, and trace element analysis.

Dr. Valković has worked as a professor of physics at Rice University, Houston, Texas; at IAEA, Vienna, Austria, as a head of the Physics-Chemistry-Instrumentation Laboratory; and at the Institute Ruder Boskovic, Zagreb, Croatia, as laboratory head and scientific advisor. He coordinated many national and international projects, including U.S.-Croatia bilateral, NATO, IAEA, EU-FP5, FP6, and FP7 projects. He was the founder and the director of the company A.C.T.d.o.o., Zagreb, implementing environmental security projects.

Dr. Valković has a lifelong experience in the study of nuclear reactions induced by 14 MeV neutrons and their analytical applications. Having experience in projects dealing with the subjects of port security, container inspections, and detection of threat materials, and keeping in mind the limited space in container terminals, he has developed the idea of autonomous ship container inspection, which has been recently prepared as a project proposal, ASCIS, with him being a scientific coordinator of the project.

Chapter 1

Introduction

1.1 HISTORY: FROM DISCOVERY TO PRESENT

In the year 1923, Rutherford had discovered the proton but, in his subsequent experiments, the atomic number was consistently less than the atomic mass. This could not be due to the number of electrons, so Rutherford hypothesized that there was another type of particle in the atomic nucleus—one with mass but no charge. In another development in 1925, physicists proposed the idea of nuclear spin to explain the Zeeman effect (shifts in atomic energy levels in a magnetic field), but this did not seem to fit the prevailing model for the atomic nucleus, believed to contain just protons and electrons. The additional hint of the neutron's existence came in 1930, when Walther Bothe and H. Becker found that when alpha radiation fell on elements like lithium and boron a new form of radiation was emitted. Initially, this radiation was believed to be a type of gamma radiation, but it was more penetrating than any known gamma radiation. Work by Irene Joliot-Curie and Frederic Joliot in 1932, though not disproving the gamma radiation hypothesis, did not particularly support it either.

Chadwick replicated one German experiment in which radiation from polonium struck a beryllium target, and the unusual form of radiation produced provided evidence of a new kind of chargeless particle. Encouraged, Chadwick devoted all his energy to further experiments, often working through the night, and his efforts paid off within weeks. In February 1932, he submitted a letter to the journal *Nature* detailing his experimental results as evidence for the existence of a neutron. He followed with a second paper in May of the same year, providing more of the technical details (Chadwick 1932a,b). The discovery was quickly championed by Niels Bohr and Werner Heisenberg, and Chadwick worked with Maurice Goldhaber to measure the mass of the neutron and concluded that neutron is a nuclear particle rather than a proton–electron pair.

Chadwick won the Nobel Prize for this discovery in 1935 just after he left the Cavendish lab, run by Rutherford, and moved to less prestigious University of Liverpool, where he spent most of his prize money to build the cyclotron

Rutherford opposed in Cavendish lab. Chadwick's discovery made it possible to create elements heavier than uranium in the laboratory via capture of slow neutrons and beta decay—a critical breakthrough for the eventual development of the nuclear bomb. The neutron proved to be an ideal "bullet" for penetrating nuclei, thanks to its lack of charge.

For more reading on Chadwick's life and discoveries, see American Institute of Physics Oral History transcript (Weiner 1969).

A neutron is one of the fundamental particles that make up matter. This uncharged particle exists in the nucleus of a typical atom, along with its positively charged counterpart, the proton. Protons and neutrons each have about the same mass, and both can exist as free particles away from the nucleus. In the universe, neutrons are abundant, making up more than half of all visible matter.

Neutrons have many properties that make them ideal for certain types of research. Because of their unique sensitivity to hydrogen, neutrons can be used to precisely locate hydrogen atoms, enabling a more accurate determination of molecular structure, which is important for the design of new therapeutic drugs. Neutrons scattered from hydrogen in water can locate bits of moisture in fighter jet wings—signs of microscopic cracking and early corrosion that pinpoint the part of the wing that should be replaced.

Besides hydrogen, neutrons can locate other light atoms among heavy atoms. This capability is helping physicists to open the field of quantum superstates, such as superconductivity and superfluidity. Researchers are making measurements of the atomic momentum distribution of liquid ^{4}He and are looking for a ^{4}He supersolid state. They have determined the critical positions of light oxygen atoms in promising high-temperature, superconducting materials.

Because the energies of thermal neutrons almost match the energies of atoms in motion, neutrons can be used to track molecular vibrations and movements of atoms of a protein during catalytic reactions. Recent studies with neutrons have revealed the earliest structural formation of the disease type of the protein huntingtin, and that research is moving forward to study protein malformation responsible for Alzheimer's and Parkinson's diseases.

The neutron is composed of three quarks, one upquark (u) and two downquarks (d). The mass of the neutron is slightly larger than that of the proton. The neutron does not exist long outside of the atomic nucleus, only some 885 s (about 15 min) on average. A neutron (udd) decays to a proton (uud), an electron, and an antineutrino. This is called neutron beta decay. (The term "beta ray" was used for electrons in nuclear decays because they did not know they were electrons!) Neutron beta decay can be represented as

$$n \rightarrow p e^- \bar{\upsilon}.$$

The fundamental properties of neutron are mass: $1.67492729 \times 10^{-27}$ kg; charge: 0; and spin: 1/2; because the neutron has a 1/2 spin, it is a fermion.

The mass of the neutron can be measured using thermal neutron capture reaction:

$$p + n_{th} \rightarrow d + E_{\gamma} + E_{recoil}.$$

Following neutron capture (in hydrogen) a deuteron, d, is formed and the excess in binding energy is released by the emission of a γ-ray that causes recoil of deuteron with energy of

$$E_{recoil} = \frac{E_{\gamma}^2}{2m_d}.$$

The rest mass of both the proton and the deuteron was determined directly by charged particle mass spectrometry; the gamma-ray energy was determined with high-precision gamma-ray spectroscopy, for details see Börner and Gönnenwein (2012). The mass of the neutron was obtained to be 939.565360(81) MeV/c^2 and exceeds that of the proton by 1.2933317(5) MeV.

1.2 14 MeV NEUTRONS AND NUCLEAR WEAPONS

Any nuclear reaction in which there is a net decrease of mass, that is, in which the total mass of the products is less than that of the interacting nuclei or nuclear particles, will be accompanied by liberation of energy. The amount of energy released E (in J) is related to the net decrease in mass m (in kg) by the Einstein equation

$$E = mc^2$$

where c is the velocity of light, $c = 3.00 \times 10^8$ m s^{-1}. In the study of nuclear reactions, it is common practice to state energies on electron volt (eV) or million electron volt (MeV) units, 1 MeV being equivalent to 1.60×10^{-13} J. Nuclear masses are generally expressed in atomic mass units (amu), these being masses on a scale in which the mass of the common isotope of carbon, ^{12}C, is taken to be precisely 12.0000. In terms of familiar mass units, 1 amu = 1.66×10^{-27} kg. Upon making the appropriate substitutions in the given equation, it follows

$$E(\text{MeV}) = 931 \times m\ (\text{amu})$$

meaning that the energy in MeV accompanying a nuclear reaction is equal to the decrease in mass in amu multiplied by a factor of 931.

Two types of nuclear reactions, in which there is a decrease in mass, are used for the large-scale release of energy in weapons. These are as follows:

1. Fission, that is, the splitting of a heavy nucleus into a pair of lighter ones
2. Fusion, that is, the combination of two very light nuclei to form a somewhat heavier one

The underlying reason why these processes are accompanied by a liberation of energy (and decrease in mass) is that in each case the total energy of attraction (or binding energy) among the constituent protons and neutrons, that is, the nucleons, is smaller in the initial nucleus (or nuclei) than it is in the products of the reaction. It is a fundamental law of nature that the rearrangement of a system from a weakly bound state to a more tightly bound state must be accompanied by a release of energy. In both processes, neutrons play a key role.

The overall design of a nuclear weapon is a complex operation involving a cooperative effort among mathematicians, scientists, and engineers in several areas of specialization. In spite of the complexity of these weapons, their design and operation are based on certain fundamental physical principles. A description of these principles in fairly elementary terms was provided in several reports originating from Los Alamos Laboratory. In them, the mathematical aspects of weapon design have been omitted or oversimplified, since the emphasis is placed on the physical phenomena rather than on the mathematical techniques used in creating them. The available ones have passed classification levels from being secret to confidential to restricted, distribution limited to finally unclassified.

Among several authors we should mention Samuel T. Glasstone, who wrote with others the unclassified *The Effects of Nuclear Weapons* from 1950 to 1977 (Glasstone 1950; Glasstone and Dollan 1977), and who also authored several other books for the U.S. Military, which remain tightly classified, even to this day. From May 1999 to December 2002, a Department of Energy declassification review was conducted on one of his secret works: *An Introduction to Nuclear Weapons*, which was published in 1962, 1963, and 1972. These reviews kept most parts of the books classified. Here, we list the editions that were reviewed and were allowed to be released in a sanitized form.

In a document (Glasstone 1962) that was written for the AEC's Division of Technical Information, Chapters 3 through 8 were withheld in their entirety, along with the index. Significant bits of information in what was released were sanitized, like the minimum critical masses of reflected spheres.

However, one can learn the following:

- As a result of spontaneous fission, ^{235}U emits an ordinary about 0.85 neutrons/kg/s, while ^{238}U emits 17 neutrons/kg/s. So, the neutron background of ordinary oralloy (93.5% weight ^{235}U) is about

2 neutrons kg/s. Oralloy is an acronym for "oak ridge alloy," which is an alloy of ^{235}U and ^{238}U. The actual concentration is classified, but generally ^{235}U is greater than 90%.
- The spontaneous fission rate of plutonium-240 is about 440 fissions/s/g, and since 3 neutrons are liberated per fission, this means that 1 g of ^{240}Pu emits well over 1000+ neutrons/s.
- In modern boosted implosion-type fission weapons, the cores are hollow subcritical shells that contain deuterium–tritium boosting gas.

In both documents (Glasstone and Redman 1972) written for the AEC's Division of Military Application, Chapters 3 through 9 were withheld in their entirety, along with the index. The title of Chapter 7 was obliterated on the table of contents as well. Again, some of information in what was released was deleted, like the minimum critical masses of reflected spheres.

In a fission weapon, the chain reaction is initiated by the introduction of neutrons into a critical or supercritical system. One of the simplest procedures for obtaining neutrons is by the action of alpha particles on certain light elements, notably beryllium; processes of this kind are referred to as (α,n) reactions. The reaction is represented as

$$^{4}He + {}^{9}Be \rightarrow {}^{12}C + n.$$

A convenient source of alpha particles, which was used extensively at one time in fission weapons, is the radionuclide ^{210}Po. It has a moderately short half-life of 138.4 days, meaning that it emits alpha particles rapidly, and a small quantity of this radionuclide can thus provide a strong neutron source in conjunction with beryllium. But the short half-life is also a serious disadvantage, because the activity falls off relatively rapidly. In 1 year, the alpha activity and hence the rate of neutron production will have decreased to 10% and in 2 years to 1% of its initial value.

Another possibility is to use d + t neutron generators as firing and neutron initiation systems for the nuclear weapons. It can be found in open-source literature that this is done in the All-Russia Research Institute of Automatics (Russian acronym—VNIIA), one of the three nuclear weapons designers in Russia. VNIIA was founded in 1954 by the initiative of academician Yuli B. Khariton, who was the immediate manager of the development of the first atomic charges and bombs in the Soviet Union. Currently, VNIIA is part of the nuclear weapons complex of the State Corporation for Atomic Energy "Rosatom."

VNIIA is the only designer of the firing and neutron initiation systems for the nuclear charges in Russia. This system is meant for initiating the charge and has a number of functions that are as follows:

- Firing the highly explosive component, which gives the energy sufficient to turn the fissile material of the nuclear charge into the condition, when the chain reaction can be started.

- Initiating the chain fission reaction by irradiating the material with the pulsed neutron flux.

The system is the most critical and the most complicated of all the nonnuclear components of a nuclear weapon. Its design requires the expertise in a number of the fundamental sciences, the application of various materials, the special elemental base, the maintenance of the numerous technologies and the unique equipment, and high-precision measurements.

VNIIA has designed and prepared the serial production of several generations of the firing and initiation systems. The weight and overall dimensions of the modern firing and initiation systems are over 100 times less compared to the first models, but the functionality, reliability, and lifetime are much better. Taking into account the reduction of the production quantity of these systems in the recent years and the unprofitability of their manufacturing at the existing serial production facilities, VNIIA has turned its pilot production capabilities into the serial production for the firing and initiation systems of the new generations (for details see http://vniia.ru/eng/oboron/index.html).

Fusion bombs use light materials deuterium and tritium as fuels. The process of fusion is safer and cleaner, but it can only be achieved at extremely high temperatures and pressures. This fact is responsible of calling them thermonuclear bombs. All models of fusion bombs so far use fission as part of the process; a pure fusion bomb has not been achieved yet.

There are two types of bombs that use fusion, which are as follows:

1. Boosted fusion bomb
2. Teller–Uman design

From them, only the second uses fusion as the main process and it is the one called thermonuclear. The information about fusion bombs has not been completely declassified, but it is known enough to have a fair idea of how the original model works. The boosted fusion bomb is essentially a fission bomb that uses fusion as part of the igniting process.

In the shell design the double spherical model still applies, but the core is filled by fusible material while the fission fuel is in the surrounding chamber. There is a set of processes that trigger each other in the following:

1. Detonator lenses start the fission process
2. The fission process compresses the core giving it enough temperature and pressure to generate fusion
3. The neutrons expelled in the fusion process bombard the remaining fissile material that did not ignite in the initial reaction

This double chain reaction occurs very fast allowing better performance than a traditional fission bomb. In the basic design, the fusion fuel is tritium and the fission fuel is plutonium. There is a second model of booster that reproduces the same idea but adding more layers of fission–fusion chambers. It is known as "alarm clock" in the United States or "layer cake" in Russia. These bombs are essentially fission weapons because only 1% of the explosion is provided by the fusion part. The inclusion of the fusion process greatly increases the efficiency; almost all present fission weapons include boosting in one way or another. Its efficiency is about 40%.

Teller–Uman hydrogen bombs are also known as "staged thermonuclear weapons" or "hydrogen bombs." The main reaction in this type of bomb is fusion; it uses a fission bomb as detonator. The whole process follows a series of steps or stages to obtain larger explosions.

The bomb consists of a spherical head containing a boost bomb and a cylindrical body with ^{6}Li deuteride as fuel sandwiched between two other cylinders: ^{238}U (outer cylinder) and plutonium (inner cylinder). Both containers are swimming in polystyrene foam (another line of research considers that the foam is not necessary) inside a reflective case. Each part works as a trigger of the next step. The first trigger is the boost bomb in the head. Most of the radiation of boost bombs is in x-rays; they are reflected by the metal-casing walls and forced to converge toward the outer cylinder melting it and transforming the foam into plasma at the same time, and the outer cylinder implodes with huge strength and both fission materials enter in contact and reach critical mass detonating a second fission bomb. This explosion provides the high temperatures and pressure needed to start the fusion process of the deuterium, which transforms into helium and emits neutrons. These neutrons reinforce the fission part of the bomb lifting the temperature even more and starting another wave of fission–fusion.

It should be mentioned that there are numerous texts on web about the bomb, weapon physics, and design (see, e.g., http://www.angelfire.com/mac/nws/design.htm).

1.3 NUCLEAR PHYSICS PROBE

The nuclear reactions induced by 14 MeV, as well as their spectroscopic data, are well known since the 1970s and 1980s and summarized in the book by Csikai (1987). These 14 MeV neutron reaction cross sections have also been compiled into the JEF databases (IAEA 1994a,b) and their updates; they have two fixed points in their cross-section curves, one at the thermal energy and the other at 14 MeV. Even the interfering reactions have been studied thoroughly (Nargolwalla and Przybylowicz 1973). However, a similar research could not have been performed for the reactions with 2.5 MeV neutrons because of the relatively poor performance of the earlier versions of d–d neutron generators. The previously mentioned handbook tried to collect all the available information of the relevant analytical data on 2.5 MeV neutron generators, but their accuracy and reliability are far from enough for analytical needs (IAEA 2012).

Neutron cross-section standards are important in the measurement and evaluation of all other neutron reaction cross sections. Not many cross sections can be defined as absolute—most cross sections are measured relative to the cross-section standards for normalization to absolute values. Previous evaluations of the neutron cross-section standards were completed in 1987 and disseminated as both NEANDC/INDC (Condé 1992) and ENDF/B standards. R-matrix model fits for the light elements and nonmodel least squares fits for the heavy elements were the basis of the combined fits for all of the data. Some important reactions and constants are not standards, but assist greatly in the determination of the standard cross sections and reduce their uncertainties—these data were also included in the combined fits.

IAEA has recognized the need to reevaluate the cross-section standards at the beginning of the twenty-first century; this need is based on the appearance of a significant amount of precise experimental data and developments in the methodology of analysis and evaluation. IAEA has initiated a series of consultants' meetings in order to improve the 1987 standards' evaluation. The evaluations of the neutron cross-section standards were finalized in October 2005. The new evaluations of the cross-section standards also include covariance matrices of the uncertainties that contain fully justifiable values. A final technical report was prepared in 2006 and subsequently published (IAEA 2007). A comprehensive paper with detailed technical description of derived standards and uncertainties was also published (Carlson et al. 2009).

Chapter 5 contains discussion of different types of 14 MeV neutron–induced reactions and presentation of conclusions reached. It turns out that neutrons are a specific probe of nuclear structure and properties of nuclear forces because of its charge neutrality.

1.3.1 Nuclear Data Centers

Both the development and maintenance of nuclear technologies rely on the availability of nuclear data to provide accurate numerical representations of the underlying physical processes. Essential data include energy-dependent reaction probabilities (cross sections), the energy and angular distributions of reaction products for many combinations of target and projectile, and the properties of excited states, and their radioactive decay data. High-quality decay data are an essential input across a wide range of nuclear applications. Well-defined nuclear data are essential to ensure safe procedures within mining operations, various nuclear fuel cycles for energy generation, environmental monitoring, specific analytical techniques, and diagnostic and radiotherapeutic treatments in nuclear medicine. The Nuclear Data Section of the Division of Physical and Chemical Sciences (NAPC) within the IAEA Department of Nuclear Sciences and Applications (NA) is responsible for undertaking agency activities in the dissemination of atomic and nuclear data for applications.

IAEA Nuclear Data Section offers data center services primarily to countries which are not members of Organization for Economic Co-operation and Development, so called non-OECD countries (except Russian Federation and China). However, most products, specifically INDC reports, IAEA-NDS documents, etc., are provided upon request to customers in all countries. IAEA-NDS online services are available at web: http://www-nds.iaea.org. Users in India, China, and neighboring countries may use IAEA-NDS mirror at web address: http://www-nds.indcentre.org.in (India) and http://www-nds.ciae.ac.cn (China). Here, we should also mention a newsletter of the IAEA Nuclear Data Section, which on regular time intervals brings news and information on the activities in this field.

There are a number of national and international nuclear data centers; we shall list the major one together with descriptions of their mission. In the United States, the National Nuclear Data Center (USA NNDC) collects, evaluates, and disseminates nuclear physics data for basic nuclear research and for applied nuclear technologies. The NNDC is a worldwide resource for nuclear data. The information available to the users of NNDC services is the product of the combined efforts of the NNDC and cooperating data centers and other interested groups, both in the United States and worldwide. The NNDC specializes in the following areas: nuclear structure and low-energy nuclear reactions, nuclear databases and information technology, and nuclear data compilation and evaluation.

For services to customers in (OECD/NEA Data Bank) member countries, the OECD Nuclear Energy Agency should be contacted. Customers from countries from the former USSR should contact the Russia Nuclear Data Center at different addresses for neutron data and photonuclear data.

For potential users in these countries the Russian center might be of interest. Neutron data could be obtained from Russia Nuclear Data Center, Centr Jadernykh Dannykh (CJD), Fiziko-Energeticheskij Institut, Ploschad Bondarenko, 249020 Obninsk, Kaluga Region, Russian Federation. Web: http://www.ippe.ru/podr/cjd. Photonuclear data could be obtained from the Centre for Photonuclear Experiments Data, Centr Dannykh Fotoyadernykh Eksperimentov (CDFE), Skobeltsyn Institute of Nuclear Physics, Lomonosov Moscow State University, Leninskie Gory, 119922 Moscow, Russian Federation, web: http://cdfe.sinp.msu.ru.

In China, there is China Nuclear Data Center, China Institute of Atomic Energy, P.O. Box 275(41), Beijing 102413, China.

The computer codes of U.S. origin are available to all countries from the Radiation Safety Information Computational Center (RSTCC), Oak Ridge National Laboratory, Oak Ridge, TN. While the computer codes from non-U.S. origin are available to all countries from NEA Data Bank in France.

Let us mention some useful documents. The best known in the neutron physics field is the Computer Index of Nuclear Reaction Data, CINDA. CINDA, see for example (CINDA 2014) contains bibliographic references to measurements, calculations, reviews, and evaluations of neutron cross sections and other microscopic neutron data; it also includes index references to computer libraries of numerical neutron data available from four regional neutron data centers. Since 2005, database is extended by photonuclear and charged particle reaction data.

Of special interest are also reports prepared for Nuclear Data Section of IAEA by the International Nuclear Data Committee (INDC). The report series started the report INDC(NDS)-0001 (Lemmel et al. 1968), the recent one being INDC(NDS)-0657 (Zolotarev and Zolotarev 2013). In the list of almost 700 reports, one might find one of particular or general interest.

1.4 MODERN APPLICATIONS

Applications of analytical methods using 14 MeV neutrons produced with neutron generators can be found in almost every conceivable field of scientific inquiry. This is true in large part to a series of unique characteristics that separate the method from most others. While it is correctly identified as a method of "chemical analysis" because of the information it can generate, it is based on nuclear rather than atomic or molecular properties. It avoids the issue of chemical state in which the analyte exists in the sample, which is a common source of uncertainty in traditional analytical methods. In addition, simplicity of design and inherent safety of operation makes it an attractive alternative to the reactor-based instrumental neutron activation analysis (INAA), for those analytical problems for which it is appropriate.

The characteristics of the analytical method using 14 MeV neutrons as a probe are its nondestructive nature, multielement capability, ease of sample preparation, and freedom from reagent blank and contamination often encountered during sample preparation steps. In addition, many sample matrices are rather transparent to the incident radiation, neutrons, as well as to the gammas emitted from the reaction product that are used to generate the analytical signal (IAEA 1988, 1999). The nondestructive characteristic of the method is enhanced due to the generally lower fluxes and lower specific activities produced in isotopes with shorter half-lives likely to be produced. The small size and transportability lend the method as well to application to field measurements and implementation in site laboratories where conventional analytical instruments are unlikely to be usable (IAEA 2004).

The 14 MeV neutrons have recently been again in the spot mainly because of the development of smart analytical gadgets for explosive detection. Because of the large cross section (~5 barns) for $d + t \rightarrow \alpha + n$ reaction at low deutron bombarding energies (~100 keV) they are easily produced. Consequently, the production devices (neutron generators) can be small in size. Oil industry has recognized this advantage some time ago and 14 MeV neutrons are widely used in well-lodging activities.

Major steps forward are taken owing to both technology and politics: politics—with the end of the nuclear race, small neutron generators providing neutron excess in nuclear weapons became available to civilians; technology—physicists have learned how to put associated alpha particle detector inside the sealed tube neutron generators.

Fields of applications of neutron generators are listed in Table 1.1. The major field of application is materials analysis, namely, elemental analysis. No doubt, the most routinely useful and enduring application of 14 MeV neutron generators as an analytical tool is for the 14 MeV neutron activation analysis for oxygen. The longevity the technique has experienced is due at least in part to the fact that no competing method has been developed and it is readily automated with computer control (James and Akanni 1983). In addition to oxygen, several other components present at percent levels and tenths of percent levels concentrations are readily determined using generators in many matrices. Commonly determined elements using fast neutron–induced reactions include silicon via $^{28}Si(n,p)^{28}Al$ nuclear reaction, nitrogen via $^{14}N(n,2n)^{13}N$ reaction, aluminum via $^{27}Al(n,p)^{27}Mg$ nuclear reaction, fluorine via $^{19}F(n,2n)^{18}F$ reaction, phosphorus via $^{31}P(n,2n)^{30}P$ nuclear reaction, and iron via $^{56}Fe(n,p)^{56}Mn$ reaction. In addition, thermalized neutrons from the same generator are used in capture reactions to measure magnesium, manganese, and others if present in sufficient quantities. For example, samples returned from lunar exploration in the 1970s were routinely measured for major/minor components using generator-based neutron activation analysis (Butterfield and Lovering 1970).

Online analyzers are very popular in coal and cement industry. The experts find the online analyzers indispensable tools for raw material production in cement factories (Endress et al. 2004). More than 400 online analyzers are operating in cement factories to monitor the composition of the raw material on the conveyor belts; source IAEA (2012). They are commercially available from several manufacturers. A typical analyzer (sometimes called continuous neutron analyzers) is built up from a 14 MeV neutron generator, a moderator (e.g., polyethylene), the belt with the material on it, and the detectors (typically scintillators) above it. A large housing accommodates the instrument, which serves as a shielding against the radiation. Measurements of 20 s are made and the spectra are collected by a multi channel analyzer (MCA) card in a computer. The spectra are compared to those of pure materials (SiO_2, Al_2O_3, $CaCO_3$, and Fe_2O_3), and the composition is determined from linear combination of the library spectra. The great advantage of this method is that the irradiation averages about 90% of the raw material, while conventional sampling covers about 10%–14%.

TABLE 1.1 SELECTED FIELDS OF APPLICATIONS OF NEUTRON GENERATORS

Field	Category	Energy Required	Existing Applications
Security	Explosives detection	14 MeV	Cargo/luggage inspection, suspicious objects inspection
	Chemical weapon detection	14 MeV	N.P. CW inspections
	Contraband detection	14 MeV	Narcotics, C, N, O
Safeguards	Nuclear material detection	2.5 MeV 14 MeV	Fission product nuclides
Industrial	Online analyzers	2.5, 14 MeV	Cement process monitoring, coal and mining industries, Ca, Si, Fe, Al
	Metal cleanliness	14 MeV	Oxygen in Mg, Al, steel
	Raw minerals	2.5 MeV, thermalized and not thermalized	Purity, contaminants
	Al-based catalysts	14 MeV	F
	Energy production	thermalized	H in fuel cell technology
Research	Cross-section measurements	2.5, 14 MeV	
Educational	Nuclear parameter training, inelastic neutron scattering, nuclear physics laws	2.5, 14 MeV	
Medical	Body screening imaging, radiography	Pulsed 14 MeV	C, O, N, Ca, Na, Cl, P
Nutrition	Protein content of food	14 MeV	N
	Animal fodder	2.5 MeV thermalized	Trace elements

(*Continued*)

TABLE 1.1 (*Continued*) SELECTED FIELDS OF APPLICATIONS OF NEUTRON GENERATORS

Field	Category	Energy Required	Existing Applications
Environmental	Recycled material	2.5 MeV thermalized	Cd, Hg, Br, Cl
	Waste material	2.5, 14 MeV	
	Radiography	2.5 MeV	Water content of plants (in vivo)
	Pesticides	2.5 MeV	Halogens (Br, Cl)
		14 MeV	Halogens (F)
Geological	Online analyzers		Ash value of coal, Si, Al, C, H, Fe, Ca, S; calorific value
	Exploration		Oil wells, minerals, kimberlite diamonds

1.4.1 In Fight against Terrorism

In today's society acts of terrorism involve, in some stages, the illicit trafficking either of explosives, chemical agents, nuclear materials, and/or humans. Therefore, the society must rely on the antitrafficking infrastructure that encompasses responsible authorities: their personnel and adequate instrumental base.

Border management system must keep pace with expanding trade while protecting from the threats of terrorist attack, illegal immigration, illegal drugs, and other contraband. The border management of the future must integrate actions abroad to screen goods and people prior to their arrival in the country, and inspections at the border and measures within the country to ensure compliance with entry and import permits.

Border control agencies must have seamless information-sharing systems that allow for coordinated communication among themselves and also the broader law enforcement and intelligence-gathering communities. This integrated system would provide timely enforcement of laws and regulations. Agreements with neighbors, major trading partners, and private industry with all extensive prescreening of low-risk traffic, thereby allowing limited assets to focus attention on high-risk traffic, are essential part of the system.

The use of advanced technology to track the movement of cargo and the entry and exit of individuals are essential to the task of managing the movement of hundreds of millions of individuals, conveyances, and vehicles.

The list of materials that are subject to inspection with the aim of reducing the acts of terrorism includes explosives, narcotics, chemical weapons, hazardous chemicals, and radioactive materials. To this we should add also the illicit trafficking of human beings.

The application of neutron generators to security operations has been intensively studied and at least provisionally implemented over the last two decades. Detection of explosive material, illegal drugs, and chemical and nuclear weapons is all potentially possible using generator-produced neutrons. Interrogation of luggage or cargo containers for hidden explosives or illegal drugs is an area of extreme interest. In addition, monitoring and identification of high explosive weapons sometimes buried or underwater are an application that requires nonintrusive methods development. The problem of material (explosive, drugs, chemicals, etc.) identification can be reduced to the problem of measuring elemental concentrations. Nuclear reactions induced by neutrons can be used for the detection of chemical elements, measurements of their concentrations, and concentration ratios or multielemental maps. Neutron-scanning technology offers capabilities far beyond those of conventional inspection systems. This highly sophisticated equipment is now ready to be deployed as part of a country-wide system of deterrence. The unique automatic, material-specific detection of terrorist threats can significantly increase the security at ports, border-crossing stations, airports, and even within the domestic transportation infrastructure of potential urban targets as well as protecting forces and infrastructure wherever they are located.

Thermal neutron analysis uses thermal neutrons as the analytical probe, so the method can make use of isotopic sources, as well as d + t or d + d generators. This can also be done using a pulsed generator that allows for the measurement of both prompt capture and decay gammas. This is sometimes combined with fast neutron analysis, which adds the capability of using inelastic scattering and other particle-emitting reactions as well.

1.4.2 In Control of Illicit Trafficking

Modern personnel, parcel, vehicle, and cargo inspection systems are noninvasive imaging techniques based on the use of nuclear analytical techniques. The inspection systems are using penetrating radiations (neutrons, gamma, and x-rays) in the scanning geometry, with the detection of radiation transmitted or produced in investigated sample.

Homeland security in today's world depends on our ability to monitor our borders for the potential trafficking of special nuclear material. In addition, there is a need for sophisticated methods to locate hidden caches of nuclear material in locations being monitored to avoid the proliferation of nuclear weapon technology. Since most materials of interest are radioactive, passive measures are effective but suffer from the possibility of the material being easily hidden and long measurement times. Shielding readily reduces or eliminates characteristic radioactive emissions. The use of active neutron probes to induce fission reactions that result in delayed neutron emission or gammas from fission product decay has a distinct advantage. Most neutron generator systems being studied for use in such a system are based on the d + t reaction (Rooney et al. 1998).

REFERENCES

Börner, H. G. And Gönnenwein, F. 2012. *The Neutron, A Tool and an Object in Nuclear and Particle Physics.* World Scientific Publishing Co., Singapore.

Butterfield, D. and Lovering, J. F. 1970. Neutron activation analysis of rhenium and osmium in Apollo 11 lunar material. (Proc. Conf. Houston, 1970). *Geochim. Cosmochim. Acta* 2: 1351–1355.

Carlson, A. D., Pronyaev, V. G., Smith, D. L. et al. 2009. International evaluation of neutron cross section standards. *Nucl. Data Sheets* 110 (12): 3215–3324.

Chadwick, J. 1932a. Possible existence of a neutron. *Nature* 129 (3252): 312–313.

Chadwick, J. 1932b. The existence of a neutron. *Proc. R. Soc. A: Math. Phys. Eng. Sci.* 136 (830): 692–708.

CINDA. 2014. Computer index of nuclear reaction data. Database Version of January 27, 2014. Software Version of 2014.04.08; https://www-nds.iaea.org/exfor/cinda.htm.

Condé, H. 1992. Nuclear data standards for nuclear measurements. 1991 NEANDC/INDC nuclear standards file, Nuclear Energy Agency Organisation for Economic Co-operation and Development.

Csikai, J. 1987. *CRC Handbook of Fast Neutron Generators.* CRC Press, Boca Raton, FL.

Endress, M., Heuschkel, S., Keydel, A., Knadel, S., and Sosna, I. 2004. Automated cement plant quality control: In-situ versus extractive sampling and instrumentation. *Cement Int.* 2: 38.

Glasstone, S. (Exc. Ed.). 1950. *The Effects of Nuclear Weapons.* U.S. Department of Defense and U.S. Atomic Energy Commission, Washington, DC.

Glasstone, S. 1962. An introduction to nuclear weapons. WASH-1038, Unique Document #: SAC200117630000, written for the AEC's Division of Technical Information.

Glasstone, S. 1963. An introduction to nuclear weapons. Unique Document #: SAC200117610000, written for the AEC's Division of Technical Information.

Glasstone, S. and Dolan, P. J. 1977. *The Effects of Nuclear Weapons,* 3rd edn. U.S. Department of Defense and the Energy Research and Development Administration, Washington, DC. http://www.angelfire.com/mac/nws/design.htm. Nuclear Weapon Physics and Design.

Glasstone, S. and Redman, L. M. 1972. An introduction to nuclear weapons. WASH-1038 Revised, Unique Document #: SAC200117640000 and SAC200117620000, written for the AEC's Division of Military Application.

IAEA. 1988. IAEA-TECDOC-459. *Nuclear Analytical Techniques for On-line Elemental Analysis in Industry.* IAEA, Vienna, Austria.

IAEA. 1994a. Table of simple integral neutron cross section data from JEF-2.2. ENDF/B-VI, JENDL-3.2, BROND-2, and CENDL-2, JEF Report 14, OECD Nuclear Energy Agency, Paris, France.

IAEA. 1994b. Thermal neutron cross-sections, resonance-parameters, resonance integrals. JEF 2.2 INTER, IAEA Nuclear Data Section, Vienna, Austria.

IAEA. 1999. IAEA-TECDOC-1121. *Industrial and Environmental Applications of Nuclear Analytical Techniques*. IAEA, Vienna, Austria.

IAEA. 2004. Special report series STI/PUB/1181. Nuclear analytical techniques. IAEA, Vienna, Austria.

IAEA. 2007. *International Evaluation of Neutron Cross-Section Standards*. IAEA, Vienna, Austria.

IAEA. 2012. Neutron generators for analytical purposes. IAEA radiation technology reports series no. 1. IAEA, Vienna, Austria.

James, W. D. and Akanni, M. S. 1983. Application of on-line laboratory computer analysis to fast neutron activation oxygen determinations. *IEEE Trans. Nucl. Sci.* 302: 1610–1613.

Lemmel, H. D., Attree, P. M., Byer, T. A. et al. 1968. INDC(NDS)-001. *Neuron Data Compilation at the International Atomic Energy Agency*. IAEA, Vienna, Austria.

Nargolwalla, S. S. and Przybylowicz, E. P. 1973. *Activation Analysis with Neutron Generators*. John Wiley & Sons, New York.

OECD/NEA Data Bank: NEA Data Bank, OECD Nuclear Energy Agency, Le Seine Saint-Germain, Issy-les-Moulineaux, France.

Rooney, B. D., York, R. L., Close, D. A., and Williams III, H. E. 1998. Active neutron interrogation package monitor, in *Proceedings of Nuc1ear Science Symposium, Toronto*, Los Alamos Nat. Lab., Los Alamos, NM, Vol. 2, pp. 1027–1030.

USSR-former-countries: *Neutron data:* Russia Nuclear Data Center, Centr Jadernykh Dannykh (CJD), Fiziko-Energeticheskij Institut, Ploschad Bondarenko, Obninsk, Kaluga Region, Russia, http://www.ippe.ru/podr/cjd; *Photonuclear data:* Centre for Photonuclear Experiments Data. Centr Dannykh Fotoyadernykh Eksperimentov (CDFE), Skobeltsyn Institute of Nuclear Physics, Lomonosov Moscow State University, Leninskie Gory, Moscow, Russia, http://cdfe.sinp.msll.ru.

USA NNDC: For services to customers in USA and Canada: US National Nuclear Data Center, Brookhaven National Laboratory, Upton, NY, http://www.nndc.bnl.gov.

Weiner, C. 1969. Interview with Sir James Chadwick, Cambridge, England, April 17, 1969. http://www.aip.org/history/ohilist/3974_3.html.

Zolotarev, K. I. and Zolotarev, P. K. 2013. INDC(NDS)-0657, Evaluation of Some (n,n′), (n,γ), (n,p), (n,2n) and (n,3n) Reaction Excitation Functions for Fission and Fusion Reactors Dosimetry Application. IAEA, Vienna, Austria.

Chapter 2

Nuclear Reaction $d + t \rightarrow \alpha + n$

2.1 PROPERTIES OF $^3H(d,n)^4He$ NUCLEAR REACTION

Nuclear reaction $d + t \rightarrow \alpha + n$ or $^3H(d,n)^4He$ has a Q value Q = 17.590 MeV. The energy released in this reaction appears in the form of kinetic energy of the ^4He and the neutron, which can be converted into heat and subsequently electrical energy. In terms of per gram of starting material, this reaction is one of the most efficient sources of nuclear energy in the universe.

The ^3H + d reaction (^3H being the target particle and d being a beam particle) initial state leads to ^5He compound nucleus whose properties are described in a compilation by Lauritsen and Ajzenberg-Selove (1966). As shown in Figure 2.1 ^5He is particle unstable even in its ground state and decays into ^4He + n. Initial state ^3H + d can result into three final states following nuclear reactions:

1. $^3H(d,n)^4He$ Q = 17.590 MeV
2. $^3H(d,2n)^3He$ Q = –2.988 MeV
3. $^3H(d,pn)^3H$ Q = –2.225 MeV

Excitation curves and angular distributions for reaction (a) measured in the bombarding energy interval of deuteron beam from E_d = 8 keV to 19 MeV are summarized in the following papers: Fowler and Brolley (1956), Jarmie and Seagrave (1957), Bransden (1960), Brolley and Fowler (1960), Stewart et al. (1960), Goldberg and Le Blanc (1961), Brill et al. (1964), and Paulsen and Liskien (1964). Below E_d = 100 keV the cross section follows the Gamow function, $\sigma = (A/E) \exp(-44.40E^{-1/2})$; see Jarvis and Roaf (1953) and Arnold et al. (1954). A strong resonance, σ_{peak} = 5.0 b, appears at E_d = 107 keV; see Figure 2.2a. In the paper Bame and Perry (1957), the differential cross sections for neutron production in the $^3H(d,n)^4He$ reaction were measured over the deuteron bombarding-energy range 0.25–7.0 MeV. Legendre polynomial fits to the data are presented and compared with existing data. The total cross section plotted as a function of deuteron energy has an anomalous behavior that may indicate a broad level in ^5He near 20 MeV excitation energy.

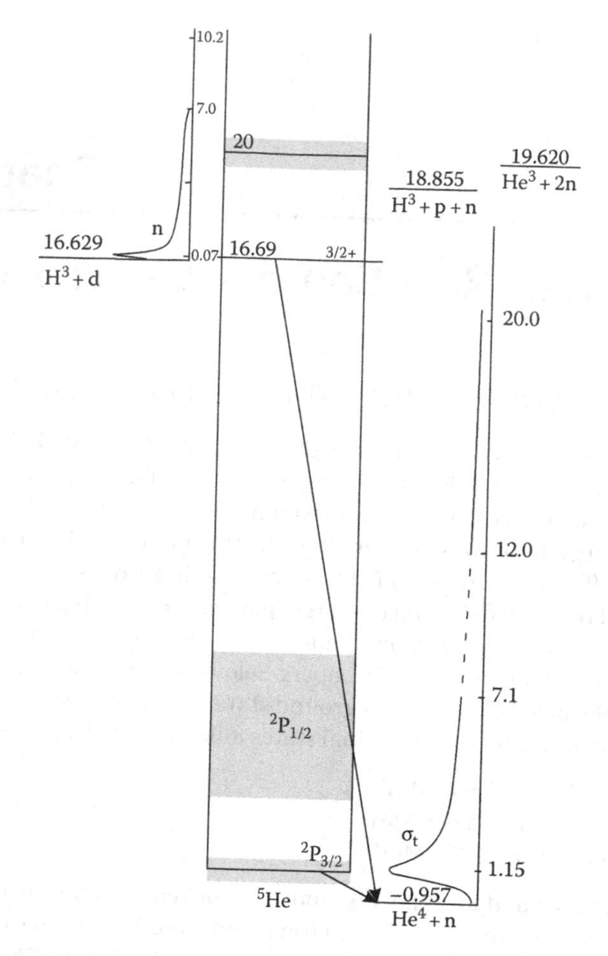

Figure 2.1 Energy diagram for ^{5}He nucleus and position of initial and final states of d + ^{3}H nuclear reactions.

In the region E_d = 10–500 keV, the cross section is closely fitted with the assumption of s-wave formation of J = 3/2$^+$ state with the parameters given in Table 2.1 (after Kunz 1955). The angular distribution of neutrons is isotropic at and below resonance and shows increasing forward peaking at higher energies; differential cross section (mb/sr) at 0° is shown in Figure 2.2b while the relative angular distribution of d + t neutrons for four deuteron energies (E_d = 50, 100, 200, and 300 keV) is shown in Figure 2.3, according to (Csikai 1987). According to Goldberg and Le Blanc (1961), the distributions in the range E_d = 6.2–11.4 MeV are all picked forward with a second maximum at about 65°, which becomes more pronounced with increasing energy and a rise at back angles. It does not

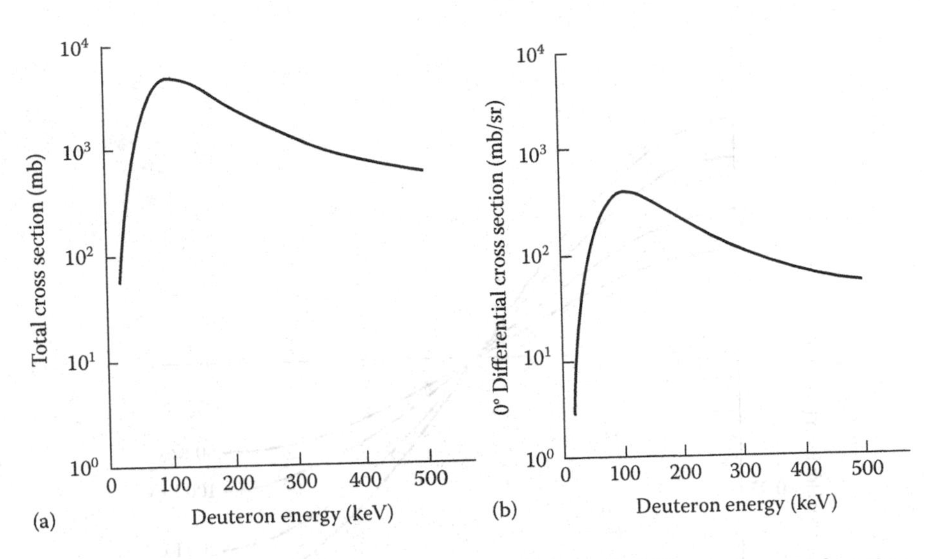

Figure 2.2 (a) Total laboratory cross section for the $^3H(d,n)^4He$ reaction as a function of deuteron bombarding energy and (b) 0° differential laboratory cross section for the $^3H(d,n)^4He$ reaction as a function of deuteron bombarding energy.

TABLE 2.1 RESONANCE PARAMETERS FOR $^3H(D,N)^4HE$ REACTION

E_R (keV)	Γ_{Lab} (keV)	l_d	J^π	$L_{n,p}$	R (fm)	E_λ (keV)	γ_d^2 (keV)	$\gamma_{n,p}^2$ (keV)	E_x (MeV)
107	135	0	$3/2^+$	2	5.0	−464	2000	56	16.70
					7.0	−126	715	17	

appear that plane wave stripping theory, including heavy-particle stripping can account for the observed distributions.

At low deuteron energies, the polarization of outgoing neutrons has been studied at $E_d = 0.1$ and 0.17 MeV by Rudin et al. (1961), at 0.6 and 1.2 MeV by Boreli et al. (1965), and from 0.1 to 7.7 MeV by Perkins and Simmons (1961) and others.

Since reactions (b) and (c) may occur only for higher E_d because of negative Q values, we shall not discuss them here.

Until 1980, the $^3H(d,n)$ low-energy fusion data used came from three main references. The first one, already mentioned, Arnold et al. (1953 and 1954) done at Los Alamos reported the measurement of σ (90°) down to $E_d \sim 10$ keV laboratory bombarding energy, claiming 2% accuracy. Because the reaction is isotropic in the center of mass system below several hundred keV, the σ (90°) is easily converted to the total cross section, σ_T. In another work, Conner et al. (1952) at

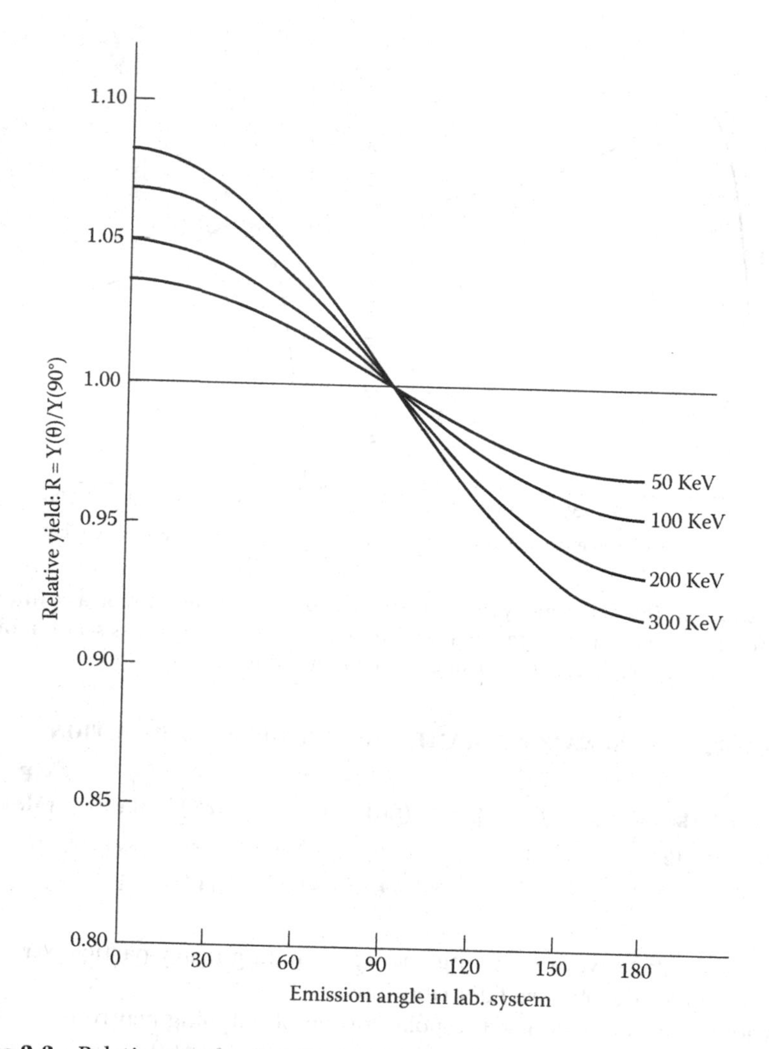

Figure 2.3 Relative angular distribution for the $^{3}H(d,n)^{4}He$ reaction as a function of deuteron bombarding energy.

Rice University, Houston, Texas, measured σ (90°) down to 10 keV with 3% accuracy. Often neglected is the work by Katsaurov (1926) from Lebedev Institute in Russia who measured σ_T down to 45 keV claiming 2%–3% accuracy.

It was Jarmie (1980) who investigated the accuracy of basic fusion data including $^{3}H(d,n)^{4}He$ reaction. He concluded that systematic errors up to 50% are possible in the reactivity values for the 10–100 keV deuteron energy region of that time $^{3}H(d,n)^{4}He$ database, most likely because of energy scale errors in the experiments. The reactivity errors would propagate proportionately into fusion probability errors in reactor calculations. The investigations of these discrepancies are reported in Jarmie et al. (1983); they have developed a low-energy fusion cross-section facility (LEFCS) to remeasure the $^{2}H(t,\alpha)n$ and other few nucleon reactions. The experimental equipment featured a windowless cryogenic target, a precision beam intensity calorimeter, a 10–120 keV accelerator producing negative tritium ions, an accurate target gas flow and temperature system, and a tritium gas handling system. The target density and geometry were calibrated with the $^{2}H(p,p)^{2}H$ reaction at 10 MeV. They have determined 18 total cross-section values for $^{2}H(t,\alpha)n$ reaction in the energy range 8.3–78.1 keV equivalent deuteron energy with a relative error of 0.5% and an absolute error of 1.4% for most energies, except for the lowest energy where both uncertainties rise to about 4.7%.

2.2 D–T FUSION REACTION: POWER PRODUCTION

The $^{3}H(d,n)^{4}He$ and $^{3}He(d,p)^{4}He$ reactions are leading processes in the primordial formation of the very light elements (mass number, $A \leq 7$) affecting the prediction of big bang nucleosynthesis for light nucleus abundances (Serpico et al. 2004). With its low activation energy and high yield, $^{3}H(d,n)^{4}He$ is also the easiest reaction to achieve on Earth. The fusion process is well understood, and the cross section is known with small uncertainties. In a work by (Navratil and Quaglioni 2012), the cross section is calculated using ab initio no-core shell model combined with the resonating group method approach. They start from a selected similarity-transformed chiral nucleon–nucleon interaction that accurately describes two-nucleon data and perform many-body calculations in order to predict the S-factor for the reaction. The results reported are in satisfactory agreement with experimental data and pave the way for microscopic investigations of polarization and electron-screening effects, of the $^{3}H(d,\gamma n)\,^{4}He$ bremsstrahlung and other reactions relevant to fusion research.

A number of experiments around the world have verified the principles of d + t fusion; however, it still has to be demonstrated that a gain in energy can be achieved. A short history of the quest for fusion and some of the interesting details (and personalities) are described in the recent book by Sheffield (2013). There are two main approaches trying to accomplish this. One is the magnetic confinement of deuterium–tritium plasma and the other is laser compression of a cryogenic layer of deuterium and tritium in a pellet. These two directions are pursued by several research facilities, namely, International Thermonuclear Experimental Reactor (ITER) (http://www.iter.org) and Joint European Torus (JET), which, in the year 1999, came under European Fusion Development Agreement (EFDA) established with the responsibility for the future collective use of JET, directed toward developing fusion power by magnetic confinement, and National Ignition Facility (NIF) (https://lasers.llnl.gov) by inertial confinement.

The cross section for the d + ^{3}H fusion is by now well known experimentally, while more uncertain (Cecil and Wilkinson 1984; Kammeraad et al. 1993) is the situation for the branch of this reaction, ^{3}H(d,γn)^{4}He, that is being considered as a possible plasma diagnostic in modem fusion experiment (Murphy et al. 2001). One alpha particle is created for each neutron in the fusion reaction:

$$d + t \rightarrow \alpha\,(3.5\text{ MeV}) + n\,(14\text{ MeV}).$$

In d–t fusion, 80% of the energy released goes into 14 MeV neutrons and only the remaining 20% into charged particles. A tokamak reactor will use the alpha particle power for heating and will ignite only when the alpha power being transferred to the plasma exceeds the plasma energy loss rate. To ensure ignition in a tokamak, the alpha particles must be well confined during their thermalization, for example, for a timescale on the order of 1 s in ITER (Zweben et al. 1995).

Unlike the charged particles, the uncharged neutrons cannot be confined by a magnetic field, and for this reason cannot be used for a direct conversion into electric energy. Instead, the neutrons have to be slowed down in some medium, heating this medium to a temperature of less than 10^3 K, with the heat removed from this medium to drive a turbo generator. This conversion of nuclear into electric energy has a Carnot efficiency of about 30%. For 80% of the energy released into neutrons, the efficiency is therefore no more than 24%. While this low conversion efficiency cannot be overcome in magnetic confinement concepts, it can be overcome in inertial confinement concepts, by surrounding the inertial confinement fusion target with a sufficiently thick layer of liquid hydrogen and a thin outer layer of boron, to create a hot plasma fireball. The hydrogen layer must be chosen just thick and dense enough to be heated by the neutrons to 100,000 K. In this way, generated, fully ionized, and rapidly expanding fireball can drive a pulsed magnetohydrodynamic generator at an almost 100% Carnot efficiency or possibly be used to generate hydrocarbons (Winterberg 2013).

The use of neutrons is discussed in a presentation by Baluc (2006) and schematically presented in Figure 2.4. In the breeding blanket, the kinetic energy of 14 MeV neutrons is transformed into heat that is transferred to a coolant. The produced vapor is then used to run a turbine that produces electricity. The neutrons produced in fusion generate atomic displacement cascades and transmutation nuclear reactions within the materials. As a result, the materials can become radioactive. Transmutation nuclear reactions yield the formation of impurities (e.g., H and He atoms). Atomic displacement cascades produce point structure defects (vacancies and interstitials). The

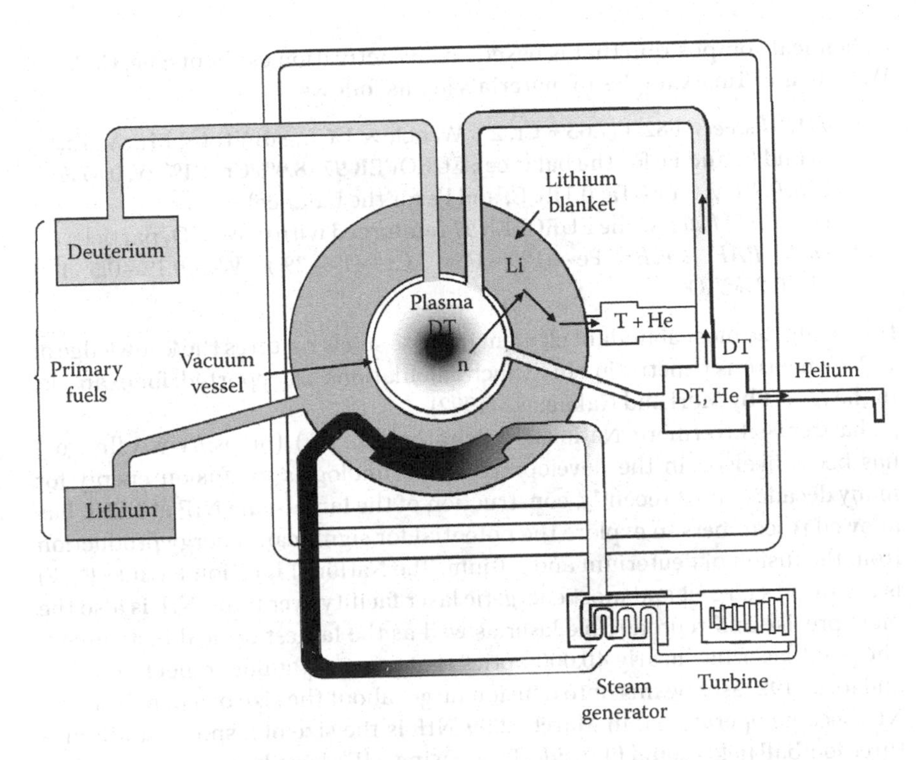

Figure 2.4 Schematic presentation of electrical power generation by d + t fusion reaction. (From Baluc, N. L., Reduced activation materials, report centre of research in plasma physics. Swiss Federal Institute of Technology, Lausanne, Switzerland, http://gcep.stanford.edu/pdfs/qa4ScQIicx-kve2pX9D7Yg/baluc_fusion_05_06.pdf, 2006.)

properties of irradiated materials are discussed in detail in this presentation by Baluc (2006). Plasma-facing components are so-called first wall and magnetic diverter. The first wall forms the plasma chamber and serves as part and protection of the breeding blanket. The first wall consists of a structural material attached to a plasma-facing material. The breeding blanket absorbs the 14 MeV neutrons, transforming their energy to provide most of the reactor output; shields the superconducting coils and other components; and allows for neutron multiplication and breeding of tritium to fuel the reactor. A breeding blanket consists of a tritium breeding material, a neutron multiplier, a coolant, and a structural material (to separate and contain the different materials). The magnetic diverter is minimizing the impurity content of the core plasma, removing the alpha particle power and pumping the He ash. It consists of a structural material (heat sink) that contains a coolant and supports a plasma-facing armor material. The main candidate materials have a chemical composition that is based on low activation elements: Fe, Cr, V, Ti, W, Si, C, and Ta. Examples of materials are as follows:

- *RAFM steels*: F82H (7.65% Cr, 2% W, below 1% in sum total Mn, V, Ta, Si and C, and Fe for the balance); EUROFER 97 (8.9% Cr, 1.1% W, 0.47% Mn, 0.2% V, 0.14% Ta, 0.11% C, and Fe for the balance)
- *ODS RAFM steels*: The EUROFER 97 reinforced with 0.3% Y_2O_3 particles
- *ODS RAF steels*: Fe—(12%–14%) Cr—(1%–3%) W—(0.1%–0.5%) Ti—0.3% Y_2O_3

Iron being the most abundant element in these steels requires the knowledge of 14 MeV neutrons penetration in it. Such calculations are reported, for example, in the report by Mori and Nakagawa (1992).

Lawrence Livermore National Laboratory (LLNL) (https://www.llnl.gov) has been involved in the development of technologies for fusion energy for many decades. Most recently, construction of the large-scale NIF at LLNL has allowed researchers to explore the potential for significant energy production from the fusion of deuterium and tritium. The National Ignition Facility (NIF) is the world's largest and most energetic laser facility ever built. NIF is also the most precise and reproducible laser as well as the largest optical instrument. The giant laser has nearly 40,000 optics that precisely guide, reflect, amplify, and focus 192 laser beams onto a fusion target about the size of a pencil eraser. NIF became operational in March 2009. NIF is the size of a sports stadium—three football fields could fit inside. By focusing NIF's laser beams onto a variety of targets, scientists create extreme states of matter, including temperatures of 100 million degrees and pressures that exceed 10^{10} times Earth's atmosphere.

While deuterium is readily available from water, tritium has to be manufactured and recycled due to its inherently short lifetime. Technologies are thus required for the production, extraction, purification, storage, and delivery of tritium. These processes need to be as efficient as possible to effectively eliminate any losses to the environment and to minimize the overall inventory of tritium material in the system. Present-day technologies are sufficient for facilities such as the NIF, but new processes will need to be developed to allow construction of next-generation facilities, as proposed by many groups around the world.

Similarly, there is a need to develop appropriate structural materials for the systems that manage the tritium production, due to the requirement to operate at high temperature and to withstand an intense neutron flux over extended periods. Conventional metals are inadequate, either due to poor mechanical properties under neutron bombardment (e.g., embrittlement) or due to unacceptable activation. Advanced materials and material processing techniques (e.g., bonding/welding) are required. Achieving ignition, characterized by the release of energy in excess of that required to initiate the fusion reaction, is still behind the schedule. So far, (mid-2014) the produced fusion energy is equal to about 1% of the laser's 1.8 MJ input (Kramer 2014).

At the Geneva Superpower Summit in November 1985, following discussions with President Mitterand of France and Prime Minister Thatcher of the United Kingdom, General Secretary Gorbachev of the former Soviet Union proposed to U.S. President Reagan an international project aimed at developing fusion energy for peaceful purposes. The initial signatories of the ITER project: the former Soviet Union, the United States, the European Union (via EURATOM), and Japan were joined by the People's Republic of China and the Republic of Korea in 2003 and India in 2005. Together, these six nations plus Europe represent over half of the world's population.

Conceptual design work for the fusion project began in 1988, followed by increasingly detailed engineering design phases until the final design for ITER was approved by the members in 2001. Further negotiations established the Joint Implementation Agreement to detail the construction, exploitation, and decommissioning phases, as well as the financing, organization, and staffing of the ITER Organization.

In ITER, the world has now joined forces to establish one of the largest and most ambitious international science projects ever conducted. ITER, which means "the way" in Latin, will require unparalleled levels of international scientific collaboration. Key plant components, for example, will be provided to the ITER organization through in-kind contributions from the seven members. Each member has set up a domestic agency, employing staff to manage procurements for its in-kind contributions.

Selecting a location for ITER was a long process that was finally concluded in 2005. In Moscow, on June 28, high representatives of the ITER Members unanimously agreed on the site proposed by the European Union—the ITER installation would be built at Cadarache, near Aix-en-Provence in Southern France. Construction work began in 2010 on the ITER site in Saint Paul-lez-Durance, France.

ITER is the culmination of decades of fusion research: more than 200 tokamaks were built worldwide, the smallest the size of a compact disc, the largest as high as a five floors the building. The objective of the ITER project is to gain the knowledge necessary for the design of the next-stage device: a demonstration fusion power plant. In ITER, scientists will study plasmas under conditions similar to those expected in a future power plant. ITER will be the first fusion experiment to produce net power; it will also test key technologies, including heating, control, diagnostics, and remote maintenance.

It is also of importance to understand the effects of 14 MeV neutron irradiation on optical components of fusion diagnostic systems that may be located in the 14 MeV neutron environment of a fusion reactor. Let us mention a report by Iida et al. (1990) in which various kinds of optical fibers and related optical components for fusion diagnostics were irradiated with 14 MeV neutrons and the degradation of important performances of the order of 10% has been observed. Their, and other similar, data are useful for the estimation of lifetimes and background levels of optical fusion diagnostic systems.

The World Nuclear Association (WNA) is the international organization that promotes nuclear energy and supports the many companies that comprise the global nuclear industry. WNA membership encompasses virtually all world uranium mining, conversion, enrichment, and fuel fabrication; all reactor vendors; major nuclear engineering, construction, and waste management companies; and the majority of world nuclear generation. Today, WNA serves its membership, and the world nuclear industry as a whole, through actions to

- Provide a global forum for sharing knowledge and insight on evolving industry developments
- Strengthen industry operational capabilities by advancing best-practice internationally
- Speak authoritatively for the nuclear industry in key international forums that affect the policy and public environment in which the industry operates

WNA represents the industry in key world forums that shape the nuclear industry's regulatory and policy environment, such as

- IAEA and NEA advisory committees on transport and all aspects of nuclear safety
- United Nations policy forums focused on sustainable development and climate change
- ICRP and Ospar deliberations on radiological protection

The WNA website (http://www.world-nuclear.org) serves as the world's leading information source on nuclear energy including information about current and future generation, nuclear fusion and power, etc. The site lists all ongoing fusion projects and active experimental sites.

2.3 ASSOCIATED ALPHA PARTICLE–TAGGED NEUTRON BEAMS

The nuclear reaction $^{3}H(d,n)^{4}He$, Q value Q = 17.590 MeV, is illustrated schematically in Figure 2.5. Usually, a deuteron beam, kinetic energy around 100 keV in the laboratory system, is bombarding the tritium target (usually tritium is implanted into some backing). The energy released in the nuclear reaction is divided so that the neutron gets E_n = 14.1 MeV while α-particle has 3.5 MeV. These facts generate some observable difficulties. It is obvious that to every produced

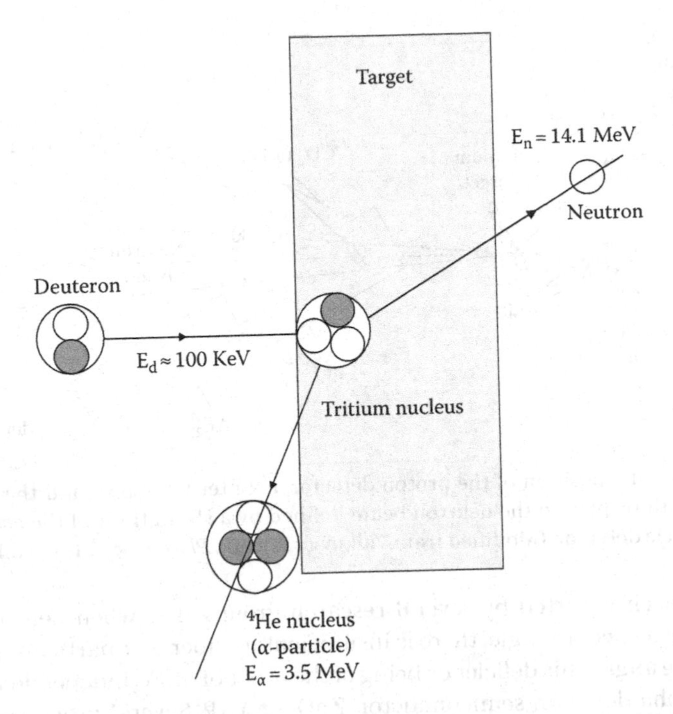

Figure 2.5 Schematic illustration of the nuclear reaction $^3H(d,n)^4He$, Q value Q = 17.590 MeV.

neutron in the nuclear reaction there is an associated α-particle traveling in the opposite direction. The detection of the α-particle at a given laboratory angle, or cone, defines a neutron direction or neutron beam. It looks simple and obvious; however, measurements have shown the smaller number of α-particles then expected one-to-one ratio. This is especially of importance if the reaction is used as a neutron source for the study of neutron-induced reactions with detection of α-particles for the purpose of neutron beam intensity monitoring.

The use of associated alpha particles to define electronically the neutron beam has been described in the literature as early as 1969 (Valkovic et al. 1969, 1970). The experimental setup is shown in Figure 2.6. The measurements were performed with an associated alpha particle detector. The arrangement was as follows: neutron source—target distance 10 cm, neutron detector—target distance 25 cm, and E detector—target distance 15 cm. The neutron detector was a cylindrical NE218 liquid scintillator, 7.6 cm in diameter and 7.6 cm thick coupled to photomultiplier. All measurements were performed in open geometry. This was used to reduce the background in the measurements, usually generated by neutrons coming from different directions. In such a way, a minimum measurable cross-section value was significantly reduced.

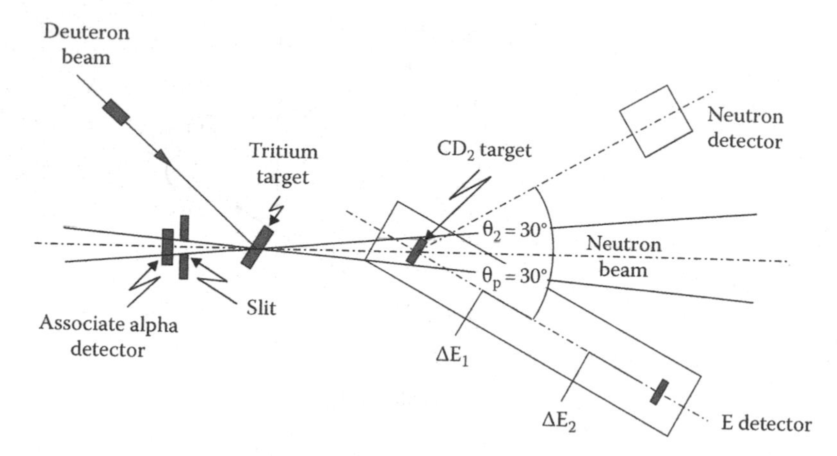

Figure 2.6 The position of the proton detector, counter telescope, and the neutron detector with respect to the neutron beam defined by a slit in front of the associated alpha particle detector. (Modified from Valkovic, V. et al., *Phys. Rev. C*, 1, 1221, 1970.)

It has been reported by several research groups that when neutrons are detected at a specific angle, there is insufficient number of α-particles in a corresponding angle, this deficiency being of the order of ~15%, independent of the type of alpha detector, semiconductor, ZnO, or YAP. Several processes could contribute to the "loss" of the α-particle of detected neutron at one particular angle, including the following:

- α-particle energy loss in leaving the tritium target. This value can be big because of two reasons: low α-particle energy (3.5 MeV) and part of the deuterons interacting with tritium nuclei toward the end of deuterium range in target generating a larger exit path for generated α-particles.
- α-particles being subject to magnetic fields in the generator resulting in the modification of their path.
- Deuterium buildup on the tritium target and increasing influence of ^{3}He (from d + d → n + ^{3}He reaction) on α-particle energy spectrum.

Here, we shall present the work done by Sudac et al. (2014) studying this phenomenon. Their experimental setup is presented in Figure 2.7. The neutron source is API-120 (manufacturer: ThermoElectron) with alpha particle detector YAP(Ce) crystal with thickness d = 0.5 mm. The inner side of the crystal is protected by evaporated silver layer, 1 mg/cm^2, in order to prevent detection of elastically scattered deuterons. The outer side of the crystal is connected by fiber optic plate (FOP) with a photomultiplier. This assembly serves as the detector of associated alpha particles.

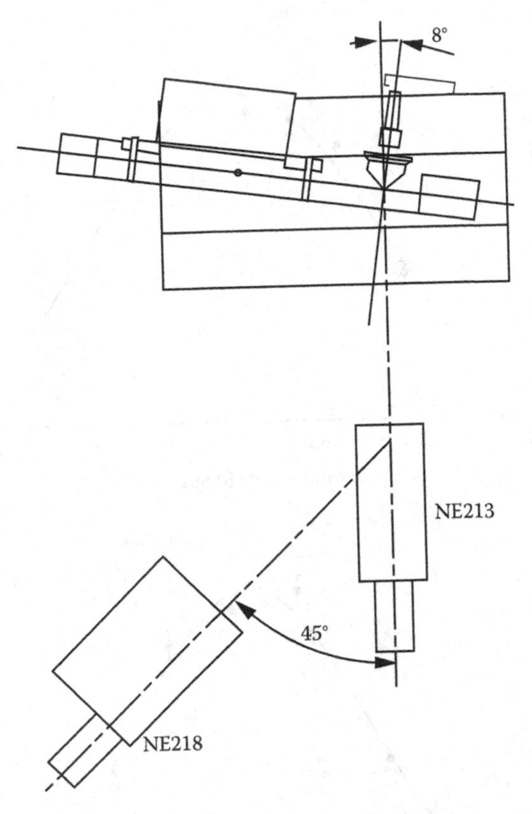

Figure 2.7 Experimental setup used in the measurements reported by Sudac et al. (2014).

The cone of tagged neutrons is determined experimentally by the rotation of neutron generator, API-120, vertically and horizontally and by measurement of α-n coincidences for a fixed neutron detector, NE213, position. Figure 2.8 shows the number of coincidences as a function of rotation angle. In order for the neutron detector, 3″ × 3″ NE218, to be inside the cone of the tagged neutrons, at the position of maximum intensity, the axis of neutron generator tube has been tilted for 8° with respect to NE213 axis. Another neutron detector, NE218, is placed 50 cm from NE213, acting as a hydrogen target to tagged neutrons, at an angle $\Theta = 45°$ with respect to NE213 axis (see Figure 2.7). From the kinematics of elastic n–p scattering the neutrons scattered at $\Theta = 45°$ have the energy 7 MeV. The electronics setup is shown in Figure 2.9 and it is measuring twofold coincidences between two neutron detectors and threefold coincidences between two neutron detectors and the alpha detector.

In conclusion, a deficit of triple coincidences (n–n-α) with respect to number of double coincidences (n-α) has been observed. The deficit is smallest for zero

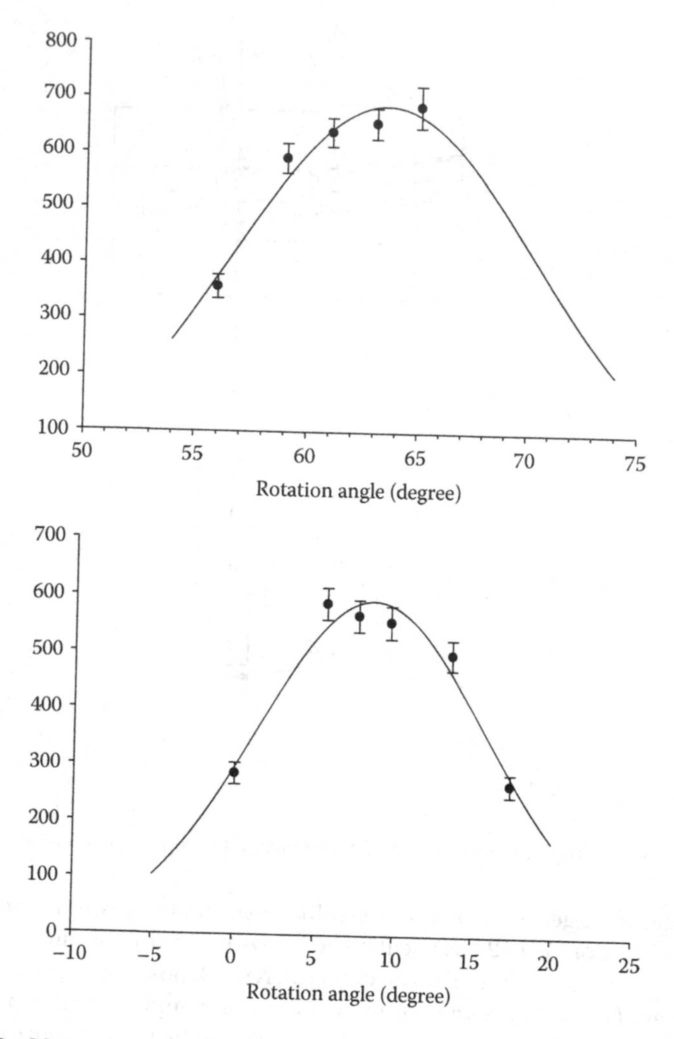

Figure 2.8 Measurements of horizontal (upper) and vertical (lower) profile of tagged neutron beam.

threshold on constant fraction discriminator (CFD) and a maximal allowed voltage (−1.7 kV) on α-detector, being (13 ± 1)%. It follows that the deficit in the number of α-particles is simply the consequence of the fact that some α-particles are not detected. Some of the alpha particles lose so much energy on their way from the point of generation to the point of detection that they "fall" below detection threshold. The number of nondetected α-particles is proportional to the threshold of α-detector CFD, αCFD, as shown in Figure 2.10 and inversely proportional to the voltage applied to alpha detector (Figure 2.11).

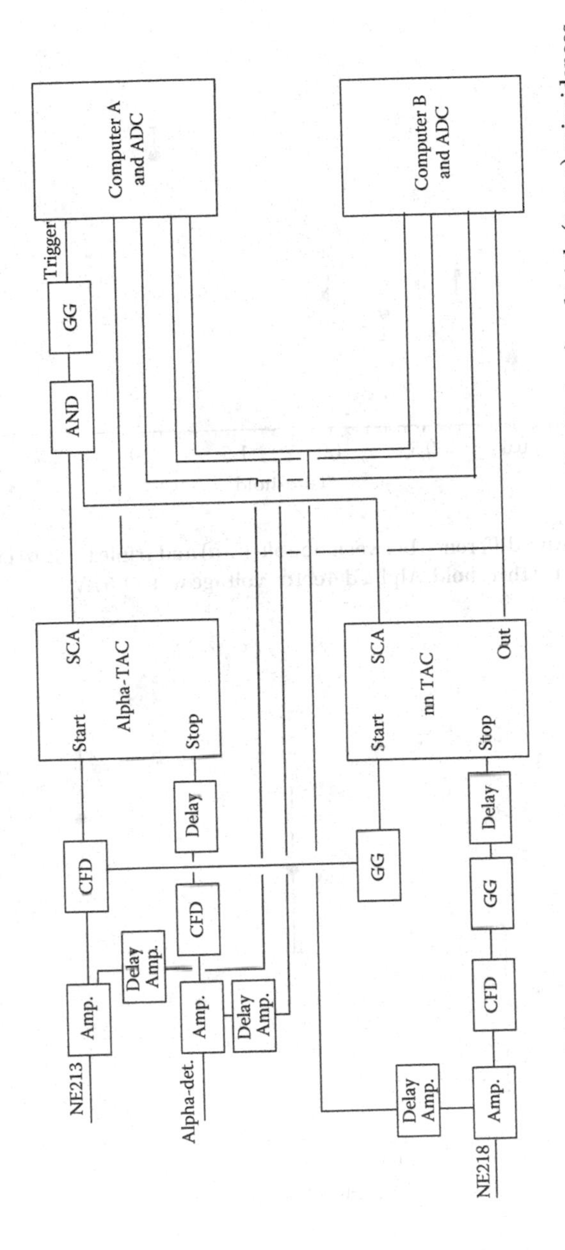

Figure 2.9 The electronics block diagram used for simultaneous measurement of double (n-α) and triple (n–n-α) coincidences.

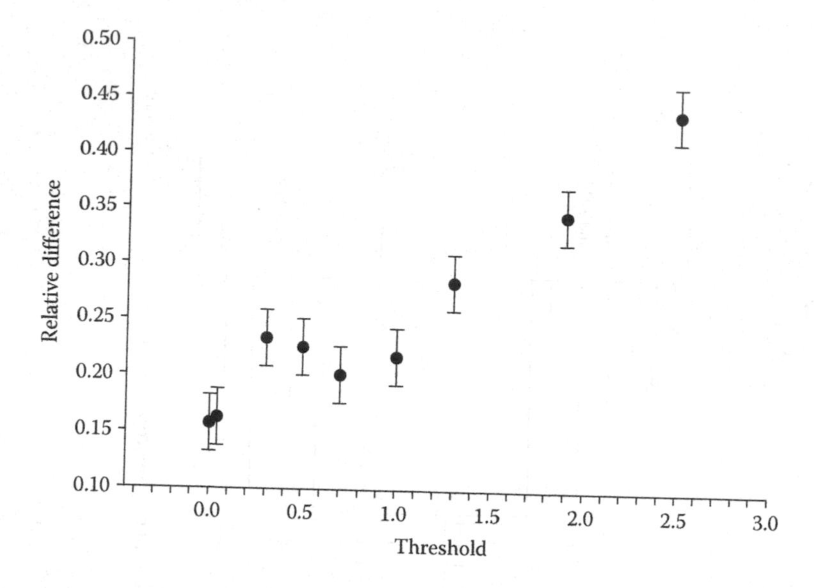

Figure 2.10 Relative difference between double (n-α) and triple (n–n-α) coincidences as a function of αCFD threshold. Alpha detector voltage was −1.6 kV.

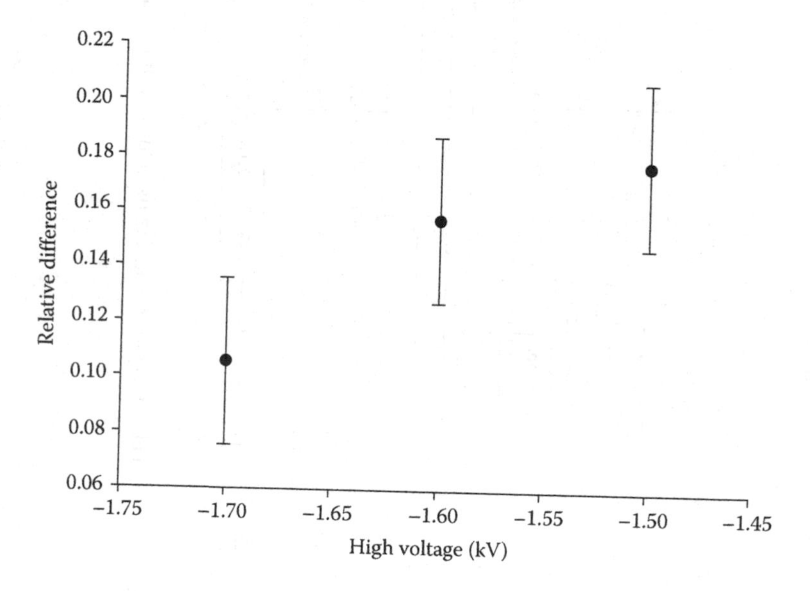

Figure 2.11 Relative difference between double (n-α) and triple (n–n-α) coincidences as a function of alpha detector voltage; the αCFD threshold is zero.

REFERENCES

Arnold, W. A., Phillips, J. A., Sawyer, G. A., Stovall, E. J., and Tuck, J. L. 1953. Absolute cross section for the reaction T(d,n)^{4}He from 10 to 120 keV. Los Alamos Laboratory Report LA-1479.

Arnold, W. A., Phillips, J. A., Sawyer, G. A., Stovall, E. J., and Tuck, J. L. 1954. Cross sections for the reactions D(d,p)T, D(d,n)^{3}He, T(d,n)^{4}He and ^{3}He(d,p)^{4}He below 120 keV. *Phys. Rev.* 93: 483–497.

Baluc, N. L. 2006. Reduced activation materials, report centre of research in plasma physics. Swiss Federal Institute of Technology, Lausanne, Switzerland, http://gcep.stanford.edu/pdfs/qa4ScQIicx-kve2pX9D7Yg/baluc_fusion_05_06.pdf.

Bame, S. J. and Perry, J. E. 1957. T(d,n)^{4}He reaction. *Phys. Rev.* 107: 1616–1620.

Boreli, F., Lazarevic, V., and Radisic, N. 1965. Polarization of neutrons from the T(d, n)He^4 reaction at 0.6 and 1.2 MeV. *Nucl. Phys.* 66: 301–304.

Bransden, B. H. 1960. *Nuclear Forces and the Few Nucleon Problem.* Pergamon, NY, 527 p.

Brill, O. D., Pankratov, V. M., Rudakov, V. P., and Rybakov, B. V. 1964. The cross section of the T(d,n)^{4}He and D(d,n)^{3}He reactions in the deuteron energy range from 3 to 19 MeV. *Atomn. En. (USSR)* 16: 141–143.

Brolley, J. E. and Fowler, J. L. 1960. in Marion, J. B. and Fowler, J. L. (Eds.) *Fast Neutron Physics.* Interscience Publishers In., New York, p. 101.

Cecil, F. E. and Wilkinson III, F. J. 1984. Measurement of the ground-state gamma-Ray branching ratio of the dt reaction at low energies. *Phys. Rev. Lett.* 53: 767–770.

Conner, J. P., Bonner, T. W., and Smith, J. R. 1952. A Study of the ^{3}H(d,n)^{4}He reaction. *Phys. Rev.* 88: 468–473.

Csikai, J. 1987. *Handbook of Fast Neutron Generators.* CRC Press, Inc., Boca Raton, FL.

Fowler, J. L. and Brolley, J. E. 1956. Monoenergetic neutron techniques in the 10 to 30 MeV range. *Revs. Mod. Phys.* 28: 103–134.

Goldberg, M. D. and Le Blanc, J. M. 1961. Yield of Neutrons from the T(d,n)^{4}He reaction for 6 to 11.5 MeV deuterons. *Phys. Rev.* 122: 164–168.

Iida, T., Kosuga, M., Sumita, K., and Heikkinen, D. W. 1990. Effects of 14 MeV neutron irradiation on optical components for fusion diagnostics. *J. Nucl. Sci. Technol.* 27(7): 651–662.

Jarmie, N. 1980. Low energy nuclear fusion data and their relation to magnetic and laser fusion. Los Alamos National Laboratory, Report LA-8087, UC-34c.

Jarmie, N., Brown, R. E., and Hardekopf, R. A. 1983. D(t,α)n reaction cross sections at low energy, in Bockhoff, K.H. (Ed.) *Nuclear Data for Science and Technology, Proceedings of the International Conference,* Antwerp, Belgium, September 6–10, 1982: pp. 318–325.

Jarmie, N. and Seagrave, J. D. 1957. Charged particle cross sections. Los Alamos National Laboratory, Report LA-2014.

Jarvis, R. G. and Roaf, D. 1953. Comparison of D-T and D-^{3}He at low energies. *Proc. Roy. Soc.* 218: 432–438.

Kammeraad, J. E., Hall, J., Sale, K. E., Barnes, C. A., Kellogg, S. E., and Wang, T. R. 1993. Measurement of the cross-section ratio 3H(d,γ)5He/3H(d,α)nat 100 keV. *Phys. Rev. C* 47: 29–35.

Katsaurov, L. N. 1962. Investigation of the reaction D(t,n)^{4}He by the thin-target method in the energy region from 70 to 750 keV. Akad. Nauk, USSR, Trudy, Fizicheskii Inst. 14: 224.

Kramer, D. 2014. Livermore ends LIFE. *Phys. Today* 67(4): 26–27.
Kunz, W. E. 1955. Deuterium + He^3 reaction. *Phys. Rev.* 97: 456–462.
Navratil, P. and Quaglioni, S. 2012. Ab Initio Many-Body Calculations of the $^3H(d,n)4He$ and $^3He(d,p)^4He$ Fusion Reactions. *Phys. Rev. Lett.* 108: 042503.
Lauritsen, T. and Ajzenberg-Selove, F. 1966. Energy levels of light nuclei, A = 5. *Nucl. Phys.* 78: 1; revised manuscript August 11, 2008.
Mori, T. and Nakagawa, M. 1992. Benchmark calculation for deep penetration problem of 14 MeV neutrons in iron. *J. Nucl. Sci. Technol.* 29(11): 1061–1073.
Murphy, T. J., Barnes, C. W., Berggren, R. R. et al. 2001. Nuclear diagnostics for the National Ignition Facility. *Rev. Sci. Instrum.* 72: 773–779.
Paulsen, A. and Liskien, H. 1964. Angular distribution for the $T(d,n)He^4$ reaction at 1 and 3 MeV deuteron energy. *Nucl. Phys.* 56: 394–400.
Perkins, R. B. and Simmons, J. E. 1961. $T(d,n)He^4$ reaction as a source of polarized neutrons. *Phys. Rev.* 124: 1153–1154.
Rudin, H., Streibel, H. R., Baumgartner, E., Brown, L., and Huber, P. 1961. Eine Quelle Polarisierter Deuteronen und Nachweis der Polarisation durch die (d,t) Reaktion. *Helv. Phys. Acta* 34: 58–84.
Serpico, P. D., Esposito, S., Iocco, F., Mangano, G., Miele, G., and Pisanti, O. 2004. Nuclear reaction network for primordial nucleosynthesis: A detailed analysis of rates, uncertainties and light nuclei yields. *J. Cosmol. Astropart. Phys.* 12: 010.
Sheffield, J. 2013. *Fun in Fusion Research.* Elsevier Inc., Philadelphia, PA.
Stewart, L., Brolley, J. E., and Rosen, L. 1960. Interaction of 6 to 14 MeV Deuterons with helium three and tritium. *Phys. Rev.* 119: 1649–1653.
Sudac, D., Nad, K., Obhodas, J., Bystritsky, V. M., and Valkovic, V. 2014. Loss of the associated α-particles in the neutron generator API-120. private communication, unpublished.
Valkovic, V., Furic, M., Miljanic, D., and Tomas, P. 1970. Neutron-proton coincidence measurements from neutron-induced breakup of the deutron. *Phys. Rev. C* 1: 1221–1225.
Valkovic, V., Miljanic, D., Tomas, P., Antolkovic, B., and Furic, M. 1969. Neutron-charged particle coincidence measurement from 14.4 MeV neutron induced reactions. *Nucl. Instr. Meth.* 76: 29–34.
Winterberg, F. 2013. Efficient energy conversion of the 14 MeV neutrons in DT inertial confinement fusion. *J. Fusion Energy* 32: 117–120.
Zweben, S. J., Darrow, D. S., Herrmann, H. W. et al. 1995. Alpha particle loss in the TFTR DT experiments. *Nucl. Fusion* 35: 893–917.

Chapter 3

Sources of 14 MeV Neutrons

3.1 INTRODUCTION

In 1920, Rutherford postulated that there were neutral, massive particles in the nucleus of atoms. James Chadwick, a colleague of Rutherford, discovered the neutron in 1932. He bombarded a beryllium target with alpha particles producing neutrons that recoiled into a block of paraffin. Chadwick's apparatus, the first neutron generator (NG), is schematically described in Figure 3.1.

By measuring protons emerging from the paraffin with a Geiger counter, Chadwick inferred that the neutron had a mass comparable to that of the proton. Soon Enrico Fermi recognized that neutrons could be used to produce radioactive nuclides, which sent off Emilio Segre to "get all of the elements in Mendeleev's table." Together they bombarded over 60 elements discovering, for example, a 1–2 day activity from bombarding gold. Fermi was awarded the 1938 Nobel Prize for this work. George de Hevesy helped Fermi obtain samples of rare earths, and he too became interested in activating these materials. His assistant Hilde Levi irradiated dysprosium with neutrons producing a large amount of radioactivity that decayed with time (Hevesy 1962). This episode triggered the realization that the half-lives and magnitudes of induced activities could be used to identify and quantify trace elements and led to the invention of neutron activation analysis (NAA).

A summary of commonly used neutron production reactions is given in Table 3.1. The total yield F of neutrons produced in a nuclear reaction is given by

$$F = N\sigma\varphi$$

where

F is the neutron yield per second

N is the number of target nuclei per square centimeter of target

σ is the reaction cross section in cm^2

φ is the incident particle rate per second

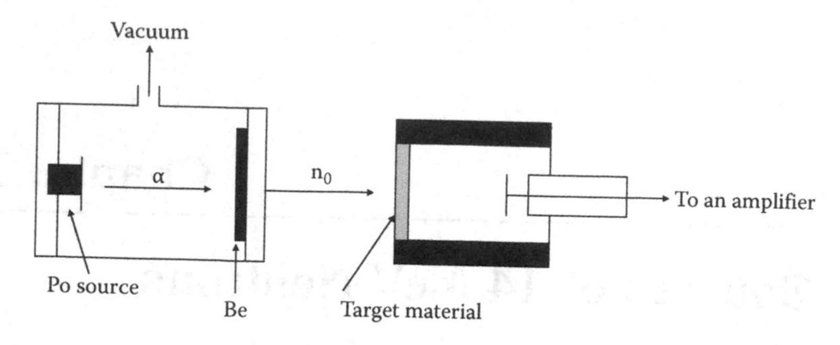

Figure 3.1 Detector used by James Chadwick to discover the neutron. Inside the detector, particles from a radioactive source hit a beryllium target. The reaction $^{9}Be + \alpha \rightarrow {}^{12}C + n$ produced neutrons that were detected when they knocked protons of paraffin wax.

TABLE 3.1 NEUTRON-PRODUCING NUCLEAR REACTIONS

Reaction	Threshold (MeV)	Q Value (MeV)	Threshold Neutron Energy (MeV)
$^{2}H(d,n)^{4}He$	0	+3.266	2.448
$^{3}H(p,n)^{3}He$	1.019	−0.764	0.0639
$^{3}H(d,n)^{4}He$	0	+17.586	14.064
$^{7}Li(p,n)^{7}Be$	1.882	−1.646	0.0299
$^{9}Be(\alpha,n)^{12}C$	0	+5.708	5.266
$^{12}C(d,n)^{13}N$	0.328	−0.281	0.0034
$^{13}C(\alpha,n)^{16}O$	0	+2.201	2.07

The incident particle rate can be expressed in terms of beam current by

$$\varphi = \frac{6.25 \times I}{C \times 10^{8}}$$

where

I is the beam current in amperes
C is the charge of the particle in Coulomb

A neutron is considered monoenergetic when the energy spectrum consists of a single line with an energy width that is much less than the energy itself. Most measurements to date have been done using the "big four" reactions $^{3}H(p,n)^{3}He$, $^{7}Li(p,n)^{7}Be$, $^{2}H(d,n)^{3}He$, and $^{3}H(d,n)^{4}He$ utilizing cyclotrons or electrostatic accelerators (Van de Graff, Cockcroft–Walton accelerators). Some of the properties of these reactions are summarized in Table 3.1. In the case of d–d and d–t reactions, the large positive Q-values and the low atomic numbers make it possible to produce high yields of fast neutrons even at low incident deuteron energies. The relative intensity of d-d,

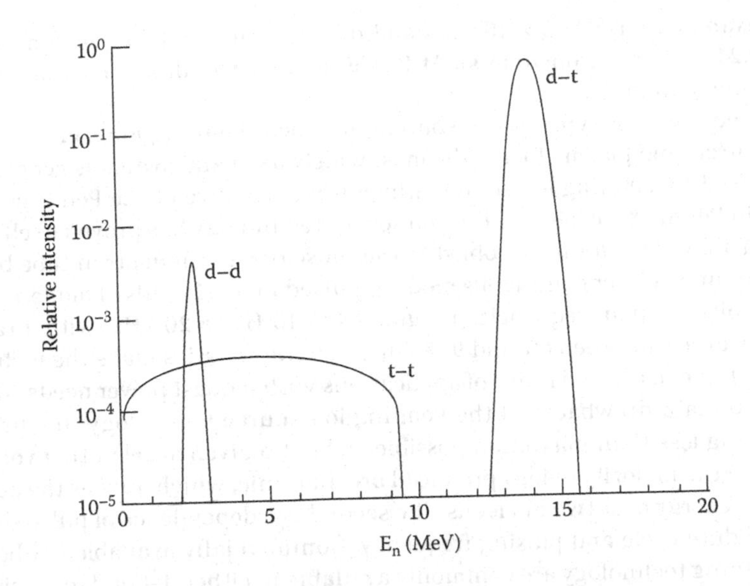

Figure 3.2 The relative intensity of d–d, d–t, and t–t fusion neutron energy spectra. (After Leung, K.-N. et al., Compact neutron generator development and applications, in *16th WCNDT 2004—World Conference on NDT*, Montreal, Canada, August 30–September 03, 2004.)

d-t and t-t fusion neutron energy spectra are shown in Figure 3.2 (after Leung et al. 2004). Another requirement for the high yield is a small energy loss of the projectile in the target and a high neutron production cross section. Since the $^3H(d,n)^4He$ reaction shows a broad resonance with a maximum value of 5 barns at $E_d = 107$ keV (see Figure 2.2), this reaction allows the production of fast neutrons with energies around 14 MeV already with small accelerators.

Neutrons produced from the d–t reaction are emitted isotropically from the target. Neutron emission from the d–d reaction is slightly peaked in the forward (along the axis of the ion beam) direction. In both cases, the He nucleus (α particle) is emitted in the exact opposite direction of the neutron. In order to estimate neutron flux density n (neutrons per square cm per second) we can use a simple relation:

$$n = \frac{N}{4\pi R^2}$$

where

N is the total neutron output from neutron generator (in neutron per second)

R is the distance from the deuterium or tritium target inside neutron tube to the location where we measure neutron flux

For example, for N = 3 × 10^{10} n/s and distance of R = 1 m neutron flux is n = 0.24 × 10^6 neutrons/(cm^2s). At R = 50 m, neutron flux is only about 10^2 neutrons/(cm^2s).

Three different types of ion sources are used: Penning ion source, spark ion source, and plasma focus. The most widely used and available generators are based on Penning ion source, which is a derivative of the Penning trap, used in Penning ion gauges. This ion source technology has proven itself over the past few decades to be robust in the sense of a given neutron tube being able to run in either continuous mode or pulsed mode. In pulsed mode, a tube can typically span frequencies ranging from 10 Hz to 20 kHz, with a range of duty cycles between 5% and 95%. These characteristics make the technology suitable to a broad range of applications with modest power needs. There are two main drawbacks of the Penning ion source technology. The first is achieving less than maximum possible yield at a given accelerating voltage because the majority of ions produced are diatomic, which divides the acceleration energy over two nucleons. The second is a dependence of pulse shape on the duty cycle and pulsing frequency. Commercially available NG based on Penning technology are commonly available in either d–t or d–d versions. Those with d–t range in yield from 10^7 to 10^{10} n/s, while those with d–d range from 10^5 to 10^8 n/s.

The basic design of a modern compact Penning diode NG (see Figure 3.3) consists of a source to generate positively charged ions and a diode structure to accelerate the ions (usually up to ~110 kV) to a metal hydride target loaded with either deuterium, tritium, or a mixture of the two. A gas control reservoir,

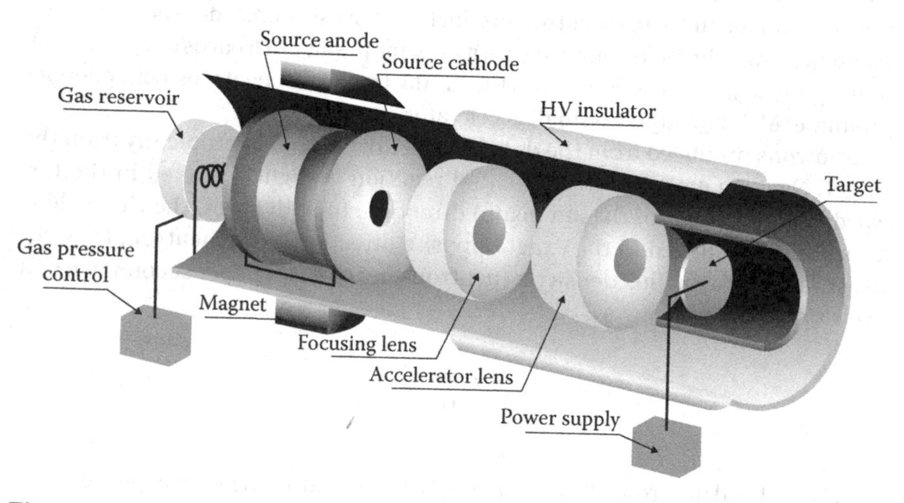

Figure 3.3 A schematic illustration of the internal structure of a compact neutron generator.

attached to the tube and made of a metal hydride material, is used to store the gas. The most common ion source used in NGs is a cold cathode or Penning ion source. This simple ion source consists of a hollow cylindrical anode (usually biased 1–2 kV) with cathode plates at each end of the anode, usually at ground potential. An external magnet is arranged to generate a coaxial field of several hundred gauss within the ion source.

When deuterium and/or tritium gas is introduced into the anode at a pressure of about 0.1 Pa (10^{-3} Torr), the electric field between the anode and cathodes ionizes the gas. Electron confinement is established in this plasma because of the orientation of the electric and magnetic fields, which forces the electrons to oscillate back and forth between the cathode plates in helical trajectories. Although some low energy electrons are lost and strike the anode, which creates more secondary electrons, most remain trapped and ionize more gas molecules to sustain the plasma. The ions are not similarly trapped, and when they strike the cathodes, they also release secondary electrons, which enter the plasma and help sustain it. Ions, however, can escape the chamber into the acceleration section of the tube through a hole at the center of one of the cathodes, called the exit cathode.

Construction methods used in building sealed neutron tubes are based on joining techniques such as welding, metal brazing, ceramic to metal brazing, and glass to metal seals. Materials used in accelerator neutron systems include glass, ceramics, copper, iron, different alloys of stainless steel, and Kovar. Most compact accelerator neutron tubes are loaded with approximately 10^{10} Bq of tritium; for comparison, a typical tritium exit sign used in an airplane or hotel might contain as much as 10^{11} Bq of the isotope.

3.2 LABORATORY SIZE ACCELERATORS

Neutron generators are small accelerators consisting of vacuum, magnetic, electrical, and mechanical components; radiation sources; cooling circuits; and pneumatic transfer systems. There are various types of ion sources, beam accelerating and transport systems, targets, high voltage and other power supplies, neutron and tritium monitors, and shielding arrangements. The operation, maintenance, and troubleshooting of a neutron generator require well-trained technicians who can successfully undertake not only preventive maintenance of the machine but also its upgrading (IAEA 1996). The principle of the neutron generator is shown in Figure 3.4.

The ion source, the extraction, the focusing, and the gas supply units placed on the High Voltage (HV) terminal need electric power, cooling, and insulated remote control systems. The ion source is RF or Penning type for standard (commercial or medium size) neutron generators while it is a duoplasmatron

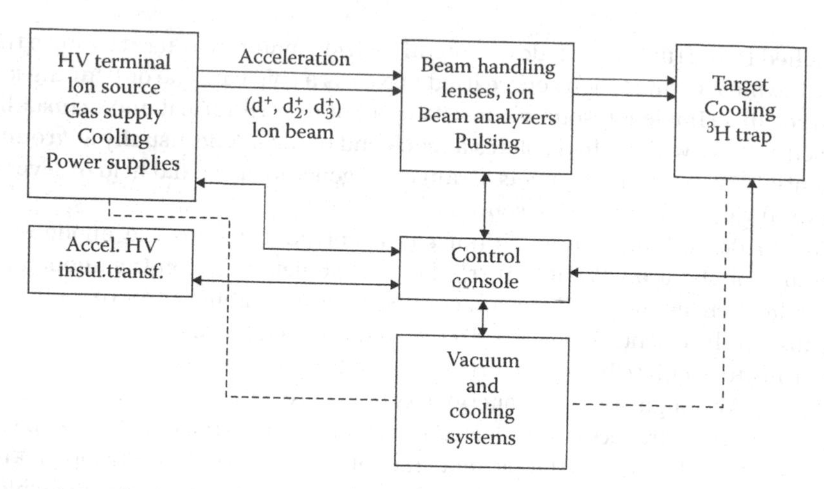

Figure 3.4 Block diagram of neutron generator.

(duopigatron) if high currents are required. The deuterium gas flow is regulated into the ion source by needle, thermomechanical, or palladium leaks. As the ion beam analyzer requires relatively high power, the separation of the d beam is made usually after acceleration.

The pulsing of the ion source allows the production of a pulsed beam, that is, 14 MeV neutron burst, in the μs range, while the nanosecond pulsing can be made before and after acceleration using special bunching units.

The acceleration tube is generally a homogeneous field type allowing a high pumping speed for the vacuum pumps situated at the ground potential. The intense neutron generators (deuteron beam current over 10 mA) utilize strong focusing single-gap or two- and three-gap tubes by which the space charge effects are taken into account.

The high voltage power supplies are usually Cockcroft–Walton voltage multipliers, Felici-type electrostatic machines, medium frequency Allibone-type voltage multipliers, insulated core transformer HV supplies, or parallel-powered Dynamitron voltage multipliers. The commercial neutron generators are modest machines with production yields of about 10^{11} and 10^{9} n/s for d–t and d–d reactions, respectively. They are utilized in basic nuclear research, education, and technology, for measurement of nuclear data, and in laboratory exercises to study the different interactions of neutrons and detection of charged particles and neutrons. They are also used in accelerator technology for activation analysis, prompt radiation analysis, irradiation, effects of fast neutrons, neutron dosimetry, etc. Practically no technological development in the field of commercial neutron generators has been reported until recent years and only a few companies were producing these moderate (150–300 kV, 1–2 mA) machines.

Here we intend to present some recent development in neutron generator described by Hardik et al. (2012). In the generator described in his work, a high beam current is extracted from an RF-driven ion source and accelerated toward the target. The generator is operated with a ^{2}H–^{3}H gas mixture and a beam-loaded target, that is, the d/t beam is driven into the target matrix for achieving acceptable lifetimes. The requirements of a sealed tube for tritium operation posed several design challenges including ultra high vacuum (UHV) compatible construction, low pressure ion source operation, and the need to limit the tritium inventory (Ludewigt et al. 2007). Neutron generator components listed are a gas reservoir and pressure regulator for supplying the ^{2}H/^{3}H gas mixture and keeping a stable pressure inside the tube, and an ion getter pump for the removal of helium and other contaminant gases (Reijonen et al. 2007). The expected 14 MeV neutron yield for a 100 keV, 50% t+/50% d+ beam impinging on a fully loaded Ti-target (Ti-^{2}H-^{3}H) was about 2.2×10^{13} n/C. However, extrapolation from their experimental results for beam-loaded d–d generators gave about half that value, likely due to an incomplete beam loading of the titanium layer.

The absolute neutron yield of the 14 MeV Frascati neutron generator (FNG) was routinely measured by means of the associated α-particle method with a silicon surface barrier detector, SSD. In the paper by Angelone et al. (1996) the work carried out to characterize the neutron source in terms of absolute intensity and angle-energy distribution of the emitted neutrons is described. The development of the measuring setup and the assessment of the measurement results are also reported. A complementary calibration procedure for validating the SSD results, based on the use of fission chambers and the activation technique, is also reported. An accurate analysis of the system has been performed via the Monte Carlo neutron and photon MCNP transport code. A detailed model of the neutron source that includes ion slowing down has been inserted into the MCNP code to permit a numerical calibration of the neutron source for comparison with the experimental results. The resulting agreement among the various methods is very good considering the uncertainties, and an accuracy of ±2% is achieved for the measurement of the 14 MeV neutron yield of the FNG (Angelone et al. 1996).

The absolute fast neutron spectrum generated by a 14 MeV neutron generator was determined by means of a liquid scintillation recoil-proton spectrometer together with threshold foil detectors. The liquid scintillation spectrometer provides the relative neutron energy spectrum. Knowing this relative spectrum, one threshold detector alone is enough to convert the relative spectrum into an absolute spectrum. In practice, more than one threshold detector is used to confirm the result. The method of obtaining the absolute spectrum is described in paper by Chuang et al. (1979), and it is estimated to have an uncertainty of about ±8%.

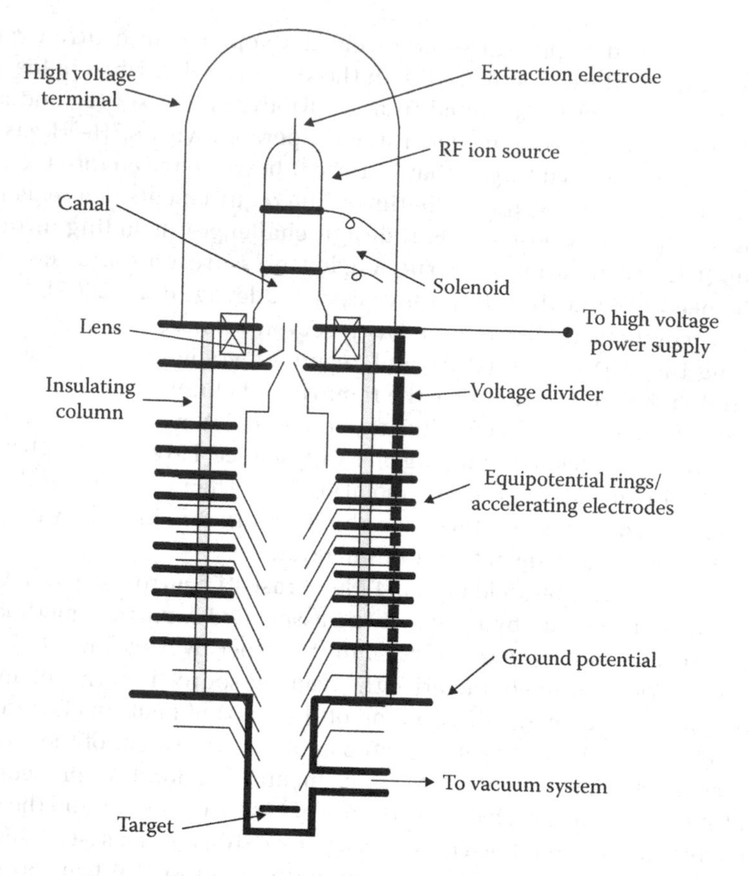

Figure 3.5 Potential drop neutron generator.

A typical laboratory-scale d+t neutron generator is shown in Figure 3.5. This will occupy a small room and can be used to produce neutrons fluxes in the range of 10^{11} n/s. Here we describe the end part of the Cockcroft–Walton accelerator at the Institute Ruder Boskovic (IRB) in Zagreb, Croatia, which is used as a neutron generator. The deuteron beam hits the tritium target placed inside the beam tube where the α-particle detector is also placed at some distance from t-target. This arrangement allows the use of electronically collimated neutron beam (Figure 3.6).

The tritium target is attached to the copper substrate for heat dissipation purpose, while the tritium is implanted into titanium, as shown in Figure 3.7. In situ measurements of depth profiles of tritium in a titanium tritide target have been reported by Okuda et al. (1986). The method used was the ion beam analysis using the $^{3}H(d,\alpha)n$ nuclear reaction. The initial distribution of tritium

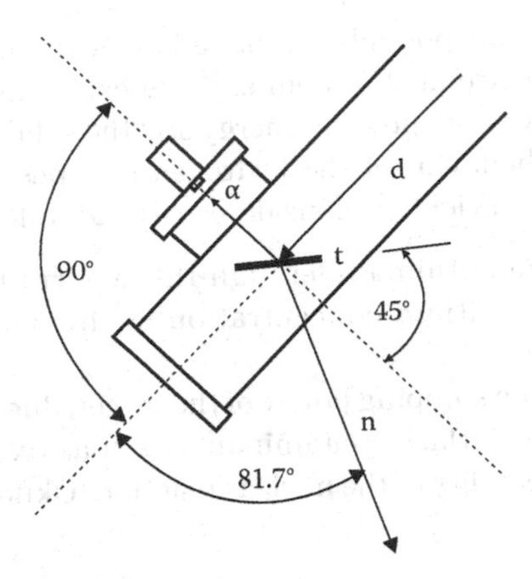

Figure 3.6 Neutron production geometry.

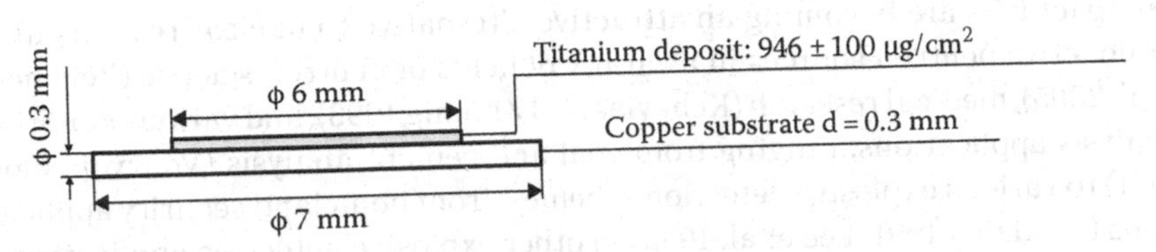

Figure 3.7 Tritium target on copper substrate.

in the unirradiated target has been observed to be nearly uniform over the depths. After the irradiation of 390 keV $^{2}H_{3}^{+}$ ions at a temperature of about 10°C, a dip has been found in the depth profile around the depth of the projected range of the ions. By the successive isochronal annealings at temperatures below 130°C, the tritium has been uniformly redistributed.

The depth profile of tritium in the tritium target should be known in evaluating the source energy spectrum of the neutrons; the concentration of tritium gives the stopping power for the incident deuterons and the reaction probability. In usual evaluation the distribution of tritium is assumed to be uniform in the target. This depth profile is approximately valid except for the near-surface region. The profile near the surface strongly depends on the contamination with oxygen. For the thin target that is used to produce monoenergetic neutrons, the surface conditions greatly influence the real neutron spectrum. These facts indicate the importance of the depth profiling of tritium for the evaluation of the source neutron spectrum, especially in the near-surface region. The depth resolution determines the energy resolution of the evaluated neutron spectrum. The depth resolution near the surface, 50 nm in full width at half maximum (FWHM), corresponds to the energy resolution of about 20 keV evaluated for the neutrons emitted in the forward direction to the incident deuterons.

The depth resolution can possibly be made higher by improving the energy resolution of the measurement system and the experimental conditions, for example, the accuracy of the probing energy and the solid angle of detection.

According to the behavior of the hydrogen isotopes, main causes of the reduction of the neutron yield are considered to be as follows:

1. The migration of tritium under high-fluence irradiation of deuterons reduces the tritium concentration in the neutron-generating region.
2. The increase of the stopping power of the target, due to the deuterium implanted or the surface contaminants such as oxygen and carbon, reduces the probability of the nuclear reaction (Okuda et al. 1986).

3.3 SMALL AND SEALED-TUBE NEUTRON GENERATORS

Compact NGs are becoming an attractive alternative to nuclear reactors and radioactive neutron sources in a variety of fields of neutron science (Reijonen et al. 2005), medical research (Kehayias and Zhuang 1993), and various material analysis applications, ranging from coal and cement analysis (Vourvopoulos 1991) to various explosive detection schemes, from homeland security applications (Grodzins 1991; Lee et al. 1995) to other explosive detection applications, like land mine detection (Womble et al. 2003). Traditionally, compact NGs have been used in oil well logging industry, using d+t reaction for high energy 14 MeV neutron productions. Currently the main commercial manufacturers of compact NGs are Thermo Electron, Sodern EADS, and Schlumberger. These NGs generate d+t neutron yield in the range of 10^8–10^{11} n/s. There are yet others, which are made by universities or national laboratories, like VNIIA-made generators (sold in the United States by Del Mar Ventures) and Lawrence Berkeley National Laboratory's Plasma and Ion Source Technology (P&IST) group (generators commercialized by Adelphi Technology Inc.). The sealed tube neutron generators or neutron tubes are 14 MeV neutron sources used in in situ geological measurements (bore-hole logging), in hospitals, and in chemical, biological, and industrial laboratories. Most neutron tubes have a Penning ion source, a one- or two-gap acceleration section, a tritium target, and deuterium–tritium gas mixture filling system. The pressure of the gas is controlled by a built-in ion getter pump and/or gas storage replenisher. The tritium displaced from the target during the operation is absorbed by the gas occlusion elements and is accelerated later back into the target, resulting in an extended target life. The advantage of the neutron tubes is their small size that makes them suitable for geological bore-hole logging or isocentric cancer therapy in hospitals. The disadvantages are their limited lifetime and the relatively high neutron production cost.

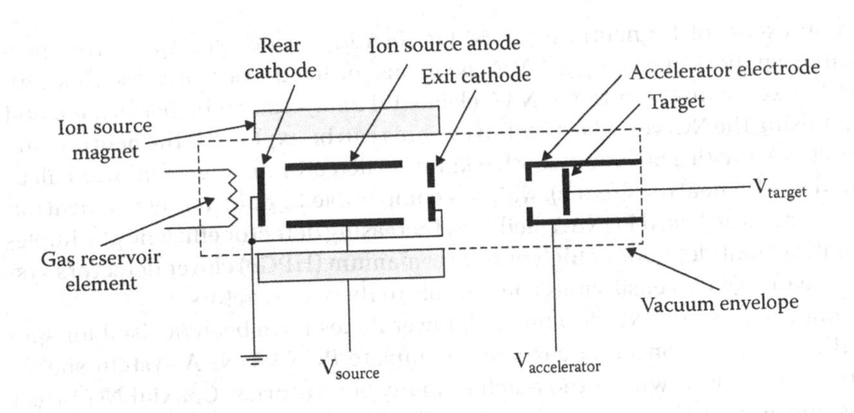

Figure 3.8 Generic neutron tube schematic.

The compact accelerator is composed of the (1) ion source, (2) accelerating region, and (3) target section. The ion source is often a low pressure gas, cold cathode source (Penning trap) requiring crossed electric and magnetic fields to contain the ions until they are accelerated from the cathode region. The schematic of a generic neutron tube is illustrated in Figure 3.8. The target is usually composed of a thin film made of a metal such as titanium, scandium, or zirconium, which is deposited on a copper or molybdenum substrate (Whittemore 1990). Titanium, scandium, and zirconium form stable chemical compounds called metal hydrides when combined with hydrogen or its isotopes. These metal hydrides are made up of two hydrogen (deuterium or tritium) atoms per metal atom and allow the target to have extremely high densities of hydrogen. This is important to maximize the neutron yield of the neutron tube. The gas reservoir element also uses metal hydrides as the active material. Many neutron tubes are designed such that the gas reservoir element and the target each incorporate equal amounts of deuterium and tritium. In these mixed gas tubes, both the ion beam and target contain 50% deuterium and 50% tritium. This allows the tubes to have very stable neutron yields over their operational life.

Recently, a new generation of NG has been developed in different laboratories that overcome many of the deficiencies of early ones. In particular at Lawrence Berkeley National Laboratory (LBNL/USA) compact NGs have been designed to produce up to at least 10^{11} n/s. The open design allows the continuous regeneration of the target so that these generators should be able to operate indefinitely at maximum neutron fluency. These generators can be continually pulsed on a very short time scale (<1 μs) allowing applications that could not be envisioned for most nuclear reactors. A tabletop NG suitable for prompt gamma neutron activation analysis (PGNAA) and NAA analysis, with moderator and shielding, has been conceptually designed and presented in LBNL report. Neutron fluxes, calculated with the General Monte Carlo N-Particle Transport Code (MCNP), are in excess of 10^8 neutrons/(cm^{2}s) for NAA in this design. Apart from particular applications using 2.5 MeV neutrons for irradiation at these thermalized fluxes, many traditional NAA elemental analyses can be performed, and by pulsing the NG, very short half-lives <1 ms can be exploited. The neutron flux for PGNAA with a favorable, low background detector arrangement was calculated as >10^6 neutrons/(cm^{2}s), which is comparable to guided thermal neutron flux at reactor-based PGNAA facilities. Increasing detector efficiency by implementing multidetector or high purity germanium (HPGe) clover detectors systems can provide sensitivities comparable to those at reactors.

Many additional NG designs with lower fluxes have been devised for specialized applications, and costs of a complete PGNAA/NAA system should run <300 k$, well within the reach of many laboratories. Coaxial NG target designs provide the maximum neutron flux, axial designs allow sample positioning within <1 cm from the generator, and low flux NGs can be used for teaching and handheld operation. Even the largest NG can be installed for bench top applications or in vehicles for transportable operation. The new generation of high performance, inherently safe NG can provide research quality facilities anywhere in the world at low cost without the concern for nuclear proliferation. It is also worthwhile mentioning new developments of d+t NGs in the field of associated particle imaging (API) technology. This technology uses the associated particles, helium nuclei, to add spatial resolution by time-of-flight and neutron trajectory computation to conventional PGNAA (IAEA 2012).

In the search for the best neutron source tailored for an application needs one has to consider several companies that could offer the style of source one is looking for. Following are the companies that are considered to offer the best neutron sources.

3.3.1 VNIIA

VNIIA or the All-Russia Research Institute of Automatics is a Russian company that produces approximately 10 models of neutron generators. These vary greatly in flux, size, and cost. However, their operating life is very short compared to other companies and their models are not easily refuelable as one would have to send them to Russia to have the service done, an almost impossible task to perform. VNIIA supplies neutron generators to the United States, the United Kingdom, Germany, and China, as well as to civilian customers. Some of the VNIIA neutron tubes are shown in Figure 3.9, while VNIIA plasma focus chambers are shown in Figure 3.10.

VNIIA is the only company in Russia that has the full-scale research and design, and fabrication capabilities to produce serially the portable neutron generators for a variety of applications with the neutron radiation of a wide

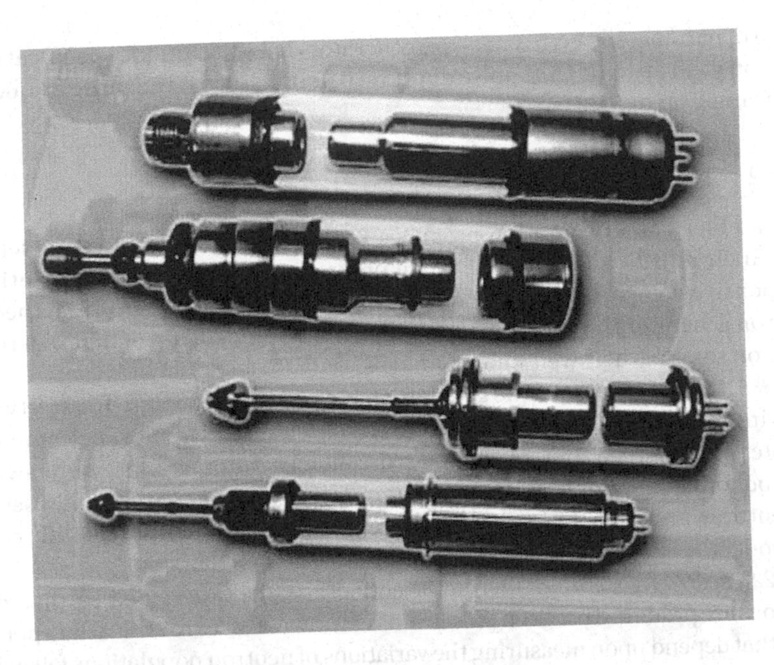

Figure 3.9 VNIIA neutron tubes.

Figure 3.10 VNIIA plasma focus chambers.

range of parameters. VNIIA neutron generators have the unique specifications: neutron yield up to 10^{12} n/s, pulse duration 0.01 μs to 10 ms, and pulse frequency from single up to 10 kHz.

3.3.2 Thermo Electron

Thermo Electron Corporation produces several neutron generator models as well. Their refueling and maintenance programs are very good and relatively inexpensive compared to some of the other companies. Most of the Thermo neutron generators have their power supplies mounted on top of them. Here we mention some of their well-known models:

P 385 neutron generator: The generator has a yield of 3×10^8 n/s. The safety features include key lock: on/off; emergency: on/off; one on electronics enclosure, one remote; normal-open and normal-closed contacts; automatic pressure switch interlock on accelerator head; warning lamp flashes when operating; and it senses current draw and light production. See more at http://www.thermoscientific.com/en/product/p-385-neutron-generator.html#sthash.0JYw70bf.dpuf.

P 221 neutron generator: Thermo Scientific™ P 211 is designed so that zero neutrons are produced between pulses. The features of P 211 are critical for applications that depend upon measuring the variations of neutron populations following the source burst. The P 211 neutron tube was originally designed by Sandia National Laboratories and incorporates the Sandia-designed "Zetatron" neutron tube. Pulse width: ≃10 μs; 1.0×10^6 n/pulse. See more at http://www.thermoscientific.com/en/product/p-211-neutron-generator.html#sthash.rD0R3QhI.dpuf.

MP 320 neutron generators: Thermo Scientific™ MP 320 is a lightweight, portable neutron generator suited for most demanding field or laboratory applications. See more at http://www.thermoscientific.com/en/products/neutron-generators.html#sthash.lgORtVNi.dpuf.

Main characteristics: typical lifetime—1200 h @ 1×10^8 n/s, pulse rate—250 Hz to 20 kHz continuous, pulse rise and pulse fall time—less than 1.5 μs, and minimum pulse width—5 μs. *Thermo Electron* and *Fisher Scientific* companies have merged into a company called Thermo Fisher Scientific Inc., http://www.thermo.com.

3.3.3 NSD Fusion

The NSD fusion neutron generator family includes a range of performance and configurations. The NSD NG 1E7 DDC unit has a dc high voltage power supply rated at 120 kV, 15 mA maximum (1.8 kW). This power is put into a glow discharge plasma load contained within the reaction chamber.

The NSD neutron generator family has three basic forms: a short electrode reaction chamber with 1.5 kW high voltage input power, a long electrode version with up to 18 kW input power, and a pulsed version with a 3 kW input rating. Each of these variants has a specific performance that is about ten times better than the generators with a solid target. This has resulted in a 24 kW input power at 160 kV neutron generator system with a maximum neutron output rate of more than 10^{10} n/s using deuterium–tritium for 14 MeV emission. Other machines are sealed d–t neutron generators with a maximum of 10^{10} n/s but with lifetimes of only a few hundred hours at this level and a few thousand hours if set to 10% of maximum with comparable manufacturing costs. The NSD devices are projected to have servicing intervals of 20,000 h of operation for d–d and 5 years for d–t due to the decay of tritium. A versatility advantage of the NSD neutron generator is that it provides a linear geometry or line source emission zone within the reaction chamber. The near field attenuation of a line source is approximately proportional to distance while a point source suffers the inverse square of distance.

3.3.4 EADS Sodern

EADS Sodern is a French company that produces three different models of neutron generators. They have excellent operating life and neutron fluxes. One of the marketing points EADS uses is how well their adaptive software works. They have made all of their neutron generators very easy to work with.

The SODITRON neutron tube is a small sealed d–t accelerator being used for neutron production. It can work either in continuous or pulsed mode. This ceramic neutron tube utilizes a Penning type ion source, a d–t gas reservoir, and a tritiated target. Extraction and acceleration of ions is possible up to 100 kV. Neutron emission at 14 MeV is adjustable up to 2×10^8 n/s in average value or up to 10^5 n/μs during 10 μs pulses. Life testing has demonstrated an average tube lifetime between 500 and 4000 h or more, depending on working conditions. Tube lifetime is strongly related to neutron output, duty cycle, high voltage level, temperature, housing, etc. The use of the SODITRON tube in industrial applications such as online inspection of materials is now possible as an alternative to isotopic neutron sources (Bach et al. 1997).

3.3.5 Adelphi

Adelphi is a new company to the neutron generator market and should have several models of generators available the near future (Figure 3.11). They are using a new technology based on research done by Lawrence Berkley National Labs. Their models will be able to produce neutrons with a flux of several orders of magnitude higher than any of the other models available presently on the market. They do have working high-flux d–d models and their d–t models are also available. They use RF plasma to generate the neutrons.

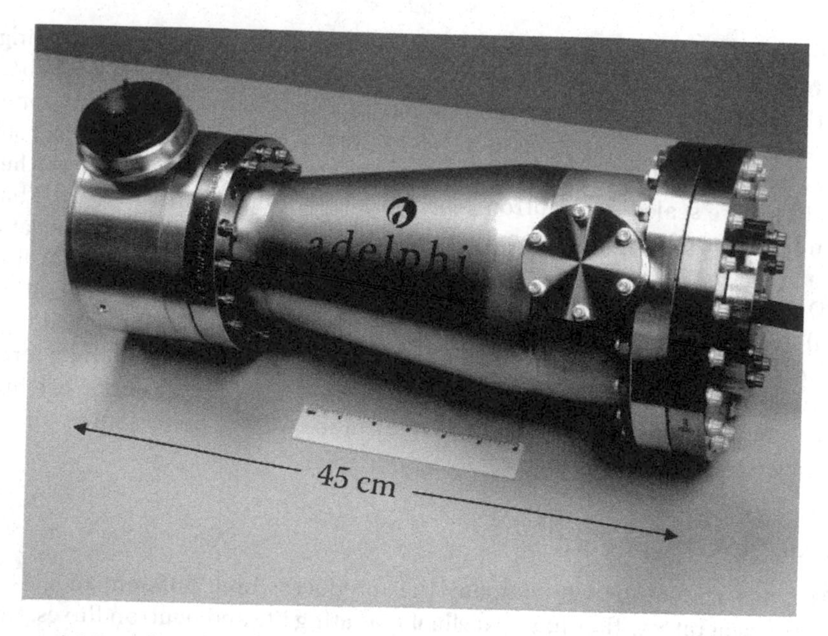

Figure 3.11 Adelphi model DD-108 neutron generator. The components at high voltage potential are shielded by placing the entire accelerator structure and its components inside modified conflate stainless steel vacuum pipe, for details see http://www.adelphitech.com.

The Adelphi generator is approximately 18 in. in length and 10 in. in diameter; it emits 1×10^{13} n/s. Some customers are concerned with the high cost of approximately $300,000; however, this cost is acceptable considering the source strength one can get for it.

The Berkeley Lab neutron generators consist of RF-driven plasma ion sources, extractors of various designs, acceleration electrodes, and titanium-covered targets. Conventional generators are usually short-lived because the target's isotopes are quickly consumed. The target in the Berkeley Lab generators, however, is constantly replenished by ions from the plasma source. These devices may last thousands of hours longer than conventional generators.

The Berkeley Lab portfolio includes a multiple beam system for imaging luggage, small neutron tubes for oil well logging while drilling (LWD), as well as designs suitable for cargo interrogation, tumor therapy, and structural inspection. Because neutron generators can be used for imaging and interrogating so many materials, the applications listed for each invention are not exhaustive.

The group led by Dr. Leung has developed a coaxial RF-driven plasma ion source for a compact cylindrical neutron generator (Leung 2002) that operates with high current density and produces high atomic deuterium or tritium ion species. The plasma and extraction electrodes electrostatically control the passage of ions out of the ion source. These electrodes contain longitudinal slots along their circumferences so that the ions radiate out of the source in a full 360° pattern. The single target coaxial neutron generator with dimensions of 26 cm in diameter and 28 cm in length is expected to generate a 2.4 MeV d–d neutron flux of 1.2×10^{12} or a 14 MeV d–t neutron flux of 3.5×10^{14} n/s.

An adaptation of the coaxial generator design uses a versatile nested configuration with multiple plasma and target layers where ion beams can impinge on both sides of the targets to enhance neutron yield. A generator with this nested design and dimensions of 48 cm in diameter and 35 cm long should generate a neutron output 10 times higher than the single target generator described earlier. Thus, d–d neutron output higher than 10^{13} n/s and d–t neutron output of 10^{15} n/s should be attainable.

The Berkeley Lab radiography system (Berkeley Lab 2009) is compact and is expected to be less expensive to construct than conventional accelerator systems. It is well suited for homeland security luggage inspection systems because the coaxial design produces multiple beams of fast neutrons, allowing several pieces of luggage on a carousel to be imaged at the same time. Neutrons produced from these t–t fusion reactions range in energy from 1 to 9.4 MeV and therefore can image a wide range of content in a single scan.

Dr. Leung's generator employs a focused ion beam accelerator with a titanium tubing target that oscillates to minimize heat buildup while maximizing neutron output. The 2 mm neutron source can also be designed to generate monoenergetic d–d or d–t neutrons or large fluxes of thermal neutrons by incorporating a moderating device. This potentially low cost device uses a simple cooling system and can be portable. Other Berkeley Lab designs can be coupled with elements of this invention for t–t, d–d, or t–d spectroscopic interrogation of luggage or other samples.

There is a need for calibration service for the many 14 MeV neutron generators now being employed for security and industrial applications. A new calibration service for 14 MeV neutron generators is being developed by the National Institute of Standards and Technology (NIST) (Heimbach et al. 2011). The calibration is done by the activation of standardized aluminum ring, with NaI gamma-ray spectrometry on the activated ring at NIST.

Flux measurement of an accelerator-based d–t neutron generator was achieved by the activation technique (Jakhar et al. 2008). The neutron generator can produce maximum neutron yield of 1.9×10^{10} n/s at the tritium target. For the measurement of 14 MeV neutron flux, $^{63}Cu(n,2n)^{62}Cu$ and $^{27}Al(n,p)^{27}Mg$ reactions are most suitable because of their higher threshold energies 11.03 and 3.25 MeV, respectively. Measurement of γ-ray activities from ^{62}Cu and ^{27}Mg is used to measure 14 MeV neutron flux. NaI(Tl) scintillation and HPGe detectors were used for the γ-radiation measurement from ^{62}Cu and ^{27}Mg, respectively. In the paper by Jakhar et al. (2008), the 14 MeV neutron flux measurement by ^{63}Cu and ^{27}Al activation was discussed. The method could also serve as a diagnostics tool for the measurement of high-energy neutron flux.

3.4 WELL LOGGING SEALED-TUBE ACCELERATORS

The neutron generator, a key part of pulse neutron logging instruments, is a neutron source based on mini-type accelerator. It consists of three sections: neutron tube, high voltage power supply, and control circuit. The desired pulse neutron beams may be provided by controlling the anode pulse frequency and the pulse width according to deuterium–tritium nuclear reaction principle. 14 MeV fast neutrons produced from this reaction interact with mediums around the borehole, and then neutrons or gamma rays are measured, which are produced from a series of reaction processes. The corresponding data are processed, and the relevant information of the medium around the tools is obtained and the theoretical foundations are provided to geologists.

There are several manufacturers of sealed-tube neutron generators, some fabricating only sealed-tube generators for oilfield services and others produce also units for different commercial and/or military applications. They are listed in Table 3.2. This variety of neutron generators is being used for the production of fast neutrons in the well. They all use the nuclear reaction $d+t \rightarrow \alpha+n$ within a sealed glass–steel tube as shown in Figure 3.12. The dimensions of the neutron emission unit (including power supply and electronics) are around 3.5 cm in diameter and 130 cm in length.

Next, we shall briefly describe SODERN Sodilog tube shown in Figure 3.13. The replenisher function is to create and regulate the right hydrogen pressure inside the tube. The ion source function is to create, when switched on, hydrogen ions that will be accelerated to the target. The very high voltage (VHV) function is to accelerate the beam in the gap in order to make the fusion reaction more likely. The target function is to have hydrogen (tritium and deuterium) atoms fixed and ready in front of the beam. Today SODERN only sells the Sodilog tube to customers who have developed their own neutron generator (housing and hard and soft electronics) based on this tube or are willing to develop such equipment.

Main characteristics of the tube are pulsed 14 MeV neutrons, 10^8 n/s typical average flux, and 10^9 n/s typical peak neutron yield. Duty cycle: 5% to continuous. Life time: more than 1000 h in oil logging environment; switchable and pulsable neutron source. More convenient to handle and export/import/store than a radioactive source! Safety: sealed tube with 120 GBq tritium inside. Mechanical characteristics: external diameter—25.4 mm (1 in.); length—224 mm; weight—<250 g and insulation suitable for SF_6 gas. Operating

TABLE 3.2 MANUFACTURERS OF COMPACT ACCELERATOR NEUTRON TUBES AND SMALL GENERATORS WITH ALPHA PARTICLE DETECTION

Company	Place, Country	Web	Tubes	Small NG with α-Detector	Tubes with α-Detector
VNIIA	Moscow, Russia	www.vniia.ru	Yes	Yes	No
Baker Hughes, Inc.	Houston, TX, USA	www.bakerhughes.com	Yes	No	No
EADS, SODERN	Paris, France	www.sodern.com	Yes	Yes	No
Halliburton Co.	Houston, TX, USA	www.halliburton.com	Yes	No	No
Schlumberger Ltd.	Princeton, NJ, USA	www.schlumberger.com	Yes	No	No
Thermo Electron Corp.	Colorado Springs, CO, USA	www.thermo.com	Yes	Yes	No

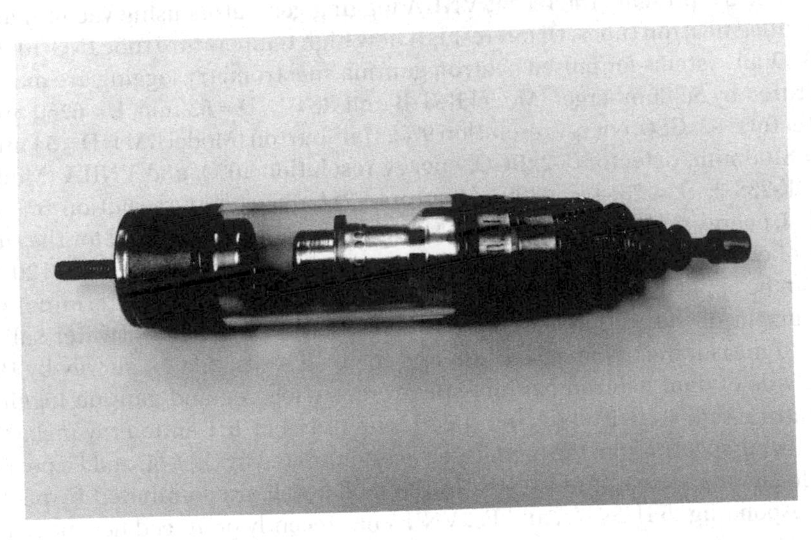

Figure 3.12 Sealed tube used in fast neutron generation.

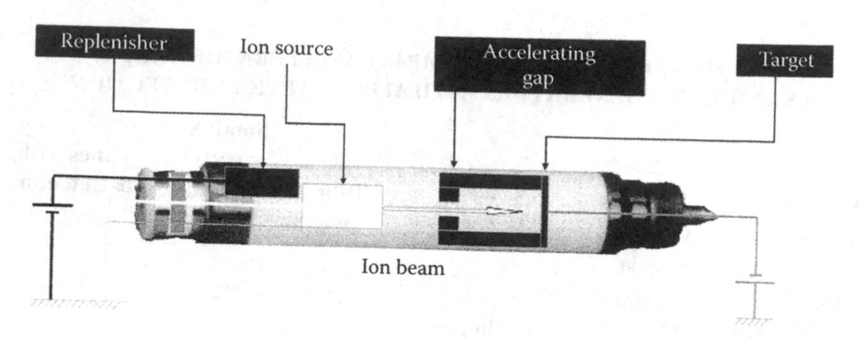

Figure 3.13 SODERN Sodilog tube.

temperature: up to 150°C. Electrical characteristics: from 50 to 100 kV, nominal is 85 kV; target current—from 10 to 100 μA. Ion source voltage: 0–2 kV. Ion source current: 0–5 mA.

Replenisher voltage: 0–5 V, nominal: 1 V. Replenisher current: 0–1.5 A.

Neutron generator output expressed in an average number of neutrons in a second can result in different peak neutron yield depending on pulse duration. Figure 3.14 shows two examples of pulsed mode resulting in the same average neutron output (10^8 n/s in this case) with very different peak output (10^9 n/s or 2×10^8 n/s). Actually, there are two limits in the neutron outputs: (1) the peak neutron yield is limited to about 10^{10} n/s, and (2) whatever your time mode is, one cannot have more than about 2×10^8 n/s in average.

Table 3.3 presents the recent VNIIA logging generators using vacuum and gas-filled neutron tubes. Of interest is a new high temperature tube ING-10-20-175. Dual systems for pulsed neutron gamma spectrometry logging are manufactured by Schlumberger (Model RST-B and RST-D, D = 63 mm, L = 6760 mm, detectors—2xGSO, energy resolution 9%); Halliburton (Model RMT, D = 54 mm, L = 8100 mm, detectors—2xBGO, energy resolution 10%), and VNIIA (Model AINK-73S-2, D = 73, L = 4300, detectors—2xLaBr, energy resolution 3%, or 2xCsI, energy resolution 9%). VNIIA Model AINK-73S-2 can be used for the survey of cased and open wells with maximum operating temperature of 120°C. It can be applied for (1) determination of rock composition, (2) determination of current oil saturation of reservoirs irrespective of formation water salinity, (3) measurements in cased side and inclined wells, and (4) survey by the methods of dual neutron gamma-spectrometry logging and gamma logging of natural radioactivity. Logging speed is up to 150 m/h. Gamma ray inelastic scattering spectra from the model well are dominated by C, O, Si, and Fe peaks, while gamma ray capture spectra in the model well are dominated by peaks corresponding to H, Si, Na, and Fe. VNIIA has recently produced new neutron tubes for logging generators as shown in Table 3.4.

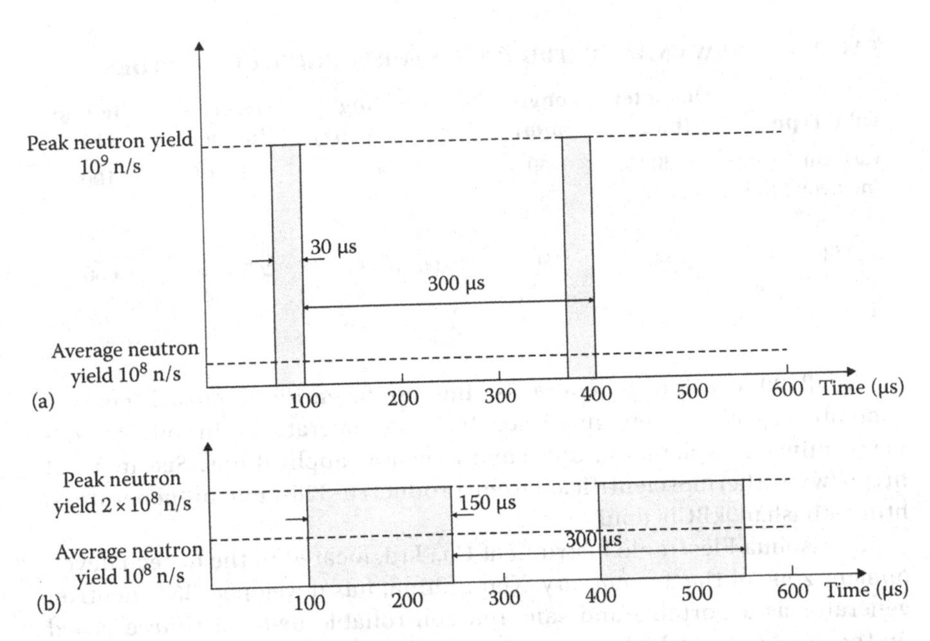

Figure 3.14 Two modes of operation with an average output of 10^8 n/s, (a) peak neutron yield of 10^9 n/s, pulse width 30 μs and (b) peak neutron yield of 2×10^8 n/s, pulse width 150 μs.

TABLE 3.3 VNIIA LOGGING GENERATORS USING VACUUM (VNT) AND GAS-FILLED (GFNT) NEUTRON TUBES

Type	Marked	Features	Flux (n/s)	Lifetime (h)	Diameter (mm)	Length (mm)	Frequency (kHz)
VNT	ING-10-20-120	General	1.0×10^8	150	34	1360	—
VNT	ING-10-20-150	High temperature	1.0×10^8	40	34	1360	—
VNT	ING-10-20-175	New—high temperature	1.0×10^8	100	34	1360	—
VNT	ING-11	Small	0.5×10^8	40	30	1720	—
GFNT	ING-08	Small	1.0×10^8	200	36	1360	0.05–10
GFNT	ING-063	For spectrum	1.5×10^8	300	60	930	10
GFNT	ING-062	High	2.5×10^8	300	70	900	0.4–10
GFNT	ING-061	General	1.0×10^8	300	70	1220	0.4–10

TABLE 3.4 NEW VNIIA NEUTRON TUBES FOR LOGGING GENERATORS

Tube Type	Diameter (mm)	Length (mm)	Operating Frequency (Hz)	Neutron Flux (n/s)	Lifetime (h)
Vacuum neutron tube VNT1-20	20	150	0–20	0.4×10^8	100
Gas-filled neutron tube GNT3-42	43	180	400–10,000	2.5×10^8	600

The B320 neutron generator is Thermo Scientific™ pulsed neutron generator specially configured for borehole operations in oilfield services, mineral exploration, and environmental applications. See more at http://www.thermoscientific.com/en/product/b-320-neutron-generators.html#sthash.ul9kBGbj.dpuf.

Xian Aohua Electronic Instrument Co., Ltd, located in the hi-tech development zone of the ancient city, Xian, China, has developed BNG neutron generator as a portable and safe and controllable neutron source based on the accelerator, which may operate in the hostile environment such as high temperature and high voltage and vibration of petroleum well logging (shown in Figure 3.15). It generates 14 MeV high energy fast neutrons by d–t reaction. Its neutron yield is controllable and its operating mode is selectable on requirements. It may generate neutrons continuously or in a pulsed beam whose frequency and width are adjustable. Specifications of the neutron generator are as follows: outside diameter—72 mm, maximal operating temperature 150°C for 2 h, and the cumulative total operating time of the neutron tube—300 h. The operating mode of ion source: dc or pulsed (100 Hz to 20 kHz). The operating mode of high voltage: dc or intermittent (this means it may increase abruptly or decrease abruptly); permissible maximum target voltage: −130 kV. Neutron yield: more than 1.5×10^8 n/s. The stability of neutron yield is greater than 10% within the whole operation temperature.

There are other reports on the development of neutron generators for logging. For example, Polosatkin et al. (2013) reported a development of miniature neutron generator for well logging. The compact neutron generator described in their presentation is composed of Cockcroft–Walton multiplier inside a gas-filled neutron tube.

A small neutron generator has been developed for meeting the need of well logging in oil fields. The miniature cold cathode Penning ion source and single electrode ion optics system are used in neutron generator. The good performance of the generator has been proved in the laboratory test and well logging

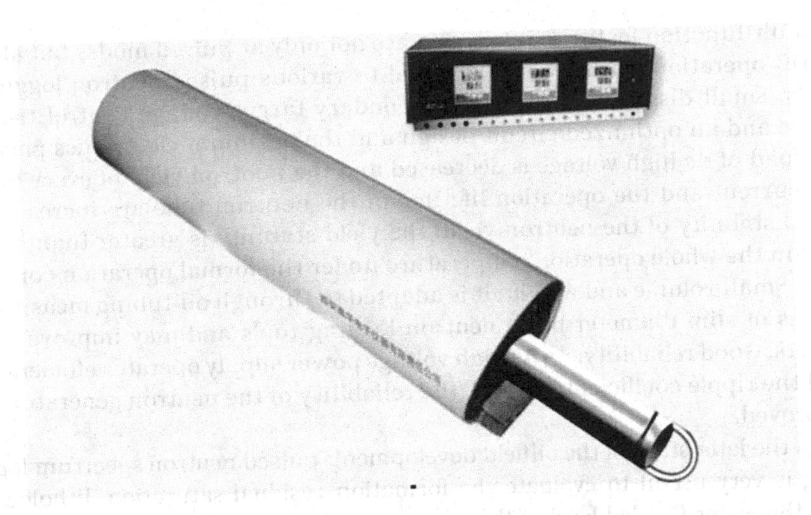

Figure 3.15 Neutron generator produced by Xian Aohua Electronic Instrument Co., China.

in oil field. The generator has good characteristics, such as small diameter, high temperature resistant, high neutron output, and good stability. At present, the neutron generator is the smallest logging generator in China, which has been used in oil field generally for conforming the geological data (Xiao et al. 2012).

There are five manufacturers of neutron generators in China: (1) Xi'an Petroleum Exploration Instrument Complex, (2) China Institute of Atomic Energy, (3) Northeast Normal University, (4) Lanzhou University, and (5) China Academy of Engineering Physics. Each manufacturer has one group related to the production and/or research about different intensity 14 MeV neutron generators.

SNG, slim diameter neutron generator for logging, produced by Xian Aohua Electronic Instrument Co., Ltd, located in Xian, China, is shown in Figure 3.31.

Specifications: outside diameter: Φ 43 mm, Tool length: 1250 mm. Maximum pressure of the housing: 80 MPa. Maximum operating temperature: 135°C. Neutron yield, at the earlier period: the neutron yield is 1.5×10^8 n/s when $70\ \text{kV} \leq V \leq 80\ \text{kV}$; at the latter period: while increasing the target voltage, the neutron yield may continuously be ensured to be 1.5×10^8 n/s.

The target voltage of the neutron tube is not greater than 95 kV. The operation mode of ion sources: dc or the pulsed mode (100 Hz to 20 kHz). The operation mode of high voltage: dc or intermittent status (i.e., it is permitted to increase high voltage or decrease high voltage instantaneously). The stability of neutron yield: the change of neutron yield ≤10% within the whole temperature range while the parameters such as VA and VT do not vary.

Multifunction features: it may operate not only at pulsed modes but also at DC operation status so it is adapted to various pulsed neutron logging tools. Small dissipation: using the secondary target voltage control technique and an optimized circuit design and the optimum electronics parts, the load of dc high voltage is decreased and the neutron yield of every target current and the operation lifetime of the neutron tube are increased. Good stability of the neutron yield: the yield stability is greater than 10% within the whole operation temperature under the formal operation condition. Small volume and weight: It is adapted to through oil-tubing measurements of slim diameter pulse neutron logging tools and may improve log effects. Good reliability: As its high voltage power supply operates efficiently and the ripple coefficient is small, the reliability of the neutron generator is improved.

In the later stage of the oilfield development, pulsed neutron spectrum logging is very useful to evaluate the formation residual saturation, lithology, and the water-flooded level of the reservoir in metal-cased boreholes. Pulsed neutron full-spectrum saturation logging (PSSL) tool is an integration of carbon/oxygen spectrum log, carbon/hydrogen spectrum log, chlorine spectrum log, gadolinium tracer spectrum log, and neutron lifetime log. This tool may efficiently measure the residual oil saturation in the formation outside of the medium such as the casing and the cement sheath under the condition of more than 10% porosity and several of salinities. And these log methods may alternately be used so that PSSL is capable of greatly improving measurement accuracy and the coincidence rate of interpretation.

The products of this Chinese company are used to

- Determine the residual oil saturation of the formation
- Explore and evaluate those omitted oil-bearing layers and gas-bearing layers in abandoned wells and those wells which need overhauling
- Measure and evaluate reservoirs and oil–water contact in the formation containing fresh water or water with lower salinity and unknown salinity where lithology is complex and mixed
- Evaluate the oil-containing reservoir where logging data provided by open hole logging and ordinary cased hole logging are not consistent with the behaviors of the reservoirs

In the water-drive reservoir project the products are used to evaluate oil saturation and fresh water migration of the formation in the oilfield.

- Using logging data, identify water layers and potential pay zones, evaluate the water -flooded level of pay zones, and identify the water-flooded mechanism of the reservoir and find water-bearing layers and provide evidences of controlling water and increasing oil production.

- Inspect cementing bond quality, check plugging effects, and seek for bypass channels and omitted layers.
- In the injection wells contaminated by isotopes, provide accurate water absorption behavior of a single layer; quantitatively evaluate fracturing effects.
- Find coal layers and evaluate their levels.

Specifications: Max. pressure: 80 MPa. Max. temperature: 150°C/4 h. Diameter: Φ90 mm. Tool length: 5.8 m. Neutron yield: ≥1.5 × 10^8 n/s. Operation lifetime of neutron tube: ≥300 h. Recommended logging speed: 50 m/h. Vibration: 29.4 m/s^2, three-dimension, 0–55 Hz, sweeping frequency for three times. Stability and linearity of energy gain within the whole temperature range: ≤0.5%. Measurement accuracy: neutron lifetime log—3% (in standard fresh water and salt water).

Spectrum logging: 5% (in saturated water sand and saturated oil sand with the porosity of 35%).

Benefits and features: Using preprocesses techniques of full-spectrum log, the quality of logging curves and the measurement accuracy of residual oil saturation are enhanced. Measure various formation parameters in a single logging run and improve the coincidence rate of interpretation. A wide dynamic range, suitable for residual oil saturation measurement with porosity over 10% and various salinities. The automatic peak stabilization for downhole tools and the drift correction technique for the surface spectrum have greatly improved the quality of logging raw data ensuring fine interpretation. The various parameters of the neutron generator are optimized and controlled to obtain a high neutron yield and good yield stability and a long operation lifetime and improve measurement accuracy of PSSL.

3.5 NEUTRON GENERATORS WITH THE DETECTION OF ASSOCIATED ALPHA PARTICLES

3.5.1 Introduction

Fast neutrons, energy of ~14 MeV, can be produced by d+t → α+n nuclear reaction in the sealed-tube generator where deuterons are accelerated by a voltage of ~100 kV toward tritium target. In this process, neutrons and α-particles are generated both emitted into 4π. It has been shown long time ago that the background in many measurements can be significantly reduced by the detection of associated α-particles. The cone of detected α-particles defines the cone of neutrons, "neutron beam," and if the neutron produced γ-rays are detected in coincidence with α-particles, the volume where they are produced is defined by overlap of "neutron beam" cone and γ-detector solid angle.

A method for measuring neutron-charged particle coincidences from the 14.4 MeV neutron-induced reactions has been developed in 1969–1970 in the Nuclear Reaction Laboratory, IRB, Zagreb, Croatia (Miljanic et al. 1969; Valkovic et al. 1969, 1970). Detection of associated alpha particles has been used to "electronically collimate" the produced neutron beam in order to reduce the experimental background. In the performed measurements the n+d → n+n+p reaction has been studied by the coincidence detection of the outgoing proton and neutron. As the neutron source a Cockcroft–Walton accelerator has been used.

The experimental setup shown in Figure 2.6 is the physical principle of all API techniques used nowadays to detect threat or illicit material in different containers. The new hardware is based on the use of sealed-tube neutron generators with detection of associated alpha particles.

In this chapter, the sealed tube generators used in the projects in which the author of this book has participated will be described in some detail with the aim to facilitate the decision on the choice of the generator for the new API systems to be developed.

3.5.2 API-120 Neutron Generator by THERMO

The API-120 is a portable neutron generator designed specifically for API applications. The API-120 is a continuous output system that generates approximately 1×10^7 n/s. In addition, the digital interface and advanced safety features make it ideal for remote operation. The system can be completely controlled from a standard RS232 port with a simple terminal program or with a LAB View TM GUI interface. Figure 3.16 shows Thermo Electron Co.—API-120 neutron generator, while its main characteristics are presented in Table 3.5. Thermo Electron has introduced additional models: API-220 and API-920, the latter model having the maximal yield of 10^9 n/s.

A simple LAB View interface is provided to run the generator if the terminal mode is not used. The LAB View interface implements the protocol necessary to use the serial interlock as well as providing rudimentary graphing and data logging capability; see Figure 3.17. The LAB View program is simply a shell program that implements and connects the buttons to specific commands sent to the generator while periodically monitoring the output. It is not required to run the generator nor is it controlling it (i.e., implementing the control loops or detecting faults) other than to send any user commands and monitor the outputs using a GUI. To run this GUI you should have a PC capable of at least a 1024 × 768 monitor resolution and obviously an available serial port. The version for API-120 requires revision 1.4 or greater of the Lab View DNC application.

Figure 3.16 Thermo Electron Co. API 120 neutron generator.

Next, we shall describe the preparation and testing of an API-120 used at Ruder Boskovic Institute in the implementation of a NATO Science for Piece project, SfP-980526 (Nebbia and Valkovic 2007). The API-120 neutron generator shown in Figure 3.18 was purchased and afterward the procedure followed was

1. Purchase of appropriate flange and installation of alpha particle detector
2. Shipment of installed detector back to Thermo Electron
3. Manufacture of API-120 at Thermo Electron and shipment to the user

The detector manufactured by Crytur Ltd., Turnov, Czech Republic, in the form of a disc of 0.5 mm thickness and 40 mm diameter was used. This was a crystal of YAP:Ce ($YAlO_3$ cerium loaded, yttrium aluminum perovskite) with the following properties:

1. Density: 5.37 g/cm^3
2. Refraction index: 1.95
3. Melting point: 1875°C
4. Nonhygroscopic
5. Linear coefficient of thermal expansion: 0.4–1.1 (10^{-5}/K)
6. Integrated light output (% NaI:Tl): 70
7. Wavelength max emission: 370 nm decay constant—25 ns

TABLE 3.5 CHARACTERISTICS OF API-120

Function	Parameter	Spec	Comments
Power	Input	24 V ±10% @ 2–5 A	Typically 2–3 A when running up to 5 A with a 2 A external warning lamp
Generator	Neutron output	~1 × 10^7 N/s	Typical at approximately 90 kV (with 60 μA of beam current)
	High voltage	70–90 kV	Limited in software to 95 kV
	Target current	20–60 μA	Limited in software to 60 μA
Source	Duty cycle	Continuous (100%)	
Interlock	Interlock current	50 mA max	Internally limited to protect operation during short circuit
Neutron lamp	Voltage out	+24 V (same as input)	Protected to take continuous short circuit
	Max. current	2 A	
Neutron lamp relay—spare contacts	Max. switched voltage	250 VDC, 230 VAC	
	Max. switched power	60 W DC or 120 VA AC, resistive load	
	Max. switched current	2 A	
Tube	SF_6 Pressure	120 psig nominal, 80 psig min., 140 psig max.	Pressure relief valves opens at 140 psig. Electrical pressure switch opens at 80 psig

To read out the crystal of YAP:Ce, a series R1450 photomultiplier tube (PMT) manufactured by Hamamatsu Inc. with the following properties was used:

1. Spectral response: 300–600 nm
2. Wavelength max. response: 420 nm
3. Structure: linear focused, 10 dynodes
4. Size: ¾ in. diameter
5. Window material: borosilicate glass
6. Voltage max.: –1800 V
7. Gain: 1.7×10^6
8. Rise time anode signal: 1.8 ns

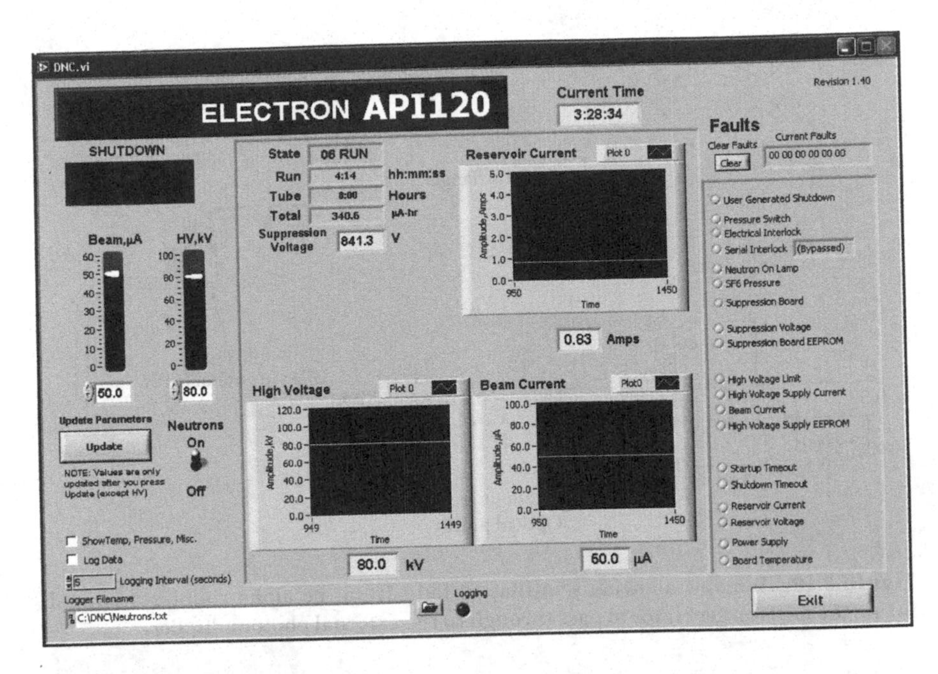

Figure 3.17 LAB View interface screen.

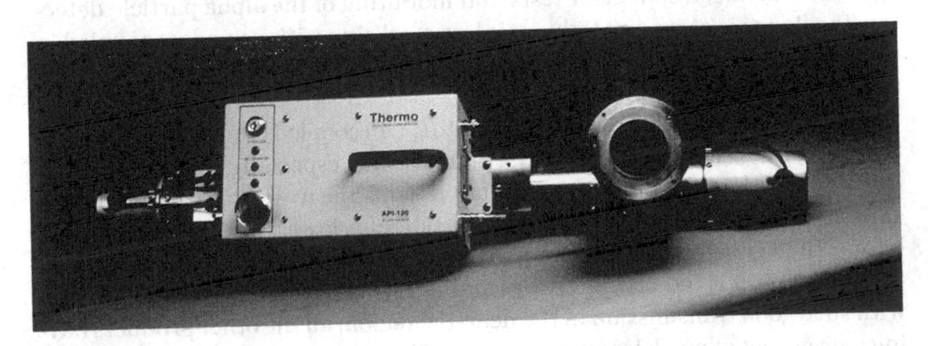

Figure 3.18 API-120 used in the implementation of SfP-980526 project.

The detector has been mechanically coupled inside a standard CF63 flange made of 316L stainless steel. The flange has been equipped with a sapphire window of 3 mm thickness, 48 mm diameter silver-copper brazed onto a ring of stainless steel and then tungsten inert gas (TIG) welded to the CF63 flange; the manufacture was performed by VLT s.r.l., Frascati, Italy. The sapphire window has been chosen for robustness and ease in the brazing process, the light transmission in the relevant wavelength region (around 370 nm) is about 80%.

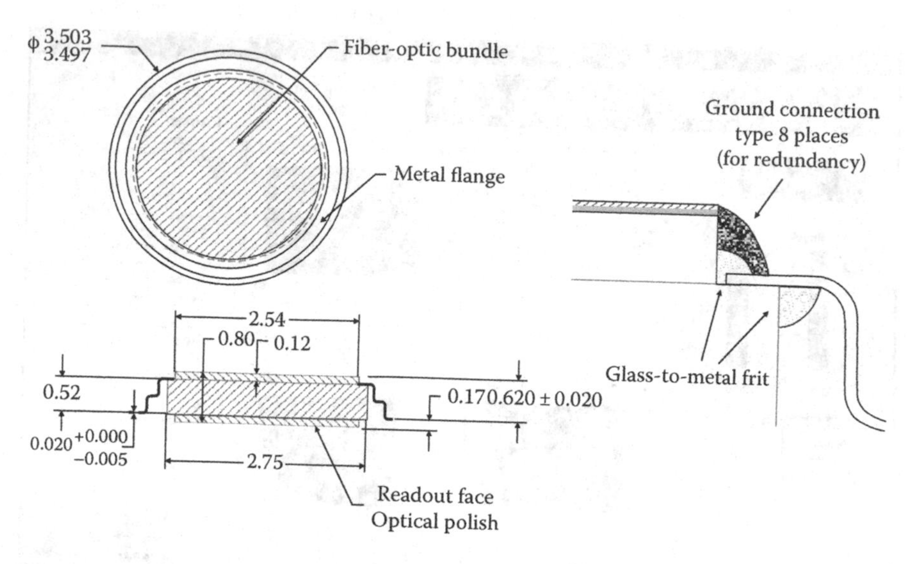

Figure 3.19 Window allowing scintillation light from the alpha counter mounted inside the neutron generator to pass through to the external photomultiplier.

A flange compatible with the API-120 neutron generator has been bought from Thermo Electron Inc. for tests and mounting of the alpha particle detector. The flange is a vacuum tight stainless steel ring with a window coupled in the middle to allow scintillation light from the alpha counter to pass through to the external photomultiplier.

The window is made of a fiber optic plate (FOP) coupled to the metallic flange by "fritting" as shown in Figure 3.19. The FOP was especially chosen to match the wavelength of the YAP(Ce) scintillation light. The wavelength of maximum emission for YAP(Ce) is about 370 nm, considerably lower than all other scintillators. The FOP chosen for this application has transmission values between 80% and 90%. It is interesting to notice that this is the only FOP in the market with such good transmission in the near-UV region, all the other products having a severe cut at much larger wavelength. The use of an FOP instead of a standard quartz or sapphire window allows to channel the light produced in the YAP crystal to the photomultiplier without further optical dispersion, therefore it opens the range of possible variations on the readout system applied outside the neutron generator.

The flange has been equipped with a disk of YAP(Ce) of 0.5 mm thickness and 40 mm diameter. The crystal has been fixed in the inner side of the flange by a small stainless steel holder "tig welded" to the sides of the flange itself. The YAP(Ce) disk has been coated with 1 mg/cm^2 of natural silver by vacuum deposition in order to guarantee light tightness and protection from the side-scattered deuteron ions inside the neutron generator. The pictures of assembled flange and of the assembled flange with alpha detector are shown in Figures 3.20 and 3.21, respectively. The flange thus assembled has been sent to Thermo Electron Inc. for final mounting on the neutron generator API-120 by laser welding.

The scintillator has been mounted on a standard stainless steel CF63 flange equipped with a sapphire window (3 mm thick, 48 mm diameter) without any optical grease, as required in the sealed neutron generator manufacturing procedure. The surface of the YAP:Ce crystal has been coated with a layer of 1 mg/cm^2 of metallic Ag to maximize the light collection, stop the elastically scattered deuterons, and protect the crystal from the UV glow inside the neutron generator. Since the counting rate capability and the time resolution were the major goals of the application, a small diameter fast PMT Hamamatsu R1450, without light guide, was used. The tagging system (i.e., the YAP:Ce mounted on the optical flange with the PMT and the associated front-end electronics) has been tested with neutrons produced in the d+t reaction making use of the 150 kV electrostatic neutron generator at the IRB in Zagreb.

The neutron-tagging detector was placed inside the reaction chamber at approximately 8.8 cm from the target at 90° with respect to the beam direction. For this test, a 5.5 mm diameter collimator has been positioned in front of the YAP:Ce detector in order to limit the size of the tagged neutron beam at the inspection plane, which is defined by the gamma detector. As a test to

Figure 3.20 A picture of the assembled flange.

Figure 3.21 A picture of the assembled flange with alpha detector.

measure the effectiveness of the method on the signal-to-noise ratio, we have irradiated a graphite sample of 5 × 5 × 5 cm^3 located at about 90 cm from the source of neutrons. The detection of the 4.4 MeV γ-rays emitted by the ^{12}C first excited level populated by neutron inelastic scattering by means of a 4 × 4 in. NaI scintillation detector has been used to determine the improvement of the spectrum quality.

The results of the irradiations are shown in Figure 3.22a is shown the γ-ray spectrum detected without requiring a strict coincidence between the alpha particle and the associated neutron hitting the sample. One can see a peak between 5 and 6 MeV due to neutrons hitting directly the NaI counter, and the 4.4 MeV and "first-escape" peaks associated to the decay of ^{12}C in the sample. In Figure 3.22b is shown the same spectrum but with a strict condition (of about 2–3 ns) on the alpha–gamma coincidence time.

One can notice a dramatic improvement of the signal-to-noise ratio (about a factor 40) due to both the "geometrical" and "time–related" reduction of the background contribution.

3.5.3 VNIIA Neutron Generator

ING-27 neutron generator is VNIIA generator with built in alpha particle detector. Standard configuration has

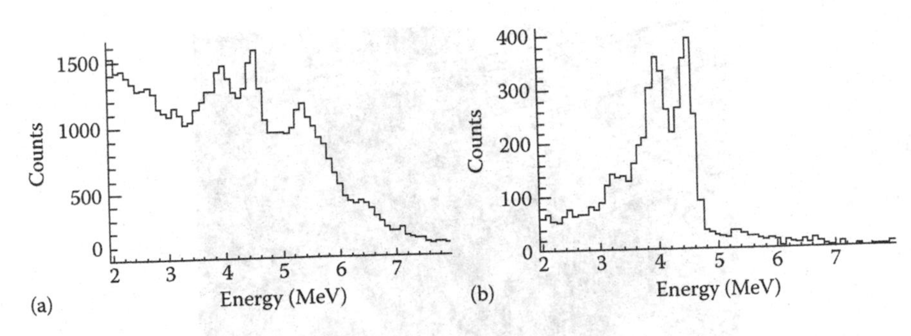

Figure 3.22 Gamma ray spectra resulting from the 14 MeV neutron irradiation of carbon, ^{12}C, target. (a) Without coincidences with associated alpha particles and, (b) with coincidences.

- Neutron unit with add-in nine-segment matrix semiconducting alpha detector; dimension of each pixel is $10 \times 10\ mm^2$
- Maximum neutron output: not less than 5×10^7 n/s by 4π
- Distance between target and alpha detector is 62 mm
- Life time: not less than 800 h at neutron output of 5×10^7 n/s by 4π with warranty period of 300 h
- Computer control
- Alpha detector is complemented with nine self-dependent preamplifiers for the generating fast analog signal, which corresponds to the moment of the registration of alpha
- Generator is powered from 220 V ± 10%

Figure 3.23 shows the photo of the generator, while Figure 3.24 shows the photo of generator control units.

3.5.4 SODERN Neutron Generators

3.5.4.1 EURITRACK Generator

The core of the EURITRACK portal is the tagged neutron–producing system made of an alpha particle detector array designed and manufactured by INFN (Italy), coupled to a 14 MeV neutron generator manufactured by EADS-SODERN (France). The compact neutron generator has been designed for an output of 10^8 n/s. The chosen interface between the tube and the tracking detector is a standard flange CF-100 (100 mm internal diameter).

As shown in Figure 3.25, the neutron tube has been designed with the target electrically grounded in order to avoid any disturbing electrical field between the alpha detector and the target itself. This field could accelerate secondary electrons coming from the target (under the interaction of the beam) into the

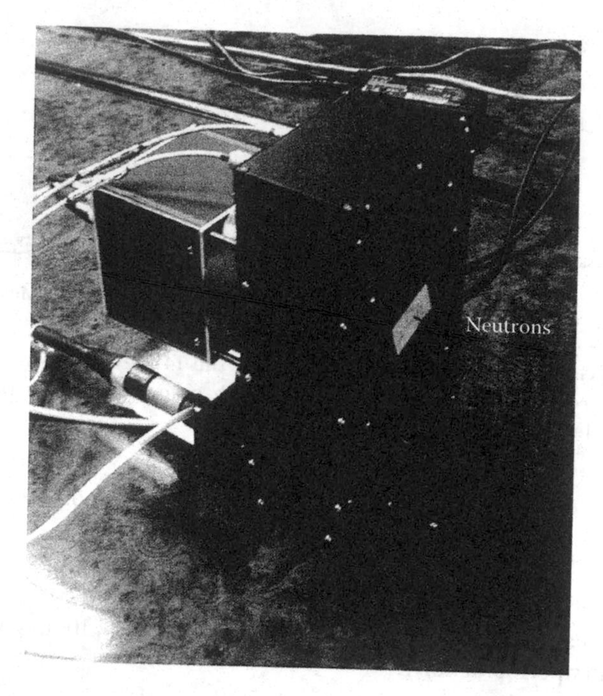

Figure 3.23 ING-27 neutron generator.

Figure 3.24 ING-27 neutron generator controls, front view.

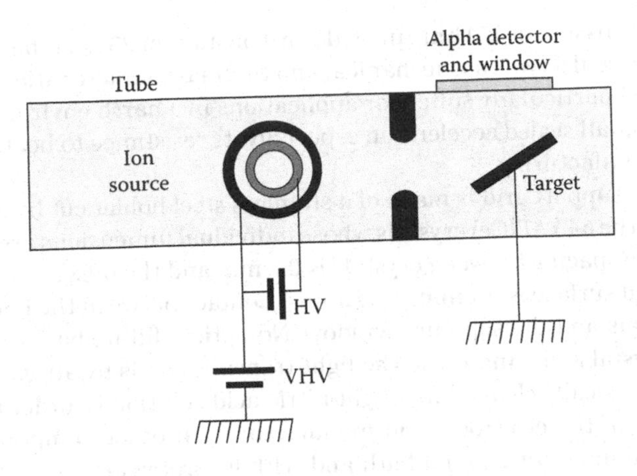

Figure 3.25 Schematic layout of the neutron generator.

YAP(Ce) scintillators constituting the alpha detector, producing a noise able to hide the light coming from alpha particles. The consequence of that choice is that the Penning ion source has to be at the high voltage HV, which is not conventional because in standard sealed-tube neutron generators the source block is grounded to facilitate the biasing of the Penning source. Biasing the ion source anode and cathode, which are working with an alternative HV of 2 kV with a VHV of about 100 kV requires ceramics insulation (usually used for the target) and a specific circuitry and supplies.

The system design implies also to have a distance of 150 mm between the tritium target center and the alpha detector. This distance defines, together with the alpha pixel size, the angular resolution of the system. The selected acceleration scheme implies the increase of the tube volume and internal tube surfaces. Consequently, specific getters have been added inside the tube to absorb the potential increase in degassing rate.

A stainless steel CF-100 high-vacuum flange with a 4 mm thick sapphire window brazed in the middle has been made in order to hold the tracking detector. The flange is necessary to separate the active part of the alpha scintillation detector, located inside the neutron generator, to the external photomultiplier, while maintaining a good optical coupling. The alpha particles produced in the $^{3}H(d,\alpha)n$ reaction are detected in an array of 64 YAP(Ce) (cerium loaded yttrium aluminum perovskite) crystals arranged in a square stainless steel grid mechanically fixed to the sapphire window.

YAP(Ce), see, for example, Baccaro et al. (1995), has the following properties: a density of 5.37 g/cm^3, a refraction index of 1.95, and a melting point of 1875°C; non hygroscopic; a linear coefficient of thermal expansion in the range 0.4–1.1 × 10^{-5} K^{-1}; a 70% integrated light output with respect to NaI(Tl); a wavelength maximum emission of 370 nm; and a decay constant of 25 ns. It has been chosen for its optical as well as mechanical and thermal characteristics that make this material particularly suited for applications in a harsh environment such as inside a small sealed accelerator, especially its resistance to heating during the tube manufacturing.

The array support grid is made of a stainless steel holder cut by electroerosion to host the 64 YAP(Ce) crystals whose individual dimensions are 5.8 × 5.8 × 0.5 mm^3. The spacing between crystals is 0.2 mm and the offset of the crystals from the grid surface is 0.4 mm. This allows contact between the back surface of the crystals and the sapphire window. No optical filling has been applied between crystal and window, so the light transmission is by air gap. The grid has been chemically cleaned in a light nitric acid solution in order to remove the residues of the electroerosion manufacture. All other components have been cleaned in an ultrasound bath and with isopropyl alcohol before being mounted on the flange. In Figure 3.26a is shown the inside of the flange with the crystal's grid mounted; in Figure 3.26b is the outside view of the flange with the appropriate location for the photomultiplier tube. The flange has been assembled in a clean room and then 1 mg/cm^2 layer of silver has been evaporated on the internal face of the crystals.

The silver coating serves several purposes: (1) to stop the side-scattered deuterons coming from the primary beam, (2) to screen against the light glow produced inside the neutron generator, and (3) to ground the internal face of the crystals to the metallic holder in order to eliminate static accumulation of charge and possible sparking. The deposition has been performed in high vacuum (about 10^{-7} torr) using a standard crucible bombarded by an electron gun.

The flange has been subsequently mounted on a high vacuum system and has undergone heating cycles for degassing during about 48 h. The heating has been pushed up to 350°C in steps of 50°C, the typical vacuum value being about 10^{-8} torr. The total degassing rate was measured by closing the pump

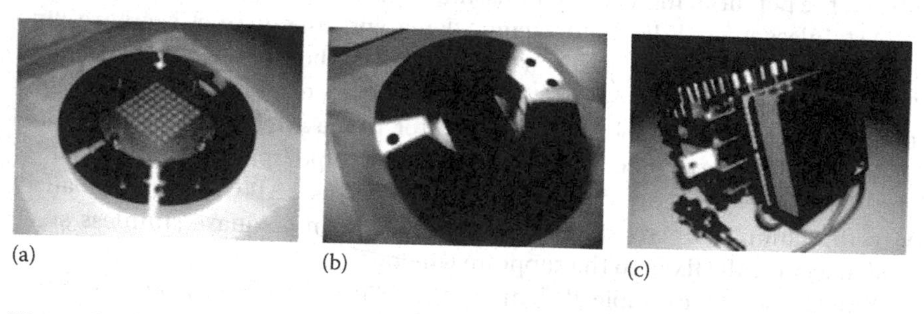

(a) (b) (c)

Figure 3.26 The alpha tracker mounted on the CF100 flange. Photograph (a) shows the detector side (i.e., inside the sealed tube), (b) shows the PMT side (i.e., outside the sealed tube), and (c) shows the H8500 PMT.

Figure 3.27 The alpha detector mounted on the sealed-tube neutron generator.

valve and by letting the system degas up to about 10^{-7} torr. The value of vacuum as a function of time was recorded for time intervals of about 30 min. The net leak rate of the flange turned out to be about 10^{-11} torr L/s.

Figure 3.27 shows the alpha detector mounted on the sealed-tube neutron generator. The detectors readout consists of a Hamamatsu H8500 PMT. The photocathode is segmented in 64 independent sections of 6×6 mm^2 thus delivering 64 separate signals to the front-end electronics. The geometry of the setup is such that each YAP(Ce) crystal is exactly facing one sector of the H8500 PMT. Nevertheless, since the 4 mm sapphire window is interposed, the light signals originating from one crystal may diffuse through the window to the adjacent sectors of the PMT. Tests were performed in order to verify the possibility of identifying the alpha particle hit position on the crystal array by a simple energy threshold discrimination on the output signal of each PMT sector. To this end, we compared the amplitude of the light signal recorded in different sectors of the PMT due to the irradiation of a single YAP(Ce) crystal by a collimated alpha source. The level of "decoupling" of the amplitude distribution spectra gives the amount of effective "optical crosstalk" between adjacent elements of the YAP (or PMT) matrix. The laboratory tests showed complete decoupling of the signals by simple threshold adjustment, thus allowing a precise determination of the YAP(Ce) hit by the alpha particle, and consequently a good spatial resolution (direction of the tagged neutron).

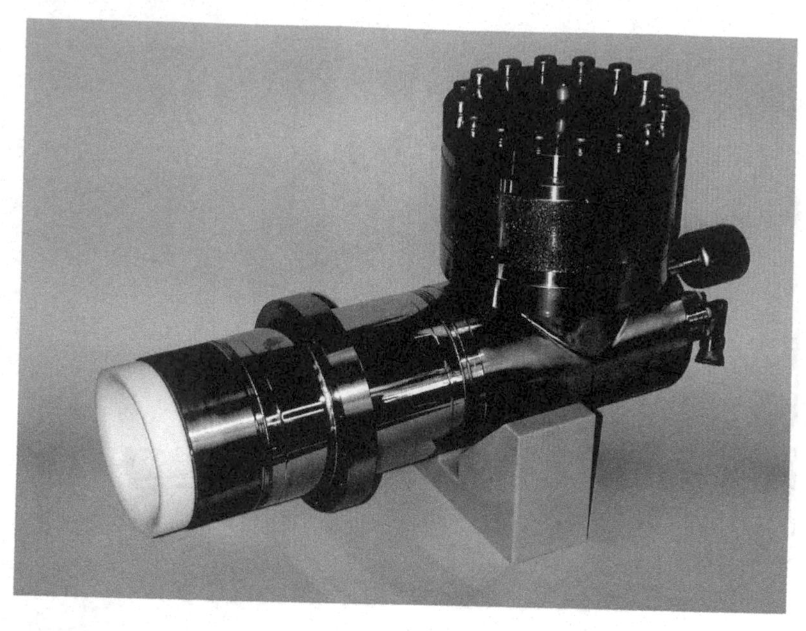

Figure 3.28 The sealed tube neutron generator with the position sensitive alpha particle detector mounted on the right and the read-out cables of its 64 individual pixels.

Finally, the sealed neutron generator with the embedded alpha tracker mounted in the EURITRACK portal is shown in Figure 3.28. It is worth noting that this is the first case of a sealed-tube neutron generator with an embedded alpha particle tracker with 64 independent crystals.

3.5.4.2 Other SODERN Neutron Generators

SODERN has discussed (Bach and Cluzeau 1997) three different 14 MeV neutron tubes specially designed for operating in bulk material analyzer or object inspection security equipment based on the time-of-flight technique. These tubes include a detector of the alpha particle emitted at the same time as the fast neutron but, unlike the tubes for already existing neutron diagnostic probes, their ion beams are not focused on a small spot of their target but are spread over a rather large area. Such a feature makes possible thousands of hours life time for the tube. When accurate imaging ability is needed they include a screen with one or many small apertures between target and alpha particle detector and thus the system can work as a pinhole camera. Advantages and drawbacks of this new concept were discussed by comparison with the already existing probes in paper by Bach and Cluzeau (1997).

In the EADS stand at the Paris Air Show 2007, EADS SODERN presented a mock-up of its portable neutron interrogation system for the detection and identification of explosives, chemicals, and nuclear and radioactive materials possibly hidden inside suspicious objects called ULIS (unattended luggage inspection system).

In EADS SODERN's current products, the neutron interrogation technology is employed by machines which analyze—in real time—the exact composition of minerals such as coal and cement when these materials pass through them on a conveyor belt. The company is currently working on adapting this proven technology to a number of security applications, particularly the ULIS system (to be discussed in some detail in Chapter 7).

The neutron interrogation technology enables threat detection even through steel walls 2 cm thick, making it suitable for the screening of cargo containers. When screening trucks, the combination with x-ray technology ensures that humans or animals are detected before the neutron flux is activated. EADS SODERN is also developing a neutron interrogation system for the detection of land mines.

It should be noted that SODERN is not selling neutron generators only anymore.

3.6 PLASMA DEVICES

A dense plasma focus, DPF (Mather 1965), is a plasma accelerator consisting of two electrodes in a coaxial configuration. The inner electrode is usually the anode; its one end is connected with the cathode via an insulator sleeve that covers part of the anode's lower part, while the other end is kept open. Depending on the purpose of the experiment, the empty space in between the electrodes is filled with a certain gas. The DPF is connected to a capacitor bank whose electrostatic energy is transferred rapidly to the gas via a fast switch. A current sheath forms at the base of the DPF, which moves upward due to a $J \times B$ force. When the sheath reaches the open end of the anode, it implodes radially, producing a short-lived column of hot dense plasma, which is finally disrupted by MHD instabilities. For systematic reasons, the time evolution of the phenomenon is divided into three successive stages: the dielectric breakdown of the gas, the axial phase, and finally the radial phase, during which the focusing of the plasma into a miniature z-pinch occurs. Being able to produce hot dense plasma, a DPF offers the possibility to investigate plasma dynamics for a broad range of applications. Moreover, DPFs with appropriate filling gases may constitute excellent sources of x-rays, EUV, and beams of energetic particles such as protons, neutrons, ions, and electrons.

A DPF is usually characterized by the energy that can be stored in the capacitor bank and the peak current between the electrodes. Historically, the trend was to construct and operate large DPFs, with energies ranging from kJ to MJ, aimed mainly at massive radiation production. Their large size, however, poses certain problems such as nonportability, considerable cost, and manpower needed for operation, and low repetition rate, if any. These, along with the plasma scalability, have led to the development of smaller DPF devices with energies of tens of joules, the so-called miniature DPFs. It is interesting to note that extensive research has proved that there are certain plasma parameters whose value remains within a narrow interval over a wide range of DPF energies: the electron density (~1024–1026 m^{-3}), the electron temperature (~200 eV to 2 keV), the plasma temperature (~1 keV), and the axial/radial velocity (~10^5 m/s). As we shall see in the following, these plasma parameters are relevant to certain DPF parameters, thus providing a useful tool for designing smaller devices with performance comparable to their larger counterparts.

An interesting aspect of the DPF is that when deuterium is used as a filling gas, fast neutrons with energy distribution well-centered around 2.45 MeV are produced. The neutron emission has a 4π distribution, and its duration is of the order of tens of nanoseconds. In the past, it was thought that these neutrons were of purely thermonuclear origin, that is, thermal collisions between deuterons in the plasma bulk, resulting in isotropic neutron radiance. However, simultaneous measurements of the neutron yield (Y_{en}) along the axial (0°) and the radial (90°) directions has revealed an anisotropy Y_{en} (0°)/Y_{en} (90°) of ~1.2–3, ruling out the possibility of a pure thermonuclear mechanism. Moreover, the average energy of neutrons in the axial direction was found to be greater than the energy of thermonuclear neutrons (2.45 MeV), providing more evidence that there is a nonthermal component contributing to the total neutron yield. Later, it was proposed that the nonthermal mechanism is the so-called beam–target effect, according to which accelerated deuterons collide with thermal deuterons and neutral atoms. To date, it is widely accepted that both thermonuclear and beam–target effects contribute.

In their work, Skoulakis et al. (2014) reported the design and construction of a pulsed plasma focus device to be used as a portable neutron source for material analysis such as explosive detection using gamma spectroscopy is presented. The device was capable of operating at a repetitive rate of a few Hz. The goal was to construct a tabletop DPF (~100 J) that would serve as a pulsed neutron source. The desirable characteristics are mainly portability and a decent repetition rate in order to compensate for the necessarily low neutron yield that a miniature DPF has. The main advantage of a DPF neutron source over continuous sources with similar spectrum such as AmBe or ^{252}Cf is that it is of the on/off type; consequently, it is easier to handle and store. Moreover, the short temporal duration of the neutron burst allows time-resolved activation analysis whenever this is necessary.

3.6.1 Nonlinear Devices, Dense Plasma Focus

The creation and acceleration of dense plasmas became a "hot" topic in the 1950s for such diverse applications as rocket propulsion, aerodynamics, chemical synthesis, and power generation. Gow and Ruby (1959) made two devices that could be operated as a steady state devices (typically 100 kV and 2–3 mA) as long as the electrodes were water cooled. Otherwise, it would be operated in a pulse mode using an impulse transformer. Impulse currents could be as high as 15 A at an indicated (ion gauge) pressure of 12 milliTorr. The other configuration was double-ended with a cathode on each side of the anode. Only one cathode served as a neutron-emitting target. This configuration minimized heavy electron bombardment at the end of the anode.

Gow and Ruby (1959) noted that the device would not pass appreciable currents until a pressure of about 9 milliTorr was reached, whereupon the current would rise steeply. Best operating conditions occurred from about 10 to 18 milliTorr. With a deuterium fill and a deuterated target, the source would produce about 10^8–10^9 n/s at peak voltages from 70 to 140 kV, see Figure 3.29.

In 1964, a mode of operation of the coaxial plasma gun called the dense plasma focus was discovered (Mather 1964). The focus is achieved at the muzzle end of a fast coaxial accelerator (sometimes called a Mather device in honor of the discoverer) where plasma densities can exceed $10^{19}/cm^3$ and temperatures may be in excess of a few thousand keV. The process is fast, on the order of 100 ns. The focus is actually a pinch effect (in this case a z-pinch as the pinch lies along the z [length] axis of the device). The basic elements of a plasma focus device are a capacitor bank, a fast spark gap switch, a pair of coaxial electrodes, and a vacuum tank. The design has to ensure a fast current rise (low inductance). Figure 3.30 is a simplified illustration of the plasma focus device. A number of factors have to come together to make the device work properly, including the geometry of electrodes, the geometry of insulator sleeve, type of fill gas, pressure of the fill gas, rate of the rise of the discharge current, and electrode polarity (Hansen 1997).

3.6.2 NSD Neutron Generators

NSD-Gradel-Fusion, Luxemburg, has developed a different neutron generation technology to industrial maturity. The reactor chamber is an improvement to the traditional spherical inertial electrostatic confinement (IEC) device. It provides a linear geometry neutron source that has great flexibility to be designed to optimize industrial neutron applications systems. The NSD neutron generator is a relatively simple way to achieve sustained nuclear fusion. While the possibility to scale the simple technology up to achieve more power output than input can be debated, there is no doubt that the technology provides

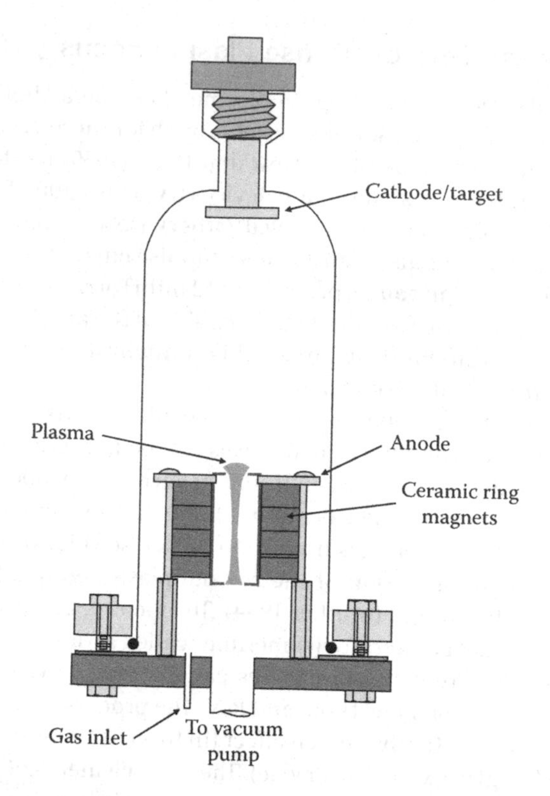

Figure 3.29 Neutron source based on crossed-field trapping. (After Gow, J.D. and Ruby, L., *Rev. Sci. Instrum.*, 30(5), 315, 1959.)

a credible neutron source without the disadvantages of a life-limiting solid target as found in common neutron tubes.

Inertial electrostatic confinement fusion can be briefly described as follows. A hollow transparent grid cathode (−) is surrounded by anode (+) at ground potential. Applying a VHV to a low pressure gas will induce a glow discharge. The positive ions are attracted to the cathode. Ideally, an ion will be channeled by the electrostatic focusing to pass through the cathode grid apertures. If there were to be no collisions with other ions and neutral particles the ion may oscillate. In reality the low pressure gas–plasma is still very dense so that there will be collisions of various types. Despite such energy loss mechanisms sufficient ions do get accelerated to kinetic energy levels over ~15 keV where fusion collisions can occur. The greater the applied voltage on the glow discharge, the greater is the probability of a fusion collision. With sufficient numbers of ions, charge space effects can have an influence on the ion trajectories in the central volume. As a result, a virtual anode forms by the self organization of the high density ion

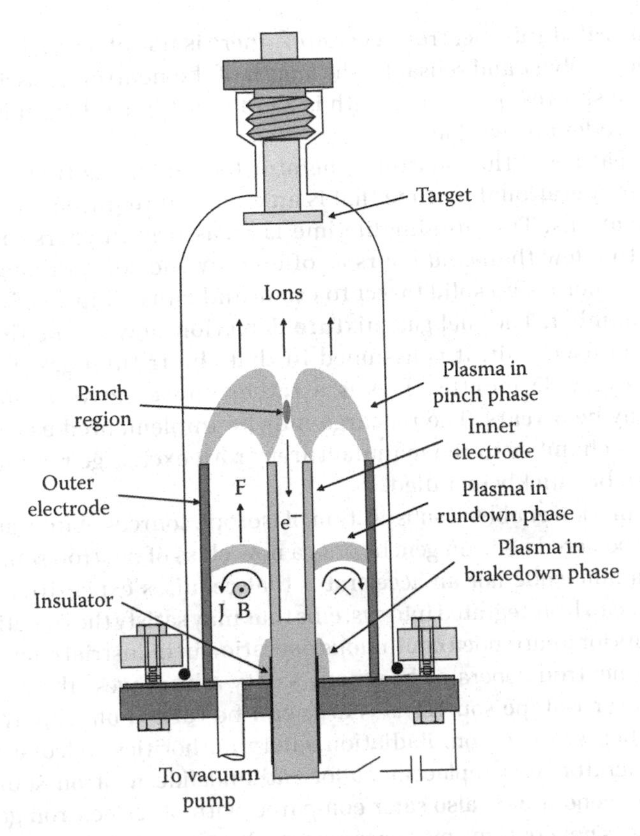

Figure 3.30 A simplified illustration of the plasma focus device. (After Hansen, S.P., *Bell Jar*, 6(3/4), 1, 1997.)

population. The virtual anode has a positive charge that is great enough to deflect the ions and effectively keep them in the central region. These ions contribute their positive charge to the charge space that is self-distributing into a shell of positive charge space. This plasma structure confines the ions at kinetic energies where fusion collisions may occur. The longer confinement time increases the fusion rate more than a simple increase of the applied electric current. With pulsed current at high voltage the super-linear scaling becomes most apparent with much greater neutron emission rates. Fusion of the hydrogen isotopes deuterium or tritium produces a charged particle and a neutron. The neutron is able to escape from the reaction chamber. Note that this process is occurring 10^6–10^{11} times per second in typical NSD neutron generators.

NSD has developed a unique linear geometry neutron generator. This may be envisaged as a fluorescent light tube but emitting neutrons. It is not a linear accelerator. It is not a derivative of a nuclear weapon "trigger." It is not "another short-lived sealed-tube neutron generator." There is no solid target. The sealed device is very robust and reusable. The length of the neutron emission zone is variable; the shortest practical length is 25 mm. A typical length is 350 mm; longer electrodes are feasible.

A vital feature of this neutron generator technology is that it has radically longer operational lifetime that is an essential requirement for industrial applications. The running lifetime is measured in years rather than hundreds to a few thousand hours as offered by the conventional neutron generators. There is no solid target to erode and cause failure of the sealed reaction chamber. The fuel gas mixture depletion may be the first reason for servicing used unit. It is assumed in that the tritium gas mix will be recharged every 36 months. This is a rather conservative estimate. The interval may be 5 years. The recharge may be implemented as a return of the reaction chamber to the manufacturer or an exchange reaction chamber that can be quickly installed.

The automation enables a substitution of isotope sources. A fully automated electrically powered neutron generator is a new class of neutron source that is neither a radionuclide nor an accelerator that requires expensive technician operators. It can be integrated into systems that may satisfy the radiation safety requirements for unattended continuous operation in industrial environments.

The NSD neutron generator has many safety advantages. The main safety advantage over isotope sources is that it can be turned off. It only radiates neutrons when switched on. Radiation safety authorities welcome the NSD neutron generator as a replacement for radionuclide neutron sources. The NSD neutron generator is also safer compared with other neutron generators because it does not contain any environmentally harmful SF_6-gas or any breakable glass parts.

NSD-350 DT neutron generator, shown on Figure 3.31, has the following technical characteristics:

- ~5×10^9 n/s with an electrode of at least 350 mm length
- 400 VAC or 200 VAC three-phase power input
- Water cooled with separate water/air heat exchanger
- Longer line source neutron emission zones are feasible
- Continuous dc or macropulsed
- The central controller and HV dc power supplies are in a 19 in. rack housing
- AC mains power (230/120 VAC)
- Short electrode = short neutron emission zone
- Chamber operational lifetime between DT gas refill servicing will be approximately 5 years

Figure 3.31 Water cooled NSD-350 chamber inserted into PE shielding block. Note also the segmented lead sleeves as shielding for Bremsstrahlung.

- At least 25,000 h operational life but probably far more due to the absence of the old beam on solid target technology
- The total lifecycle cost economics of the products compared to other neutron source technologies is more attractive

The neutron generator head is able to be fully submersed, for example, for use underwater; see Figure 3.32. The connecting cables are contained within water-proof tubes and the feedthrough connect to the generator head. In this situation the generator is cooled by a closed internal water-cooling loop with an external air–water heat exchanger.

The NSD NG has a flexible configuration intended to be adaptable to industrial requirements. The sealed reaction chamber can be configured for different applications. The minimum electrode length and fusion zone is 50 mm. A longer "line source" neutron emission zone can be specified, that is, of length ~830 mm. Figure 3.33 shows the schematic of reaction chamber intended for neutron radiography.

NSD-350-24-C neutron generator d–d 2.5 MeV or d–t 14 MeV or t–t 0.5–9.5 MeV options:

Figure 3.32 NSD-35 chamber submerged in water tank.

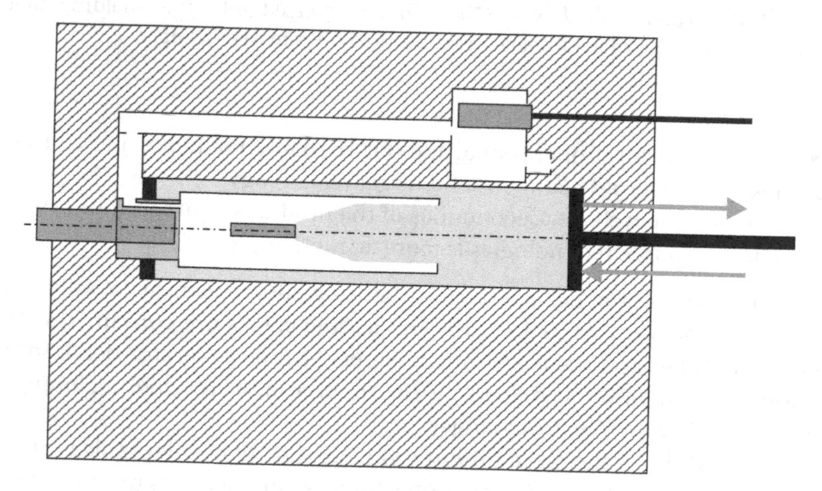

Figure 3.33 End view reaction chamber for neutron radiography.

Electrode and fusion emission zone length in the middle of reaction chamber: 350 mm

d–d 2.5 MeV neutron yield: >2 × 10^8 n/s output

d–t 14 MeV neutron yield (50× d–d): >1 × 10^{10} n/s

t–t 0.5–9.5 MeV neutron yield: ~3× d–d (TBC)

Maximum voltage 160 kV and maximum current 150 mA
Greater than 10,000 h lifetime
Nonpulsed continuous operation
"Macro pulse" >100 ms pulse ~50 Hz maximum
Quick start to near full output
Stable and repeatable operations and starting from lowest (100 n/s DD) to highest output
Outer casing is at ground potential
Reaction chamber: 1300 mm long; 135 mm OD
Neutron emission module ~30 kg
Input power 380/220VAC ± 10% or 415/240VAC ± 10% by three-phase configuration to HV PSU
VNC control HMI over LAN via any PC
OPC or MIDAS for integration into systems
Automated unattended operation for industrial applications
Submersible option
Neutron count rate (accepts neutron detector pulses) and availability monitor for gating of momentary outages

Industrial lifetime: Reaction chamber neutron emitter life is greatly extended because there is no solid target. Deuterium refill is ~10 years due to helium gas buildup.

Operation endurance >10,000 h: probably 2× more. For tritium versions the manufacturer recommends a refill after 5 years. Similarity to the Mk0 life test provides high confidence for longevity of our Mk1 product. The primary reason that the manufacturer has developed the NSD neutron generator is economic industrial lifetime. A few thousand hours of operational life of a solid target neutron generator and then a "repair" charge for replacement cartridges of almost half the purchase price is not economic. The NSD neutron generator is reusable after an eventual relatively low-cost servicing.

The continuously renewed gas–plasma target means that there is no solid target failure through thermal cycles causing metal fatigue or gas depletion. So, frequent on–off deep thermal cycle operation will not reduce operational lifetime. The NSD-NG can be refueled: old gas out—new gas mixture in. Ultimately replacement and cleaning of internal components is also possible at moderate cost.

It is possible to switch on the NSD-NG and achieve almost immediately the desired maximum neutron output. There is no slow run-up to maximum output. This suits scanning or screening applications. The sealed vessel gas pressure getter pump system must, of course, be warmed up first over 20–30 min. The getter pump can remain heated in continuous standby for near instant operation.

3.6.3 Berkeley Laboratory Compact Neutron Generators

The P&IST group at LBNL has developed various types of advanced d–d (neutron energy 2.5 MeV), d–t (14 MeV), and t–t (0–9 MeV) neutron generators for a wide range of applications. These neutron generators utilize RF induction discharge to ionize the deuterium/tritium gas. This discharge method provides high plasma density for high output current, high atomic species from molecular gases, and long life operation and versatility for various discharge chamber geometries.

This RF technology with unique impedance-matching circuitry and source designs capable of high power operation is now being used in various neutron generators. RF induction discharge ion source has unique features that make it an excellent source type for high power, high yield neutron generators. These features include the high plasma density generation for high beam current extraction, extremely high atomic ion generation from molecular gases like hydrogen, and reliable long-life operation. These ion sources are currently used in three different types of neutron generators: axial, single ion beam neutron generator; co-axial, multi-beam, high current neutron generator; and point neutron generator with high instantaneous neutron yields. The latest addition to the line of neutron generators is a subcompact, low-yield neutron generator for subsurface water/hydrogen analysis applications. These different types of neutron generators are described in the paper by Reijonen (2005).

Let us mention subcompact miniaturized neutron generator being one of the latest fields of neutron generator technology, which the P&IST group is addressing, involves miniature, low-yield neutron generator for applications, such as subsurface ice/water and mineral analysis for planetary exploration or miniature high-energy gamma calibration device for high-energy gamma detector applications. The emphasis in these applications is the low power consumption, low weight, ruggedness, and portability. To achieve these goals, some compromises have to be made. The miniaturized neutron generator would be operated at considerably lower neutron yield than the regular axial neutron generator. The neutron yield is in the order of 10^4–10^5 n/s for d–d operation and in the order of 10^6 n/s for d–t at ~40 kV of accelerator voltage and with a few microamperes of extracted beam current. The power consumption would be in the order of few watts. This enables fairly long operation time by using batteries or solar cells. The neutron generator would not be actively cooled. The diameter of the device is ~25 mm.

The ion source of this kind of neutron generator could be cold or hot cathode type, using high voltage discharges or electron emission from a filament. In certain applications, where the packaging of the neutron generator is less of a concern, the ion source could also be driven with RF-induction discharge.

Scientists at Berkeley Lab have developed innovative neutron generators that can be tailored to meet a variety of specifications. The generators invented by Leung (2012) and colleagues are unusual because they are compact, designed to be long-lasting, and inexpensive to construct yet capable of using safe d–d reaction to produce a high neutron yield or flux. They can also be designed to use t–t reaction to generate neutrons across a broad energy spectrum or d–t reactions to produce higher energy neutrons.

The radiation-producing devices developed by P&IST group at LBNL are utilizing powerful RF-induction discharge, low-energy accelerator structures, and actively water-cooled beam targets. The RF-induction discharge method provides high plasma density for high ion output current, high atomic species from molecular gases, such as hydrogen, deuterium, and tritium, long life operation with stable operation characteristics and versatility for various discharge chamber geometries enabling the development of devices of diverse geometries. DC beam accelerators are utilized either in single gap diode or multigap accelerator column configurations. The actively water-cooled target provides high beam power handling capability with various target materials. Three devices are discussed in this presentation: two neutron generators, one for neutron imaging applications and one for pulsed fast neutron transmission spectroscopy (PFNTS)–based cargo interrogation system and a gamma generator for active interrogation of nuclear materials based on photofission and photoneutron detection (Kalvas et al. 2006; Reijonen et al. 2007).

The Berkeley Lab neutron generators consist of RF-driven plasma ion sources, extractors of various designs, acceleration electrodes, and titanium-covered targets. Conventional generators are usually short-lived because the target's isotopes are quickly consumed. The target in the Berkeley Lab generators, however, is constantly replenished by ions from the plasma source. These devices may last thousands of hours longer than conventional generators. Let us describe some of the developed generators:

1. *Cylindrical neutron generator with nested option, IB-1764* Dr. Leung has developed a coaxial RF-driven plasma ion source for a compact cylindrical neutron generator that operates with high current density and produces high atomic deuterium or tritium ion species. The plasma and extraction electrodes electrostatically control the passage of ions out of the ion source. These electrodes contain longitudinal slots along their circumferences so that the ions radiate out of the source in a full 360° pattern.

   The single target coaxial neutron generator with dimensions of 26 cm in diameter and 28 cm in length is expected to generate a 2.4 MeV d–d neutron flux of 1.2×10^{12} or a 14 MeV d–t neutron flux of 3.5×10^{14} n/s. An adaptation of the coaxial generator design uses a versatile nested configuration with multiple plasma and target layers where ion beams can impinge on both sides of the targets to enhance neutron yield. A generator with this nested design and dimensions of 48 cm in diameter and 35 cm long should generate a neutron output 10 times higher than the single target generator described earlier. Thus, d–d neutron output higher than 10^{13} n/s and d–t neutron output of 10^{15} n/s should be attainable (Leung 2002).

2. *Mini neutron tube, IB-1793a* The same group has designed a very compact d-d or d-t fast neutron generator in a capsule configuration that is less than 8 mm in diameter and only 2 cm long. This generator is portable and is ideal for insertion into an oil well logging drill. A larger variation of this invention with a longer target is designed to produce d–d neutron flux higher than 10^{11} n/s with a modest length and diameter (Leung et al. 2008).
3. *Ultrashort ion and neutron pulse production, IB-1707* Dr. Leung has developed a neutron generator that produces ultrashort neutron pulses by pulsing the ion beams instead of using a mechanical chopper (Hahto et al. 2005).
   When pulsed neutron beams are used to examine a sample, it is possible to separate the different types of neutron–nuclei interactions after each neutron pulse. For many applications, pulsing provides more complete imaging or spectroscopic information (Leung et al. 2006; Leung 2012).
4. *Mini neutron generator, IB-1793b* Dr. Leung has designed a compact d–d or d–t neutron generator specifically for treating tumors less than 5 cm in diameter. The generator is composed of a 2 × 2.5 cm² plasma source from which deuterium or tritium ions are extracted and then accelerated. The accelerated ion beam is then directed down a 10 cm long needle-like tube (diameter ~3 mm) with a titanium target at the end.
   The neutron source can be inserted directly into the tumor. This should minimize damage to healthy tissue while delivering a potent dose of radiation to the tumor. In contrast to high-dose rate ^{252}Cf sources, Berkeley Lab's neutron generator should not generate any dose to clinical personnel when the source is turned off.
5. *Compact spherical neutron generator, IB-1675* Dr. Leung has developed a compact spherical neutron generator that can use safe deuterium–deuterium (d–d) instead of radioactive deuterium–tritium (d–t) or tritium–tritium (t–t) reactions, and still provide a high neutron flux. This generator is designed with a small, spherical shell-shaped RF-driven plasma ion source surrounding a multihole extraction electrode that in turn surrounds a spherical target. Ions passing through the holes in the extraction electrode are focused

onto the target, which is loaded with d or t by the impinging ion beam. Neutrons are emitted at all angles from the spherical generator (Leung 2006).

3.6.4 Adelphi

A new commercially available NG that also uses a simple diode to generate neutrons is the radio frequency induction (RFI) plasma NG. In this case, the deuterium ions are generated in RFI plasma and accelerated across a DC potential toward a Ti target. The main components are located along a single axis and tubes with this design are referred to as axial generators. Like the Penning diode, the RFI plasma NG is fundamentally a diode, but with a RFI plasma ion source instead of a Penning ion source. Also, like the Penning diode, the RFI source uses a titanium target; however, the titanium is beam loaded in the initial few seconds of operation. Deuterium ions are generated in the RFI plasma and are accelerated across a voltage potential to the titanium target. The $^{2}H^{+}$ ions strike the titanium and in the first few seconds of operation they form titanium hydride in the first few microns of the target surface. Subsequent deuterium ions fuse with ^{2}H atoms of the hydride resulting in the d–d reaction and emitting 2.5 MeV neutrons. Thus the generator's neutron emission is self-sustaining and the sealed Penning diode that uses an impregnated target have the same properties. On the other hand, the RFI generator needs a compressed source of deuterium gas and an active vacuum system (e.g., turbo pump).

Previous generators of this type had various components exposed that were at high voltage potential, making the generator dangerous to operate. In the new design the high voltage is shielded by placing the entire accelerator structure and its components inside conflate stainless steel piping. The target and secondary electron shielding are inside and cooled through a unique insulator feed through structure.

These generators have been shown to be capable of yields of 10^{8} n/s using an axial design and a d–d reaction. Two new axial generators manufactured by Adelphi Technology Inc. are capable of yields of up to 10^{10} n/s. A prototype sealed generator has been fabricated at Adelphi with an operated sealed tube using the d–d reaction and it can be easily converted to d–t operation (Adelphi 2013). (http://adelphitech.com/products/dt110.html).

Adelphi Technology Inc. has two products of interest:

1. *API-108 API neutron generator* Adelphi has developed a prototype high-flux, small spot-size API neutron generator. This generator has an integrated target for the production of neutrons and alpha particles. When deuterium and tritium atoms fuse on the generator's target, both an alpha particle and a neutron are produced. These are

then emitted simultaneously 180° apart so that the direction of the neutron can be determined by detecting the direction of the alpha particle. A scintillator provides an optical signal to determine the location, and thus direction, of the alpha particle. A small ion beam spot size (<2 mm) on the target is required to provide a precise origin for the neutrons and alphas and, thus, an accurate angular resolution. Adelphi's microwave ion plasma technology allows for flexible and precise ion beam optics, which in turn provide for a small spot size, even with large neutron outputs. The generator is ideal for making a mapped image of gamma emission from fast neutron scattering. Both the API neutron generator and a gamma detector for detection of the activated or scattered gamma radiation (e.g., HPGe) can be located on one side of an object to be imaged, greatly simplifying the setup. The combination of neutron, alpha, and gamma detection allows for the possibility of the imaging and identification of distant and concealed objects such as contraband and special nuclear materials.

2. *DT110–14 MeV neutron generator* The DT110 is a new sealed, high output neutron generator using the d–t reaction to produce 14 MeV neutrons (see Table 3.6). The expected output is 10^{10} n/s. The generator will

TABLE 3.6 GENERAL SPECIFICATIONS OF ADELPHI D–T GENERATOR

DT neutron yield	1×10^{10} n/s maximum
Neutron energy	14.1 MeV
Neutron source size	1 cm × 4 cm (side view: standard)
Small source size (option)	≥4 mm diameter (axis view)
Generator head lifetime	2000 h guaranteed at maximum yield
Servicing	Generator head serviced at Adelphi
Operating mode	Continuous
Pulse on demand (option)	≥100 μS, to 100% duty factor
Accelerator voltage	60–100 kV
Operating beam current	3–5 mA
Control interfaces	Computer with graphical user interface; RS-232 optional
Generator head weight	50 lb (with turbo pump)
Control and power rack	180 lb
Heat exchanger weight	57 lb
Total weight	287 lb
Power requirements	120 VAC, 1 phase; 20 A circuit
Safety features	Both personnel and machine safety interlocks included

have applications in security for baggage and container screening for the detection of explosives or contraband and for the detection of special nuclear materials in cargo or luggage. The d–t generator is based on a nuclear reaction between two of hydrogen isotopes, deuterium and tritium. Tritium is a radioactive gas, so d–t generators are sealed; also, U.S. Nuclear Regulatory Commission (NRC), see U.S. NRC (2014), regulations require that U.S. customers of d–t source possess a Type A broad scope specific license authorizing possession of tritium. The U.S. NRC website has further details of licensing requirements.

As already said, the d–t neutrons have energy 14.1 MeV, whereas neutrons originating from the d–d reaction have only 2.45 MeV. Thus, the d–t neutrons will penetrate further material and induce a different range of activation than d–d neutrons. The output energy is an important consideration in shielding. The 14.1 MeV neutrons are much more energetic and require more material in order to moderate and absorb these neutrons. Shielding designs are therefore much more significant. The d–t reaction is also more efficient (100×) than the d–d reaction, requiring less acceleration power than the d–d fusion reaction. This is reflected in the modest size of the total power required (about 500 W). Since the power requirements are modest, compact power supplies that can be operated using ac or battery power can be designed. A prototype sealed generator has been fabricated at Adelphi and should be available by now.

3.7 HIGH FLUX SOURCES

Recently, 14 MeV neutron generators with a yield higher than $>10^{12}$ n/s were developed for fusion-related applications, neutron cross section measurements, production of long-lived isotopes, investigations on radiation effects, cancer therapy, and elemental analysis of small samples. The original dc-intense neutron generators were equipped later with pulsing systems to measure the secondary and leakage neutron spectra. Such experiments require about 100 times higher neutron yields than that of today's intense neutron generators to achieve the required accuracy of data for fusion reactor design. These programs became more important after the first successful experiment with d–t fusion in the Joint European Torus in Abingdon, United Kingdom, in 1991. The design of a fusion reactor needs accurate data for tritium breeding, nuclear heating, bulk shielding, secondary reactions, and gas production. The accuracy of the energy and angular distributions of secondary neutrons—double differential cross sections (DDX) measurements—for blanket and other structural materials should also be increased.

For calculations of the induced activity, more accurate activation cross sections are needed not only for the materials of the major components but also for their impurities. Precise activation data are required mainly for the dosimetry reactions and for unfolding the neutron field in different parts of the fusion reactor.

During the early intense neutron generator period (second half of the 1970s) when several such machines had been constructed for fusion-oriented research and neutron therapy, it became clear that these originally dc generators would be more suitable for fusion studies if a pulsing system helped the DDX and neutron transport measurements. The Osaka OCTAVIAN generator and the JAERI FNS were equipped with nanosecond pulsing units as the first intense neutron generator, the RTNS-I at LLL.

The paper by Fleischer et al. (1965) describes some aspects of a feasibility study on a high flux, large area 14 MeV neutron generator. The purpose of the study was to provide a preliminary design of a 14 MeV neutron generator capable of uniformly delivering a 1000-rad dose on a 6 × 6 ft target. A 14 MeV source of about 10^{13} n/s for 4 h yields a 1000-rad dose with a uniformity of 10% to a 6 ft target located 10 ft from the source. There are numerous charged particle reactions capable of yielding significant numbers of 14 MeV neutrons, but the requirement for monoenergetic 14 MeV neutrons precluded all reactions but the (d–t) reaction. The 14 MeV neutrons are produced by a 250 mA deuteron beam with 250 keV energy impinging on an 18 in. diameter tritiated-titanium target. The use of a large area tritiated-titanium target makes target cooling and depletion problems manageable.

There are a large number of intense neutron generators around the globe. Here we shall describe only some efforts.

A cooling system of a stationary target has been designed some time ago; see Seki et al. (1979) to satisfy the structural, thermal, and hydraulic requirements. Two square tubes for ion beam and for cooling water were concentrically placed and the target was mounted on the top of the beam guide tube. The end plate of the outer tube was devised to be removable for easier replacement of the target. In order to test the cooling capability of the system, dummy target assemblies with electrical heaters were used in the experiment of heat transfer in place of using an accelerator. Correlations of heat transfer and head loss were obtained experimentally as a function of Reynolds number. The extrapolation of the data has shown that for the present target system, about 2.3 kW is the maximum power for the beam in diameter of 15 mm. This value was sufficiently large compared with the required heat load of Fusion Neutronics Source (FNS) in JAERI.

The P&IST group at the LBNL has been engaging in the development of high yield compact neutron generators for the last 20 years (Leung et al. 2004). Because neutrons in these generators are formed using either d–d, t–t, or d–t fusion reaction, one can produce either monoenergetic (2.4 or 14 MeV) or white neutrons. All the neutron generators being developed by this group utilize 13.5 MHz RF induction discharge to produce a pure deuterium or a mixture of deuterium–tritium plasma. As a result, ion beams with high current density and almost pure atomic ions can be extracted from the plasma source. The ion beams are accelerated to ~100 keV and neutrons are produced when the beams impinge on a titanium target. Neutron generators with different configurations and sizes have been designed and tested at LBNL. A novel small point neutron source has been also developed for radiography application. The source size can be 2 mm or less, making it possible to examine objects with sharper images.

The RF-driven ion sources developed at LBNL are simple to operate, have long lifetimes and high gas efficiencies, and provide high-density plasmas with high monatomic species yields. These characteristics make the RF-driven ion source a viable candidate for the next generation of compact, high-output, sealed- tube neutron generators, utilizing the d–d, t–t, or d–t fusion reactions. Described in the following are three different types of neutron sources developed at LBNL for luggage and cargo container inspection.

Compact axial extraction neutron generator: The compact axial extraction neutron generator is approximately 40 cm in length and 15 cm in diameter. The ion source is a quartz-tube incorporating external antenna. The back plate has the deuterium gas and pressure readout feedthrough. The target is housed in an aluminum vacuum vessel and it is insulated from the ground potential with an HV insulator. Figure 3.34 shows a drawing of the axial extraction neutron

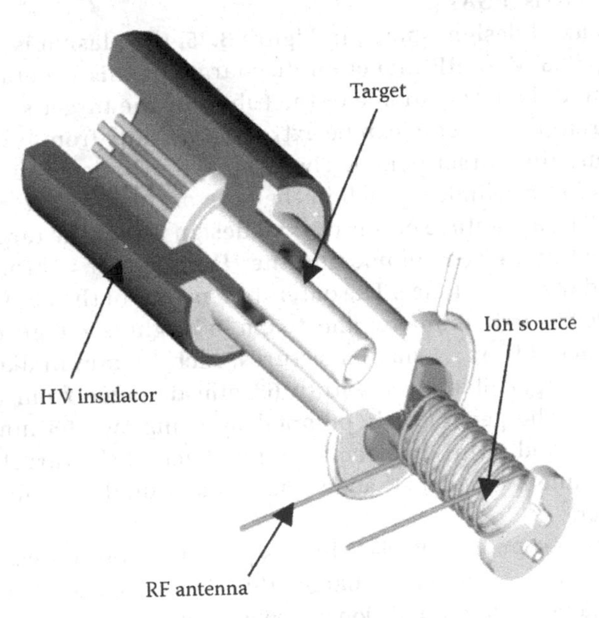

Figure 3.34 The axial extraction neutron generator.

tube. The ion beam is extracted from a 3 mm diameter aperture and accelerated to a target at 100 kV. The target is water-cooled and is made of explosive-bonded titanium-on-aluminum material.

The axial neutron generator is mainly used in pulsed mode experiments. The pulsing is performed by switching the RF power on and off. The current RF power supply is limited in pulse length of 2 ms. The rise time of the plasma can be as short as a few microseconds, which is achievable using a fast rise-time RF generator. The compact axial extraction neutron generator is constantly being evacuated by a turbo molecular pump. This is possible because no radioactive tritium gas is used. Some applications of the neutron generator require portability and/or the use of tritium for 14 MeV neutron production. In these cases, the generator has to be operated in sealed (no pumping) condition. The 10^9 d–d n/s flux corresponds to 10^{11} n/s for the more widely used d–t reaction because of the much higher cross section of the later reaction. The flux is limited mainly by the available high voltage and for the maximum duty factor that the generator can be operated.

High yield radial extraction neutron generator: For applications that require high neutron output from a relatively compact dimension, a new type of coaxial, radial extraction neutron generators have been developed. In this generator, the main task has been to produce high neutron flux in cw mode for applications such as medical treatments (BNCT), NAA, and prompt gamma activation analysis (PGAA).

In this coaxial design, shown in Figure 3.35, the plasma is also formed by utilizing 13.5 MHz RF induction discharge. In this generator, the ion source chamber is in the middle of the tube and the target surrounds the source. Therefore, the beam can be extracted radially from the ion source to the surrounding target panels. These coaxial cylinders are surrounded by a HV insulator cylinder, which in the case of this generator is made of Pyrex glass. The advantage of this coaxial design is that the target area can be maximized in a given volume and the HV insulator is protected from the sputtered target particles. The outer dimensions of the coaxial neutron generator are 30 cm in diameter and 40 cm in height (see Figure 3.35). The beam is extracted from 24 small apertures, each 1.5 mm in diameter. The water-cooled target plates are within 63 mm distance from the plasma chamber wall. The generator is pumped by using two 63 mm diameter pumping ports and a turbomolecular pump. Each of the target plates has a pair of permanent magnets that acts as an individual secondary electron emission filter.

Point neutron source for PFNTS: In this point neutron source, the plasma is produced by RF induction discharge. This type of discharge can provide high current density with atomic ion concentration greater than 90%. The gas employed will be a mixture of 50% deuterium and 50% tritium for 14 MeV d–t

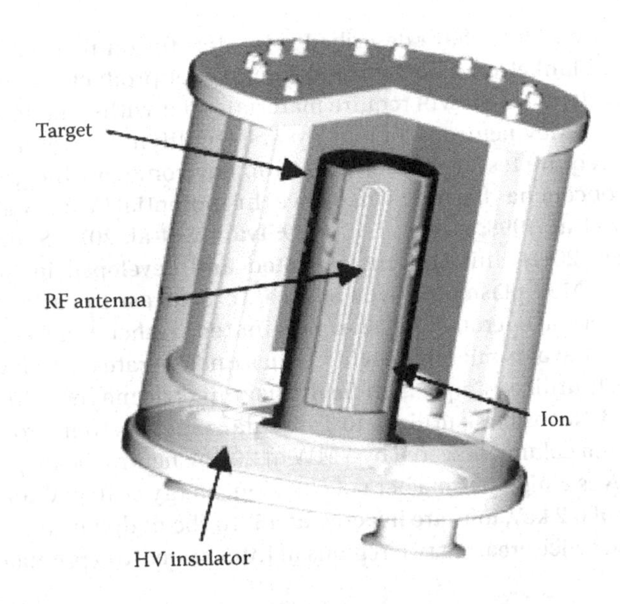

Figure 3.35 The coaxial type neutron generator.

neutron production and pure tritium gas for "white" (0–9 MeV) neutron production. The discharge will be operated in pulsed mode with a 3 kW, 13.5 MHz RF generator. Multiple beamlets will be extracted from the plasma source. The ions of each beamlet will be accelerated to 100 kV and focused down to a beam spot size of ~2 mm diameter. These beamlets will irradiate the Ti target uniformly. The Ti target surface will be loaded with deuterium or tritium atoms. When the incoming ions impinge on these atoms, neutrons will be generated via the d–t or d–t fusion process. The neutrons produced will appear to come from a small "point source" with diameter no larger than 2 mm. They will then be directed to the target object for interrogation study. The power density deposited by the focused ion beams on the target surface can be maintained at ~500 W/cm^2. This modest heat load can be easily removed by using an oscillating, water-cooled target stage. The entire neutron generator is about 20 cm diameter by 20 cm high. The deuterium–tritium plasma is first generated in a toroidal quartz or ceramic chamber by means of 13.5 MHz RF induction discharge. The antenna is a copper or aluminum coil wrapped around the plasma chamber. Since the antenna is located outside the plasma chamber, there is no limitation on its lifetime.

In order to achieve 10^{12} d–t n/s of 10^{10} t–t n/s, the total beam power deposited on the target is 120 kV, 10 mA. The target is Ti tubing with an outer diameter of 2 mm. Water is running through the center for heat removal. Ti is used as the target material because it can absorb deuterium or tritium atoms efficiently.

Simonen et al. (2013) have described a 14 MeV fusion neutron source for material and blanket development and fission fuel production. Economical fusion energy deployment will require materials that withstand intense bombardment by 14 MeV neutrons for many years. Qualifying materials and components will require testing in such a neutron environment. The gas dynamic trap (GDT) concept has been shown to have this potential (Mirnov and Ryutov 1979; Ryutov et al. 1998; Fisher et al. 1999; Ivanov et al. 2010; Simonen et al. 2010; Simonen 2011). This concept, invented and developed in Novosibirsk Russia, is a 14 MeV plasma neutron source. In contrast to earlier magnetic mirror concepts it operates collision dominated, rather than collisionless. Consequently it avoids microinstability issues and operates with low electron temperature. It utilizes simple high-field axisymmetric magnets. The existing GDT device at Novosibirsk utilizes 10 T circular magnets to mirror confine a 7 m long plasma column powered by 5 MW of 20 keV neutral beam power. GDT plasma achieves a high plasma beta of 60%, ion energy of 10 keV, and electron temperature of 0.2 keV. Ions are injected at 45° to the midplane magnetic field to be reflected twice creating two regions of intense neutron production at the midplane.

The neutron source concept considered in Simonen et al. (2013) is based on these GDT results. Dimensionless parameters such as beta and Te/Ti are identical while the magnetic field, neutral beam energy, and power are increased by a factor of four (Simonen 2011). In this case, the neutron flux is 2 MW/m^2 and meets the material science community's spatial uniformity and environmental requirements. The GDT-type source would produce a much larger test zone than accelerator-based sources but considerably smaller than tokomak systems; so tritium consumption is small (~100 g/year), alleviating the need to breed tritium. Simonen et al. (2013) evaluated methods to control temporal variations of neutron flux output and also described breeding blanket concepts that could be employed to utilize the GDT concept for fissile fuel production.

Extrapolation of the GDT database to the fuel production mission requires demonstrating MHD stability with lower end losses with axisymmetric end plugs (Simonen et al. 2010) and showing that the electron temperature can significantly exceed 1% of the ion energy. Initial feasibility tests of these issues can be performed with modifications to GDT.

Burning fusion plasmas will have small fluctuations in neutron output associated with normal variation of plasma properties arising from various plasma relaxation and control phenomena. There are many sources of this variability with a broad range of timescales, such as burning plasma thermal variations associated with plasma shape, position, heating, and fueling systems as well as variations due to numerous plasma equilibrium relaxation phenomena. The entire blanket structure will immediately feel these temporal variations through the change in volumetric neutron heating. The heating is necessarily nonuniform and leads to the appearance of mechanical stresses driven by thermal expansion and contraction. Estimates (ANL 2005) for typical designs of fusion blankets with 10% flux variations at ~1 Hz will experience temperature excursions as high as 10°C. In the mechanically complex structure of the blanket, 30 million cycles per year may lead to unanticipated early failure.

In their paper Simonen et al. (2013) illustrated that the GDT-type neutron source is capable of producing controlled temporal variations of neutron flux in order to address the effects on material structural properties. Four methods (Ryutov 2012) of modulating the neutron output of the GDT-type neutron source are evaluated.

High intensified neutron generator (HINEG) is an accelerator-based d–t fusion neutron source with steady line and pulse line, which is designed and under construction by the Institute of Nuclear Energy Safety Technology, Chinese Academy of Sciences (Song et al. 2013). The main features of HINEG are the two lines shared the same ion source and accelerating tube and the pulse beam was realized by chopper and buncher after preaccelerating. The design index of steady neutron source strength is reaching 3×10^{13} n/s and of the FWHM for pulse neutrons is less than 1.5 ns. The design work has been completed, including the ion source, accelerator components, rotating target, and auxiliary facilities. The results of computer simulation for the design confirmed that design index was reached. The details of design and computer simulation were introduced in the contribution. HINEG will be widely used in the areas of advanced nuclear energy technology such as the study of fusion reactor, subcritical reactor and accelerator-driven subcritical system. It will also be a vital experimental platform for nuclear technology application such as neutron radiography, neutron irradiation breeding, and cancer therapy by fast neutron (Song et al. 2013).

3.8 LASER-GENERATED NANOSECOND PULSED NEUTRON SOURCES

The ability to induce a variety of nuclear reactions with high-intensity lasers has been demonstrated recently in several laboratories (Umstadter 2000; Ledingham et al. 2003). Laser-induced activation (Ledingham et al. 2000; Ewald et al. 2003; Liesfeld et al. 2004), transmutation (Magill et al. 2003), and fission (Schwoerer et al. 2003) and fusion (Karsch et al. 2003; McKenna et al. 2003) have been demonstrated without recourse to reactors or other traditional neutron sources. These results came as a consequence of several technological breakthroughs in the field of high-intensity lasers in the last decade, which made laser intensities in the focal spot above 10^{19} W/cm^2 possible. Under these conditions, the laser beams generate plasmas with a temperature greater than 10 billion degrees (10^{10} K). The interaction of high-intensity laser light with this plasma on a surface of a thin solid target generates collimated jets of high-energy electrons and protons (Clark et al. 2000). Experimental studies on thin foil targets with thicknesses greater than the wavelength of the laser light have demonstrated the production of beams of protons with energies up to 58 MeV, using a large single-shot kilojoule laser with a wavelength of ~1 μm and an intensity of 3×10^{20} W/cm^2. Two aspects are of particular interest: the proton energy attained and the high-conversion efficiency between the laser energy and the proton beam. In the aforementioned studies, a conversion efficiency of 12% was obtained (Snavely et al. 2000).

Taking into account the high-proton energies, high laser to proton energy conversion efficiency, and short proton pulse length, the (p,xn) reactions in high Z materials seem to be a promising option for a novel fast pulsed neutron source. Recently, McKenna et al. (2005) showed that a beam of protons accelerated in an experiment on the large single-shot Vulcan laser closely resembles the expected energy spectrum of evaporative protons (below 50 MeV) produced in GeV-proton-induced spallation reactions. In their work, McKenna et al. showed that laser-generated proton beams could be used for the study of residual isotopes in lead spallation target. In their paper, they presented a new analysis of laser-generated proton reactions on lead with a view to characterizing the neutron emissions in both the Vulcan experiment and in experiments with high-repetition terawatt table-top lasers. This characterization of the neutron production capability is significant due to the possible application of this technology for compact generation of pulsed neutrons with table-top lasers. Due to the relatively high-proton generation efficiency on solid targets such sources could be significantly stronger than other neutron sources generated with the interaction of femto second-laser pulses with deuterium clusters (Zweiback et al. 2000) where neutrons are generated directly from deuterium fusion or deuterium–tritium fusion reactions in laser-heated plasma. Such cluster fusion sources are capable of yields up to 10^5 fusion neutrons per joule of incident laser energy (Hartke et al. 2005).

The highly efficient conversion of laser light into fast protons, achieved with irradiation of thin solid targets, has opened up the possibility of generating pulsed neutrons in a completely new way. These sources are based on proton to neutron conversion in thick converter materials. Using a phenomenological approach based upon nuclear cross section data and proton range calculations, Žagar et al. (2005) have estimated proton to neutron conversion efficiencies for different materials. They have shown that low Z materials like lithium can be used as proton to neutron converters in current laser systems with correspondingly low proton energies (which has also been demonstrated by Lancaster et al. (2004) and Yang et al. (2004)). However, high Z materials, such as lead, are the materials of choice for efficient proton to neutron conversion for laser systems in the near-term future, when much higher proton energies will be available. The future laser systems might reach laser intensities well beyond 10^{21} W/cm^2/μm^2, even up to 10^{24} W/cm^2/μm^2, in the next decade. At these intensities proton energies will be high enough to initiate spallation reactions in high Z solid targets. Spallation reactions are even more efficient in neutron production than fission reactions, making for higher efficiency neutron generation. The characteristics of these neutron sources are an anisotropic neutron flux, a continuous energy spectrum, and an extremely short pulse width.

By using a combination of experimental data and phenomenological analysis, it has been demonstrated that the Vulcan laser is capable of producing 2×10^9 neutrons per laser shot. Using the model developed for the Vulcan analysis, it has been predicted that current state-of-the-art table-top lasers can generate neutron pulses with a rate of 10^6 n/s by using lithium as a proton to neutron converter. It was further predicted that with the table-top lasers currently under construction, pulsed neutron sources with 5×10^9 n/s (in pulses smaller than 1 ns) would be readily achievable. The extremely short neutron pulse length means that this could be a very promising neutron source for applications where precise time determination in nuclear reactions is important (e.g., pulsed fast neutron activation methods or prompt neutron material damage repair studies). Moreover, the short pulse length and strength of such neutron sources is also of interest for pulsed fast neutron interrogation methods. Finally it could be concluded that, given the current rate of evolution in laser technology, fast, cheap, flexible, pulsed neutron sources on the table-top scale would become available in the next few years.

3.9 CALIBRATIONS OF 14 MeV NEUTRON GENERATORS

Small d–t neutron generators are now routinely used in a variety of applications, including geophysical analysis and online process control. Obtaining satisfactory results from these measurements often requires a specific neutron output level. However, the output level from a neutron generator can vary, depending upon such factors as its previous operating history, the ambient temperature and the pulsing scheme. Thus, it is desirable to have a method that can determine the neutron output level in the actual conditions under which the neutron generator is to be used. This method should also be insensitive to the flux of scattered, low energy neutrons and gamma rays that are produced in the measurement environment. Also, a standard method of demonstrating the neutron output level is essential as a quality control indicator in the manufacturing of d–t neutron generators. Liberman et al. (1993) proposed a standard method of establishing the absolute neutron output from small, d–t, 14 MeV neutron generators.

This method uses a copper activation measurement in a configuration that they have calibrated with fission ionization chambers from NIST. The absolute uncertainty in this calibration is less than ±7%. The copper activation method is insensitive to backgrounds from low-energy scattered neutrons because it uses the $^{63}Cu(n,2n)^{62}Cu$ reaction that has a 12 MeV threshold. With this calibration method, measurements of absolute neutron output are possible under a variety of experimental conditions, including those simulating nuclear well logging. In addition, the configuration of the copper samples gives high counting rates so that the statistical precision of the measurement of neutron output, depending upon the generator voltage and beam current, is on the order of 1%.

Flux measurement of an accelerator-based d–t neutron generator was achieved by the same activation technique by Jakhar et al. (2008). The neutron generator can produce maximum neutron yield of 1.9×10^{10} n/s at the tritium target. For the measurement of 14 MeV neutron flux, $^{63}Cu(n,2n)^{62}Cu$ and $^{27}Al(n,p)^{27}Mg$ reactions are most suitable because of their higher threshold energies 11.03 and 3.25 MeV, respectively. Measurement of γ-ray activities from ^{62}Cu and ^{27}Mg is used to measure 14 MeV neutron flux. NaI(Tl) scintillation and HPGe detectors were used for the γ-radiation measurement from ^{62}Cu and ^{27}Mg, respectively. In this chapter, the 14 MeV neutron flux measurement by ^{63}Cu and ^{27}Al activation is discussed. The method could also serve as a diagnostics tool for the measurement of high-energy neutron flux (Jakhar et al. 2008).

A new calibration service for 14 MeV neutron generators is being developed. The calibrations may be done at NIST or at a customer site, by activation of a standardized aluminum. NIST plans to provide a needed calibration service for the many 14 MeV neutron generators now being employed for Homeland Security and industrial applications. A summary of the new NIST capabilities was presented at the *21st International Conference on the Application of Accelerators in Research and Industry*. A survey of techniques was done to investigate providing calibration services for 2.5 MeV neutrons, but no NIST capability is currently in place. These generators, using the deuterium–deuterium reaction, are of some interest as they allow facilities access to neutrons without special nuclear material or tritium. They also produce no neutrons when turned off, as opposed to isotopic sources, with NaI gamma-ray spectrometry on the activated ring at NIST (NIST 2012).

NBS-1 is the U.S. national neutron reference source. It has a neutron emission rate (June 1961) of 1.257×10^6 n/s with an uncertainty of 0.85% (k = 1). Neutron emission-rate calibrations performed at the NIST are made in comparison to this source, either directly or indirectly. To calibrate a commercial 14 MeV neutron generator, NIST performed a set of comparison measurements to evaluate the neutron output relative to NBS-1. The neutron output of the generator was determined with an uncertainty of about 7% (k = 1). The 15-h half-life of one of the reactions used also makes off-site measurements possible. Consideration is given to similar calibrations for a 2.5 MeV neutron generator (see Heimbach 2011).

3.10 NEW DEVELOPMENTS

Commercial requirements for compact, low-power electronic neutrons sources became conventional in the 1970s through the commercial requirements of mining and logging. Then, as the state of the art increased, increasing reliability and software tools for efficient analysis of composition and density were developed. Great savings came from identifying the composition of coal as it is mined from the ground (along the conveyer belt) to evaluate grade and composition so that a fair price could be determined. In addition to the proper evaluation for price, the buyer now had a tool to identify the quality of coal during delivery. This permitted efficient burning and real-time mixing of a variety of coal qualities within a single shipment.

Materials analysis for scientific activities was the next venue that gained from the increased analytical tools and databases of information that were being developed as a result of the expanded use of the neutron technique. After 9/11, cargo inspection applications grew. The ever-expanding requirements for deeper penetration into cargo packaging and higher speed analysis/results still exceed the capabilities of the neutron activation technique. However, cargo inspection by neutron analysis is being performed on a regular basis, as a secondary tool, after x-ray analysis provides a visual image. When questionable x-ray images result, specific areas for neutron probing are identified.

The two areas for improvement would be in compact accelerator reliability/fabrication technology and detector collection efficiency. Unfortunately, compact accelerator technology rarely has high priority. Investigation of microlaser-powered dielectric particle accelerators are underway primarily for medical applications but with interesting applications to remote sensing (Yoder et al. 2007). Approaches to increasing flux are being investigated using micromachined multitip cathode arrays to reach neutron flux densities of 10^{10} n/cm^2 (Reichenbach et al. 2008). Additional techniques that have developed in high flux sources developed for boron neutron capture therapy (BNCT) was found to be an experimental success in cancer treatment.

3.10.1 Computer Chip–Shaped Neutron Source

Technological advances in the field of neutron generators are being made at Sandia National Laboratory (SNL). By thinking outside the box, scientists have come up with a computer chip-shaped neutron source that is inexpensive and may have commercial uses including treating cancer patients from home instead of at the hospital (SciTechDaily 2012).

There are two main problems with accelerator-based neutron generators—their size and their cost. There are applications for which a 3 in. (7.5 cm) cylinder is too large, either physically (implanted neutron cancer therapy) or when a point source of neutrons is desired (e.g., for neutron inspection of weld flaws). Also, accelerator-based generators start at about a hundred thousand U.S. dollars, which is too large a price for some uses. For example, a neutron generator is needed for NAA, a technique for rapidly identifying the composition of a sample. This is the sort of technique that would be amazing to incorporate in a Star Trek-style tricorder, but has been far too large and expensive.

Recently Dodson (2012) reported the development of a new type of neutron generator that solves many of these problems by putting a particle accelerator on a chip. As seen in Figure 3.36, the neutristor is layered in ceramic insulation because of the large voltages being used. The unit shown here produces neutrons through d–d fusion. The d–t reaction is easier to initiate, but the decision was made to require no radioactive materials in the design of the generator.

A voltage is applied between the ion source and the deuterium target so that the deuterium ions from the source are attracted to the deuterium target. The ions accelerate in the drift region between the source and the target. The drift region must be in vacuum so the ions do not scatter from the air molecules. When the energetic ions hit the target, a small fraction of them will cause d–d fusion, thereby generating a neutron. Sandia did not announce typical acceleration voltages used with the neutristor, however, commercial neutron generators use around 100 kV, but significant neutron yields can be obtained at voltages under 10 kV.

The ion lens modifies the electric field between the ion source and the target so that the accelerated ions are concentrated on the region of the target loaded with deuterium. The SNL disclosure does not mention how the deuterium gas

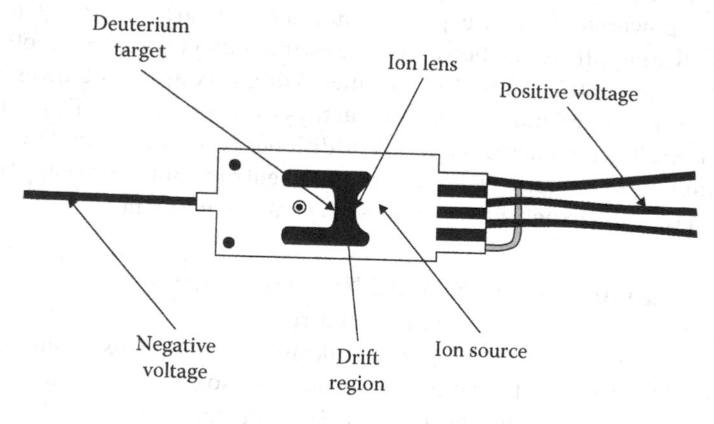

Figure 3.36 Smallest neutron generator called Sandia neutristor.

is stored, but one common approach is to coat the ion source and/or the target with palladium or some other metal that readily forms hydrides, or in this case, deuterides. For example, a palladium coating can store nearly one deuterium atom for each palladium atom. The ion current is sufficiently low that even these small amounts of deuterium will last a very long time in the completed neutristor. Neutristors can be operated in continuous or pulsed mode as required. Current neutristors have a drift region a few millimeters across, forming a sufficiently small package for many new applications. The estimated production cost for neutristors is around U.S.$2000, about a fiftieth of the cost of current accelerator-based neutron generators. The next generation of entirely solid-state neutristors will not require a vacuum for operation, thereby reducing the cost and increasing the durability of the device. In addition, SNL is working on neutristors two to three orders of magnitude smaller that would be fabricated using microelectromechanical systems, MEMS, technology.

3.11 SAFETY AND REGULATIONS

3.11.1 Introduction

Neutron generators have in the general public a reputation of hazardous material that they do not deserve. They are seen as far more dangerous than familiar x-ray instruments of which everybody has received radiation at his dentist or at the hospital. In reality, NGs are not so dangerous; however, they are radiation sources and must be operated in ad hoc installations by specifically skilled people.

The use of NGs with the knowledge of and in compliance with the safety rules and regulations is extremely important. It will allow correcting the wrong image of NGs, and help to develop the use of neutron technology in all industrial applications where it can be beneficial due to its great analytical or imaging capabilities. The willingness of making NG users be aware of safety aspects in order to develop the neutron technology justifies the presence of this chapter, which is not so obvious.

The goal of this text is not to be a reference or applicable document. Users are strongly encouraged to refer to official documents provided by the Nuclear Safety Agency of their country. Therefore, this text will not be completely exhaustive and moreover not "official" on this safety subject. The goal is to give a safety overview through the topics "technical aspects," "general rules to follow," and "regulation aspects" in such a way that allows one not to be completely lost in complicated official documents. For example, U.S. NRC regulations require that buyers have a Type A broad scope specific license authorizing possession of tritium in the applicable quantities before the purchase of tritium neutron generators or you must apply for an engineering safety sealed source and device (SS&D) review and have a specific license authorizing the possession of these devices.

3.11.2 Technical Aspects of the Neutron Generator Safety

There are two main hazards with NGs: the electrical hazard and the radiation hazard.

3.11.2.1 Electrical Hazard

As seen in previous chapters, NGs generate neutrons thanks to a nuclear reaction obtained in a target on which ions have been accelerated by an electrical field. To obtain a high level electrical field (typically 100 kV/cm), a high voltage power supply (typically 100 kV) is needed. However, if the generator is designed according to the norms of the manufacturer, there are no naked pieces under voltage that are accessible. If the connections (mains and ground) are done as indicated in the user manual, and if the generator elements are not modified or opened, there is no electrical hazard. However, in compliance with the legislation (depending on the country), the user will discover legal labels on some parts of its generator warning him against VHV hazards.

3.11.2.2 Radiation Hazards

Neutron generator radiation hazards can be divided into three different categories:

1. Radiation hazard due to the fact that 14 MeV neutron tubes contain tritium, a radioactive gas, on the contrary, 2.5 MeV neutron tubes only contain deuterium, which is not radioactive.
2. Radiation hazard due to neutron and γ emission of a NG during normal operation.
3. Radiation hazard due to the neutron activation phenomena that makes objects and materials radioactive, including the NG itself.

Radiation Hazard due to the Fact That 14 MeV Neutron Tubes Contain Tritium

Tritium is a hydrogen isotope that consists of one proton and two neutrons. It is a β emitter (emission of a β particle from one of the neutrons, which becomes a proton) with a half-life of 12.33 years; the maximum energy of the β particle is 18 keV and the average energy is 5 keV. No radiation is measurable outside the tube because the electrons do not have enough energy to penetrate the wall of the tube.

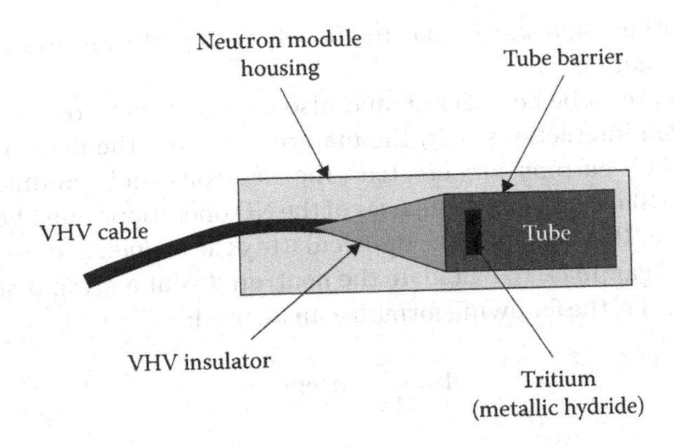

Figure 3.37 Scheme of the tritium containment of a 14 MeV NG. The containment consists of three barriers: housing barrier, tube barrier, and getter barrier.

Tritium is known for the manufacturing of objects that are self-luminous in the dark, for example, exit signs for hotels or airplanes and watches. Tritium is present with deuterium in 14 MeV neutron tubes. The order of magnitude of the quantity is between 10 and 1000 GBq for specific and high flux NGs.

Generally three barriers are present to prevent the tritium from escaping the tube (see Figure 3.37):

1. The barrier of the emission module housing. This housing contains the tube, the VHV connection, and also the VHV insulator between the tube's VHV metallic parts and the metallic grounded housing.
2. The tube barrier itself.
3. The getter barrier (zirconium, titanium, vanadium, erbium, etc.) that stores the gas inside the tube as metallic hydride.

The escape of tritium is very rare. Moreover, if a leak occurs because of a mechanical problem, air immediately starts to flow inward, because the pressure inside the tube is generally very low (~0.1 Pa [10^{-3} Torr]). If a leak occurs, the right course of action is to put the tube in a plastic bag and close it. No radiation will come out then.

The toxicity of tritium is very low, but it has to be considered. The transient time in the human body is just 10 days.

Radiation Hazard due to Neutron and γ Emission of a NG during Normal Operation

The main hazard of NGs is obviously the fact that they are radiation emitters, which is what they are built for. The user must take that into consideration before operating them, and he has to know how to protect himself and others from their radiation.

Neutrons are to be considered, and also gammas that are a consequence of the neutron interactions with the materials around the neutron tube target where the neutrons are created. The neutrons and gammas are produced in 4πsr. Biological consequences of the NG operations must be evaluated before any use. This must be done by calculating the biological dose rate due to neutrons and gammas. To calculate the neutron flux at a given distance from the neutron tube, the following formula can be used:

$$\Phi = \frac{S}{4\pi r^2} \, n/cm^2/s$$

where

Φ is the neutron flux at a given distance r from the source
S is the neutron emission per second

The formula is valid only if there are no obstacles between the source and the location at distance r. Taking into account the energy of the neutrons of 14 MeV, the biological dose rate can be calculated with appropriate conversion factors. For instance a flux of 17×10^8 n/cm²/h gives 1 Sv/h, which is 10^6 μSv/h. So 1 μSv/h corresponds to $17\times10^8/(10^6 \times 3600) = 0.47$ n/cm²/s.

In most cases, distance alone is not sufficient to lower the dose rate to an acceptable level and the NG must be operated with shielding to reduce the dose rate to operators. These shielding could be the walls of a lab room or shielding put directly around the neutron source. Typical thickness of shielding to reduce the neutron flux and dose rate by a factor of 10 is about 38 cm of water or concrete. Highly sophisticated materials are developed to optimize the shielding, namely, for fusion reactor research purpose. Figure 3.38 shows attenuations of fast neutron fluxes in 0.7 m thick shields made from various materials. The hydrogen-rich hydrides show superior neutron shielding capability compared to the conventional materials. Neutron transport calculations of the 0.7 m thick outboard shields indicated that $Mg(BH_4)_2$, TiH_2, and ZrH_2 can reduce the thickness of the shield by 23%, 20%, and 19%, respectively, compared to the combination of steel and water. NGs generally must not be directly accessible while operational. If accessibility is possible by opening a door or moving a mobile shielding, the NG must be switched off. NGs are usually provided with a safety loop that can be connected to the safety circuit of the lab.

Radiation Hazard due to the Neutron Activation Phenomena

Neutron activation is usually the required physical effect of neutron irradiation, as it allows performing measurements on the sample being

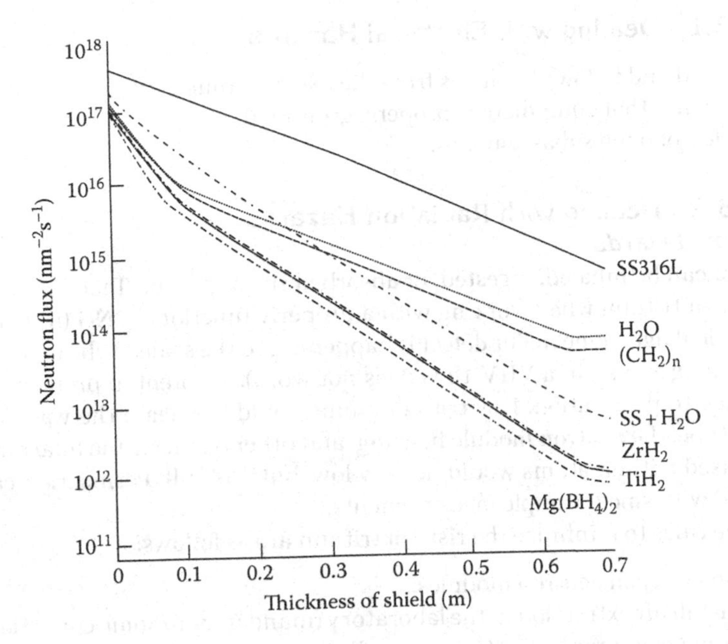

Figure 3.38 Attenuations of fast neutron fluxes in 0.7 m thick shields made from various materials.

irradiated. It is also a drawback as the sample becomes immediately active. Consequently, ad hoc operations have to be considered for the management of the activated samples. The activation hazard risk is low with standard NGs, and most importantly one has to deal with the regulations aspects. Levels of activation can be very different depending on the considered object. The final level of activation depends on the mass of the object, the flux it receives, and the duration of irradiation. Ranking the activated objects regarding their radioactivity from most activated to least activated generally gives the following list:

1. NG emission module
2. Lab structures supporting the NG emission module
3. Fixed samples exposed to flux
4. Mobile samples

3.11.3 General Rules

The rules mentioned in this section are just common sense rules. They are not part of a safety procedure. One needs to check local and official rules for safety procedures.

3.11.3.1 Dealing with Electrical Hazards

- Read and follow all notices from the user's manual.
- Ensure that equipment is properly grounded.
- Do not open subassemblies.

3.11.3.2 Dealing with Radiation Hazards

Tritium Hazards

Tritium can be inhaled, ingested, or absorbed through skin. There is no major leakage of tritium when working with a properly functioning NG (if there is a tritium leakage, some major defect is happening to the sealed tube and the NG can no longer sustain a VHV, thus does not work). A potential problem could arise due to permeation. Few tritium atoms could permeate the walls of the sealed tube, the neutron module housing, and other barriers. The final amount of released tritium atoms would be very low, but not null. Permeation can be checked with smear sample measurements.

Some rules to minimize the risks of tritium are as follows:

- Do not open neutron modules.
- Install air extraction in the laboratory (mandatory in some countries).
- Emergency protocol exits. In case of hazards enclose NG in plastic bags.

Radiation Hazards due to Neutron and Gamma Emission of the NG during Normal Operation

Some rules to minimize the risks due to radiation emission during normal operation are as follows:

- Design shielding and walls according to the expected maximal flux (or limit the flux according to the existing facility).
- Map the expected dose rate through calculations and compare it with measurements.
- Close the active area access with a safety loop. If safety loop is broken, NG will automatically be stopped.
- Follow local rules to train people and ensure utilization of individual dosimeters.

Radiation Hazards due to Activation Phenomena

Some rules to minimize the risks of activation radiation are as follows:

- Keep in mind the order of magnitude of level of activation for activated objects.
- Measure the activated objects when samples have to be moved.
- Use labels to store activated objects.
- Follow local rules.

3.11.4 Regulations

Regulations strongly depend on national rules and policies. Moreover, it depends on national regulations of the NG supplier and on the national regulations of the NG customer—buyer and user.

3.11.4.1 Administrative Work before Buying a Neutron Generator

Two important topics:

1. Obtaining authorization according to radiological safety
2. Obtaining authorization according to antiproliferation rules

3.11.4.2 Radiological Aspects

The philosophy is that according to national safety organization rules, NG suppliers can only send NGs to customers proven authorized.

The usual procedure is as follows:

1. Customers must ask their local authorities for the right to obtain and use NGs (NG suppliers are asked to supply a NG brochure for local authority information purposes).
2. Customers must send their certificate to NG suppliers.
3. Based on the customer's certificate, NG suppliers ask their local authorities for the right to sell and supply customers with the requested NGs.

3.11.4.3 Antiproliferation Rules That Apply for Any NG

The philosophy is that the NG supplier's national administration has (likely) committed to the United Nations about controlling the export of "dual use" products (civilian and military).

Based on a filed form by the customer, NG suppliers must get a license from its administration before delivering the NG to a customer.

1. Customers must provide information for the NG provider. This information could be defined by specific forms that the NG provider can send to his prospect. Typically the minimum information required to fill in these forms is the identification of end users, typical use of NG, and the place where the NG will be used and stored. Information about final use should also be a commitment to use it as described.
2. NG suppliers provide the information and commitment to their national authorities to get the license.

When the NG supplier has all authorizations, NG can be shipped. The administrative procedure typically takes several weeks to months.

3.11.4.4 Regulations for Operation

Only national regulations apply. Users must receive authorization from their administration to operate NG. This process could be simple or complicated, be short or time consuming, depending on the level of details that has to be provided, which depends highly on the neutron flux that has to be considered.

3.11.4.5 Regulations about NG's End of Life

Regulations about NG's end of life depend also only on the supplier's and customer's national regulations. Some countries require the neutron tube to be sent back to NG suppliers after 10 years. NG suppliers usually have the necessary equipment to recycle NGs at their end of life. It must be noticed that sending back NGs at their end of life (or at any time) must be done according to the process described in the previous sections, where customer and NG provider switch roles (IAEA 2012). Note: This text is following the IAEA publication *Neutron Generators for Analytical Purposes*, IAEA Radiation Technology Reports No. 1, IAEA, Vienna, Austria, 2012.

REFERENCES

Adelphi. 2013. DT110 – 14 MeV Neutron Generator. http://adelphitech.com/products/dt110.html. Accessed November 10, 2014.

All-Russia Research Institute of Automatics (VNIIA). http://www.vniia.ru/eng/index.html.

Angelone, M., Pillon, M., Batistoni, P., Martini, M., Martone, M., and Rado, V. 1996. Absolute experimental and numerical calibration of the 14 MeV neutron source at the Frascati neutron generator. *Rev. Sci. Instrum.* 67: 2189–2196.

ANL. 2005. Fusion blanket design and optimizations techniques, Argonne National Laboratory, ANL/NE-15-05.

Baccaro, S., Blazek, K., de Notaristefani, F., Maly, P., Mares, J. A., Pani, R., Pellegrini, R., and Soluri, A. 1995. Scintillation properties of YAP:Ce. *Nucl. Instrum. Methods* A361: 209–215.

Bach, P. and Cluzeau, S. 1997. New concept of associated-particle neutron generator sealed-tube, CP392 in Duggan, J. L. and Morgan, I. L. (Eds.) *Application of Accelerators in Research and Industry*. pp: 899–904. AIP Press, New York.

Bach, P., Bernardet, H., and Stenger, V. 1997. Operation and life of SODITRON neutron tube for industrial analysis, CP392, in Duggan, J. L. and Morgan, I. L. (Eds.) *Application of Accelerators in Research and Industry*. pp: 905–908. AIP Press, New York.

Berkeley lab. 2009. U.S. Compact Neutron Generator IB-1764. Patent # 6,870,894.

Chuang, L. S., Wong, H. K., and Wong, K. C. 1979. Absolute determination of 14 MeV neutron spectrum by means of liquid-scintillation spectrometer and threshold foil-detectors. http://dx.doi.org/10.1016/0029-554X(79)90204-0. Accessed October 10, 2014.

Clark, E. L. et al. 2000. Measurements of energetic proton transport through magnetized plasma from intense laser interactions with solids. *Phys. Rev. Lett.* 84: 670–673.

Dodson, B. 2012. World's smallest neutron generator—It's just for nukes anymore, Sandia National Laboratories, SNL, Press Release August 24, 2012.

Ewald, F. et al. 2003. Application of relativistic laser plasmas for the study of nuclear reactions. *Plasma Phys. Control. Fusion* 45: A83–A91.

Fisher, U., Moslang, A., and Ivanov, A. A. 1999. The gas dynamic trap as neutron source for materials irradiations. *Fusion Technol.* 35: 160–164.

Fleischer, A. A., Jorgensen, N. E., Nissen, R. F., Wells, D. K., and Younger, F. C. 1965. A 14 MeV neutron source capable of delivering a 1000 rad dose uniformly over a 6 × 6-foot area. *IEEE Trans. Nucl. Sci.* 12(3): 262–265.

Gow, J. D. and Ruby, L. 1959. Simple, pulsed neutron source based on crossed-field trapping. *Rev. Sci. Instrum.* 30(5): 315–317.

Grodzins, L. 1991. Nuclear techniques for finding chemical explosives in airport luggage. *Nucl. Instrum. Methods Phys. Res. B* 56/57: 829–833.

Hahto, S. K., Hahto, S. T., Leung, K.-N., Reijonen, J., Miller, T. G., and Van Staagen, P. K. 2005. Fast ion beam chopping system for neutron generators. *Rev. Sci. Instrum.* 76: 23304.

Hansen, S. P. 1997. Neutrons and neutron generators. *Bell Jar* 6(3/4): 1–9.

Hardik, D. P., Darji, P. H., Makwana, R., Abhangi, M., Sudhirsinh Vala, S., and Rao, C. V. S. 2012. 14 MeV neutron generator—Literature review. *Int. J. Sci. Eng. Technol.* 1: 16–20.

Hartke, R. et al. 2005. Fusion neutron detector calibration using a table-top laser generated plasma neutron source. *Nucl. Instrum. Methods Phys. Res. A* 540: 464–469.

Heimbach, C. R. 2011. Calibration of a 14 MeV neutron generator with reference to NBS-1, in McDaniel, F. D. and Doyle, B. L. (Eds.) *Application of Accelerators in Research and Industry: Twenty-First International Conference. AIP Conference Proceedings*, Vol. 1336, pp. 437–443. American Institute of Physics, College Park, MD.

Heimbach, C. R., Dewey, M., and Gilliam, D. M. 2011. Presentation at *the 21st International Conference on the Applications of Accelerators in Research and Industry*, Fort Worth, TX, August 8–13, 2010.

Hevesy, G. 1962. *Adventures in Radioisotope Research*, Vol. 1, pp. 47–62. Pergamon Press, New York.

IAEA. 1996. Manual for troubleshooting and upgrading of neutron generators. IAEA-TECDOC-913, IAEA, Vienna, Austria.

IAEA. 2012. Neutron generators for analytical purposes. IAEA Radiation Technology Reports No. 1, IAEA, Vienna, Austria.

Ivanov, A. A. et al. 2010. Results of recent experiments on GDT device after upgrade of heating neutral beams. *Fusion Sci. Technol.* 57: 320–325.

Jakhar, S., Rao, C. V. S., Shyam, A., and Das, B. 2008. Measurement of 14 MeV neutron flux from D-T neutron generator using activation analysis, in *Nuclear Science Symposium Conference Record, 2008, NSS '08, IEEE*, October 19–25, 2008, pp. 2335–2338.

Kalvas, T., Hahto, S. K., Gicquel, F., King, M., Vainionpaa, J. H., Reijonen, J., Leung, K.-N., and Miller, T. G. 2006. Fast slit beam extraction and chopping for neutron generator. *Rev. Sci. Instrum.* 77: 03B904-03B904-3.

Karsch, S. et al. 2003. High-intensity laser induced ion acceleration from heavy-water droplets. *Phys. Rev. Lett.* 91: 015001.

Kehayias, J. J. and Zhuang, H. 1993. Use of the zetatron D-T neutron generator for the simultaneous measurement of carbon, oxygen, and hydrogen in vivo in humans. *Nucl. Instrum. Methods Phys. Res. B* 79: 555–559.
Lancaster, K. L. et al. 2004. Characterisation of $^{7}Li(p,n)^{7}Be$ neutron yields from laser produced ion beams for fast neutron radiography. *Phys. Plasmas* 11: 3404–3408.
Ledingham, K. W. D. et al. 2000. Photonuclear physics when a multiterawatt laser pulse interacts with solid targets. *Phys. Rev. Lett.* 84: 899–902.
Ledingham, K W. D., McKenna, P., and Singhal, R. P. 2003. Applications for nuclear phenomena generated by ultra-intense lasers. *Science* 300: 1107–1111.
Lee, W. C. et al. 1995. Thermal neutron analysis explosive detection based on electronic neutron generators. *Nucl. Instrum. Methods Phys. Res. B* 99: 739–742.
Leung, K-N. 2002. Formed with coaxial radio frequency driven plasma ion source; detecting explosives and nuclear material; simplicity, high flux, compactness; high throughput. U.S. Patent # 6,907,097.
Leung, K.-N. 2006. Spherical neutron generator. U.S. patent #7,139,349.
Leung, K.-N. 2012. Fast pulsed neutron generator. U.S. patent appl. no. 20120213319.
Leung, K.-N., Reijonen, J., Gicquel, F., Hahto, S., and Lou, T. P. 2004. Compact neutron generator development and applications, in *16th WCNDT 2004—World Conference on NDT*, Session: Radiography, paper code 706, 6 pages, Montreal, Canada, August 30–September 03, 2004.
Leung, K.-N., Barletta, W. A., and Kwan, J. W. 2006._Ultra-short ion and neutron pulse production. U.S. patent #6,985,553.
Leung, K.-N., Lou, T., and Reijonen, J. 2008. Neutron tubes. U.S. patent application #20080080659.
Liberman, A. D., Albats, P., Pfutzner, H., Stoller, C., and Gilliam, D. M. 1993. A method to determine the absolute neutron output of small D-T neutron generators. *Nucl. Instrum. Methods Phys. Res. B* 79: 574–578.
Liesfeld, B. et al. 2004. Nuclear reactions triggered by laser-accelerated electron jets. *Appl. Phys. B: Lasers Opt.* 79: 1047.
Ludewigt, B. A., Wells, R. P., and Reijonen, J. 2007. High-yield D–T neutron generator. *Nucl. Instrum. Methods Phys. Res. B* 261: 830–834.
Magill, J. et al. 2003. Laser transmutation of iodine-129. *Appl. Phys. B: Lasers Opt.* 77: 387–390.
Mather, J. W. 1964. Investigation of the high-energy acceleration mode in the coaxial gun. *Phys. Fluids* Suppl. 7: 5–28.
Mather, J. W. 1965. Formation of a high-density deuterium plasma focus. *Phys. Fluids* 8: 366–377.
McKenna, P. et al. 2003. Demonstration of fusion-evaporation and direct-interaction nuclear reactions using high-intensity laser-plasma-accelerated ion beams. *Phys. Rev. Lett.* 91(7): 075006.
McKenna, P. K. et al. 2005. Broad energy spectrum of laser-accelerated protons for spallation-related physics. *Phys. Rev. Lett.* 94: 084801.
Miljanic, D., Antolkovic, B., and Valkovic, V. 1969. Applications of time measurements to charged particle detection in reactions induced by 14.4 MeV neutrons. *Nucl. Instrum. Methods* 76: 23–28.
Mirnov, V. V. and Ryutov, D. D. 1979. Linear gas-dynamic trap for plasma confinement. *Sov. Tech. Phys. Lett.* 5: 678.

Nebbia, G. and Valkovic, V. 2007. NATO Science for Piece Project, SfP-980526, Control of illicit trafficking of threat materials. Final Report, Zagreb and Padova.

NIST. 2012. A new calibration service for 14 MeV neutron generators. http://www.nist.gov/pml/div682/grp03/14mev.cfm. Accessed November 14, 2014.

NSD-Gradel-Fusion, Luxemburg. http://www.nsd-fusion.com.

Okuda, S., Yamamoto, T., and Fujishiro, M. 1986. Tritium depth profiling study on titanium tritide target for generating 14- MeV neutrons. *J. Nucl. Sci. Technol.* 23: 667–672.

Polosatkin, S. V. et al. 2013. Development of miniature neutron generators for science and well logging. Presentation by Grishnyaev, E. S. at *4th Asian Forum for Accelerators and Detectors (AFAD-2013)*, Novosibirsk, Russia, February 25–26, 2013.

Reichenbach, B., Solano, I., and Schwoebel, P. R. 2008. A field evaporation deuterium ion source for neutron generators. *J. Appl. Phys.* 103: 94912.

Reijonen, J. 2005. Compact neutron generators for medical, home land security, and planetary exploration, in *Proceedings of 2005 Particle Accelerator Conference*, Knoxville, TN, May 16–20, 2005, pp. 49–53.

Reijonen, J. et al. 2005. DD neutron generator development at LBNL. *Appl. Radiat. Isot.* 63: 757–763.

Reijonen, J. et al. 2007. Development of advanced neutron/gamma generators for imaging and active interrogation. Applications, in Saito, T. T., Lehrfeld, D., and DeWeert, M. J. (Eds.) *Optics and Photonics in Global Homeland Security III, Proceedings of SPIE*, Vol. 6540, p. 65401P. SPIE Press, Bellingham, WA.

Ryutov, D. D. 2012. Modulating the neutron flux from a mirror neutron source. *FUNFI Workshop*, Varenna, Italy, September 12–15, 2011; *AIP Conf. Proc.* 1442: 247–254.

Ryutov, D. D., Baldwin, D. E., Hooper, E. B., and Thomassen, K. I. 1998. A high-flux source of fusion neutrons for material and component testing. *J. Fusion Energy* 17: 253–257.

Schlumberger Limited, http://www.slb.com.

Schwoerer, H. et al. 2003. Fission of actinides using a tabletop laser. *Europhys. Lett.* 61: 47–52.

SciTechDaily. 2012. Computer chip-shaped neutron source, April 17, 2012, by Staff.

Seki, M., Ogawa, M., Kawamura, H., Maekawa, H., and Saokawa, K. 1979. Water-cooled target of 14 MeV neutron source, design and experiment. *J. Nucl. Sci. Technol.* 16(11): 838–846.

Simonen, T. C. 2011. Extrapolation of GDT results to a neutron source for fusion materials testing. *Fusion Sci. Technol.* 59: 36–38.

Simonen, T. C., Moir, R., Molvik, A., and Ryutov, D. 2010. A DT neutron source for fusion materials development, in *Proceedings of the 23rd International Conference on Fusion Energy 2010*, Daejeon, South Korea, October 11–16, 2010. IAEA, Vienna, Austria. CD -ROM file FTP/P7–02 and http://www.naweb.iaea.org/napc/physics/FEC/FEC2010/html/index.htm.

Simonen, T. C., Moir, R. W., Molvik, A. W., and Ryutov, D. D. 2013. A 14 MeV fusion neutron source for material and blanket development and fission fuel production. *Nucl. Fusion* 53: 063002 (5 pp.).

Skoulakis, A. et al. 2014. A portable pulsed neutron generator. Applications of nuclear techniques (CRETE13). *Int. J. Mod. Phys.: Conf. Series* 27: 1460127 (8 pp).

Snavely, R. A. et al. 2000. Intense high-energy proton beams from petawatt-laser irradiation of solids. *Phys. Rev. Lett.* 85: 2945–2948.

Sodern, http://www.sodern.fr.

Song, G., Song, F., Zhu, Q., Liao, Y., Wu, Y., and FDS Team. 2013. Progress in design and development of high intensified neutron generator. http://www.icenes2013.org/ViewFile.aspx?codReg = 29. Accessed October 20, 2014.

Thermo Fisher Scientific Inc., http://www.thermo.com.

Umstadter, D. 2000. Photonuclear physics - Laser light splits atom. *Nature* 404: 239–239.

U.S. NRC. 2014. U.S. Nuclear Regulatory Commission, http://www.nrc.gov. Accessed May 05, 2014.

Valkovic, V., Miljanic, D., Tomaš, P., Antolkovic, B., and Furic, M. 1969. Neutron-charged particle coincidence measurements from 14.4 MeV induced reactions. *Nucl. Instrum. Methods* 76: 29–34.

Valkovic, V., Furic, M., Miljanic, D., and Tomaš, P. 1970. Neutron-proton coincidence measurement from the neutron induced break-up of deuteron. *Phys. Rev. C* 1: 1221–1225.

Vourvopoulos, G. 1991. Industrial on-line bulk analysis using nuclear techniques. *Nucl. Instrum. Methods Phys. Res. B* 56/57: 917–920.

Whittemore, W. L. 1990. Neutron radiography. *Neutron News* 1: 24–29.

Womble, P. C., Vouvopoulos, G., Paschal, J., Novikov, I., and Chen, G. Y. 2003. Optimizing the signal-to-noise ratio for the PELAN system. *Nucl. Instrum. Methods Phys. Res. A* 505: 470–473.

Xian Aohua Electronic Instrument Co., Ltd, http://www.chinaogpe.com/showroom/605/html/product_SNG_Slim_Diameter_Neutron_Generatorfor_Logging_10_8421.html.

Xiao, K.-X., Ai, J., Shi, G.-J., Xiang W., Liang, C., and Mei, L. 2012. Development of small neutron generator for well logging. *Nucl. Phys. Rev.* 29(1): 81–84.

Yang, J. M. et al. 2004. Neutron production by fast protons from ultraintense laser-plasma interactions. *J. Appl. Phys.* 96: 6912–6918.

Yoder, R. B., Travish, G., and Rosenzweig, J. B. 2007. Laser-powered dielectric structure as a micron-scale electron source, in *Particle Accelerator Conference, 2007 IEEE*, Albuquerque, NM, June 25–29, 2007. pp: 3145–3147.

Žagar, T., Galy, J., Magill, J., and Kellett, M. 2005. Laser-generated nanosecond pulsed neutron sources: Scaling from VULCAN to table-top. *New J. Phys.* 7: 253.

Zweiback, J. et al. 2000. Characterization of fusion burn time in exploding deuterium cluster plasmas. *Phys. Rev. Lett.* 85: 3640–3643.

Chapter 4

Detectors

4.1 CHARGED PARTICLE DETECTION

The charged particles from the neutron-induced nuclear reactions could be detected by a counter telescope similar to the one described by Kuo et al. (1961), consisting of three proportional gas counters followed by a scintillation CsI(Tl) counter. The scintillation counter was used as the energy counter. Its resolution for 14 MeV protons was ~4%. The proportional counter closest to the E counter was a dE/dx counter. The width at the half-maximum of the distribution of the energy losses for 12.8 MeV deuterons in the dE/dx counter was ~24%. The first proportional counter rejected the pulses produced by particles originating in the front wall of the telescope. The second counter served for collimation and considerably reduced the background. Each of the four pulses from the telescope, after passing the amplification stage, was led to the coincidence–anticoincidence gating unit, from which two pulses were derived: one proportional to the energy loss of the particle in the dE/dx counter and the other proportional to the residual energy of that particle in the scintillation counter. These pulses were analyzed by a 2D analyzer. The resulting 3D graphs show proton, deuteron, and triton dE versus E spectra simultaneously. A typical distribution of energy losses in the dE/dx counter is shown in Figures 4.1 and 4.2.

Applications of time measurements to charged particle detection in reactions induced by 14.4 MeV neutrons have also been described by Miljanic et al. (1969). The time-of-flight measurements of a charge particle together with the measurement of its energy and energy loss allow the study of neutron-induced reactions with low cross sections down to about 10 μb/sr.

The main problems in the study of neutron-induced reactions arise from the very low intensity of the incident neutron beam. These difficulties have been tried to overcome by increasing the dimensions of the target and of the detectors.

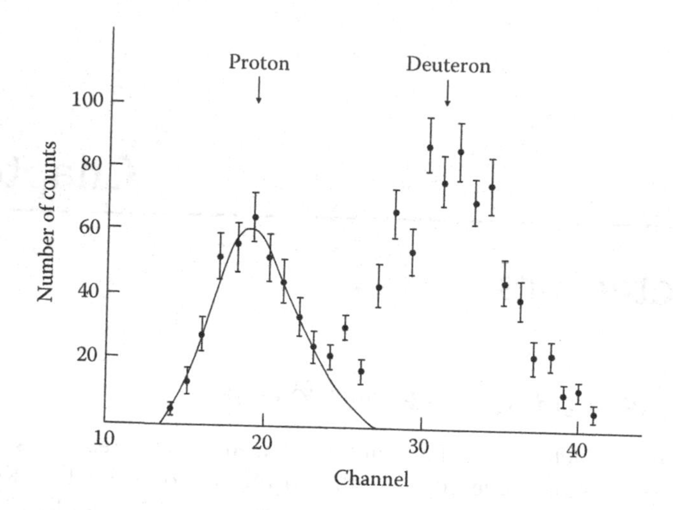

Figure 4.1 Spectra of proton and deuteron energy losses in the dE/dx counter. The curve is calculated from the Landau–Symon theory.

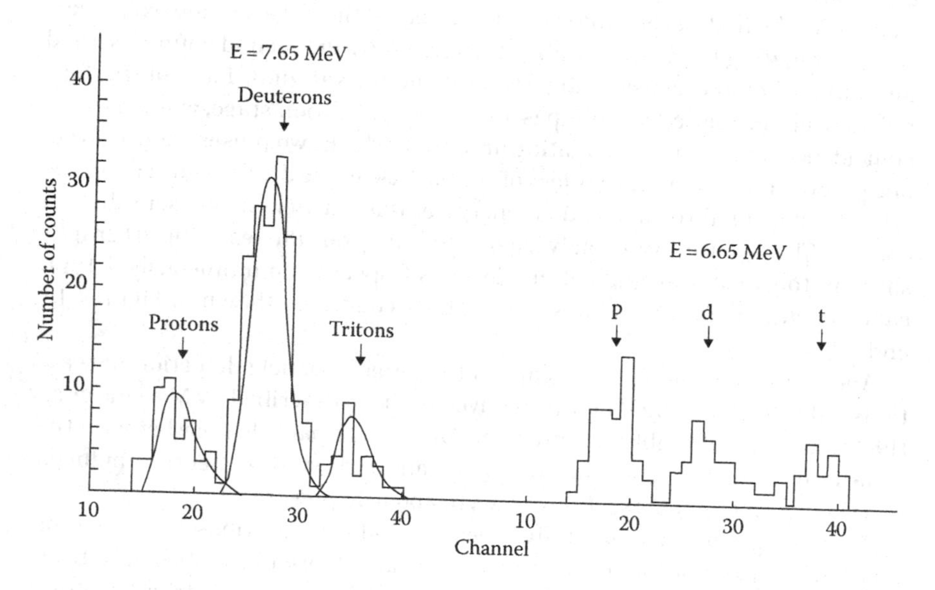

Figure 4.2 Data from the E-dE/dx telescope; the typical discrimination between protons, deuterons, and tritons. The solid curves are calculated using the Landau–Symon theory.

Most of the measurements in neutron-induced reactions are therefore characterized by poor geometry. The angular resolution of most of the counters used is of the order of 10°. Attempts to improve the angular resolution by reducing the sizes of the detectors and of the targets would result in an extremely low counting rate requiring long measurements. Most of the data on the energy spectra and angular distributions of charged particles from neutron-induced reactions have been obtained by using the counter telescope technique. The most frequently used telescope is a combination of an E detector (surface barrier or scintillator) and several gas proportional detectors. One of the gas proportional counters is usually used for measuring the energy loss dE/dx. The background reduction is achieved by two additional dE/dx counters: one in front of the target and the other placed immediately behind it. Separation of heavier charged particles, like Li, is also possible (Figure 4.3).

A natural approach to the problem of improving the angular resolution might be the improvement of the counter telescope technique. Such a position sensitive counter telescope for the study of neutron-induced reactions has been described as shown in Figure 4.4 (Valkovic et al. 1970; Valkovic and Tomas 1971).

The battery of the gas proportional counters was used to "cut" the E detector into slices (see Figure 4.4). The geometry of the system could be improved by using a rectangular E detector. The battery was employed to generate a pulse indicating the two wires between which the detected particle had passed. The battery was made from stainless steel wires, 2×10^{-3} cm in diameter, with a wire separation of 0.5 cm. The wires were stretched between two planes of stainless steel mesh made from wires of 5×10^{-3} cm in diameter, 0.2 cm apart. The distance between the mesh and the wires was 1.5 cm.

The experimental setup used in the measurements with a ^{7}LiF target is shown in Figure 4.4. The operating conditions of the long dE/dx counter were chosen such that the alpha particle from the ^{7}LiF target could be detected. The data were recorded in three-parameter format (E, ΔE, T). The simultaneous measurement of the energy and the energy loss (E and ΔE) allows the identification of particles, while the measurement of the wire carrying the pulse (T) allows the determination of the position. In such a way, the data can be collected simultaneously at several angles with improved angular resolution. The data obtained by neutron irradiation of the ^{7}LiF target are shown in Figure 4.5. The distribution of the events in the E–ΔE plane is shown on the left side of the figure. The data are distributed along the hyperbola corresponding to the alpha particles. The tail of the triton hyperbola is seen in the corner. The distribution of the events from the alpha particle hyperbola in the E–T plane is shown on the right. The projection on the T-axis is shown at the bottom of Figure 4.5. Good separation between the wires has been achieved.

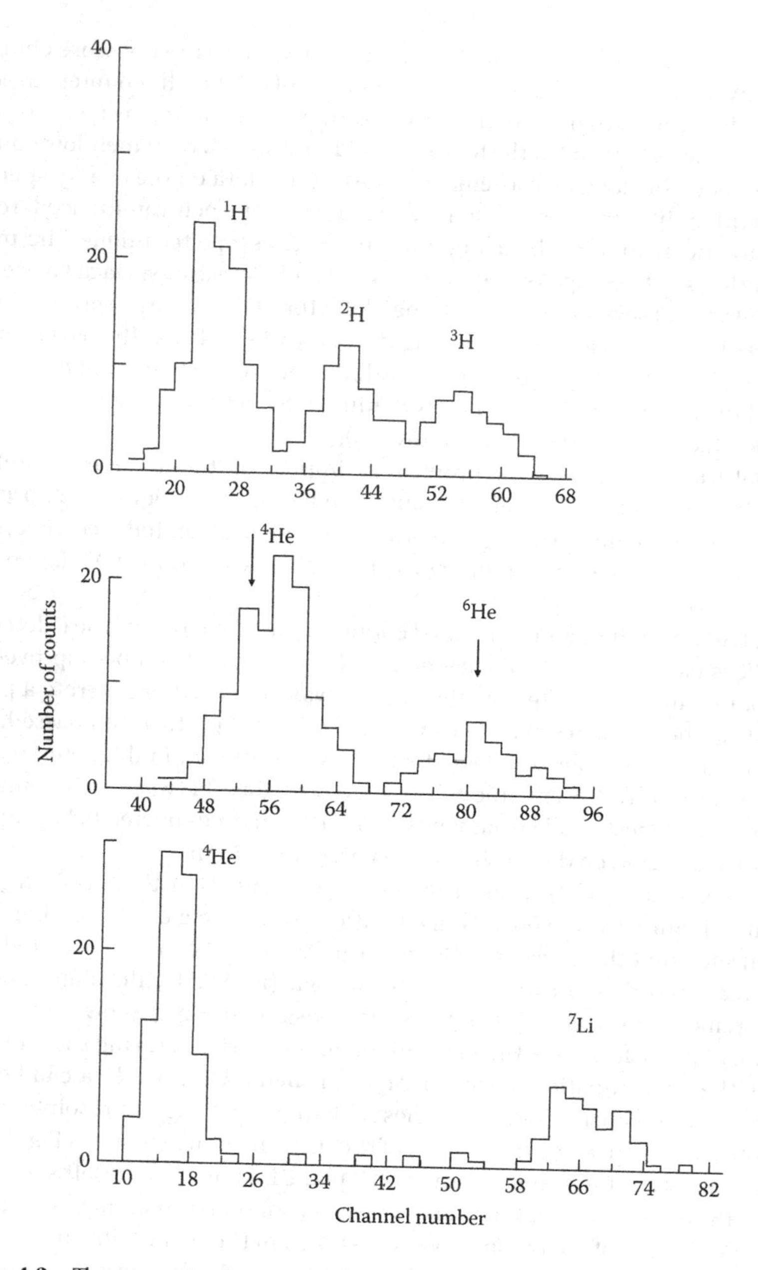

Figure 4.3 The separation between different charged particles obtained by the measurement of the energy loss in the ΔE_2 counter. For p, d, and t the ΔE_2 counter was filled with CO_2 gas, while for ^{4}He, ^{7}He, and ^{7}Li the counter was filled with H_2 gas.

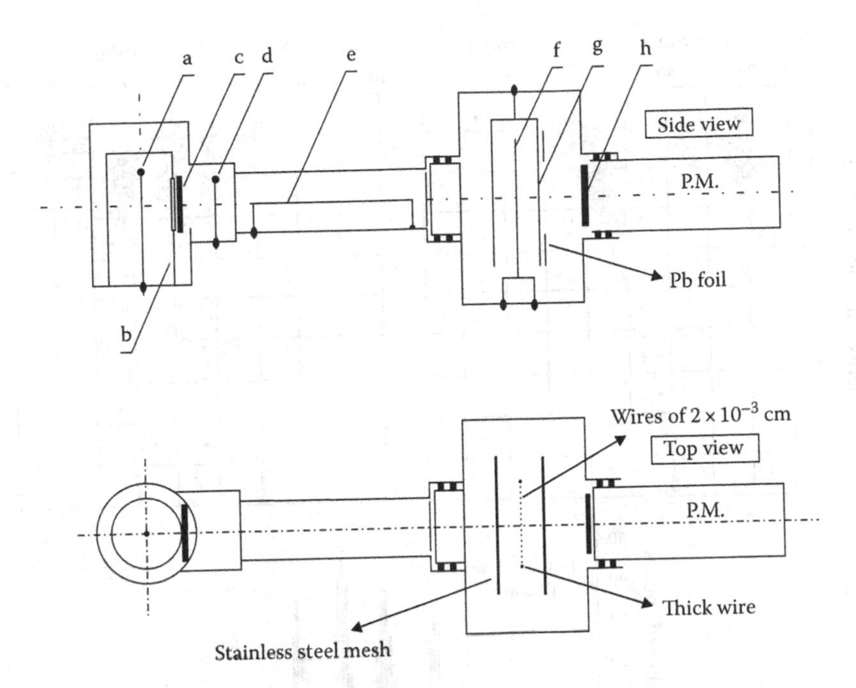

Figure 4.4 Counter telescope: (a) dE/dx gas proportional counter, (b) carbon and gold lining, (c) target, (d) dE/dx gas proportional counter, (e) long dE/dx gas proportional counter, (f) battery of the gas proportional counters—position sensitive part, (g) stainless steel mesh, and (h) scintillation. Both top and side views are shown.

4.2 SCINTILLATION DETECTORS

Many inorganic and organic materials have scintillation properties. Here, we make reference to the website (Derenzo et al. 2014) providing the measured properties of many inorganic materials and citations to the published papers in which the measurements were reported. It is a live document that grows as additional measurements are published. The document is intended for two main uses: (1) as a web accessible reference to useful scintillation detector materials and (2) as an aid in developing fundamental theories or empirical relations between basic material properties and scintillation performance. The table has 592 entries presently, each listing material formula, emission mechanism, density (g/cm^3), luminosity (photons/MeV), time (ns), emission peak (nm), energy resolution (% FWHM @ 662 keV), comment on material properties, and reference to the published measurement. All measurements are at room temperature, unless otherwise noted. Only measurements made with ionizing radiation (e.g., electrons, x-rays, and gamma rays) have been included. Measurements of crystalline powders have been included because many

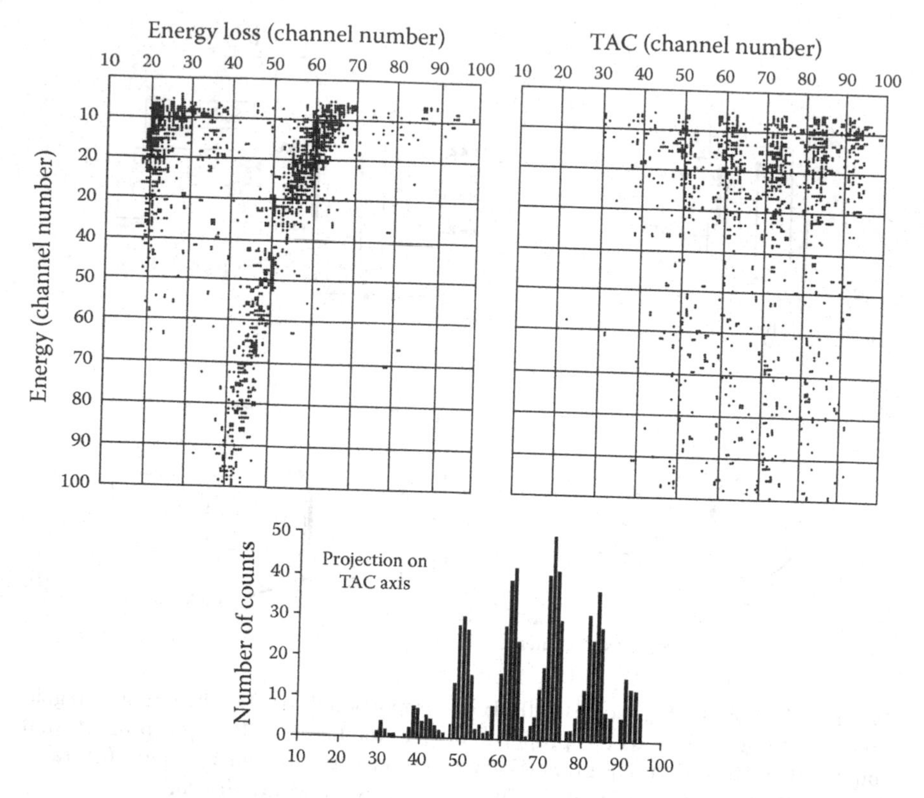

Figure 4.5 Data obtained from neutron irradiation of the ^{7}LiF target. Only the events from the alpha particle hyperbola in the E–ΔE plane are presented in the E·T plane. The projection on the time-to-amplitude converter (TAC)-axis is shown.

scintillation materials are useful in powder form and material in powder form can provide a qualitative measurement of the luminosity and good measurements of the time response and the scintillation emission spectrum.

Many scintillating materials have been used as radiation detectors; here we show characteristics, see Table 4.1, of some often used in experimental work and different applications. Let us describe some of them. High flashpoint EJ-309 liquid scintillator is an alternative to the commonly used EJ-301 (= NE213). EJ-309 has a flashpoint of 144°C and is not listed as dangerous goods material. Its pulse shape discrimination (PSD) properties are just slightly inferior to EJ-301. See Table 4.2 for the properties of EJ-309 and EJ-309:B5 scintillators. To increase the neutron sensitivity, EJ-309 can be doped with boron up to a weight percent of 596 of natural boron. This material is called EJ309:B5. EJ-309 can be encapsulated in a variety of geometries and can be read out

TABLE 4.1 DATA FOR SOME SCINTILLATION MATERIALS

Physical Properties	NaI(Tl)	YAG(Ce)	YAP(Ce)	BGO	CaF(Eu)	CsI(Tl)
Density (g/cm^3)	3.67	4.57	5.37	7.13	3.18	4.51
Hardness (Mho)	2	8.5	8.6	5	4	2
Index of refraction	1.85	1.82	1.95	2.15	1.44	1.78
Crystal structure	Cubic	Cubic	Rhombic	Cubic	Cubic	Cubic
Melting point (°C)	651	1970	1875	1050	1360	621
Hygroscopic	Yes	No	No	No	Yes	No
Linear coeff. of thermal expansion (10^{-5}/K)	5.75	0.8–0.9	0.4–1.1	0.7	1.95	5
Cleavage	Yes	No	No	No	Yes	No
Chemical formula	NaI	$Y_3Al_5O_{12}$	$YAlO_3$	$Bi_4(GeO_4)_3$	CaF_2	CsI
Luminescence properties						
Integrated light output (% NaI(Tl))	100	40	70	15–20	50	45
Wavelength of max. emission (nm)	415	550	370	480	435	550
Decay constant (ns)	230	70	25	300	940	900
Afterglow (% at 6 ms)	0.5–5	<0.005	<0.005	<0.005	<0.3	<2
Radiation length (cm)	2.9	3.5	2.7	1.1	3.05	1.86
Photon yield at 300°K (10^3 photons/MeV)	38	40–50	25	8–10	23	52

TABLE 4.2 PROPERTIES OF EJ-309 AND EJ-309:B5 SCINTILLATORS

Properties	EJ-309	EJ-309:B5
Light output (rel. to anthracene)	75%	
Photon yield/MeV electrons	11.500	
Maximum of emission wavelength	424 nm	
Density (15°C)	0.964 g/cc	
H:C ratio	1.25	
No. of C atoms/cc	4.37×10^{22}	4.13×10^{22}
No. of H atoms/cc	5.46×10^{22}	5.34×10^{22}
No. of electrons/cc	3.17×10^{23}	3.16×10^{23}
No. of ^{10}B atoms/cc	—	5.34×10^{23}
Flash point	144°C	144°C
Decay time short component	Approx. 3.5 ns	Approx. 3.5 ns
Refractive index	1.57	1.57
Light attenuation coefficient	>1 m	>1 m
Chrome	5	BB

with suitable photomultiplier tubes (PMT) to obtain the optimum timing and neutron gamma separation via PSD; see Figure 4.6.

All materials suffer to some extent from radiation damage from γ-rays and neutrons. In scintillators, radiation damage manifests itself by the formation of absorption bands in the material (coloration), which negatively affects the energy resolution and the light output of the scintillator. The onset of radiation damage in NaI(Tl) or Bismuth Germanate (BGO) starts at doses of approx. 500–1000 Gray. Part of the damage is self-annealing. The effect of fast neutrons on the performance of germanium detectors is in degradation of HPGe coaxial and planar detectors' energy resolution when bombarded with fast neutrons.

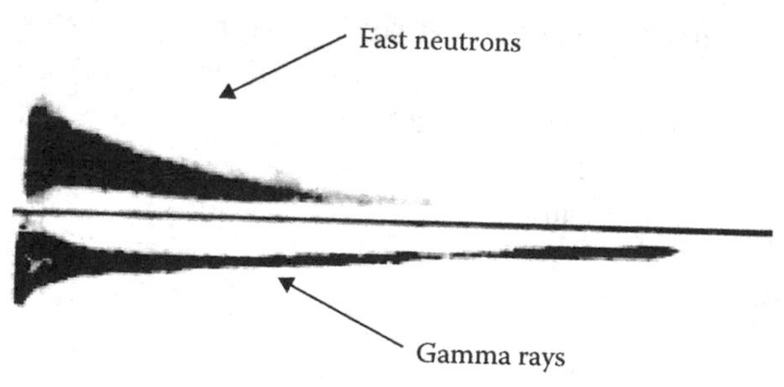

Figure 4.6 Two-dimensional scatter plot showing neutron–gamma separation in EJ-309.

Evidence suggests that in order to achieve the highest possible resistance to neutron damage, the detector should operate at a temperature as close to 77 K as possible. Streamline HPGe detector cryostats may be equipped with an internal heater and companion hardware to allow in cryostat neutron damage repair at the place of detector use. Typical annealing temperatures and duration of the annealing cycle for 30% relative efficiency detectors with severe damage are 120°C and 168 h for P-type and 100°C and 24 h for N-type. Larger volume HPGe detectors start showing energy resolution degradation at lower fluences than smaller detectors.

The increase in resistance to neutron damage of N-type detectors versus P-type can be as high as a factor of 20. However, this increase depends on a variety of factors including detector size, configuration of electric field internal to the detector, energy spectrum of the neutrons, and probably other variables. When P-type detectors are annealed, there is a significant loss in relative efficiency because of inward diffusion from the lithium outer contact. Such loss is negligible for N-type detectors because the lithium diffusion is at the inner contact where an increase in the lithium diffusion depth has little effect on the detector's active volume (IAEA 2012). There is no known limit to the number of times that a neutron-damaged detector can be annealed if proper precautions are taken. There is no evidence of a radiation hardening effect. Slightly neutron-damaged N-type detectors show some improvement in energy resolution when cycled to room temperature. However, severely neutron-damaged N-type detectors show a catastrophic deterioration of energy resolution when cycled to room temperature. As a general rule, it is best to maintain neutron-damaged detectors at LN_2 temperature until they are annealed (IAEA 2012).

4.2.1 Gamma Detection

The gamma-ray detectors must be suitable for operation in mixed radiation fields where neutrons and gamma rays are present. The detector material must have a high Z value to effectively detect characteristic photons with energies up to 10.8 MeV. The detection medium must also provide the energy resolution that allows resolving peaks of interest. Ideally, the detector should provide minimum interference with the signal emitted from a sample when the detector material is irradiated with neutrons. Thus, if possible, the detector material should avoid isotopes that are anticipated in the analyzed samples. The neutron-induced gamma-ray peaks for elements of the detector material should not interfere with the sample's spectral signatures. In addition, neutrons may produce radioactive activation products with the time-delayed decay inside the detector volume. These decays (e.g., the beta decay) may produce photons or charged particles that interfere with the characteristic gamma-ray spectra adding the noise and overloading the data acquisition electronics. It is a complicated task to satisfy all these requirements, especially with added cost limitations. Usually, the trade-off between various detector parameters including its cost is considered for a particular application (Barzilov and Womble 2006).

Standard gamma-ray detector solutions for spectroscopy are high-purity germanium (HPGe) detectors with liquid nitrogen or mechanical cooling subsystems and scintillation detectors such as NaI(Tl), $Bi_4Ge_3O_{12}$ (Feng et al. 1992), $LaBr_3$(Ce) (van Loef et al. 2001), etc. Noble gas scintillation or ionization detectors and gamma-ray telescopes can also be used for neutron-induced photon measurements in the MeV energy range.

The NaI(Tl) scintillator material has the light yield 38 photons/keV, 1/e decay time 250 ns, and density 3.67 g/cm^3. Atomic numbers are 53 and 11 for iodine and sodium, respectively.

Under neutron irradiation, the NaI(Tl) scintillator is activated by neutrons showing the delayed beta decay spectral continuum with the end point energy ~2 MeV. The BGO scintillator has the light yield 9 photons/keV and 1/e decay time 300 ns. Due to the high atomic number of bismuth, 83, and the crystal's high density of 7.13 g/cm^3, the BGO scintillator is very effective for the detection of high-energy photons. Its energy resolution is lower than NaI(Tl) resolution: ≈10% FWHM versus ≈7% FWHM for 662 keV γ-rays. The BGO demonstrates excellent behavior under neutron irradiation without delayed decay issues. Significant downside of the BGO detector is its sensitivity to the environmental temperature (Womble et al. 2002).

The $LaBr_3$(Ce) scintillator has the ~3% resolution of the 662 keV peak and a density of 5.08 g/cm^3. The lanthanum atomic number is 57. This scintillator has high light yield 63 photons/keV, fast 1/e decay time 16 ns, and better timing properties than NaI(Tl). The $LaBr_3$(Ce) material contains small quantities of the radioactive ^{138}La isotope ($T_{1/2}$ = 1.02 × 10^{11} years) producing the 1.47 MeV gamma-ray peak that is always visible in the spectrum; it can be used for calibration purposes. The $LaBr_3$(Ce) is affected by neutrons showing the delayed beta decay spectral continuum with end point energy ~3 MeV when irradiated with a d–t neutron source. The measured decay curve exhibits cumulative nature: two isotopes decay at the same time. The ^{80}Br decays with a half-life 17.68 min. The ^{82}Br isotope decays with the half-life of 35.28 h. Lanthanum halide demonstrates stable gamma-ray spectrum parameters in the mixed field under d–t neutron irradiation, when properly shielded. The good energy resolution under the room temperature, the high brightness, and the high scintillation decay speed pose this material as a promising candidate for active neutron interrogation applications, if the crystal's neutron activation issues are properly addressed.

The performance of two large NaI(Tl) scintillation detectors has been determined as a function of detector type and as a function of temperature (Reeder and Stromswold 2004). One detector had dimensions of 4 × 4 × 16 in.^{3} (4.194 × 10^{-3} m^3) with a stainless steel shell while the other detector had 2 × 4 × 16 in.^{3} (2.097 × 10^{-3} m^3) with an aluminum shell. Absolute counting efficiencies for photopeaks and total counts were measured at 0.46 and 2.0 m for gamma sources ranging in energy from 25 to 2500 keV. Photopeak resolutions were measured over the same energy range. The changes in pulse height and photopeak resolution were measured as a function of temperature over the range −50°C to +60°C. As expected from prior literature data, the scintillator light output decreases at both higher and lower temperatures compared to room temperature. However, the maximum peak height in this work occurred at 0°C, whereas the literature gives the maximum light output at about 40°C. This difference is attributed to the fact that in this work, the phototubes and preamplifiers were heated and cooled along with the scintillator. Both detectors continued to function successfully over the entire temperature range studied in this work. The pulse height decreased by about 33% at −50°C and about 25% at +60°C compared to its maximum at 0°C for both detectors. The resolution of both detectors degraded from about 8% to about 10% at −50°C for the 662 keV peak in ^{137}Cs, but it did not change significantly at elevated temperatures.

The effects of radiation on 4″ × 4″ NaI(Tl) detector by the 14 MeV neutrons have been studied by Sudac and Valkovic (2010).

The GEANT4 Monte Carlo simulation tool was used to model a prototype 14 MeV neutron fiber detector (Mengesha et al. 2006). Detail features of the prototype were implemented in the modeling to assess the directionality and detector performance. The prototype was built using plastic fibers consisting of a core scintillating material and an acrylic outer cladding. A total of 64 square fibers were used in parallel with a fiber pitch of 2.3 mm. Recoil protons scattered by an incident monodirectional 14 MeV neutrons were tracked to enable the reconstruction of a 2D direction of the incident neutrons. Simple kinematics of neutron scattering together with a back projection technique was implemented. Reconstructed direction has a peak with a full width at half maximum (FWHM) of 10° and rests on a pedestal of uniform counts. Results from the present GEANT4 simulation have demonstrated promising directionality of a prototype scintillating fiber detector.

The choice of gamma detector in the explosive detection is of paramount importance. The properties of a 9 × 9 × 20 mm^3 [$LaBr_3$(Ce)] detector made by Saint-Gobain company under the name BrilLanCe 380 (see http://www.detectors.saint-gobain.com) and NaI detector of the same size have been studied (Sudac et al. 2010). The neutron source was sealed tube generator ING-27. Figure 4.7 shows measured gamma-ray spectra from irradiation of RDX

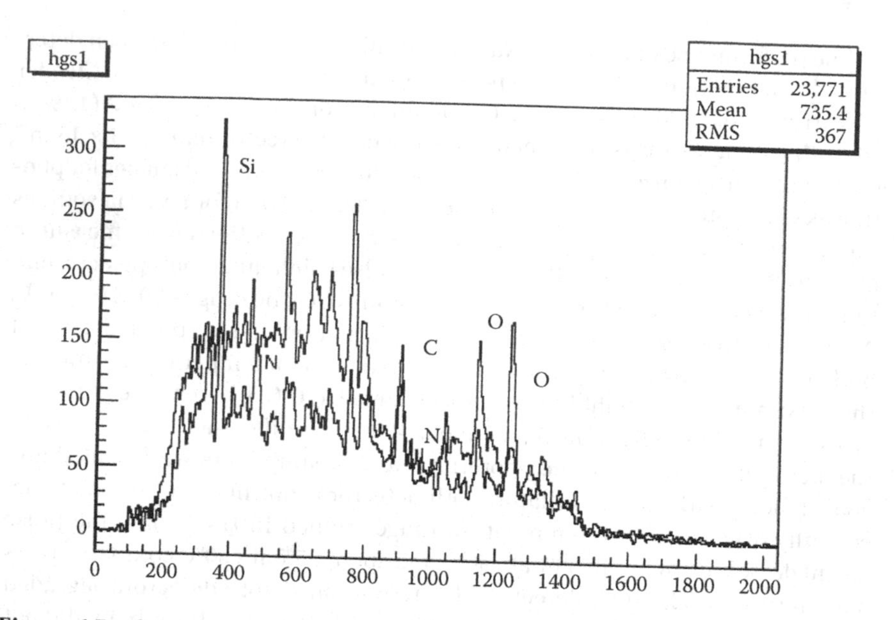

Figure 4.7 Gamma-ray spectra of RDX explosive stimulant, SiO_2 + melanin (lower) and stone (upper) as measured by a 9 × 9 × 20 mm³ [$LaBr_3$(Ce)] detector.

explosive stimulant, SiO_2 + melanin (lower) and stone (upper) as measured by a 9 × 9 × 20 mm³ [$LaBr_3$(Ce)] detector.

Measurements with 3″ × 3″ BGO detector were done on the samples of Trinitrotoluene (TNT) simulant having formula $C_7N_3O_6H_6$, water, and quartz sand. In the measurements, samples were located in the plastic container volume of 1 L. As an example, Figure 4.8 shows results of measurements for TNT simulant and plain water. Measurements lasted for about 2120 s using neutron beam of intensity ~3.6×10^7 n/s. Total number of tagged neutrons was 15×10^7.

In comparison with other scintillators, BGO spectrum does not contain second escape peak of oxygen at 6.13 MeV and the spectrum at lower energies (around 2 MeV) is cleaner.

Bystritsky et al. (2013) have designed and developed gamma detector based on the use of BGO crystal. The detector response linearity and its energy resolution in the energy range of 0.5–2.6 MeV were determined using the gamma lines of isotopes from the set of standard spectrometric gamma emitters. The widths of total gamma–quanta absorption peaks and the positions of their maxima in the amplitude spectrum were determined by fitting the peaks with a Gaussian function under a linear background description. The measured dependence of the detector response on the gamma quantum energy is represented by a straight line with an accuracy of no less than 0.4%.

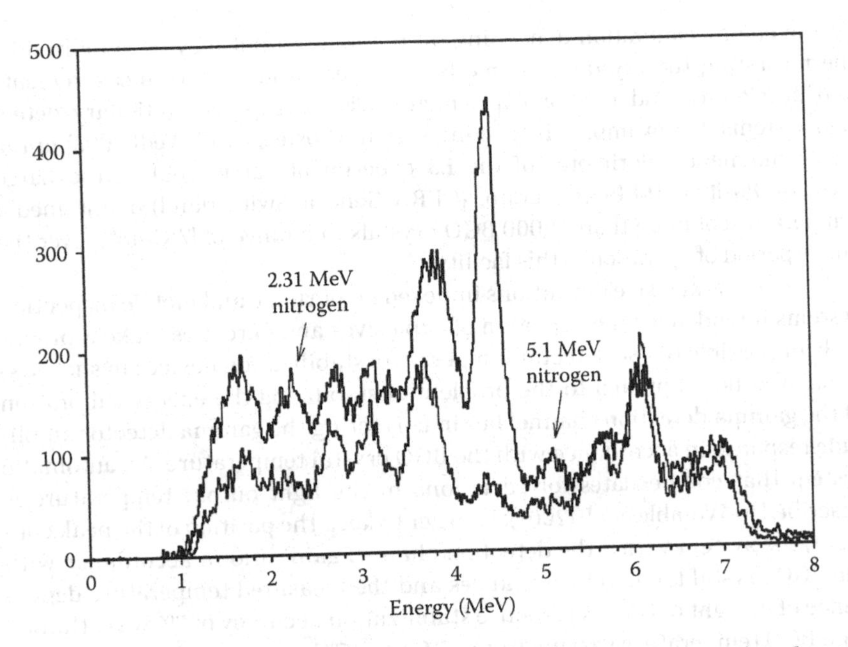

Figure 4.8 Gamma-ray spectra from TNT explosive stimulant (upper) and water (lower) samples.

The response linearity for γ quanta with energies in excess of 2.6 MeV was checked directly via the energy spectra of gamma quanta produced by the irradiation of pure carbon and melamine ($C_3H_6N_3$) samples with fast neutrons. The deviation from linearity in the energy range of 2.6–6.1 Me V does not exceed 0.35%.

The typical relative energy resolution of the detector equals 9.8% at the ^{137}Cs line and 7.8 and 7.4% at the ^{60}Co lines with energies of 1173 and 1332 keV, respectively. The dependence of the relative energy resolution on the γ-quanta energy is described fairly accurately by the following relation:

$$\frac{FWHM}{E_\gamma} = \frac{A}{E_\gamma^{1/2}} + B$$

When using γ detectors based on BGO crystals, one must take into account the fact that the temperature dependence of the light output and decay time of these crystals is rather sharp. If the temperature of the crystal increases from 0°C to 40°C, the decay time decreases from 400 to 200 ns and the light output undergoes more than twofold reduction (Globus et al. 2005). The value of the coefficient of light output change depends on the quality of the crystal and falls within interval of 1.0%–1.6%/°C.

The energy calibration depending on the light output may be stabilized by thermostating the crystal. This method is used in laboratory measurements and accelerator and reactor experiments with large-scale stationary detection systems. For example, the cooling system (Bosteels and Weill 2003) of the electromagnetic calorimeter of the L3 experiment carried out at the Large Electron-Positron (LEP) accelerator (CERN Geneva, Switzerland) maintained a temperature of more than 11,000 BGO crystals in a range of 17°C–18°C over the entire period of operation of this facility.

Strict mass and size limitations imposed on portable and mobile inspection systems intended for the detection of explosives and narcotics make it practically impossible to use detector temperature stabilization devices in such systems. Another approach to the problem of stabilizing the energy calibration of the gamma detection channel lies in correcting the gamma detector amplitude response in accordance with the BGO crystal temperature. An automatic system that compensates for variations in the light output temperature is described in Womble et al. (2001). In order to keep the position of the peak constant, the system varied the detection channel gain ratio in accordance with the BGO crystal temperature changes and the measured temperature dependence of its light output. A response stabilization accuracy of 7% was attained for a BGO temperature varying from −5°C to +45°C.

The demand for radiation portal monitor (RPM) systems has increased, and their capabilities are being further scrutinized as they are being applied to the task of detecting nuclear weapons, special nuclear material, and radiation dispersal device materials that could appear at borders. The requirements and constraints on RPM systems deployed at high-volume border crossing are significantly different from those at weapons facilities or steel recycling plants, where RPMs have been historically employed. In this new homeland security application, RPM systems must rapidly detect localized sources of radiation with a very high detection probability and low false alarm rate, while screening all of the traffic without impeding the flow of commerce. In the light of this new Department of Homeland Security application, the capabilities of two popular gamma-ray detector materials as applied to these needs are reexamined by Siciliano et al. (2005). Both experimental data and computer simulations, together with practical deployment experience, are used to assess currently available polyvinyl toluene and NaI(Tl) gamma-ray detectors for border applications. Measured and simulated spectra for current commercially available NaI(Tl) and polyvinyl toluene (PVT) detectors have been obtained for various sources. Comparisons have been made of the effectiveness of these two scintillator materials for applications such as RPMs for border screening. The main advantage of PVT is its low cost, which enables large detectors to be employed. Secondary benefits are reduced sensitivity to damage or gain drifts from temperature changes. The PVT detectors have very limited spectral capability and are best suited for efficient gamma-ray detection such as in primary portals where large numbers of vehicles need to be screened rapidly for the presence of gamma-ray sources. NaI detectors are more expensive than PVT, but they provide the benefit of isotopic identification because of their enhanced resolution. Secondary screening to determine the specific source of alarming events is one use of NaI(Tl). However, the amount of the NaI(Tl) needs to be sufficient to match the sensitivity used in the primary screening. Scattering of gamma rays in cargo degrades the quality of spectra available from any detector and limits the benefits of higher resolution in some situations (Siciliano et al. 2005).

4.3 NEUTRON DETECTION, NEUTRON–GAMMA SEPARATION

In general, neutrons are more difficult to detect than gamma rays because of their weak interaction with matter and their large dynamic range in energy. In 2D scatter plot neutron–gamma separation it can be easily seen. The boron-related neutron capture peak is located at a significant higher gamma equivalent energy (100 keV) than in boron-doped EJ-301 (60 keV).

The measurement of the neutron detector efficiency in the energy range from about 2 to 14 MeV was performed (Jackson et al. 1967) using the $^{2}H(d,n)^{3}He$ and $^{3}H(d,n)^{4}He$ reactions and elastic neutron–proton scattering. The $^{2}H(d,n)^{3}He$ and $^{3}H(d,n)^{4}He$ reactions were induced by 150 keV deuterons in a scattering chamber using deuterium and tritium targets. Charged particles were recorded by a surface barrier detector and their spectra measured with and without the coincidence requirement between the charged particle and the kinematically corresponding neutron. The ratio of the number of counts under the ^{3}He (or ^{4}He) peak, measured in coincidence and in free mode directly, gave the absolute efficiency of the neutron detector. Due to the slow variation of the neutron energy with the angle of detection in the $^{2}H(d,n)^{3}He$ and $^{3}H(d,n)^{4}He$ reactions induced by 150–200 keV deuterons, the neutron energy range that could be explored by these reactions is limited to intervals of about 2–3 and 13–15 MeV. The energy range not comprised by the aforementioned reactions was investigated by the elastic scattering of neutrons by protons.

A simple and straightforward method for experimentally determining neutron time-of-flight detector threshold and efficiency as a function of neutron energy is presented. A procedure for neutron-charged particle coincidence studies is included also (Jakson et al. 1967). For many experiments involving neutron detection, the range of neutron energies of interest is such that the detection efficiencies may vary significantly. This is particularly true for neutron energies below 5 MeV or for values near the detector threshold. When time-of-flight measurements are to be made on the neutrons, a simple and straightforward method exists for experimentally determining the detector threshold and measuring the detector efficiency as a function of neutron energy. It consists of using a gamma-ray source to establish reproducibly the discriminator setting on the detector, n–p elastic scattering to measure the relative neutron detection efficiencies over the complete range of interest, and the $^{2}H(d,^{3}He)n$ reaction to obtain several absolute neutron detection efficiencies with which to normalize the relative values. The results of the relative efficiency measurements are shown in Figures 4.9 and 4.10. Shown in Figure 4.10 are the theoretically calculated efficiency curves that best fit the experimental values. Since the counter actually detects the recoil proton resulting from an n–p collision in the scintillator, the number of such recoil protons must be calculated to obtain the detector efficiency. The presence of the carbon must be accounted for in the secondary n–p collisions; the double scattering from protons can effectively be neglected since it will only serve to increase the detector pulse height corresponding to the

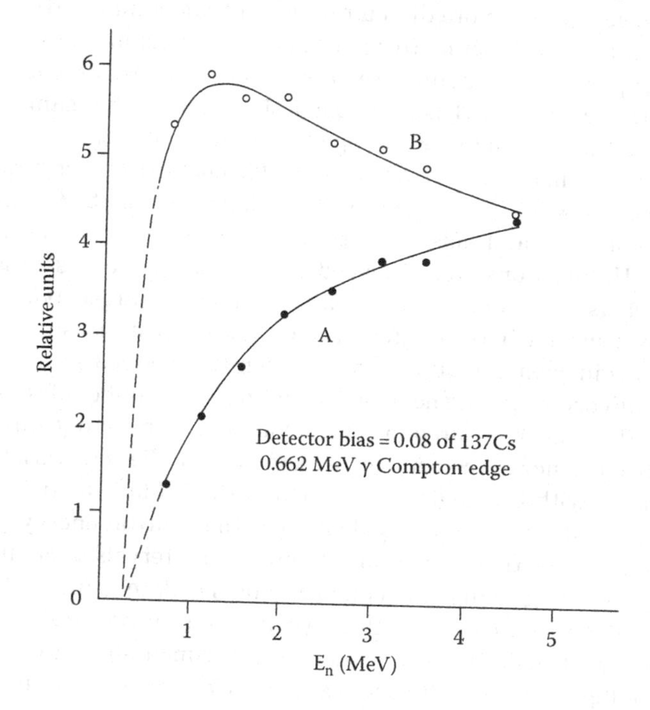

Figure 4.9 The relative efficiency obtained for the neutron detector from the n–p scattering measurements. Curve A is the ratio of the number of counts in the n–p elastic scattered peak to the number of counts in the monitor long counter. Each point measured has a 1% statistical uncertainty. Curve B is the result of applying the correction due to the geometrical factor $A(E_n,\theta)$ (attenuation in scatterer) and the variation in the laboratory differential cross section for n–p scattering. The curves have been drawn by inspection.

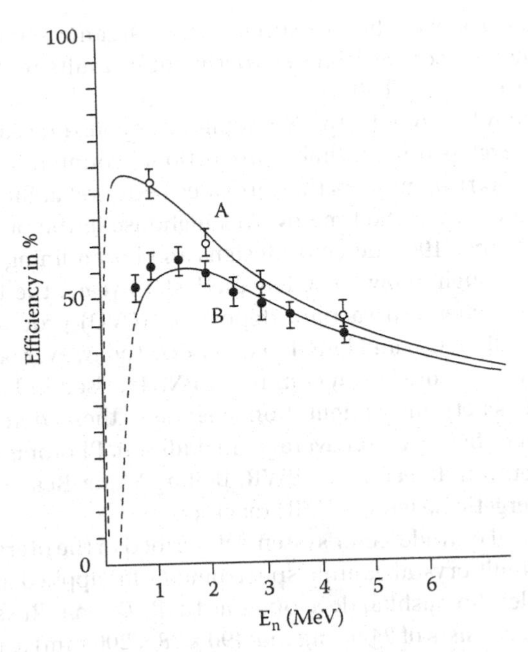

Figure 4.10 The absolute neutron detection efficiency obtained by normalizing the relative efficiency measurements by the use of absolute efficiency measurements obtained from the $^{2}H(d,n)^{3}He$ reaction as described in the text. Correction for attenuation in the chamber wall has been included. The errors indicated are the statistical uncertainties of the n–p and $^{2}H(d,n)^{3}He$ measurements. Curve A does not correspond to slow bias set, while Curve B corresponds to the slow bias conditions.

initial collision. Detector pulse heights below a certain level will not be observed due to the bias-level set. To calculate the effect this bias setting will have on the detection efficiency, it is necessary to know the relation between the neutron energy and its corresponding recoil proton pulse height spectrum. Since the pulse height for a given proton recoil is approximately proportional to the recoil energy and since the differential cross section for n–p scattering below 15 MeV is isotropic in the center of mass system, a simple calculation shows that for neutrons of incident energy E_n, the pulse height spectrum will be rectangular. In practice, the spectrum will deviate from the rectangular shape due to the resolution of the detector, nonlinearity of the scintillator response, photomultiplier tube noise, and other effects. The shape of the efficiency curve is very sensitive to variations in the parameters used for the calculations, and the fact that both experimental curves could be fitted so closely by only a variation of B attests to the validity of the approximations and methods used. Although a detailed error analysis cannot be justified under these circumstances, the success of the method indicates a confidence level of about 5% for the neutron detection efficiencies over the measured range. Organic scintillator neutron detector efficiency—a comparison of experimental results with predictions is discussed by Plasek et al. (1973).

French company Photonis (http://www.photonis.com/nuclear/applications/#research) has developed boron-lined proportional counters working in pulse mode during the start-up process that provides better reliability than the standard BF_3 counters used at the time by Westinghouse in the pressurized water reactors (PWRs). Since 1966, detector designs and boron-lining processes have been improved through many tests. For the last 15 years, the lining has been made by a unique chemical vacuum deposition (CVD) process giving excellent boron-layer adherence and long-life counters. Today, over 60 AREVA PWRs use two or four out-of-core boron counters CPNB44. Used in Rolls Royce Civil Nuclear start-up safety instrumentation channels, these detectors are now changed only every five years on average. In addition, Photonis is producing a wide range of neutron detectors for PWR, Boiling Water Reactors (BWR), and Water-Water Energetic Reactor (VVER) reactors.

There are several multidetector systems described in the literature. Here, we shall mention a multicrystal gamma spectrometer for applied and fundamental research, called Romashka, developed at JINR, Dubna, Russia (Skoy et al. 2013). Romashka consists of 24 hexagonal (90 × 78 × 200 mm) NaI detectors, as shown in Figures 4.11 and 4.12.

Figure 4.11 NaI(Tl) detectors 90 × 78 × 200 mm in Al cage, separated from Hamamatsu PMT by Al flange and glass window. (From Skoy, V.R. et al., Multi-crystal gamma-spectrometer for applied and fundamental neutron research Frank Laboratory of Neutron Physics. Report JINR Publishing Office, Dubna, Russia, 2013.)

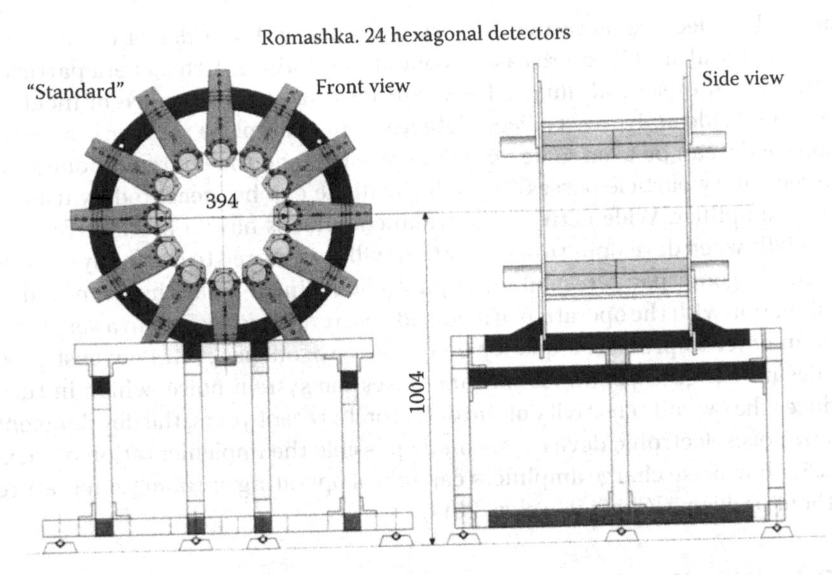

Figure 4.12 Front and side views of Romashka detectors arrangement. (From Skoy, V.R. et al., Multi-crystal gamma-spectrometer for applied and fundamental neutron research Frank Laboratory of Neutron Physics. Report JINR Publishing Office, Dubna, Russia, 2013.)

Solid-state detectors have been previously used for direct detection of neutrons; however, the detection of fast neutrons using silicon diodes has inherent problems in some of these techniques (Zhang et al. 2001; Bassini et al. 2002; Spieler 2006; Caruso 2010; Talebitaher et al. 2011). Silicon photodiodes can also detect x-rays, electrons, and protons, amongst a few types of ionizing radiations. A more common method for neutron detection is through the use of scintillators that convert the absorbed neutron energy into light that can be detected using photodiodes, photomultipliers, or charge coupled devices (CCDs). Scintillators are also sensitive to many kinds of ionizing radiations, including gamma rays, and their efficiency is relatively low and only viable with a high-gain light detector. A more interesting proposition is neutron nuclear activation of an appropriate material followed by detection (Zhang et al. 2001) of the secondary particle emission as the activated products decay. Such techniques can be tailored to eliminate contributions to the detector signal from electrons and high-energy x-rays. There is the added advantage that the measurement of the activation can take place after the neutron beam is off, which is particularly important for neutrons that are produced in devices that generate a significant amount of electrical noise.

For low-yield neutron sources, the nuclear activation technique can offer better efficiency when compared to scintillation detection techniques.

The nuclear decay of activated materials occurs on the order of the half-life of the material, and it can release secondary radiations such as beta particles in the region of several hundred per cm^2 for a moderate fluence of incident neutrons. Wide active area silicon detectors, for example, a p-i-n or avalanche photodiode, can be used to detect beta emission, and the charge produced as the secondary particle passes through the diode can be measured by using a charge amplifier. Wide active area avalanche diodes have a prohibitive cost, especially when developing a detector that will be required to use many of these to provide good spatial resolution. A photodiode linear array that is moved in conjunction with the operation of a pulsed neutron source can scan a very large area in order to produce a quality image at a reasonable cost. Low-cost p-i-n diodes have high capacitance that increases the system noise, which in turn reduces the overall sensitivity of the detector. In recent years, the development of low-noise electronic devices has made possible the implementation of inexpensive low-noise charge amplifiers capable of operating at room temperature without problems (Chatzakis et al. 2014).

4.3.1 Directional Neutron Detectors

Several directional neutron detectors have been reported in the literature (Holslin et al. 1992; Wurden et al. 1995; Singkarat et al. 1997; Miller et al. 2003). They all rely on knock-on proton-type detectors that are susceptible to x-rays and low-energy gamma rays. In the report by Mascarenhas et al. (2005), the two novel plastic scintillating fiber directional neutron detector prototypes were investigated. One prototype used a fiber selected such that the fiber width was less than 2.1 mm, which is the range of a proton in plastic. The difference in the distribution of recoil proton energy deposited in the fiber was used to determine the incident neutron direction. The second prototype measured both the recoil proton energy and direction. The neutron direction was determined from the kinematics of single neutron–proton scatters.

Both detectors were prescribed to be small, optimized for 14 MeV neutrons, and immune to x-ray background. Studies performed by Mascarenhas et al. (2005) suggest that the first detector has a poor angular resolution of around 45°. They found that 0.5 mm round fibers achieve the best resolution for this detector.

A promising type of detector is a tracking neutron detector (Peel et al. 2006). It uses an 8 × 8 array of 0.5 mm^2 fibers. This detector is sensitive to a change in incident neutron direction of better than 10°. This detector shows good directional performance in the presence of an x-ray to neutron background of 10^6:1. It is the first tracking neutron detector to be developed, which can reconstruct the direction of 14 MeV neutrons from kinematics of single neutron–proton scatters. This tracking neutron detector uses Saint

Gobain BCF-12 scintillating fiber in an 8 × 8 square array. The fibers were paced 2.3 mm apart with an air gap between fibers. The fiber dimension was 0.5 × 0.5 × 100 mm. One end of each fiber was coupled to an anode of a multianode photomultiplier tube (Hamamatsu RS900-00-M64) using optical grease. The fiber assembly was housed in a thin cylindrical aluminum shell that kept the assembly light tight. The fibers were coated with a thin black coating, and a black Mylar sheath was placed in between the aluminum shell and the fibers to reduce cross talk among the fibers. The Bicron fibers were composed of a polystyrene core with an acrylic cladding. The cladding thickness was about 4% of the core size.

The neutron detector and a neutron generator were placed in a shielded experimental cave. The shielding consisted of 20 cm thick, boron-loaded polyethylene wall to reduce the neutron flux outside the cave. The neutron generator pulsed at 100 Hz with a pulse width of 100 μs. The neutron flux was 8.2×10^6 n/s. The 14 MeV neutrons are generated isotropically in a d–t reaction. The detector was placed at 0.75 m from the source and rotated to various angles to study directionality. The energy calibration was performed using 14 MeV neutron incident along the fiber axis. The resulting proton recoil spectrum in a fiber has an end point of 14 MeV; by detecting this end point one can determine the detector energy scale. The advantage with using this technique was that the calibration was performed using high-energy recoil proton.

One can exploit the kinematics of the neutron–proton scattering to deduce the direction of the neutron source. First, the energy of the scattered proton is given by

$$E_p = \frac{4A}{(1+A)^2 E_n}\cos^2\theta \rightarrow E_p = \frac{4A}{(1+A)^2} E_n \cos^2\theta$$

E_p and E_n are the proton and neutron energies, respectively, and θ is the angle between the incident neutron direction and the direction of the scattered proton. A is the ratio of the mass of the target nucleus to the neutron mass. Since neutrons and protons have essentially the same mass (1.67×10^{-27} kg), A = 1. For neutron–proton scattering this simplifies to

$$E_p = E_n \cos^2\theta$$

If the incident neutrons are monoenergetic, E_n is known and fixed, then θ is determined from the kinematics of a single proton scatter by detecting the energy and direction of the scattered proton. This concept is used in tracking detector. Protons are detected as they slow down in the detector and produce scintillation light. The scintillation light is proportional to the energy deposited by the proton in the scintillator. Summing the energy from all the fibers hit yields the proton energy.

4.3.2 New Developments in Neutron Detection

In future, fusion tokamaks devices, detectors, and radiation monitors will operate in harsh environments characterized by high level of nuclear radiation and by temperatures higher than the standard room temperature. The present available silicon diode detectors are not suitable to withstand these working conditions despite the efforts done to develop more radiation hard devices such as those based on oxygenated silicon. For many years, scientists have been studying new materials that could substitute silicon for detecting medium in harsh environments. Currently, several candidates have been proposed as alternatives to silicon and among them polycrystalline chemical vapor deposited diamond seems to be the most promising one.

Diamond has a large band-gap energy (5.5 eV), a high breakdown voltage ($\approx 10^7$ Vcm), high radiation hardness ($>3.0 \times 10^{15}$ n/cm^2), large saturated carrier velocities, and low atomic number. Because of these outstanding electronics properties fast and low-noise diamond detectors can be obtained (Traper 2000). Indeed, the development of a new class of detectors based on polycrystalline CVD diamond films is of interest not only for fusion reactors but also for high-energy physics (Adam et al. 2003) and dosimetry (in the latter case it must also be stressed that the atomic number of carbon is very close to that of human tissue).

An Austrian company, CIVIDEC Instrumentation (http://www.cividec.at), produces diamond detectors employing electronic grade synthetic diamonds, either polycrystalline (pCVD) or single-crystal (sCVD) diamond substrates. Their detector designs are compact, RF tight, and usable in vacuum as well as at cryogenic temperatures. They provide tailored solutions for the layout of the detectors. The size and thickness of the diamonds and the electrodes can be made according to customer specifications. Their diamond detectors are proven to operate at cryogenic as well as high temperatures for the detection of fast neutrons.

In the published literature, there are an increasing number of papers concerning the use of CVD diamonds as particles detector in several fields (Hassard et al. 1998; Schmid et al. 2003), but application in a tokamak environment is missing. This is mainly due to the harsh tokamak's environment and the difficulty in locating and operating a charge sensitive preamplifier (needed by the CVD detector) close to a tokamak. Another point of concern, when using CVD diamond, is the problem linked to polarization effects that usually become evident after a short time, so that long-lasting measurement cannot be performed.

Polycrystalline CVD diamond detectors are of great interest in harsh environments due to their capability to operate even at high temperature and to withstand great neutron fluencies. Their use is of interest in many fields including the nuclear fusion reactors. In the paper by Pillon et al. (2005), the successful characterization of polycrystalline CVD diamond detectors as 14 MeV neutron monitors is reported. Two polycrystalline CVD diamond detectors of 96 and 126 μm thickness, respectively, were used. The detectors' count rate versus neutron flux was studied for various applied electric fields, showing that these detectors have a linear behavior in the flux range explored. The detector efficiency was also derived resulting in the stable behavior versus the neutron flux. The low sensitivity to gamma-ray field and the possibility to reject the counts due to gamma rays was also assessed. The time-dependent neutron emission from the 14 MeV Frascati neutron generator (FNG) was recorded with CVD diamond detectors. The data for diamonds were compared with that recorded by the standard monitors available at FNG showing good agreement. Good stability and capability to operate (with negligible pileup effect) with neutron flux up to 3×10^8 n/cm^2/s were observed. Following this characterization, one diamond detector was installed at JET and successfully operated during the trace tritium experiment (TTE) campaign as a flux monitor of the 14 MeV neutron emissions from JET.

A polycrystalline CVD diamond detector was installed on JET tokamak in order to monitor the time-dependent 14 MeV neutron emission produced by d–t plasma pulses during the TTE performed in October 2003 (Angelone et al. 2005). This was the first tentative ever attempted to use a CVD diamond detector as neutron monitor in a tokamak environment. Despite its small active volume, the detector was able to detect the 14 MeV neutron emissions ($>1.0 \times 10^{15}$ n/shot) with good reliability and stability during the experimental campaign that lasted 5 weeks. The comparison with standard silicon detectors presently used at JET as 14 MeV neutron monitors is reported, showing excellent correlation between the measurements. The results prove that CVD diamond detectors can be reliably used in a tokamak environment and therefore confirm the potential of this technology for next-step machines like ITER.

Compact pulsed hadron source (CPHS) project of Tsinghua University is a facility similar to the low-energy neutron source (LENS) of Indiana University, Bloomington, IN; a series of exchange visits and teleconferences among universities and laboratories in China, the United States, Japan, and Korea attested to the common desire for international collaborations. It was reported (CPHS monthly 2010) that the progress has been made in the development of incorporating a neutron–electron converter in the microchannel plates (MCP) to be used as a key component of a novel neutron detector. Microscopic analysis with backscattered-electron scan indicated successful coating of $^{nat}Gd_2O_3$. Further development of ^{10}B-coated MCP is planned.

The harsh radiation environments such as a nuclear reactor core, high-energy physics experiments, or outer space need detectors with radiation hard properties. Fast neutron detectors based on wide band-gap semiconductors are being developed for the quantification of fissionable materials (Dulloo et al.

2003; Rahman et al. 2004; Ruddy et al. 2006). Neutron flux measurement for in-core of nuclear reactor and detection of concealed fissionable materials are among the applications of these detectors. Fast neutrons can be detected as a result of nuclear reactions that produce charged particle products. One of the promising materials is silicon carbide (SiC), which has wide band-gap energy from 2.2 to 3.2 eV for the most common polytypes. A large energy band gap has the possibility of low leakage currents, a good radiation resistance, and a reasonable sensing capability for the charges created during an ionization process.

Commercially available single crystals of SiC are the hexagonal 4H-SiC and 6H-SiC. 4H-SiC single crystal wafer was used for fabrication and characterization of neutron detector. The present work focuses on the development of a fast neutron semiconductor detector based on a wide band-gap SiC semiconductor, operating at a zero-biased voltage by using a strong internal electric field due to a PIN structure. Ha et al. (2011) reported the fabrication of the 4H PIN SiC radiation detector with three-layer structures that were 0.5 μm of a P-layer, 2 μm of an I-layer, and 30 μm of an N-layer on a 350 μm N-type substrate. The detector was designed and fabricated to operate at zero bias voltage for neutron detection at room temperature. Electrical and radiation response properties of detector were measured such as current–voltage characteristics and alpha and neutron response. The fabricated PIN SiC detectors performance showed a possibility as a neutron detector for harsh environment by operating at a self-powered mode.

4.4 PERSONAL X-RAY, GAMMA, AND NEUTRON DETECTORS

Protection of workers in a radioactive environment requires an accurate and timely monitoring of the radiation dose equivalent received by each worker. Monitoring dose equivalents received for neutron exposures must take into account not only the radiation quantity but also the radiation quality. Unlike x, γ, or β radiations for which the hazards are substantially the same per unit of absorbed dose for the commonly encountered energies, neutron radiation can result in a hazard that increases with both increased unit absorbed dose and neutron energy. For a device to measure accurately the neutron dose equivalent received by a person exposed to unknown or varying spectra of neutron energies, the device should not only count the neutron events but also compensate properly for the variability in hazard as a function of the neutron energy. To monitor the dose equivalent in a timely manner, the device should indicate the accumulated dose equivalent at any desired time and provide a warning when the accumulated dose equivalent reaches a chosen action level.

Several devices are commonly used to measure neutron dose equivalents received by personnel in radioactive environments. One such device is a badge containing neutron sensitive film as the measurement medium. Neutrons impinging on the film may strike a hydrogen atom in the film emulsion. This hydrogen atom is ionized into a proton, which then causes an ionization recoil track in the emulsion. Development of the film forms an image of the proton recoil track that may be visually detected. Evaluation of the accumulated dose equivalent received by the person wearing the badge can be made by counting the proton recoil tracks on the film, usually manually with the aid of a microscope.

Another commonly used device is a badge containing thermoluminescent dosimeters (TLD) as the measurement medium. These TLD are crystalline materials, usually containing the isotopes ^{6}Li or ^{10}B, and have the property of luminescing when they are heated to a high temperature if they had previously been exposed to radiation. Both ^{6}Li and ^{10}B have a significant cross section for the (n,α) reaction, whereby the alpha causes ionization in the crystal, imparting energy to the electrons. A portion of these electrons are trapped until the crystal is heated to a temperature sufficient to release them. The release of a trapped electron is accompanied by a flash of light (luminescence). The neutron dose equivalent is then evaluated from the measurement of the light output when the TLD are evaluated by heating.

Although attempts to provide portable neutron counters were described in articles published in 1971 (Johnson 1971; Falk et al. 1971) the introduction of electronic neutron dosimeters started in 1980s (Falk and Tyree 1984) and the state of the art of electronic personal dosimeters for neutrons was presented two decades later by d'Errico et al. (2003). Despite a widely recognized need, then and now, electronic devices for personal dosimetry of neutrons or mixed neutron–photon fields are still far less established than systems for photon or beta radiations. Their paper describes the investigation of many commercial and laboratory systems and their response characteristics. These were determined so far with measurements using ISO standard monoenergetic beams up to 19 MeV at Physikalisch-Technische Bundesanstalt (PTB) in Braunschweig, Germany.

There is a growing need for personal radiation monitoring equipment. The growing market has resulted in some advanced companies producing sophisticated detectors. They come in various shapes and with different characteristics, see for example http://www.bing.com. We shall mention here only some without prejudices, inforcement, or recommendations. Although companies usually have different models, some with x-ray and gamma detectors only, we shall mention only some of the meters capable of measurement of neutron radiation also.

TABLE 4.3 SM100N SPECIFICATION SUMMARY

Detection capability	X-ray, gamma, and neutrons
Measurement types	Gamma: air kerma or H*10
	Neutrons: counts/s
Display and measurement range	Gamma: 0.1 μSv/h to 3 Sv/h
	Neutrons: thermal to 14 MeV
Energy range	Gamma: 50 keV to 1.5 MeV
	Neutrons: thermal to 14 MeV
Data logging	5000 data logging points dose, dose-rate, alarm, and event information
Product compliance	Designed to meet IEC 60846 and ANSI N42.17, N42.33 Type 2

4.4.1 Innovative Physics

SM100n personal gamma–neutron survey meter, made by Innovative Physics, Landguard Manor, Shanklin, Isle of Wight, PO37 7JB United Kingdom, is capable of fast and reliable measurements of gamma, x-ray, and neutron radiation. Its versatility and compact size enhance its ability to be used easily in a number of applications. Its specifications are presented in Table 4.3. In addition, the company has two other models: SN125n personal gamma-neutron survey meter and SM150n advanced gamma–neutron survey meter.

4.4.2 Thermo Scientific

The Thermo Scientific™ Radeye™ NL personal highly sensitive neutron radiation detectors complement passive and active neutron dosimeters and can be easily used by anyone working with industrial neutron sources, staff and inspectors of nuclear facilities, first responders, and law enforcement officers. See more at http://www.thermoscientific.com/content/tfs/en/product/radeye-nl-personal-highly-sensitive-neutron-radiation-detectors.html.

Their product EPD-N2, electronic personal gamma–neutron dosimeter, is well known. Its characteristics are

- Sensitive to x- and γ-radiation (E > 20 keV) and neutrons 0.025 eV < E < 15 MeV
- Direct readout of Hp(10) for neutron and photon dose
- Multiple diode detectors with converters and energy compensation shields
- Display units: Sv and rem (with prefixes μ, m) set via internal software

- Generally in accordance with ANSI standards 13.11, 13.27, and 42.20 (photons performance) and most aspects of IEC 61525 (neutrons and photons)
- Dose display and storage: 0 μSv to >16 Sv, autoranging
- Resolution for display: 1 μSv (<10 mSv/1 rem) (γ, and neutron under best conditions)
- Resolution for storage: 1/64 μSv (~1.5 μrem) (γ), 1 μSv for neutron dose under best conditions
- Dose-rate display: 0 μSv/h to >4 Sv/h (400 rem/h), autoranging, variable resolution

4.4.3 Polimaster

Dosimeter shop (http://www.dosimetershop.com) describes personal radiation detector PM1703GNM (made by Polimaster), which combines sensitivity with precision for detecting and measuring harmful radiations. The dual-purpose PM1703GNM lets you detect and locate both gamma and neutron radiation sources. The instrument is equipped with three detectors:

- CsI(Tl) scintillation detector for searching gamma radioactive sources
- LiI (Eu) scintillation detector for neutron radiation registration in the search mode
- Small-sized GM tube detector for measuring gamma radiation in the wide DER and DE range

The scintillation counters are highly sensitive for detecting even the slightest amounts of gamma and neutron radiation emitting materials. The extra GM counter facilitates accurate measurement of gamma radiation DER. The high-precision device provides measurement in a wide range of up to 10 Sv/h (1000 R/h) and DE up to 10 Sv (1000 R).

4.4.4 Mirion Technologies (MGP)

Their product DMC-2000 GN monitors gamma and neutron dose and dose rates. The full energy range of thermal, intermediate, and high-energy neutrons is covered with a high sensitivity and very good gamma rejection, tested to 6 MeV. The DMC 2000GN is a gamma and neutron detection dosimeter featuring dose rate and programmable alarms.

This instrument is compliant with IEC61526 eD2 for gamma and neutron display units: mSv or mrem. Its technical specifications for neutron measurements are

- *Dose display*: 1 μSv to 10 Sv (0.1 mrem to 1000 rem)
- *Measurement range*: 10 μSv to 10 Sv (1 mrem to 1000 rem)

- *Dose-rate display*: 10 μSv/h to 10 Sv/h
- *Energy range*: 0.025 eV to 15 MeV

Their product PDS-100 GN pocket radiation detector is a sensitive pocket-sized device designed to detect, locate, and quantify any radioactive material in a very short time. The PDS has been designed specifically for first responders, law enforcement, customs inspectors, and for personal and site security in critical infrastructures.

4.4.5 Landauer Inc.

Their product Neutrak dosimetry service provides neutron radiation monitoring with CR-39 and track etch technology (http://www.landauer.com). The Neutrak detector is a CR-39 (allyl diglycol carbonate)-based, solid-state nuclear track detector that is not sensitive to x, β, or γ radiation, and can be packaged specifically for neutron detection only, or as a component of another dosimeter to include x, γ, and β radiation monitoring. The CR-39 is laser engraved for permanent identification and wrapped with a 2D bar code to assure efficient chain of custody.

Technical specifications—energy range: 40 keV to 40 MeV; dose measurement range: 20 mrem to 25 rem (200 μSv to 250 mSv).

4.4.6 Fuji Electric Co., Ltd.

Their product NRF31 measures gamma ray and thermal and fast neutrons. Here are its characteristics (http://www.fujielectric.com/products/radiation/personal/box/doc/ECNO2310c_NRF.pdf):

Accuracy: Gamma(x)-ray ±10% (0.1 mSv to 9.999 Sv, ^{137}Cs); neutron ±15% (0.5 mSv to 9.999 Sv, ^{252}Cf)

Energy range: Gamma(x)-ray 30 keV to 6 MeV; neutron 0.025 eV to 15 MeV

Linearity for wide range of dose rate: Gamma(x)-ray ±10 (0.mSv/h to 9.99 Sv/h, ^{137}Cs); neutron ±20% (0.5 mSv/h to 9.999 Sv/h, ^{252}Cf)

So, one can read the manufacturers' catalogs on their web pages or simply visit http://www.dosimetershop.com/gamma-neutron-detector before making a decision, or even better ask for advice from somebody experienced.

REFERENCES

Adam, W., Sampietro, M., et al., Eijk, B., van, Hartjes, F., Noomen, J. 2003. The development of diamond tracking detectors for the LHC. *Nucl. Instrum. Methods Phys. Res. A* 514: 79–86.

Angelone, M., Bertalot, L., Marinelli, M. et al. 2005. Time dependent 14 MeV neutrons measurement using a polycrystalline CVD diamond detector at JET tokamak. *Rev. Sci. Instrum.* 76: 013506 (6 p.).

Barzilov, A. and Womble, P. 2006. Comparison of gamma-ray detectors for neutron-based explosives detection systems. *Trans. Am. Nucl. Soc.* 94: 543.

Bassini, R., Boiano, C., and Pullia, A. 2002. A low-noise charge amplifier with fast rise time and active discharge mechanism. *IEEE Trans. Nucl. Sci.* 49(5): 2436–2439.

Bosteels, M. and Weill, R. 2003. The cooling systems of the calorimeters and tracking subdetectors of the L3 experiment. *Nucl. Instrum. Methods Phys. Res. B* 498: 165–189.

Bystritsky, V. M., Zubarev, E. V., Krasnoperov, A. V. et al. 2013: Gamma detectors in explosives and narcotics detection systems. *Phys. Part. Nuclei Lett.* 10(6): 566–572.

Caruso, A. N. 2010. The physics of solid-state neutron detector materials and geometries. *J. Phys. Condens. Matter* 22: 443201–443233.

Chatzakis, J., Hassan, S. M., Clark, E. L., Talebitaher, A., and Lee, P. 2014. Improved detection of fast neutrons with solid-state electronics, in *Applications of Nuclear Techniques (CRETE13), International Journal of Modern Physics: Conference Series*, Vol. 27, pp. 1460138 (8pp.).

CPHS monthly. 2010. Compact Pulsed Hadron Source, CPHS, CPHS monthly, Hadron Application and Technology Complex, HATC-Tsinghua, No. 3.

Derenzo, S., Boswell, M., Weber, M., and Brennan, K. 2014. *Scintillation Properties*. Lawrence Berkeley National Laboratories, Berkeley, CA. http://www.scintillator.lbl.gov

d'Errico, F., Luszik-Bhadra, M., and Lahaye, T. 2003. State of the art of electronic personal dosimeters for neutrons. *Nucl. Instrum. Methods Phys. Res. A* 505: 411–414.

Dulloo, A. R., Ruddy, E. H., Seidel, J. G., Adams, J. M., Nico, J. S., and Gilliam, D. M. 2003. The thermal neutron response of miniature silicon carbide semiconductor detectors. *Nucl. Inst. Methods Phys. Res. A* 498: 415–423.

Falk, R. B, and Tyree, W. H. 1984. Personnel electronic neutron dosimeter. US Patent 4489315 A.

Falk, R. B., Tyree, W. H., and Johnson, V. P. 1971. A pocket-sized integrating neutron counter. Rocky Flats publication RFP-794, January 22.

Feng, X., Cheng, C., Yin, Z., Khamlary, M. R., and Townsend, P. D. 1992. Two kinds of deep hole-traps in $Bi_4Ge_3O_{12}$ crystals. *Chin. Phys. Lett.* 9(11): 597–600.

Globus, A., Grinyov, B., and Kim, J. K. 2005. Inorganic scintilators for modern and traditional applications. Report, Institute for Single Crystals, Kharkow, Polland.

Ha, J. H., Kang, S. M., Kim, H. S. et al. 2011. 4H-SiC PIN-type Semiconductor detector for fast neutron detection. *Prog. Nucl. Sci. Technol.* 1: 237–239.

Hassard, J., Liu, J. H., Mongkolnavin, R., and Colling, D. 1998. CVD diamond film for neutron counting. *Nucl. Instrum. Methods Phys. Res. A* 416: 539–542.

Holslin, D., Shreve, D., and Hagan, B. 1992. A directional fast *neutron* detector using scintillating fibers and electro-optics, in *Proc. of SPIE Neutrons, X Rays, and Gamma Rays*, Vol. 1737, pp. 14–19.

IAEA. 2012. Neutron generators for analytical purposes. IAEA Radiation Technology Reports No.1., IAEA, Vienna, Austria.

Jackson, W. R., Divatia, A. S., Bonner, B. E. et al. 1967. Method for neutron detection efficiency measurements and neutron-charged particle coincidence detection. *Nucl. Instrum. Methods* 55: 349–357.

Johnson, V. P. 1971. A Pocket-Sized Integrating Neutron Counter. Rocky Flats Publication, REP-794, January 22.

Kuo, L. G., Petravić, M., and Turko, B. 1961. AdE/dx—E counter telescope for charged particles produced in reactions with 14 MeV neutrons. *Nucl. Instr. Methods* 10: 53–65.

Mascarenhas, N., Mengesha, W., Peel, J. D., and Sunnarborg, D. 2005. Directional neutron detectors for use with 14 MeV neutrons. Sandia Report SAND2005-6255, Sandia National Laboratories, Albuquerque, NM.

Mengesha, W., Mascarenhas, N., Peel, J., and Sunnarborg, D. 2006. Modeling of a directional scintillating fiber detector for 14 MeV neutrons. *Nucl. Sci. IEEE Trans.*, 53 (4): 2233–2237.

Miller, R. S., Macri, J. R., McConnell, M. L., Ryan, J. M., Flueckiger, E., and Desorgher, L. 2003. *Nucl. Instrum. Methods Phys. Res. A* 505: 36–40.

Miljanić, D., Antolković, B., and Valković, V. 1969. Applications of time measurements to charged particle detection in reactions induced by 14.4 MeV neutrons. *Nucl. Instrum. Methods* 76: 23–28.

Peel, J., Mascarenhas, N., Mengesha, W., and Sunnarborg, D. 2006. Development of a directional scintillating fiber detector for 14 MeV neutrons. *Nucl. Instrum. Methods Phys. Res. A* 556: 287–290.

Pillon, M., Angelone, M., Lattanzi, D., Marinelli, M., Milani, E., Tucciarone, A., and Verona-Rinati, G. 2005. Characterization of 14 MeV neutron detectors made with polycrystalline CVD diamond films, in *Fusion Engineering 2005, Twenty-First IEEE/NPS Symposium on*, pp. 1–4.

Plasek, R., Miljanić, D., Valković, V., Liebert, R. B., and Phillips, G. C. 1973. Organic scintillator neutron detector efficiency: A comparison of experimental results with predictions. *Nucl. Instrum. and Methods* 111: 251–252.

Rahman, M., Al-Ajili, A., Bates, R., Blue, A. et al. 2004. Super-radiation hard detector Technologies: 3D- and widegap detectors. *IEEE Trans. Nucl. Sci.* 51(5): 2256–2261.

Reeder, P. L. and Stromswold, D. C. 2004. Performance of large NaI(Tl) gamma-ray detectors over temperature -50°C to +60°C. Pacific Northwest National Laboratory, Richland, WA, Report PNNL-14735.

Ruddy, F. H., Dulloo, A. R., Seidel, J. G., Das, M. K., Ryu, S.-H., and Agarwal, A. K. 2006. The fast neutron response of 4H silicon carbide semiconductor radiation detectors. *IEEE Trans. Nucl. Sci.* 53: 1713–1718. doi: 10.1109/TNS.2006.875151.

Schmid, G. J., Griffith, R. L., Izumi, N. et al. 2003. CVD diamond as a high bandwidth neutron detector for inertial confinement fusion diagnostics. *Rev. Sci. Instrum.* 74: 1828–1831.

Siciliano, E. R., Ely, J. H., Kouzes, R. T., Milbrath, B. D., Schweppe, J. E., and Stromswold, D. C. 2005. Comparison of PVT and NaI(Tl) scintillators for vehicle portal monitor applications. *Nucl. Instrum. Methods Phys. Res. A* 550: 647–674.

Singkarat, S., Boonyawan, D., Hoyes, G. G. et al. 1997. Development of encapsulated scintillating fiber detector as a 14-MeV neutron sensor. *Nucl. Instrum. Methods Phys. Res. A* 384: 463–470.

Skoy, V. R., Kopatch, Yu. N., and Ruskov, I. 2013. Multi-crystal gamma-spectrometer for applied and fundamental neutron research Frank Laboratory of Neutron Physics. Report JINR Publishing Office, Dubna, Russia.

Spieler, H. 2006. *Semiconductor Detectors Systems*, 2nd edn. Oxford University Press,Oxford, U.K.

Sudac, D. and Valković, V. 2010. Irradiation of 4" × 4" NaI(Tl) detector by the 14 MeV neutrons. *Appl. Radiat. Isot.* 68: 896–900.
Talebitaher, A., Springham, S. V., Rawat, R. S., and Lee, P. 2011. Beryllium neutron activation detector for pulsed DD fusion sources. *Nucl. Instrum. Methods Phys. Res. A* 659 (1): 361–367.
Trapper, R. J. 2000. Diamond detectors in particle physics. *Rep. Prog. Phys.* 63: 1273–1316.
Valković, V., Kovačević, K., and Vidić, S. 1970. Position sensitive counter telescope for the study of neutron induced reaction. *Nucl. Instrum. Methods* 79: 13–18.
Valković, V. and Tomaš, P. 1971. A position sensitive counter telescope for the study of nuclear reaction, induced by 14 MeV neutrons. *Nucl. Instrum. Methods* 92: 559–562.
van Loef, E. V. D., Dorenbos, P., van Eijk, C. W. E., Krämer, K., and Güdel, H. U. 2001. High-energy-resolution scintillator: Ce3+ activated LaBr3. *Appl. Phys. Lett.* 79(10): 1573–1575.
Womble, P. C., Vourvopoulos, G., Novikov, I., and Paschal, J. 2001. PELAN 2001: current status of the PELAN explosives detection system. In James, R. B. (Ed.) *Proceedings of SPIE: Hard X-Ray and Gamma-Ray Physics III*, San Diego, CA, Vol. 4507, pp. 226–231.
Womble, P. C., Vourvopoulos, G., Paschal, J., Novikov, I., and Barzilov, A. 2002. Results of field trials for the PELAN system. *Proc. SPIE* 4786: 52–57.
Wurden, G. A., Chrien, R. E., Barnes, C. W., and Sailor, W. C. 1995. Scintillating-fiber 14-MeV neutron detector on TFTR during DT operation. *Rev. Sci. Instrum.* 66 (1): 901–903.
Zhang, L., Mao, R., and Zhu, R.-Y. 2001. Fast neutron induced nuclear counter effect in Hamamatsu silicon PIN diodes and APDs. *IEEE Trans. Nucl. Sci.* 58 (3): 1249–1256.

Chapter 5

Nuclear Reactions Induced by 14 MeV Neutrons

5.1 INTRODUCTION

Neutrons may interact with matter in one or more of the following reactions:

1. Elastic scattering: (n,n) reaction: A neutron collides with the atomic nucleus and then loses its kinetic energy. It should be noted that the neutron loses less kinetic energy when it collides with a heavy nucleus. In contrast, it loses more kinetic energy when it collides with a light nucleus. Hydrogen (^{1}H) is, therefore, the most effective neutron moderator because it is the lightest nucleus having a mass almost the same as a neutron. Elastic scattering is most important in the production of low energy or slow neutrons from fast neutrons emitted from the source for neutron radiography. Water, paraffin, and polyethylene are common neutron moderators. In fact, hydrogen-2 (^{2}H, the so-called deuterium) is the best neutron moderator due to its extremely low neutron absorption probability. Heavy water ($^{2}H_2O$) has neutron absorption cross section only about 1/500 that of light water ($^{1}H_2O$), but heavy water is very costly.
2. Inelastic scattering: (n,n′) or (n,n′γ) reaction: This is similar to elastic scattering but when a neutron collides with the atomic nucleus it has enough kinetic energy to raise the nucleus into its excited state. After collision, the nucleus will give off gamma ray(s) in returning to its ground state. Even inelastic scattering also reduces the energy of a fast neutron but this is not preferable in neutron radiography because it increases in contamination of gamma rays to the system.
3. Neutron capture: (n,γ) reaction: A neutron can be absorbed by the atomic nucleus to form a new nucleus with an additional neutron resulting in an increase of mass number by 1. For example, when ^{59}Co captures a neutron it will become radioactive ^{60}Co. The new nucleus mostly becomes radioactive and decays to a beta particle followed by

the emission of a gamma ray. Some of them are not radioactive such as ^{2}H, ^{114}Cd, ^{156}Gd, and ^{158}Gd. A few of them only decay by a beta particle without the emission of a gamma ray, such as ^{32}P. This reaction plays a vital role in neutron radiography when a metallic foil screen is used to convert neutrons into beta particles and gamma rays.

4. Charged particle emission: (n,p) and (n,α) reactions: Most of the charged particle emission occurs by fast neutrons except for the two important (n,α) reactions of ^{6}Li and ^{10}B. The $^{6}Li(n,\alpha)^{3}H$ and $^{10}B(n,\alpha)^{7}Li$ reactions play important roles in neutron detection and shielding. In neutron radiography, these two reactions are mainly employed to convert neutrons to alpha particles or to light. The (n,p) reaction is not important in neutron radiography but it may be useful when a solid-state track detector is selected as the image recorder.
5. Neutron-producing reaction: (n,2n) and (n,3n) reactions: These reactions occur only with fast neutrons that require a threshold energy to trigger. They may be useful in neutron radiography, particularly when utilizing 14 MeV neutrons produced from a neutron generator. By inserting blocks of heavy metal like lead (Pb) or uranium (U) in a neutron moderator, the low-energy neutron intensity can be increased by a factor of 2–3 or higher from (n,2n) and (n,3n) reactions.
6. Fission: (n,f) reaction: Fission reactions are well known for energy production in nuclear power plants and neutron production in nuclear research reactors. Heavy nuclei like ^{235}U and ^{239}Pu undergo fission after absorption of neutrons. The nucleus splits into two nuclei of mass approximately one half of the original nucleus with emission of 2–3 neutrons. When uranium is used to increase neutron intensity by the (n,2n) and (n,3n) reactions mentioned earlier, the fission reaction also contributes additional neutrons to the system. The degree of contribution depends on the ratio of ^{235}U to ^{238}U in uranium.

Figure 5.1 is a schematic presentation of the areas that represent the probability that a neutron will interact with different nuclei. Scattering cross section σ varies "randomly" from element to element and even isotope to isotope. Typical σ is ~10^{-24} cm^2 for a single nucleus.

The Lawrence Livermore Laboratory initiated in their 1967 pulsed sphere program extensive measurements of the neutron spectra from spherical targets bombarded with a centered 14 MeV neutron source. Neutron emission spectra were measured for materials ranging from hydrogen to lead for thicknesses between 1 and 5 mfp. The objective of the program was to measure the neutron spectra with adequate resolution in order to make available test cases for the neutron transport codes and input neutron cross sections. The resolution was sufficiently good to enable checking not only

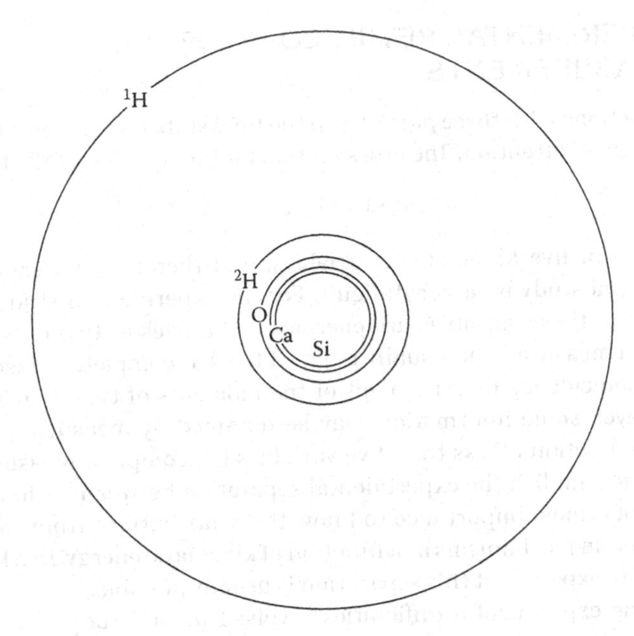

Figure 5.1 Graphical representation of the total nuclear cross sections for several isotopes.

the total cross sections as a function of energy but also the 14 MeV differential elastic and inelastic cross sections to low-lying strongly excited levels (Hansen et al. 1969, 1970, 1973). Measured spectra for materials of particular interest to controlled thermonuclear reactors, for example, ^{6}Li, ^{7}Li, C, and Fe were reported by Anderson (1974). The emitted neutron energy spectra were measured using time-of-flight techniques (2–10 ns FWHM system resolution) in a geometry where the flight path (7–10 m) is long compared to the dimensions of the spherical targets. It has been proved possible to measure emission spectra from 14 MeV (the source energy) down to 10 keV. The examples cited are chosen to emphasize the usefulness of these bulk measurements in (1) checking transport codes, (2) evaluating neutron cross sections, and (3) assessing the reliability of computational and/or reaction models.

Nuclear deformations have an effect on the total cross section; Marshak et al. (1966) measured the effect of nuclear deformation on the total cross section of ^{165}Ho using 14 MeV neutrons and an aligned target and compared it with the total cross section of unoriented ^{165}Ho. The observed difference in cross sections was reported to be $+(3.48 \pm 0.75)$, where the positive sign indicates a larger cross section for nuclei aligned perpendicular to the incident beam than for randomly oriented nuclei.

5.2 EXPERIMENTAL SETUP, COINCIDENCE MEASUREMENTS

Nuclear reactions with three particles in the final state have received a considerable amount of attention. The cross section for the reaction of the type

$$b+t \rightarrow 1+2+3 \tag{5.1}$$

is a function of five kinematical variables, and therefore, its experimental and theoretical study is rather difficult. For the experimental study, the natural choices of these variables are energies and angles of two outgoing particles. Such a measurement, usually referred to as a "complete measurement," requires a coincidence measurement of the moments of two final state particles. However, some information may be obtained by measuring the cross section as a function of less than five variables ("incomplete measurement"). In order to accomplish the experimental separation between various mechanisms, it is of utmost importance to know the contributions from each of the possible reaction mechanisms as a function of kinematic energy. In a kinematically complete experiment this separation is usually possible.

Due to the experimental difficulties involved in the study of correlation spectra in neutron-induced reactions, experimental results of complete measurements are very scarce. However, some of the fundamental information about nuclear forces could be extracted from the study of reactions such as the n + d → n + n + p reaction and neutron–proton bremsstrahlung. Besides, important spectroscopic information about light nuclei might be obtained by studying the reactions of type (1) induced by neutrons. As an example, we shall mention (n,d*) reactions (d* represents the neutron–proton final state interaction in the $^1S_0$ configuration).

Some of the information about the (n,d*) and n + d → n + n + p reaction was obtained from "incomplete measurements." However, the extracted parameters are subject to serious criticism. We shall mention that some of the measurements have been performed on the n + d → n + n + p reaction using the time-of-flight technique with the detection of two outgoing neutrons.

In the article by Valković et al. (1969, 1970), the experimental technique for detecting the outgoing neutron and the charge particle in coincidence in reactions of type (1) induced by 14 MeV neutrons is described. The possibility for the experimental study of the (n,d*) reaction, n + d → n + n + p and n–p bremsstrahlung, is discussed. In conclusion, nuclear reactions induced by neutrons resulting in three particles in the final state, one of them being a neutron, could be studied using three-parameter analysis. A usual E–ΔE measurement should be performed in order to achieve charged particle discrimination. The measurement of the neutron time of flight does not have to be performed in such a way that neutron energy could be determined. The use of associated alpha particles is necessary to eliminate background events when reactions with small cross sections are to be studied. The measurements of the neutron–proton bremsstrahlung process and the neutron-induced breakup by coincidence measurement of the outgoing neutrons and protons are feasible.

5.3 NUCLEAR REACTIONS ON LIGHT NUCLEI

The nuclear reactions on light nuclei are of interest in the study of the nucleon problem (Šlaus et al. 1964). A study of the neutron–proton bremsstrahlung could give important insight into this problem. A kinematically complete experiment on the neutron–proton bremsstrahlung at a neutron energy of 14.4 MeV (Furić et al. 1970) was performed. Protons and neutrons were detected on opposite sides of the neutron beam. Protons were identified and their energy measured. The associated particle method and the neutron–proton time-of-flight difference were used to reduce the background. An upper limit of 400 $\mu b \cdot sr^{-2}$ was found for the neutron–proton bremsstrahlung differential cross section at the detector setting $\theta_p = \theta_n = 30°$.

The effects of final state interactions are evident in the charge particle spectrum from $n + d \rightarrow n + n + p$ reaction. Furthermore, it has been demonstrated that the shape of the proton spectrum around the maximum energy is quite sensitive to the value of neutron–neutron $^1S_0$ scattering length. In the experiment done by Ilakovac et al. (1961), the sensitivity of the proton spectrum to the neutron–neutron scattering length has not been fully exploited due to an insufficient number of points determining the shape of the peak (the analysis has been based on six points) and to the large uncertainty in the rather poor energy resolution. Also, the experiment has been done with a 5 × 20 channel analyzer that allowed taking simultaneously only five points. It was, therefore, thought to be worthwhile performing a more accurate measurement of the proton spectrum from the reaction $^2H(n,p)2n$. Figure 5.2 shows the spectrum of the breakup protons at 4.8°. The errors shown are statistical errors. The data are corrected for the variation of the energy intervals arising from different energy losses at different energies. The correction is also made for neutrons degraded in energy and/or incident in other directions.

The data of Cerineo et al. (1964) were compared with those of Ilakovac et al. (1961) in Figure 5.2. The excellent agreement in the shapes of the two spectra is obvious. However, the absolute cross sections do not agree, the value from Cerineo et al. (1964) being ~25% larger. Two other measurements of the reaction $^2H(n,p)2n$ at the nominal angle of 0° and at 14 MeV were performed by Bonnel and Lévy (1961) and by Veljukov and Prokofjev (1962). Both spectra show a pronounced peak at the maximum proton energy and their shapes are in good agreement with the work by Cerineo et al. (1964). The absolute cross section

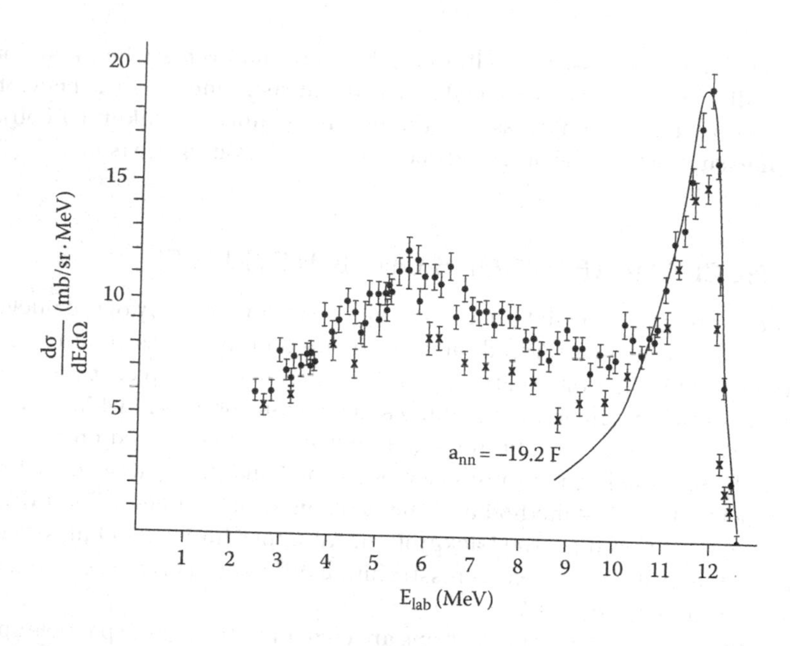

Figure 5.2 Energy spectrum of protons from deuteron breakup at 14.4 MeV at 4.80 in the laboratory system. Crosses: measurement by Ilakovac et al. (1961), dots: measurement by Cerineo et al. (1964). The curve represents the theoretical calculated spectrum for $a_{nn} = -19.2$ F after smearing and normalization.

reported by Bonnel and Levy (1961) is ~20% larger than the value reported by Cerineo et al. (1964). Veljukov and Prokofjev (1962) do not explicitly give the cross section, but an estimate made comparing their breakup protons with the elastically scattered protons from the known hydrogen contamination in their target yields a value of ~20 mb/sr in the peak, in fair agreement with the value of Cerineo et al. (1964).

By varying both a_{nn} and the normalization factor, Cerineo et al. (1964) reported the best fit to the experimental data for $a_{nn} = -21.7 \pm 1$ F. The error is only statistical. The uncertainty in the resolution, which was $(5.8 \pm 0.5)\%$, together with the uncertainty brought by the variation in the number of experimental points taken into analysis may argue for the increase of the total error to ± 2 F. The reported value is in excellent agreement with the value previously obtained by Ilakovac et al. (1961). This has prompted the study of charge dependence of nuclear forces from the breakup of deuterons and tritons (Cerineo et al., 1964).

The experimental value $a_{nn} = -21.7 \pm 1$ fm measures the effective strengths of nuclear and magnetic interactions. The correction due to magnetic interaction can be calculated, and the pure nuclear value for the neutron–neutron $^1S_0$ scattering lengths is found to be $a_{nn} = -22.5 \pm 1$ fm. This value implies the nuclear forces to be about 2%–3% charge dependent. Wong and Noyes (1962) predicted for a_{nn}, assuming the exact charge symmetry, $a_{nn} = -27.0 \pm 1.4$ fm. A noticeable difference between values was reported by Cerineo et al. (1964), and this value indicates that nuclear forces may be charge dependent.

Triton spectrum from $n + {}^7Li \rightarrow t + \alpha + n$ reaction was measured (Valković 1964a) at the neutron energy $E_n = 14.4$ MeV using E–dE/dx counter telescope. The measured triton spectrum, shown in Figure 5.3, contains an intensive peak corresponding to the neutron-α final state interaction in the ground state of ^5He. Figure 5.4 shows predictions for the simultaneous breakup process. The solid curve represents the incoherent sum of a simultaneous breakup and the sequential decay via $^5He_{g.s.}$.

The reactions $^{10}B(n,t)\alpha\alpha$, $^7Li(n,t)\alpha n$, and $^6Li(n,d)\alpha n$ have been studied at 14.4 MeV (Valković et al. 1967). The energy spectra indicate the importance of the sequential decay mechanism in these processes. The Phillips–Griffy–Biedenharn

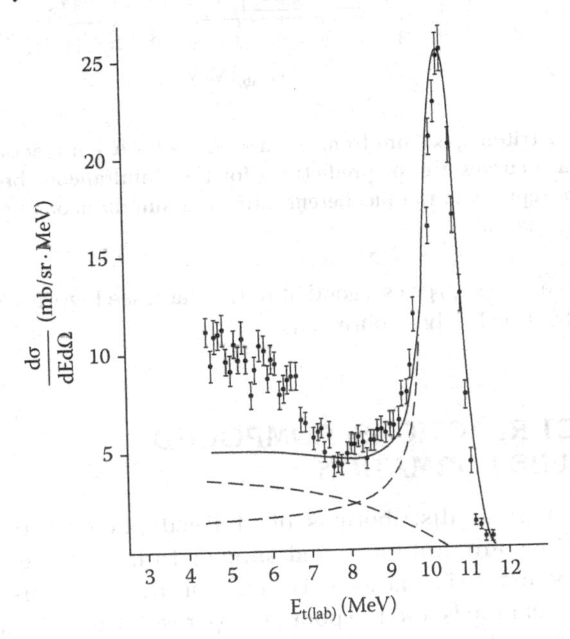

Figure 5.3 Triton spectrum from the $n + {}^7Li \rightarrow \alpha + t + n$ reaction. The errors shown are statistical. The dashed curve represents the contribution from the α + n final state interaction in the ground state of ^5He. The dashed–dotted curve represents the contribution from the α + t final state interaction in the 4.63 MeV excited state of ^7Li. The solid curve is the sum of these two contributions.

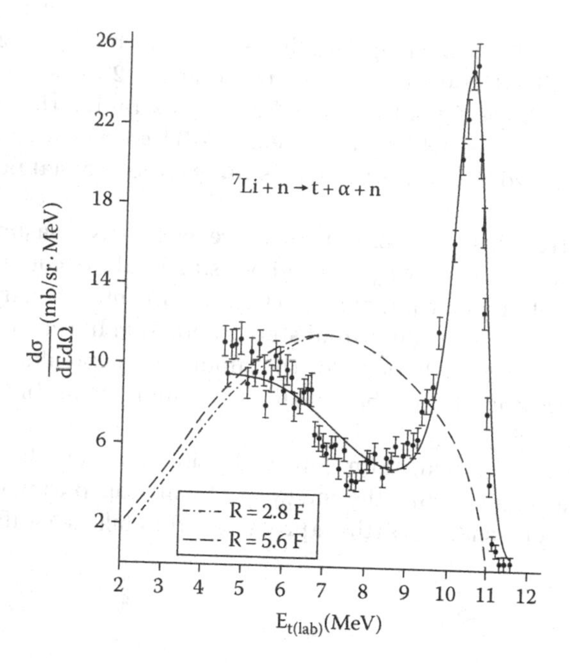

Figure 5.4 The triton spectrum from the n + ^{7}Li → t + α + n reaction. The dashed and dot-and-dash curves are the predictions for the simultaneous breakup process. The solid curve represents the incoherent sum of a simultaneous breakup and the sequential decay via $^5He_{g.s.}$.

model (Phillips et al. 1960) gives a good fit to the data; see Figure 5.5. No evidence of spatial localization has been observed.

5.4 DIRECT REACTIONS, COMPOUND NUCLEUS FORMATION

The data on angular distributions of charged particles from neutron-induced reactions are quite rare and subject to large uncertainties, due mostly to the poor discrimination between protons, deuterons, tritons, etc., in the outgoing channels and the poor energy resolution. The advent of the dE/dx–E counter telescopes and multidimensional analyzers offered the possibility of performing quantitative investigations of (neutron, charged particle) reactions.

In general, the angular distributions of charged particles leaving the residual nuclei in low-lying states exhibit pronounced forward peaks. Such peaks are usually interpreted as indication of reaction proceeding via the

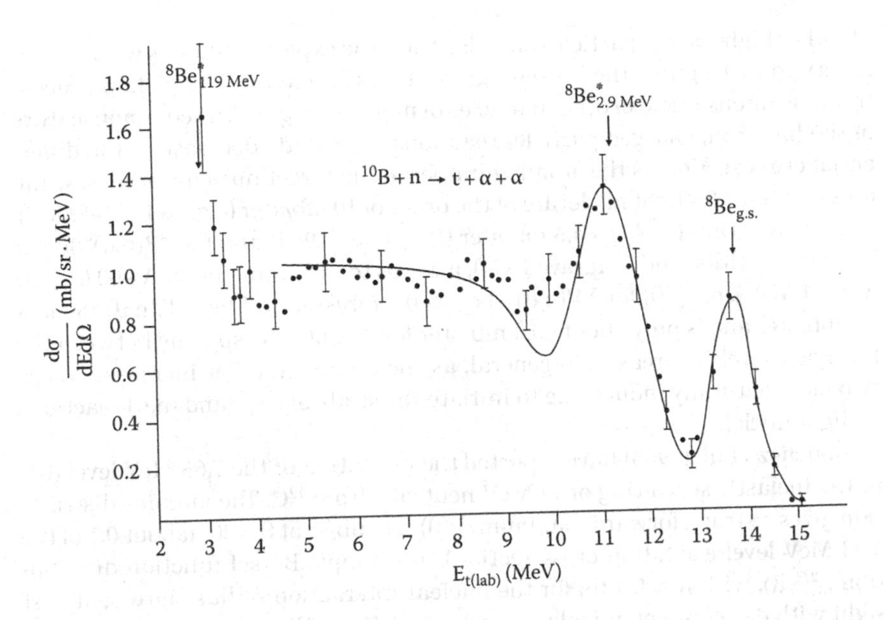

Figure 5.5 Triton spectrum from the n + ^{10}B→ t + α + α reaction at θ_{lab} = 4°. The curve represents the calculation based on the Phillips–Griffy–Biedenharn model assuming the contributions of the α–α final state interaction in the ground state and the 2.9 MeV excited state in ^{8}Be and the α–t final state interaction in the 4.63 MeV excited state in ^{7}Li.

direct reaction mechanism. Some fraction of the reaction at moderate incident energies (~5–10 MeV) proceeds also, via the compound nucleus formation, but for E_{inc} ~ 14 MeV, the processes seem to be predominantly direct interactions.

For instance, for a (p,p′) reaction the compound nucleus formation is appreciable up to Ep ~10 MeV, but for higher energies, for example, between 10 and 16 MeV, the relative contribution of the compound nucleus diminishes rapidly.

Direct interactions are suitable for the study of the structure of low-lying nuclear states. Particularly, (n,d) reactions yield information about the proton configurations, and therefore, they are complementary to (d,p) or (d,t) reactions. Stripping reactions predominantly excite single particle levels. Pickup reactions, on the contrary, favor the hole levels. This distinction argues for the study of both types of processes. The existing experimental data, however, show a large disproportion. There are only few studies of pickup reactions compared to an impressive amount of data on stripping reactions. The reason mainly lies in the experimental difficulties. Pickup reactions are characterized by large negative Q values (typically ~8 MeV), which make it necessary to have a beam of rather high-energy particles and also leave the experimentalists with the difficulty to distinguish the low-energy particles formed by the pickup process from the intense yield of other particles of higher energies. The common feature of the (neutron, charged particle) reactions is a steady decrease in the differential cross section as the atomic number of the target nucleus increases. The cross sections for light nuclei are of the order of 10 mb/ster (e.g., at $E_n = 14$ MeV), σ_{max} (θ) for $^{16}O(n,d)^{15}N_{g.s.}$ is 4.5 mb/ster (Paic et al. 1964) and for $^{10}B(n,d)^{9}Be_{g.s.}$ it is 8 mb/ster (Ribe and Seagrave 1954), it drops to ~1–2 mb/ster for A ~50 (e.g., at $E_n = 14$ MeV), σ_{max} (θ) for $^{51}V(n,d)^{50}Ti_{g.s.}$ is 0.4 mb/ster and for $^{48}Ti(n,d)^{47}Sc$ it is 2.4 mb/ster and is only about <0.3 mb/ster for A ~200. The spacing between the low-lying levels decreases, in general, as the atomic number increases. These two facts naturally induce one to initiate the study of (n,p) and (n,d) reactions on light nuclei.

Bouchez et al. (1963) have reported the excitation of the 7.65 MeV level (0+) in the inelastic scattering of 14 MeV neutrons from ^{12}C. The angular distribution gives a strong forward maximum $\sigma(\theta) \approx 8$ mb/sr at $\theta = 20°$ (about 0.1 of the 4.43 MeV level excitation cross section). The simple Bessel function distribution $j_0^2(kR)$, with $R \approx 5.4$ fm for the nuclear interaction radius, agrees, at first sight with data (except at wide angles) suggesting a direct interaction process (α-cluster excitation).

5.4.1 Elastic Scattering of 14 MeV Neutrons

Early measurement of elastic scattering on hydrogen (Barschall and Taschek 1949) has shown proton recoils originating in a thin polythene radiator bombarded by 14 MeV neutrons which could pass through three proportional counters between which coincidences were observed. By rotating the foil and the counter about the center of the foil, the angular distribution of the recoiling protons could be measured. Protons corresponding to neutrons scattered through 90°, 120°, 150°, and 180° in the center of mass (CM) system were observed. Within the statistical accuracy of about 6% at each angle the angular distribution of the scattered neutrons appeared isotropic.

The nucleon–deuteron interaction has been intensively studied in the 1950s and 1960s in the hope of obtaining information about the nature of the forces between nucleons. Comparisons of p–d and n–d elastic scattering could give information on the charge symmetric nature of these forces, while studies of the inelastic interactions could shed light on the n–n interaction. It is clear that a combination of very precise measurements and equally precise methods of theoretical analysis is required to obtain this desired information; see Berick et al. (1968).

The optical model (Serber 1947; Fernbach et al. 1949) has been used for many years to explain various nuclear interactions. A most striking result was the qualitative interpretation of the experimental data for total neutron cross sections by Feshbach et al. (1954). The quantitative fitting of the experimental data was later considerably improved by the introduction of a diffuse-surface optical model with a rounded well potential (Melkanoff et al. 1957). For 14 MeV neutron elastic scattering, it appears that the best fits so far are obtained by Bjorklund and Fernbach (1958) using a rounded well potential to which is added a spin-orbit term: As was shown by (Begum and Galloway 1979) the polarization of 14.2 MeV neutrons elastically scattered through 20° by Cu and Pb is substantially larger than calculated from standard optical model potentials, a similar effect to that reported for 16 MeV neutrons (Galloway and Waheed 1979). It is proposed that these effects may be related to the geometry of the spin-orbit term in the optical model potential.

The elastic scattering of neutrons at 14 MeV is particularly interesting in that the scattering is essentially all of elastic shape, the decay of the nucleus through the compound elastic channel being relatively improbable. The absence of Coulomb interaction is also a simplifying factor. Elastic scattering of neutrons from ^{16}O compared with the predictions of the optical potential used in the calculations is shown in Figure 5.6 (after Bauer et al. 1963). In the 1950s, precise measurements of differential elastic scattering cross sections for

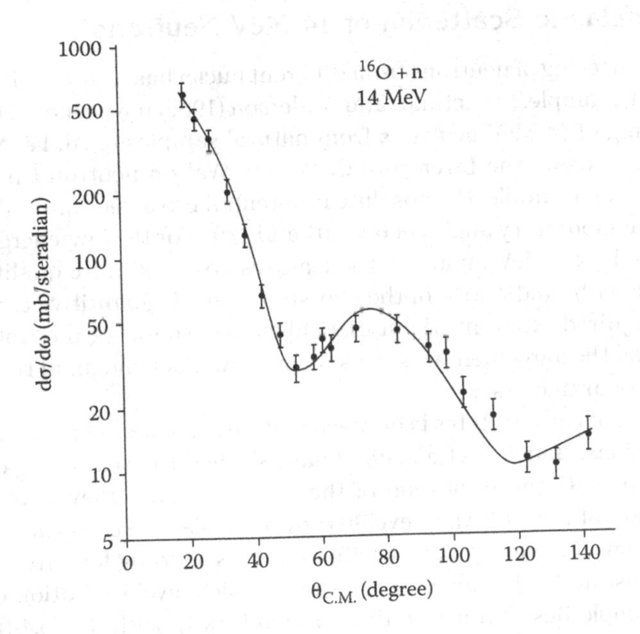

Figure 5.6 Elastic scattering of neutrons from ^{16}O compared with the predictions of the optical potential used in the calculations. (After Bauer, R. et al., *Nucl. Phys.*, 47, 241, 1963.)

several neutron energies and for various elements have been reported by many investigators; see Pierre et al. (1959), Rayburn (1959), and references therein. Angular distribution of 14 MeV neutrons scattered by tritons was measured by Coon et al. (1951).

Nuclear emulsions in conjunction with a neutron collimator have been used to measure the differential cross section as a function of angle for the elastic scattering of monoenergetic 14 MeV neutrons by protons and by deuterons. Measurements were carried out in the angular regions corresponding to neutrons being scattered at angles between 48° and 154.5° in the center-of-mass system by protons and between 46° and 176° in the same coordinate system by deuterons. The n–p scattering data are consistent with isotropic scattering in the center-of-mass system and are in excellent agreement with the data of Barschall and Taschek (1949). The n–d elastic scattering data are strongly anisotropic. The shape of the angular distribution of the neutrons elastically scattered by deuterons may be inferred from the differential cross sections in millibarns for the following laboratory angles: 2°, 435 ± 28; 10°, 342 ± 7; 20°, 78 ± 9.3; 30°, 38 ± 4.2; 40°, 59 ± 5.2; 50°, 104 ± 9.0; 60°, 136 ± 12; and 65°, 142 ± 14 (Alred et al. 1953).

5.4.2 Inelastic Scattering of 14 MeV Neutrons

Inelastic scattering of neutrons from different nuclei has been studied by many authors, for example, Schectman and Anderson (1966) have studied the inelastic scattering of 14 MeV neutrons from natural samples of Al, Fe, Ni, Cu, Cd, Sn Pb, and Bi using the Livermore Cockcroft–Walton neutron time-of-flight facility. For each sample, the absolute differential cross sections, $\sigma(\theta, E)$, were measured at laboratory angles of 60°, 90°, and 120°. For the low-energy neutron spectra ($1 < E_n < 5$ MeV), many of the features observed were qualitatively in accordance with predictions of the statistical model. Quantitive comparison, however, required extending the statistical model to describe multiple neutron emission and the agreement was not satisfactory, independent of the choice of level density parameters.

Excitation of carbon states is of special interest because of possible analytical applications. Bouchez et al. (1963) have studied the scattering of 14 MeV neutrons from ^{12}C and excitation of the 7.65 MeV level. They have observed the excitation of the 7.65 MeV level (0+) in the inelastic scattering of 14 MeV neutrons from ^{12}C. The angular distribution gives a strong forward maximum $\sigma(\theta) \approx 8$ mb/sr at $\theta = 20°$ (about 0.1 of the 4.43 MeV level excitation cross section). The simple Bessel function distribution $j_o^2(kR)$, with $R \approx 5.4$ fm for the nuclear interaction radius, agrees at first sight with data (except at wide angles) suggesting a direct interaction process (α-cluster excitation). In addition, experimental differential cross sections are also presented for elastic scattering and for the 4.43 and 9.63 MeV levels.

The $^{12}C(n,n'\gamma)^{12}C$ reaction has been studied at 15.0 MeV by Spaargaren and Jonker (1971). Absolute neutron–gamma angular correlation data were taken with the γ-detector in the reaction plane and perpendicular to that plane. The neutron spin-flip probabilities were determined. Differential elastic and inelastic neutron scattering cross sections and the differential cross section of the 4.44 MeV deexcitation γ-rays were obtained simultaneously during the angular correlation experiment. The results were compared with other experimental determinations and theoretical predictions. No significant differences between neutron and proton scattering were found.

The inelastic collision cross sections of a number of elements (B, C, N, Al, Fe, Cu, Cd, Au, Pb, and Bi) for 14 MeV neutrons have been measured (Phillips et al. 1952) using threshold detectors inside spherical shells of the scattering material. Reactions in copper, aluminum, and phosphorus with thresholds at approximately 11.5 MeV, 2.6 MeV, and 1.4 MeV were used as threshold detectors. Feshbach and Weisskopf (1949) have published a schematic theory of nuclear cross sections. Their "reaction cross section," σ_r, may be identified with the inelastic collision cross section reported by Phillips et al. (1952), measured with copper detectors. They have calculated σ_r using the values of the nuclear radius, R, obtained from the analysis of measurements of σ_{tot}. It should be pointed out that the measurements of σ_{tot} for 14 MeV neutrons are in disagreement with the values used by Feshbach and Weisskopf (1949).

Differential and total cross sections have been determined for the scattering of 14.1 MeV neutrons from several levels in carbon by Singletary and Wood (1959). The data were measurements of scattered neutrons made by means of the proton recoil reaction in nuclear emulsion plates. The inelastic cross sections were 203 mb for the 4.4 MeV level, 96 mb for the 9.6 MeV level, and 124 mb for the unresolved higher levels. The shape of the angular distribution for the 4.4 MeV level indicated the possibility of some direct interactions. No significant amount of scattering was observed due to the 7.6 MeV level. The elastic scattering angular distribution showed a forward-peaked diffraction type structure and yielded a cross section of 805 mb. Comparisons are made with other data on the elastic and 4.4 MeV level scattering.

A NaI scintillation spectrometer has been used to study the gamma rays produced by the inelastic scattering of 14 MeV neutrons in carbon and oxygen. The scattering material was placed around the neutron source with the gamma-ray detector outside the scatterer. A gamma ray from the 4.43 MeV level in ^{12}C was obtained with a cross section of about 0.3 barn. Gamma rays from the known level at 6.13 MeV and levels close to 7 MeV were obtained from ^{16}O with a total cross section of about 0.2 barn (Thompson and Risser 1954).

The gamma rays emitted by ^{12}C under 14 MeV neutron bombardment have been investigated with a three-crystal pair spectrometer. In the energy range 1.5–5.5 MeV, the observed pulse-height distribution is consistent with the assignment of a single-energy line at 4.4 MeV. The calculated value for the production cross section of 4.4 MeV gamma radiation is 245+35 millibarns (Battat and Graves 1955).

Total cross sections for 14 MeV neutrons of C, H, ^{2}H, O, and N were measured by Poss et al. (1952). The effective neutron energy was evaluated from the kinematics of the $^{3}H(d,n)^{4}He$ reaction, the dependence of the reaction's cross section on energy and angle, and the multiple scattering of the deuterons and their rate of energy loss in the target. A least squares analysis of the transmissions of six thicknesses each of carbon and polyethylene (40%–85% transmissions) yielded, for total cross sections σ in barns: $\sigma_C = 1.279 \pm 0.004$ and $\sigma_H = 0.689 \pm 0.005$. Transmissions of light and heavy water gave $\sigma^2{}_H = 0.803 \pm 0.014$ and $\sigma_O = 1.64 \pm 0.04$. Transmission of melamine gave $\sigma_N = 1.7 \pm 0.1$.

One of the methods for determining nuclear radii is based on measurements of the total cross sections of nuclei for fast neutrons. Nuclear radii are most likely to be calculable from such measurements if the neutron wavelength divided by 2π is small compared to the nuclear radius, but not small enough that the nucleus is transparent for the neutrons used. Neutrons of energies of the order of 20 MeV satisfy these conditions. Several measurements using neutrons of energies between 13 and 25 MeV have been published in the 1940s and the 1950s and have served to determine nuclear radii. Each investigation covers a relatively small number of elements. Measurements performed at different neutron energies and in different geometries are difficult to compare particularly because of the strong angular dependence of diffraction scattering about which only very limited experimental information is available. The total cross sections of over 50 elements were measured in good geometry for 14 MeV neutrons by Coon et al. (1952); see Figure 5.7. A plot of the square root of the total cross section versus the one-third power of the atomic weight shows deviations from the linear relationship predicted by statistical theory. The deviations are most pronounced for the heaviest elements.

Experiments have been performed by Scherrer et al. (1953) to observe the pulse-height distribution produced in a single crystal spectrometer by gamma radiation arising from the bombardment of various materials with 14 MeV neutrons. Various gamma-ray lines have been identified and estimates made of the total cross section for gamma-ray production on C, O, Al, Ni, Cu, Cd, and Pb.

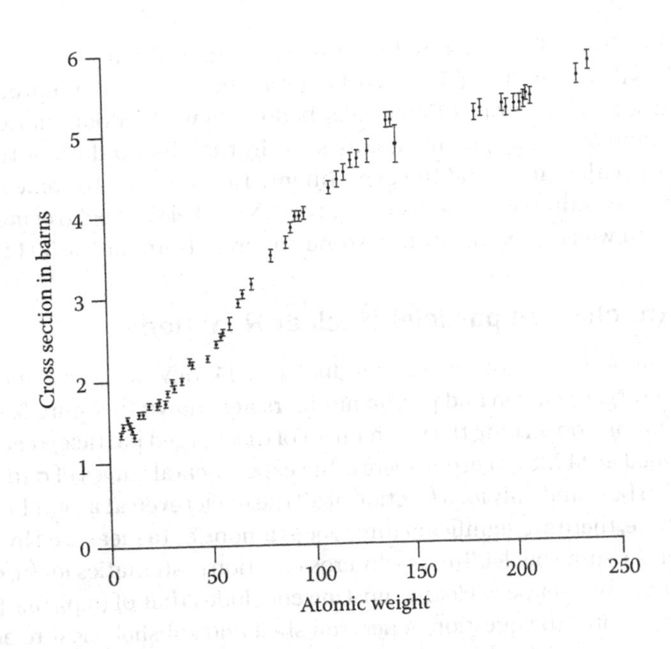

Figure 5.7 The total cross sections of the elements for 14 MeV neutrons plotted against atomic weight. (After Coon, J.H. et al., *Phys. Rev.*, 88, 562, 1952.)

Backscattering of neutrons is an important phenomenon which one often encounters in analyzing neutron streaming problems in ducts and cavities. The phenomenon is described by the albedo concept. The concept is utilized in some of the neutron transport calculations at material boundaries to make the calculation model simple. It is known that the albedos effectively save the computing time in duct streaming calculation without the expense of the accuracy of the results if the albedos are given precisely.

Systematic albedos are given by a Monte Carlo calculation or discrete ordinates codes in 1D or 2D geometry. To assure the accuracy of the albedo data obtained by the calculations, measurements that supply benchmark data for the albedo calculations are required. Especially for 14 MeV neutrons, the cross-section data of the neutrons are not always accurate enough for both (n,xn) and (n,xγ) reactions, and this may introduce unacceptable errors in the albedo data that are obtained by the calculations based on these cross-section data. In the experiments performed by Shin et al. (1988), simultaneous measurements of fast neutron and prompt γ-ray albedos were performed by 14 MeV neutrons for the type 304 SS and the limestone concrete. Both neutrons and γ-rays were detected by an NE-213 scintillator with the aid of the n–γ discrimination circuit. The time-of-flight method was used to obtain neutron spectra, whereas the unfolding method was applied to analyze γ-ray pulse-height data of the scintillator. The measured data were compared with those calculated by the MCNP code with the ENDF/B-IV data. The calculation reproduced the measured secondary γ-ray differential albedo data well for both materials. For neutrons, apparent disagreement was seen in the albedo data of the 304 SS between the calculation and the experiment. This was due to some inaccuracies in the chromium cross sections in the ENDF/B-IV. The agreement of the neutron data was better for the limestone concrete than for the 304 SS.

5.4.3 (n, charged particle) Nuclear Reactions

Approximately 20 reactions can be produced by 14 MeV neutrons. The possible changes in target neutron and proton numbers are shown in Figure 5.8. Most of the information concerning the mechanism of (n, charged particle) reactions has been obtained at 14 MeV neutron energy. The experimental values of σ (n, charged particle) can be found only for a fraction of all the nuclei even at around 14 MeV. At the same time, there are significant differences among data measured by different authors for the same nuclei. This is why cross-section systematics for (n, charged) reactions have not yet been cleared up. One concludes that of (n,p), (n,α), (n,^{3}He), and (n,t) reactions, the question of neutron shell and subshell closure effects on the cross sections arising from causes other than Q values and neutron-binding energies remains the main problem for investigation. Further studies should be performed on the formation of nucleon clusters in the nuclear surface and their effects on reaction mechanisms and cross sections. To obtain reliable information on reaction mechanisms, more complete and accurate data are needed for the excitation functions, angular distributions, and particle spectra.

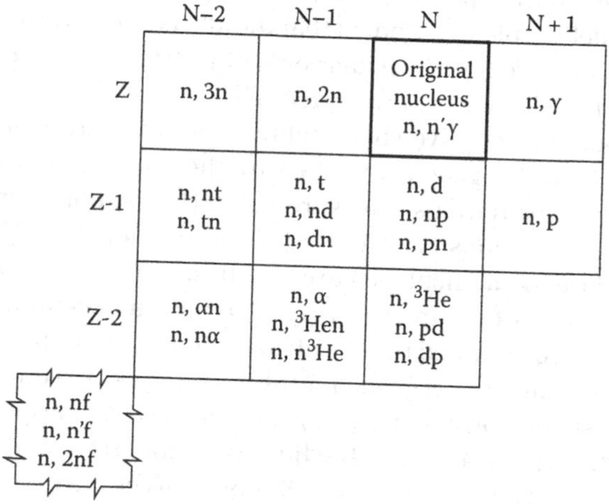

	N−2	N−1	N	N+1
Z	n, 3n	n, 2n	Original nucleus n, n′γ	n, γ
Z-1	n, nt n, tn	n, t n, nd n, dn	n, d n, np n, pn	n, p
Z-2	n, αn n, nα	n, α n, ^{3}Hen n, n^{3}He	n, ^{3}He n, pd n, dp	

n, nf
n, n′f
n, 2nf

Figure 5.8 Possible reactions produced by fast neutrons.

Cross sections for the production of charged particles in fast neutron interactions are of recent interest in view of their importance in the conceptual and technical design of fusion reactors. Cross section of the reaction $^{16}O(n,\alpha)^{13}C$ is relevant for a specific reason. An important characteristic of a fusion reactor is the neutron multiplication in the reactor wall. The reactor wall is a complex structure with different components, one of those being made of Be. Neutrons are multiplied partly in the reactor blanket via (n,2n) reaction on Be. Neutron multiplication constant is usually determined experimentally in a manganese bath. For precise determination of this constant, careful analysis of neutron interaction in the whole experimental setup is necessary. One of the important points is the neutron interaction with water, especially elastic and inelastic interaction with oxygen. Apart from elastic scattering on oxygen, neutrons undergo also (n,α) reaction with ^{16}O and are thus lost for the multiplication process. Therefore, a good knowledge of the cross section for this reaction is necessary. Requested precision of the (n,α) cross section is better than 10%.

Hlavač et al. (1994) measured discrete cross section of γ-ray production cross sections in the $^{16}O(n,\alpha\gamma)^{13}C$ reaction at 14.7 MeV. Cross sections were determined relatively to the well-known reference $^{52}Cr(n,n'\alpha)$ reaction with precision better than 10%. They found that Doppler effects play an important role in this very light system. For the short-lived 3/2″ level, they accounted for this fact and arrived at the cross section that is comparable roughly to the mean value of other known experimental data around 14 MeV. They determined the total discrete γ-ray production to be (80.9 ± 5.4) mb and deduced the total $^{16}O(n,\alpha)^{13}C$ cross section of (98.9 ± 6.4) mb. This latter value is lower than the other experimental data as well as the ENDF/B-VI evaluation; it is, however in excellent agreement with the JENDL-3 evaluation.

Reactions with three particles in the final state exibit in a very explicit way the three-body complications of the scattering theory. Such complications are also present in transfer reactions with two particles in the final state. It is extremely difficult to calculate the three-body effects in the transfer reactions since they are obscured by many other uncertainties. However, due to the small separation energies of nucleons and clusters of nucleons, transfer reactions on light nuclei offer the possibility of studying these effects. Transfer reactions to particle unstable states result in three or more particles in the final state. Measurement of the angular distribution could be performed by measuring the energy spectra (single counter experiment) at different angles only in some special cases. As an example, the deuteron spectrum from n + $^{6}Li \rightarrow d + \alpha + n$ reaction is shown in Figure 5.9. The (n,d) transition to the $^{5}He_{g.s.}$ was determined by the summation of the events under the observed peak, since the shape of the energy spectrum was well reproduced by final state interaction type theory. It is obvious that the transition rate could not be determined precisely. The situation is more complicated when a transition to a weakly excited and broad state is considered, when the contributions from other final state interactions

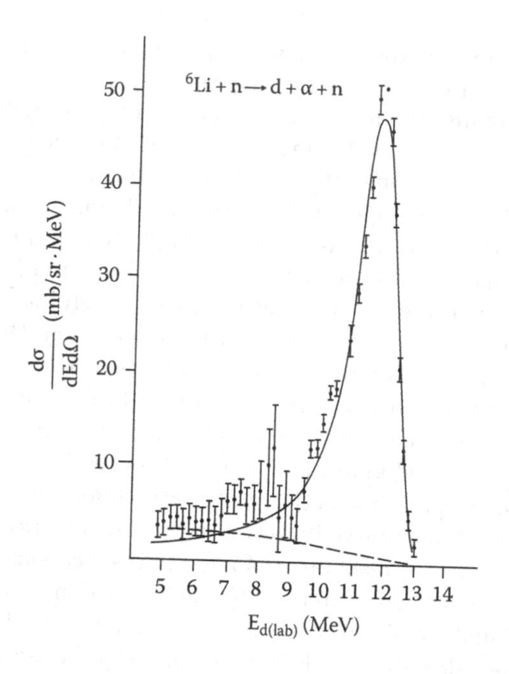

Figure 5.9 Deuteron spectrum from n + ^{6}Li → d + α + n reaction with En = 14.4 MeV at θ = 0°. The solid curve is calculated using the theory from (Phillips, Griffy and Biedenharn 1960); the dashed curve is the phase space distribution.

are simultaneously present. These have been studied as the neutron-induced nonelastic processes on light nuclei (Valković et al. 1970).

For light nuclei, distorted wave Born approximation (DWBA) does not reproduce the shape of the angular distribution although the absolute value of the cross section is fairly well predicted. As an interesting detail, the DWBA fit to the ^{3}He(n,d)^{2}H data (Antolković et al. 1967) is shown in Figure 5.10. Both shape and absolute values are well reproduced. A question should be asked as to how to interpret such an agreement. The ratio of the experimental and theoretical spectroscopic factors is shown in Figure 5.11 for each (n,d) transition. The only extreme disagreement is for the ^{6}Li(n,d)^{5}He$_{g.s.}$ reaction. In the case of the ^{9}Be* 2.43 MeV particle-unstable excited state, the disagreement between theory and experiment is within the limits imposed by the validity of the DWBA at these energies for light nuclei., However, a noticeable disagreement is present in the case of the ^{14}N(n,d)^{13}C reaction.

Our conclusion is that the existence of the deuteron cluster in the target nucleus is responsible for the deviations of S_{exp}/S_{theor} since the DWBA calculations consider only proton pickup. It is expected that a knockout-type calculation might give much better agreement with the experimental data.

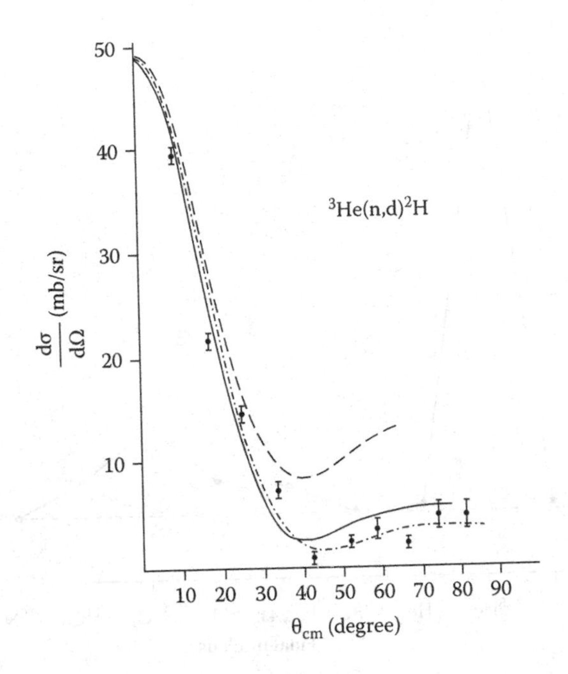

Figure 5.10 Comparison of DWBA predictions with the measured data on the $^{3}He(n,d)^{2}H$ reaction.

The situation with (n,t) reactions is even more complex. The study of angular distributions of tritons by 14 MeV neutron bombardment of light elements was initiated in the 1960s (Valković 1960). Only a few measurements have been performed since then. The shapes of the (n,t) angular distribution curve to the low-lying states reflect a direct interaction mechanism. Figure 5.12 shows the angular distribution of tritons from the reaction $^{10}B(n,t)^{8}Be_{g.s.}$.

Angular distributions of direct (n,t) reactions are characterized by the total orbital angular momentum L of the transferred deuteron; L must satisfy the relations

$$L = l_n + l_p,$$

$$S_t + J_f = L + J_i + S_n,$$

where the indices n, p, and t refer to neutron, proton, and triton, respectively. In the case of the $^{10}B(n,t)^{8}Be_{g.s.}$ reaction, with $J_i = 3^+$ and $J_f = 0^+$, the permitted values for L are 2 and 4. Assuming for ^{10}B the shell configuration $(S_{1/2})^4(P_{3/2})^6$, the only allowed total angular momentum value of the transferred deuteron is L = 2.

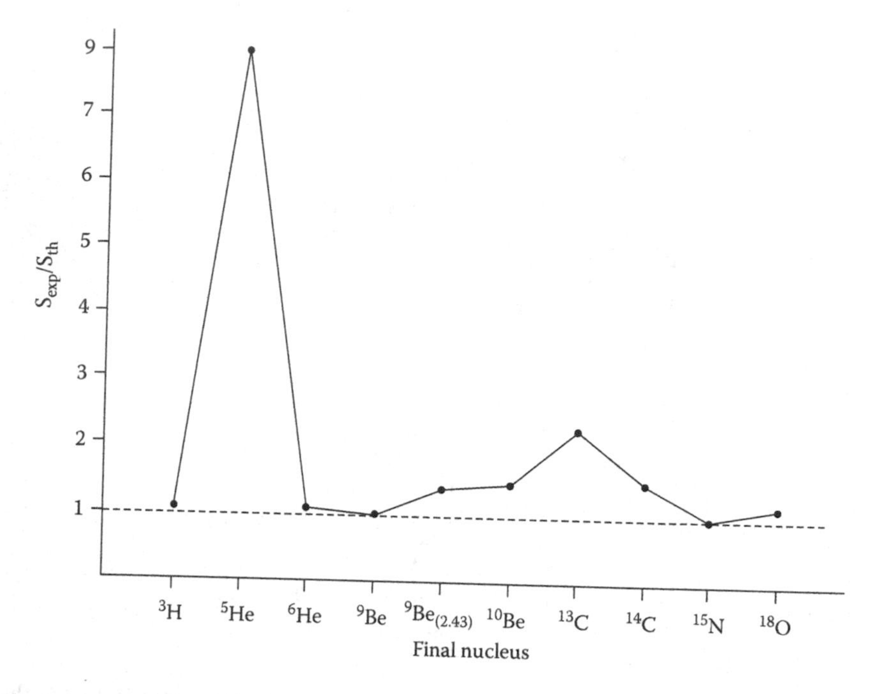

Figure 5.11 Ratios of the experimental spectroscopic factors—obtained by reasonable optical model potential parameters (Miljanić 1970)—to the theoretical ones.

The target nucleus ^{10}B can be visualized as a ^{8}Be core plus a deuteron, This suggests the use of the Butler theory. The L = 2 Butler angular distribution was calculated using several radii. The best fit was obtained with the Butler radius R = 6.5 fm and is shown in Figure 5.12. The agreement with the first maximum is quite good.

The data about (n,t) reactions are scarce due to the difficulties encountered in the study of neutron-induced reactions (Valković and Tomaš 1964). Also, the Q values (usually about −10 MeV) restrict the number of reactions to be studied with 14 MeV neutrons.

The analysis of measured angular distribution of tritons from the ^{11}B(n,t)^{9}Be reaction at 14.4 MeV (Miljanić et al. 1970a) was based on the Gledenning approach (Glendenning 1962, 1965) to the two-nucleon transfer reaction. In these calculations, the code SALLY (Bassel et al. 1962) was used.

The agreement of experimental (n,t) angular distribution with the DWBA predictions is shown in Figure 5.13. The shape of the measured angular distribution is well reproduced by theoretical calculations with one of the potential used for the triton channel. However, a serious disagreement in the absolute value of the cross section is observed with both potentials used (a factor of 12 and a factor of 4, respectively). This discrepancy is due to the approximation in the

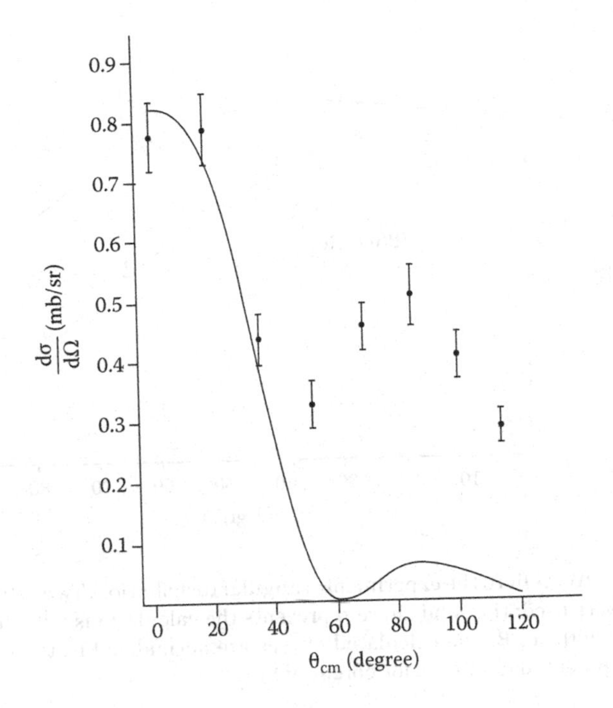

Figure 5.12 The angular distribution of tritons from the reaction $^{10}B(n,t)^{8}Be_{g.s.}$. The solid curve is the Butler curve obtained with L = 2 and R = 6.5 fm.

calculation of the form factor of the transferred pair and the lack of appropriate parameters of the triton optical model potential.

In a study of (n,d) and (n,t) reactions on ^{7}Li (Miljanić et al. 1970b; Valković and Tomaš 1971) the angular distribution of deuterons was measured from the $^{7}Li(n,d)$ reaction and it is found to be in agreement with the DWBA prediction. The absolute value of the cross section is much better reproduced than the shape of the angular distribution. More precise information about the cross section at backward angles and the transition to the first excited state in ^{6}He could be obtained using the dE/dx counter filled with a gas that does not produce deuterons when irradiated by neutrons.

The shape of the angular distribution of tritons is reasonably well reproduced by DWBA calculations. The disagreement between the absolute value of the calculated cross section assuming the pickup of two nucleons and the experimental value observed by the summation under the observed peak in the triton spectrum is significant. This should be partly attributed to the inadequacy of the DWBA and the choice of the optical model potential parameters. However, the contribution of quasifree n–t scattering appears much more important and is preferred because

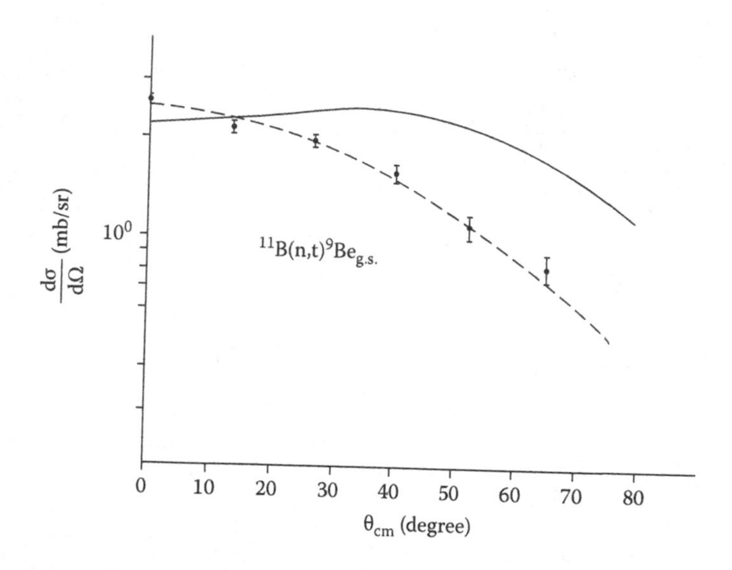

Figure 5.13 DWBA fit to the experimental angular distribution. Two sets of potential parameters were used: the solid curve represents the calculations with set A and the dashed curve with set B. The calculated curves are normalized to the experimental points (×4 for potential A and ×12 for potential B).

the structure of the nuclei is involved. The two aforementioned processes could be separated by a coincidence measurement of two outgoing particles.

Direct effects in (n,p), see Valković et al. (1964) and Paić et al. (1964), (n,d) see Miljanić et al (1967, 1968), and (n,t), see Valković (1964b) and Valković and Tomaš (1964), reactions have been studied in detail. Recently, a number of hydrogen production cross sections have been measured and calculated for elements and isotopes at around 14 MeV neutron energy, mainly for the radiation damage assessment of fusion-related materials, to complete the database of neutron activation cross sections for environmental protection and to estimate the potential radiation hazards connected with the nuclear techniques. Cross sections for the generation of long-lived radionuclides, of importance in fusion reactor technology and accelerator-based neutron sources, have a high priority. Precise data at 14 MeV are needed for validation and testing of different data libraries and for the normalization of excitation functions. Recently, on the basis of precise cross-section measurements, a number of investigations were devoted to improve the systematics based on the isotopic and asymmetry parameter dependences. The reliable theoretical and empirical formulae are indispensable to estimate cross sections if experimental data are not available.

In light of that, cross sections have been measured, deduced, and adopted for 208 (n,p) reactions at (14.7 + 0.2) MeV incident neutron energy in the report by Doczi et al. (1997). Systematics based on asymmetry parameter dependence and isotopic effects have been confirmed. Recommended formulae were fitted to the same database to be able to test their applicability. Total (n,p) cross sections are given for 54 elements and compared with the results of direct methods. Some $\sigma_{n,p}$ data both for long-lived target and for residual radionuclides were estimated, see also Levkovskii (1974).

In the work reported by Demetriou et al. (1994), the quantum mechanical theory of Feshbach et al. (1980) has been applied to analyze neutron-induced reactions, particularly nucleon emission reactions at 14.1 MeV on ^{51}V, ^{56}Fe, Zr, ^{93}Nb, Mo, and Sb. The analysis is carried out over the whole energy and angle range and aims to account for all the reaction flux. An unambiguous method of isolating the multistep direct component by taking the forward–backward differences in the centre-of-mass system has been developed and used to obtain the average value for the strength V_0 of the effective N–N interaction. This method has been applied to a wider range of nuclei in order to find the dependence of V_0 on the target mass number.

Owing to the development of counter telescopes and multidimensional analyzers, neutron-induced one-nucleon transfer reactions have been studied with increasing interest. Most of the experiments have been performed with 14 MeV neutrons and for Q values around –7 MeV. However, most of the measurements suffer from large uncertainties as a result of poor angular resolution (from 5° to 10°). Several DWBA studies of (n,d) reactions on light nuclei induced by 14.4 MeV neutrons have been published (Valković et al. 1965; Miljanić et al. 1968). Features of these measurements are good discrimination between various charged particles, low background, and the determination of the absolute cross section to an accuracy of approximately 10%. Now it becomes worthwhile to apply the distorted-wave method to the analysis of (n,d) reactions. They then become a valuable tool of nuclear spectroscopy, since they probe the proton configurations, while the commonly studied (d,p) and (d,t) reactions are concerned with the neutron configuration. In all those investigations, the optical potentials have been obtained either by adjusting the parameters to achieve the best fit to the measured (n,d) data or by fitting the corresponding elastic scattering data where available.

In the report by Valkovic et al. (1965), the reactions $^{48}Ti(n,d)^{47}Sc$, $^{16}O(n,d)^{15}N$, $^{10}B(n,d)^{9}Be$, and $^{6}Li(n,d)^{5}He$ have been studied at 14.4 MeV. The absolute differential cross sections have been analyzed using the distorted-wave method and assuming simple proton pickup. The effects of finite range, nonlocality, and radial cutoff have been investigated. Good fits obtained for the reactions $^{16}O(n,d)^{15}N_{g.s.}$, $^{10}B(n,d)^{9}Be_{g.s.}$, and $^{10}B(n,d)^{9}Be_{2.43\,MeV}$ yield spectroscopic factors in close agreement with shell-model predictions. The shape of the angular distribution of the reaction $^{48}Ti(n,d)^{47}Sc$ is well explained, but a discrepancy in absolute magnitude of about a factor four is not yet understood (see Figure 5.14). The agreement between theory and experiment for the reaction $^{6}Li(n,d)^{5}He_{g.s.}$ is not satisfactory, and this may indicate the importance of other processes such as knockon. All possible reaction mechanisms are indicated in Figure 5.15 (after Gadioli and Hodgson 1992).

The (n,d) reactions on light nuclei presented in Table 5.1 are discussed in detail (Miljanić and Valković 1971). In this chapter, the authors have tried to explain all the measured (n,d) angular distributions on light nuclei with the same three sets of deuteron optical-model potential parameters for all the reactions.

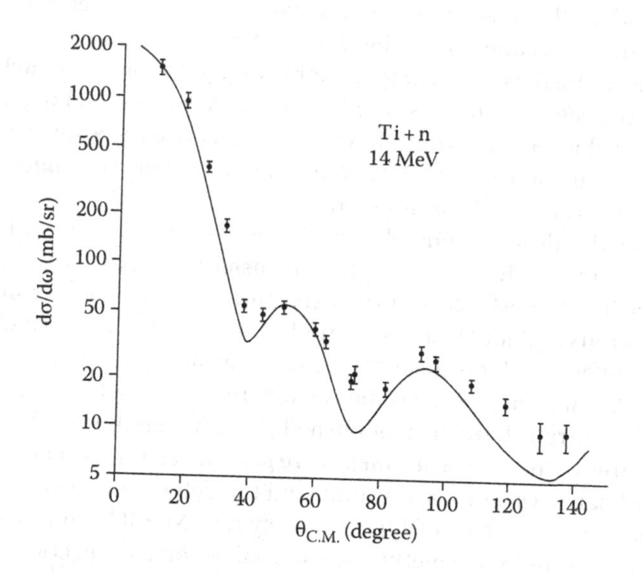

Figure 5.14 Angular distribution of Ti(n,d) reaction for E_n = 14 MeV.

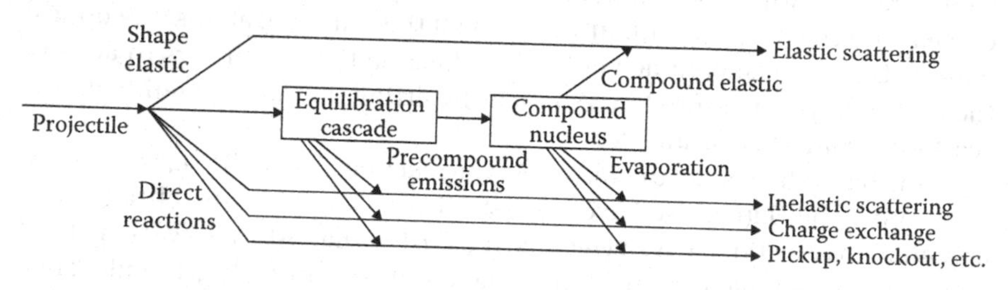

Figure 5.15 Direct, precompound, and compound nucleus processes in nuclear reactions. (After Gadioli, E. and Hodgson, P.E., *Pre-equilibrium Nuclear Reactions*, Clarendon Press, Oxford, U.K., 1992.)

TABLE 5.1 SOME PROPERTIES OF THE (N,D) REACTIONS ON LIGHT NUCLEI

Target Nucleus	Final Nucleus	Q(MeV)	Θ_{max}(Lab Degrees)	σ_{max}(mb/sr)	l
^{3}He	^{2}H	−3.27	8.5	39.6	0
^{6}Li	$^{5}He_{g.s.}$	−2.43	4	51.3	1
^{7}Li	$^{6}He_{g.s.}$	−7.76	20	2.2	1
^{10}B	$^{9}Be_{g.s.}$	−4.36	20	8.6	1, (3,5)
^{10}B	$^{9}Be_{2.43\ MeV}$	−6.79	20	5.8	1, (3,5)
^{11}B	$^{10}Be_{g.s.}$	−9.00	20	2.1	1
^{14}N	$^{13}C_{g.s.}$	−5.33	20	9.2	1
^{15}N	$^{14}C_{g.s.}$	−7.98	20	3.3	1
^{16}O	$^{15}N_{g.s.}$	−9.90	15	4.7	1
^{19}F	$^{18}N_{g.s.}$	−5.77	4	23	0

The analysis in the frame of the DWBA approach was performed assuming the proton pickup. Other terms in the transition amplitude were neglected (although this is difficult to justify for light nuclei) to avoid difficulties in the analysis, which were caused by the additional terms introduced. The obtained spectroscopic factors S were compared with the calculated values in the frame of different models. The comparison with the S values obtained from other proton-transfer reactions and neutron-transfer reactions to the states analogues of those involved in (n,d) reactions was also made.

Experimental angular distributions of deuterons from (n,d) reactions on light nuclei with neutron bombarding energy E_n = 14.4 MeV indicate the predominance of direct reaction mechanism. Most of the measured (n,d) angular distributions could be described in the DWBA frame assuming the proton pickup. This analysis shows that deuteron potentials with surface imaginary term are most suitable for the description of these processes in this energy interval and mass region. However, the cases of ^{7}Li and ^{14}N, where better results have been achieved using potential with volume imaginary term, probably indicate that there is no unique choice of the type of potential for all these nuclei.

The ratios of the experimentally obtained spectroscopic factors and the calculated values show the reasonable agreement. The only strong disagreement for the ^{6}Li(n,d) reaction can be explained by the contributions of other processes different from proton pickup. It could be concluded that the DWBA calculations and the shell-model theory give a good explanation of (n,d) data on light nuclei.

5.4.4 (n,2n) Reactions

Currently, to evaluate the cross sections of (n,2n) reactions with neutron energy near 14.6 MeV, empirical and semiempirical formulas and approaches based on the systematics of experimental data for cross section nuclear reactions are widely used. In most situations, the parameters of systematic relationships were identified involving relatively limited number of experimental data, particularly with large errors.

In spite of the lack of data and the large discrepancies both in the individual values and in the excitation functions, some rough trends can be recognized in the cross sections. The plot of (n,2n) cross sections enables establishment of the following rough trends: (1) In the mass number region $19 < A < 40$, the cross section decreases from 50 mb to 4–5 mb. (2) At $A \approx 48$, the cross section rises rapidly to ≈1000 mb and shows a trend of increasing slowly up to a maximum value of ≈2200 mb for the heaviest nuclei. (3) In the region $48 < A < 100$, strong local fluctuations exist.

The trends in (n,2n) reaction cross sections have been investigated by many authors and both isotopic and isotonic, as well as odd–even effects, have been observed (see Qaim 1974; Holub and Cindro 1976; and others).

Semiempirical and empirical formulas for $\sigma_{n,2n}$ are given by many authors, see for example (Tel et al., 2008). For example, following the idea of Levkovskii (1957) that the cross section can be factorized to the compound nucleus formation cross section and to a factor depending exponentially on (N–Z)/A, Chatterjee and Chatterjee (1969) recommended a similarly deduced formula for (n,2n) cross sections:

$$\sigma_{n,2n} \approx 45.2(A^{1/3}+1)^2 \exp[-2.60(N-Z)/A] \text{mb}$$

This formula is valid for cross sections at a 3 MeV excess energy above the (n,2n) threshold.

From the experimental $\sigma_{n,2n}$ values, Lu and Fink (1971) have given the following empirical equation as a function of (N–Z)/A obtained by least squares fitting:

$$\sigma_{n,2n} = 61.6(A^{1/3}+1)^2[1-1.319 \exp(-8.744(N-Z)/A] \text{mb}$$

This equation reproduces the $\sigma_{n,2n}$ values to about ±20%, except for the lightest stable isotopes of even Z elements. In general, this equation may be used for the quick estimation of (n,2n) cross sections at 14.4 MeV for stable nuclei from Z = 28 to 82.

By expressing the binding energy of a neutron S_n, in the target nucleus by Weizsacker's formula, and assuming the energy distribution of preequilibrium emission neutrons to be constant, Bychkov et al. (1982) obtained the following dependence of $\sigma_{n,2n}$ on N and Z of the target nucleus at 14.5 MeV:

$$\sigma_{n,2n} = \sigma_{n,M}[0.68 + (N - Z)/A - 5.2(\varepsilon/2(N - Z/A) - 1)\exp(-\varepsilon/2(N - Z/A)]\text{mb}$$

where $\sigma_{nM} = \sigma_a[1 - \exp(-33(N-Z/A))]$, ε is a parameter of Weizsacker' s formula, and σ_a is the neutron absorption or the composite formation cross section. This equation can be reduced to the form

$$\sigma_{n,2n} = \sigma_a[1 - k\ \exp(-m(N - Z)/A]$$

which can also be obtained empirically.

A new comprehensive code for nuclear data evaluation, CCONE, has been developed for the evaluation of nuclear data for actinides. Neutron-induced reaction cross sections for uranium isotopes (A = 232, 233, 234, 235, 236, 237, 238) were analyzed for incident energies from 10 keV to 20 MeV using CCONE in order to validate its capability. Reproducibility of the cross-section calculations for various reactions such as total, fission, capture, and (n,2n) was tested using simple parameterization. The calculated cross sections and neutron emission spectra show good agreement with the experimental data (Iwamoto 2007).

During the 1970s and 1980s, a number of authors (for extensive list of references see Csikai 1987) have reported results on (n,2n) cross sections for fast neutrons measured by an improved activation method. Details of precise measurements of excitation functions from threshold to 30 MeV and data in the 14 MeV region were published, for example, by the Vienna, Geel, Los Alamos, Livermore, Obninsk, Debrecen, Bratislava, Julich, Zagreb, Bruyères-le-Châtel, and Beijing groups (Csikai 1987).

Konobeyev and Korovin (1999) reported that a new formula for the (n,2n) reaction cross-sections estimation at an energy of 14.5 MeV has been obtained by employing the preequilibrium and evaporation models. The formula gives a more precise description of experimental data and represents the isotopic dependence of the cross-sections better than the systematics proposed earlier by other authors. Unlike other systematics, the formula obtained takes account of the difference in the cross section values for nuclei of different parity. Figure 5.16 shows the schematic representation of (n,2n) reaction cross sections at the energy of 14.5 MeV for 126 nuclei from ^{40}Ar to ^{209}Bi as a function of the neutron excess parameter (N–Z)/A obtained from the experimental data (see Konobeyev and Korovin 1999; Pashchenko 1990). Figure 5.17 shows the ratio of 126 cross-section values obtained from the experimental data analysis by Pashchenko (1990) to the cross-section values calculated through formula by Konobeyev and Korovin (1999).

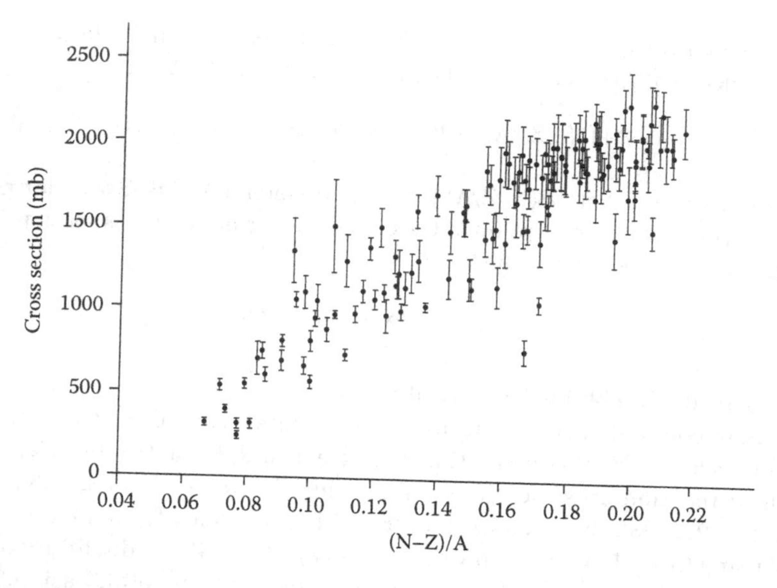

Figure 5.16 The schematic representation of (n, 2n) reaction cross sections at the energy of 14.5 MeV for 126 nuclei from ^{40}Ar to ^{209}Bi as a function of the neutron excess parameter (N–Z)/A obtained from the experimental data. (From Konobeyev A. Yu. and Korovin Yu. A., *Nuovo Cimento*, 112A, 1001, 1999.)

5.5 ACTIVATION ANALYSIS: (n,p), (n,np), (n,α), (n,nα), (n,n′γ), (n,2n), (n,t), (n,^{3}He) REACTIONS

If a nuclear reaction leads to the formation of an unstable isotope, the resulting radioactivity provides a means for determining the reaction cross section. In particular, if there are beta particles emitted from the product nucleus, they may be counted in order to determine the total number of active atoms present. The cross section for the formation of these active atoms can then be found from the neutron flux, sample weight, and the number of activated atoms. In cases involving K-capture, beta counting may still be employed if a known fraction of the disintegrations involves beta particle emission. Similarly, the cross section for the production of isomers may be determined to obtain the cross section for the production of a particular isotope provided each isomer decays at least in part by beta emission.

Figures of cross sections in JENDL dosimetry file 99 (JENDL/D-99) are published by Shibata (1999). The cross sections of following nuclear reactions (n,γ), (n,p), (n,α), (n,f), (n,2n), (n,np), (n,n′), and (n,t) are presented as a function of the neutron bombarding energy covering the neutron bombarding energy interval from reaction threshold to 20 MeV.

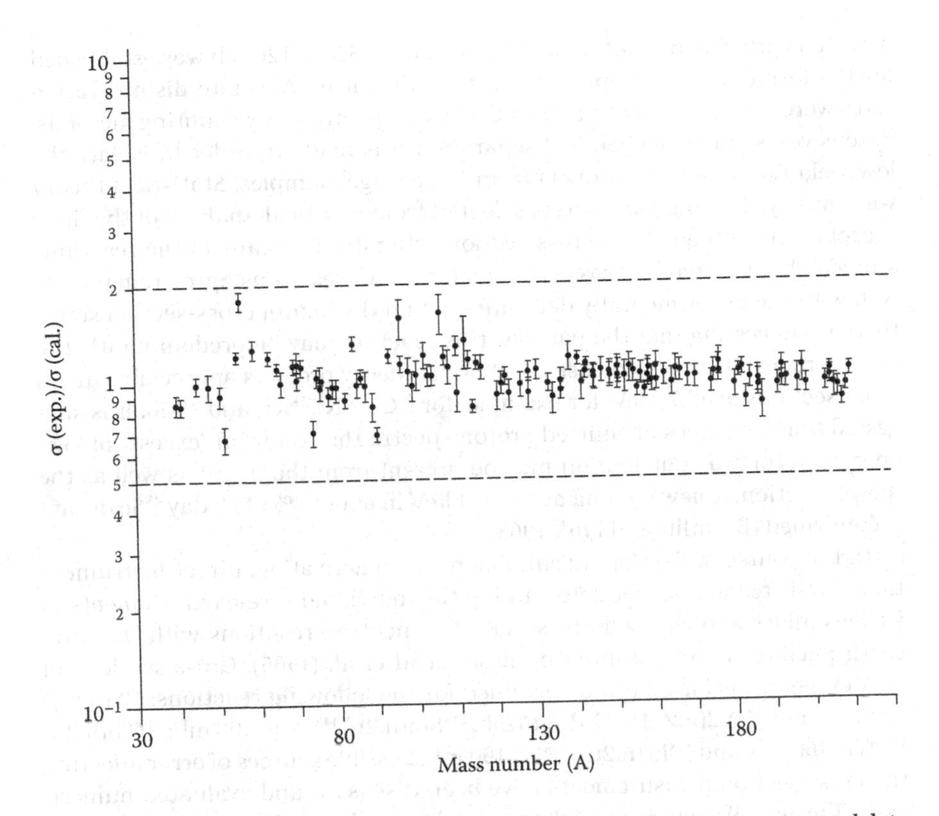

Figure 5.17 Ratio of 126 cross-section values—obtained from the experimental data analysis by Pashchenko (1990)—to the cross-section values calculated through formula by Konobeyev and Korovin (1999).

In the report by Forbes (1952), cross sections for several nuclear reactions induced by 14 MeV neutrons were measured by the activation method. The number of activated atoms was determined by absolute beta counting. Corrections for finite sample thicknesses were determined experimentally in every case. Cross sections for the following reactions were measured: (n,2n) for ^{63}Cu, ^{65}Cu, ^{107}Ag, and ^{109}Ag; (n,p) for ^{27}Al, ^{31}P, ^{56}Fe, and ^{65}Cu; and (n,α) for ^{27}Al.

Activation cross sections for 14.7 ± 0.2 MeV neutrons were measured by Bramlitt and Fink (1963) for [(n,nα) + (n,αn)] reactions on ^{65}Cu, ^{70}Zn, ^{71}Ga, and ^{93}Nb, while upper limits were set for this reaction on ^{51}V, ^{76}Ge, ^{81}Br, ^{87}Rb, ^{107}Ag, ^{109}Ag, ^{115}In, ^{197}Au, and ^{203}Tl. Cross-section limits also were set for (n,2p) reactions on ^{29}Si, ^{41}K, ^{45}Sc, ^{50}Ti, ^{51}V, ^{55}Mn, ^{75}As, ^{89}Y, ^{93}Nb, ^{133}Cs, ^{139}La, ^{141}Pr, and ^{159}Tb, and for (n,He^{3}) reactions on ^{45}Sc, ^{93}Nb, ^{197}Au, and ^{205}Tl. Cross sections were determined for (n,γ) reactions on ^{89}Y, ^{93}Nb, and ^{141}Pr, and upper limits were set for (n,3n) reactions with ^{141}Pr, ^{197}Au, and ^{203}Tl. The [(n,np) + (n,pn)] reaction was detected with ^{58}Ni, but not with ^{92}Mo. A value of 520 ± 120 mb was established for the former and an upper limit set for the latter. Absolute disintegration rates were obtained by both beta and gamma spectroscopy counting methods. Extensive use of radiochemical separation was made in order to isolate the low-yield rare reaction products from large target samples. Statistical theory was employed (using parameters selected from a critical analysis of the literature) to make theoretical cross-section estimates for many of the reactions studied. The theoretical cross sections for (n,αn) reactions agree remarkably well with the experimentally determined [(n,nα) + (n,αn)] cross-section sums, thereby suggesting that the path for this reaction may be predominantly the (n,αn) process. For certain cases, statistical theory predicts appreciable (n,2p) cross sections at 14.7 MeV; for example, for ^{50}Cr, ^{58}Ni, ^{78}Kr, and ^{92}Mo. It is suggested that in studies of emitted proton spectra that show an "excess" of low-energy protons, a contribution may be present from the (n,2p) as well as the (n,np) reaction. A new gamma at 176 ± 4 keV in about 5% of 1.8 day ^{48}Sc decays is confirmed (Bramlitt and Fink 1963).

In the course of developing suitable neutron activation, direct instrumental analysis techniques for determining the individual rare-earth elements in various alloys and compounds, several fast neutron reactions with the rare-earth nuclides were measured by Broadhead et al. (1965). Cross sections of 14 MeV neutrons have been determined for the following reactions: $^{89}Y(n,n'\gamma)Y^{89m}$, 400 mb; $^{142}Nd(n,2n)\,^{141m}Nd$, 545 mb; $^{144}Sm(n,2n)\,^{143m}Sm$, 400 mb; $^{159}Tb(n,2n)\,^{158m}Tb$, 160 mb; and $^{168}Er(n,2n)^{167m}Er$, 190 mb. Possible sources of error affecting the cross-section measurements have been discussed and evaluated numerically. The overall accuracy was determined to be about ±12%.

Neutron-induced cross sections are of interest for practical applications and for testing nuclear models. In the work, elaborated in the thesis by Reimer (2002) reactions (n,p), (n,np), (n,α), (n,nα), (n,n′γ), (n,2n), and (n,3n) on vanadium, molybdenum, technetium, and lead have been measured in the energy range of 0.5–20.6 MeV using the activation technique. The radioactive reaction products with half-lives between 58 s and 20300 years have been measured off-line via high-resolution γ-ray spectrometry and liquid scintillation counting, the latter in combination with radiochemical separation. Irradiations in the energy range from 13.4 to 20.6 MeV were performed using the $^{3}H(d,n)^{4}He$ reaction, with a solid Ti/T target. Most of the reactions were investigated using a light mass setup to minimize scattering effects, but for short half-lives a pneumatic sample transport system was used as well. A special sample holder was developed for the measurement of the $^{nat}Mo(n,x)^{94}Nb$ reaction. All cross sections were measured relative to the $^{27}Al(n,\alpha)^{24}Na$ standard cross section and all necessary corrections due to the irradiation process and the measurement of the induced activity have been applied. As a result of this work, an extended database for neutron-induced cross sections on four elements was obtained.

The China nuclear data activities consist of nuclear data measurement and related measurement methods development, data evaluation and model study, data library establishment and library management, and nuclear data benchmark testing and validation. The mainly activities are being carried out at China Nuclear Data Center (CNDC), China Institute of Atomic Energy (CIAE), and China Nuclear Data Coordination Network (CNDCN), and more than 10 institutions and universities are involved in CNDCN. Substantial progress on nuclear data measurement has been made in China in recent years. More and more needs for nuclear data measurement have been required with the progress of the ADS, TMSR, and ITER projects in China. Some new facilities such as CSNS are under construction; these facilities will greatly improve the capability of the nuclear data measurement in China in the near future (Ge and Ruan 2013).

5.6 FISSION INDUCED BY 14 MeV NEUTRONS

This process has been studied in some detail; however, most of data are still classified. As an example, we discuss the paper by Wahl (1955) on the fission of ^{235}U by 14 MeV neutrons nuclear charge distribution and yield fine structure. The results were derived from experiments in which iodine was quickly separated from irradiated uranium metal at a known time after irradiation. The general procedure followed was to irradiate for a short time a stack of two or more pieces of uranium metal in a flux of 14 MeV or thermal neutrons. One piece was dissolved immediately after the irradiation and the iodine quickly separated. The results indicate that the pronounced peak in the mass–yield curve found at mass number 134 in thermal neutron fission is nearly washed out in 14 MeV neutron fission. The higher independent yields of late members of the chains in 14 MeV neutron fission indicate a shift toward stability of the most probable initial nuclear charge. This shift is about 0.7 charge unit for the chains in the mass region studied.

The spatial and spectral distributions of the neutrons emitted when heavy elements are irradiated by fast neutrons yield information concerning the basic mechanism of such interactions. Specifically, these data shed light on (1) the type of interaction processes involved, for example, the relative probability of compound nucleus formation versus interactions involving a small number of nucleons; (2) the level density of the residual nucleus; and (3) the applicability of various nuclear models. In a paper by Rosen and Stewart (1957), the spatial and spectral distributions of the neutrons from 14 MeV neutron interactions with Ta and Bi have been obtained by using nuclear emulsion detectors in conjunction with a neutron collimator. These results indicate that roughly 85%–90% of the interactions proceed via compound nucleus formation while approximately 10%–15% proceed by "direct" interactions. The space-integrated neutron spectrum has been obtained for Be by means of a sphere experiment. Cross sections for neutron emission, nonelastic scattering, (n,n′), and (n,2n) are derived for all three elements.

These experiments on Ta and Bi were made possible by the use of a large neutron collimator of a type initially developed for observing neutron spectra from nuclear detonations and since successfully employed on a number of experiments involving 14 MeV neutrons (Allred et al. 1953). This collimator is made of iron backed by paraffin. Iron was chosen on the basis of the large value of the product of its nonelastic cross section and nuclear density and the fact that the nonelastic spectrum of emitted neutrons is much degraded. Even so, paraffin backing must be added to achieve further degradation in order to take data down to 0.5 MeV.

5.7 Neutron-Induced SEU

For over 20 years, the military, the commercial aerospace industry, and the computer industry have known that high-energy neutrons streaming through our atmosphere can cause computer errors known as single-event upsets (SEUs). These are "soft" errors—no permanent damage is done—but a single digit in computer memory suddenly changes, or a logic circuit produces an erroneous result that may hang up (or crash) an application. The neutron's head-on collision with a nucleus is what does the mischief. It produces a burst of electric charge that causes a single transistor—the basic building block of the integrated circuits patterned on the surface of a microchip—to flip from the *off* state to the *on* state.

Several incidents across many industries have been reported in recent years. Among these are the following:

- On October 7, 2008, an Airbus A330-303 operated by Qantas Airways was en route from Perth to Singapore. At 37,000 ft, one of the plane's three air data inertial reference units had a failure, causing incorrect data to be sent to the plane's flight control systems. This caused the plane to suddenly and severely pitch down, throwing unrestrained occupants to the plane's ceiling. At least 110 of the 303 passengers and 9 of the 12 crew members were injured. The injuries of 12 of the occupants were serious and another 39 occupants required treatment at a hospital. All potential causes were found to be "unlikely," or "very unlikely," except for an SEU. However, the Australian Transport Safety Board (ATSB 2011) found it had "insufficient evidence to estimate the likelihood" that an SEU was the cause.

- Canadian-based St. Jude Medical issued an advisory to doctors in 2005, warning that SEUs to the memory of its implantable cardiac defibrillators could cause excessive drain on the unit's battery.
- Cisco Systems issued a field notice in 2003 regarding its 1200 series router line cards. The noticed warned of line card resets resulting from SEUs.

The occurrence of SEU in aircraft electronics has evolved from a series of interesting anecdotal incidents to accepted fact. A study completed in 1992 demonstrated that SEUs are real, that the measured in-flight rates correlate with the atmospheric neutron flux, and that the rates can be calculated using laboratory SEU data. Once avionics SEU was shown to be an actual effect, it had to be dealt within avionics designs. The major concern is in random access memories (RAMs), both static (SRAMs) and dynamic (DRAMs), because these microelectronic devices contain the largest number of bits, but other parts, such as microprocessors, are also potentially susceptible to upset. In addition, other single-event effects (SEEs), specifically latch-up and burnout, can also be induced by atmospheric neutrons (Normand 2002).

The same microchips used in avionics are appearing everywhere in our digital world, for example, in ground-level civilian systems for banking, transportation, medicine, communication, entertainment, and more. They are critical in insulin monitors and GPS-enabled emergency response systems, in antilock brakes, and smart stoplights, smart phones, increasingly realistic video games, advanced audio systems, and the supercomputers that forecast the weather.

SEUs are soft errors, meaning that the integrated circuit can be recovered to a fully functional state. SEUs usually mean a change in the logic state of a memory cell that can be corrected by rewriting the data to the cell. The probability for a device to be upset is known as the SEU cross section, and it is an important factor in predicting how a device will function in a radiation environment. The SEU cross section is traditionally considered to depend mainly on the particle LET, increasing with increasing LET. The SEU cross section can be given in units of cm^2/bit or cm^2/device. The SEU cross section relates to an actual area because at high LET values the SEU cross section usually saturates to a certain value and each particle crossing the device is considered to cause an upset, thus outlining a projected area. The mechanism behind SEUs is the following: a penetrating particle deposits energy into a semiconductor and generates electron–hole pairs (= carriers) that are then transported within the semiconductor. Before the electron–hole pairs have time to recombine, a fraction of them is separated by the electric field near a p–n junction of the semiconductor and collected at the nearby device contact. If the collected charge exceeds the critical threshold, an SEU occurs. Reverse-biased p–n junctions are the most sensitive regions to SEUs because of their large depletion region and strong electric field.

In recent years, cosmic rays–induced SEUs have been recognized as a key reliability concern for microelectronic devices used not only in space but also at the ground level or in airplanes at higher altitude. The SEU is one of the transient radiation effects by which the memory state of a cell can be flipped from a 1 to a 0 or vice versa, resulting in malfunction. The SEU is initiated by the interaction of incident cosmic-ray particles with materials in microelectronics devices. Then, light-charged particles and heavy recoils are generated via the nuclear reaction with a constituent atomic nucleus, mainly ^{28}Si, and then deposit the charge in a small sensitive volume (SV) of the device. The deposited charge is collected at one of the nodes keeping the memory information and the resulting transient current generates an SEU. Knowledge on nuclear physics and radiation physics is indispensable to understand these elementary processes in the SEU phenomena well. Particularly, nuclear reaction data play an essential role in estimating the SEU rate accurately, because the nuclear interaction takes place in the first stage of the SEU process. Watanabe (2007) examined the role of nuclear data in the study of SEU phenomena in semiconductor memories caused by cosmic-ray neutrons and protons. Neutron and proton SEU cross sections are calculated with a simplified semiempirical model using experimental heavy-ion SEU cross sections and a dedicated database of neutron- and proton-induced reactions on ^{28}Si. Some impacts of the nuclear reaction data on SEU simulation are analyzed by investigating the relative contribution of secondary ions and neutron elastic scattering to SEU and influence of simultaneous multiple ions emission on SEU.

The SEU rate per microchip depends on three things multiplied together: the neutron intensity, the intrinsic sensitivity of each transistor to neutron-induced SEUs, and the number of transistors on the microchip. Suppose the SEU rate for a particular microchip with particular transistors, used at a certain altitude, is 1 every 1000 h and there are 100 microchips in use. Then at that altitude, 1 of those 100 microchips will suffer an SEU once every 10 h. In other words, the higher the altitude, the greater the neutron sensitivity of the transistor and the larger the number of microchips in use, the higher the SEU rate.

An SEU is a bit flip in a memory element of a semiconductor device. These upsets are random in nature, do not normally cause damage to the device, and are cleared with the next write to that memory location or by power cycling the device. The result of upsets is data corruption. Many systems can tolerate some level of soft errors; for example, corrupted data in a video or audio stream may or may not be noticeable or important to the user. When charged particles strike the silicon substrate of an IC, they leave an ionization trail. Similarly, when a high-energy particle such as a neutron strikes the substrate, it collides with atoms in the substrate, liberating a shower of charged particles that then leave an ionization trail. If the resulting deposited charge from these particles is sufficient, it can change the state of a memory bit or flip-flop. Specifically, the charge (electron–hole pairs) generated by the interaction of an energetic charged particle with the semiconductor atoms corrupts the stored information in the memory cell. The flux of cosmic rays impacting the Earth's atmosphere is modulated by both the solar wind and the Earth's magnetic field. Combining these factors results in the neutron flux being a function of latitude, longitude, altitude, and solar activity, with the greatest flux occurring at high altitudes over the poles during quiet periods of solar activity. For example, a resident of a northern sea-level city such as New York experiences twice the neutron flux as someone living near the equator in Singapore, but only about one-fourth that of Denver residents. On the other hand, airplane passengers experience nearly 600 times the neutron flux of New Yorkers. Initially, the SEU issue gained widespread attention in the late 1970s as a memory data corruption issue when DRAMs began to show signs of apparently random failures. Although the phenomenon was first noticed in DRAMs, SRAM memories and SRAM-based programmable logic devices are also subject to the same effects. Unlike capacitor-based DRAMs, SRAMs are constructed of cross-coupled devices that have far less capacitance in each cell. The lower the capacitance of a cell, the greater the chance of an upset. As both the voltage and cell size are reduced with each new process generation, the SRAM cell capacitance continues to decrease, making the cell even more vulnerable to more types of particles, and to particles of lower energy. Normally, configuration memory upsets can be cleared with a reconfiguration or system power cycling and have no lasting effect. However, configuration memory upsets can create illegal conditions within the FPGA, for example, creating high-current conditions due to contentions as a result of the misconfiguration. This high current draw may damage the device or the board on which it is mounted. If not corrected, configuration memory upsets that result in simultaneously enabling pull-ups and pull-downs or serious bus contention may physically damage the FPGA.

Let us mention some interesting studies of this phenomenon as reported in the scientific literature. In the report by Tsao et al. (1988), the SEU rates due to neutron-induced nuclear recoils were calculated for Si and GaAs components using the HETC and MCNP codes and the evaluated nuclear data file (ENDF) database for (n,p) and (n,α) reactions. For the same critical charge and sensitive volume, the upset rate in Si exceeds that of GaAs by a factor of about 1.7, mainly because more energy is transferred in neutron interactions with lighter Si nuclei. The upset rates due to neutrons are presented as functions of critical charge and atmospheric altitude. Upsets induced by cosmic-ray nuclei, secondary protons, and neutrons are compared.

In the report by Arita et al. (2004), neutron SEUs induced by elastic scattering were investigated by an experiment using 2 MeV neutron beams and by a calculation based on scattering cross-section data and angular distribution data from the ENDF. The SEU rates obtained by the calculation and experiment are in fairly good agreement if the region sensitive to the SEUs (sensitive volume) is properly defined. Adopting the calculation to atmospheric neutrons, the fraction of the SEU rates induced by elastic scattering accounted for approximately 26%–32% of the total fast-neutron-induced SEUs in the atmospheric neutron environment. Reducing well depth will effectively reduce the number of SEUs.

Lambert et al. (2006) investigated the sensitivity of SOI and bulk SRAMs to neutron irradiations with energies from 14 to 500 MeV. The technology sensitivity is analyzed with both experiments and Monte Carlo simulations. In particular, simulations include the nuclear interactions of neutrons with both silicon and oxygen nuclei (n-Si and n-O), in order to investigate the influence of isolation upper layers on the device sensitivity. The device cross sections are analyzed for monoenergetic neutron irradiations and discussed in terms of nuclear interaction type (n-Si and n-O) and distribution of the secondary ion recoils. The dimensions of the interaction volume around the sensitive cell as a function of the device architecture was also investigated. See also reports by Clemens et al. (2011) and Hands et al (2011).

Palau et al. (2002) used a Monte Carlo approach to obtain statistical information on the effect of the spatial distribution of the numerous secondary ions involved in neutron-induced soft error rates (SERs). The sorting criteria for the occurrence of upset are derived from a simplification of previous work on full-cell 3D SRAM device simulation. The time thus saved allows the treatment of a wide variety of track conditions. The shape and extension of the sensitive region are explored and correlated to the secondary ion properties. Details on the variations of the sensitivity with depth into the sensitive region as well as on the geometrical conditions associated with those tracks that cause upsets are given.

In the report of Ibe et al. (2010), trends in terrestrial neutron-induced soft-error in SRAMs from a 250 to a 22 nm process are reviewed and predicted using the Monte Carlo simulator CORIMS, which is validated to have less than 20% variations from experimental soft-error data on 180–130 nm SRAMs in a wide variety of neutron fields like field tests at low and high altitudes and accelerator tests in LANSCE, TSL, and CYRIC. The following results are obtained: (1) SERs per device in SRAMs will increase ×6–7 from 130 to 22 nm process, (2) as SRAM is scaled down to a smaller size, SER is dominated more significantly by low-energy neutrons (< 10 MeV), and (3) the area affected by one nuclear reaction spreads over 1 M bits and bit multiplicity of multicell upset become as high as 100 bits and more.

Terrestrial neutron-induced soft error or SEU in semiconductor devices is one of the most crucial reliability issues in the cutting-edge memory devices (Yahagi et al. 2002; Ibe et al. 2002, 2006; Nakamura et al. 2008). In the case of DRAM, cell data inversion from high state to low state mainly occurs. Designers of electronic systems such as high-end servers, therefore, need the SER data of the electronic components in order to architect the system reliability securely. Hayakawa et al. (2010) have intensively developed the SER estimation method for memory devices mainly using monoenergetic and quasimonoenergetic neutron beams generated by accelerators. The SER of the device is estimated in the unit of FIT (failure in time; a number of errors in 10^9 h); the SER of the DRAMs was effectively suppressed as its down sizing from 220 to 150 nm. This is estimated as the effect of the shrinking of junction volumes and the securing of constant storage capacitance. However, the SER beyond 150 nm process was found constant. This is estimated the offset of the scaling effect and the gradual decreasing of storage capacitance by the lower voltage to capacitor, see also Jahinuzzaman et al. (2013). The securing of storage capacitance may be becoming difficult on advanced DRAM because of the reduction of the cell area and the lower voltage to capacitor.

SEUs, which are induced by neutron or alpha particles, have been noticeable in system-on-chip (SoC) circuits (Baumann 2002). SEU events may occur in any part of an SoC circuit such as memories, sequential logic (FF and latches), and combinational logic to cause reliability issues. To obtain chip-level SER of a SoC, first, it is necessary to estimate device-level SER for each part of the SoC separately. Based on device-level SER results, chip-level SER of the SoC is then estimated by the chip-level SER simulation tool. Device-level SER for memories and sequential logic cells can be obtained by performing the accelerated SER tests. However, with this method, it usually takes time to design the test chip, test facility booking, data collection, and data analysis. It is also difficult to cover the SER results of all kinds of logic cells used in an SoC by limited accelerated SER tests. Thus instead of accelerated SER tests, an accurate device-level SER simulator with shorter simulation duration becomes useful for SER estimation (Palau et al., 2001). Device-level neutron and alpha-induced SER simulations are presented for SRAM and logic flip-flop (FF) cells in paper by Fang et al. (2013). Compared with the accelerated SER test results, the FIT rate simulation results are close to the measured FIT rate for both SRAM and FF cells. Accurate neutron-induced multiple cell upsets (MCU) and multibit upsets (MBU) statistics for the SRAM at different test patterns are also exhibited.

The high-energy neutron source at the Neutron and Nuclear Science (WNR) facility provides a capability for accelerated neutron testing of semiconductor devices. This testing is important because neutrons have been recognized as a significant threat to semiconductor devices at aircraft altitudes and below. The shape of the neutron spectrum produced at WNR is very similar to the spectrum of neutrons produced in the atmosphere by cosmic rays, but it is more than five orders of magnitude more intense. Since 1992, many companies, including Texas Instruments, the Boeing Company, Hewlett Packard, Lockheed Martin, Digital Equipment, and Intel, continue to use the high-energy neutron source at WNR to study various failure modes caused by neutron radiation. (There has been more industrial participation in this activity than in any other user activity at LANL.) The ultimate goal of this research will be to find ways to overcome the effects of incoming galactic and solar cosmic-ray-induced neutrons.

SRAM-based FPGAs have been studied for their sensitivity to atmospheric high-energy neutrons. FPGAs with the supply voltage 5 and 3.3 V were irradiated by 0–11, 14, and 100 MeV neutrons and showed a very low SEU susceptibility (Ohlsson et al. 1998) to SEUs caused by neutrons. No permanent effects were detected and reconfiguration of the device was sufficient to regain full functionality after the occurrence of an SEU. It was concluded that these SRAM-based FPGAs can be used without limitation in the atmospheric radiation environment, contrary to SRAM memories were precaution in the use is necessary because of neutron-induced SEU.

A comprehensive text on the subject of the terrestrial neutron-induced failures in semiconductor devices and relevant systems and their mitigation techniques is presented by Ibe (2011). His attention is focused on multicell upsets (MCUs) that, defined as simultaneous errors in more than one memory cell induced by a single event, have been under close scrutiny. The concept of MCU, therefore, contains both upsets that can be corrected by error detection/error correction code (EDAC/ECC) and those that cannot. The latter is called "multiple-bit upset" or "multibit upset" (MBU) of memory cells, and can lead, for example, to hang-ups of computer systems. Though MBUs can be avoided by a combination of ECC and interleaving technique, MCUs may still be problematic in high-performance devices such as contents addressable memories (CAMs) used in network processors and routers. In the case of system design, it is, therefore, very important to evaluate MCUs as well as SERs of the device in design phase. Historically, MCUs are understood as taking place when two or more storage nodes are hit by one secondary ion from nuclear spallation reaction in a device. As device scaling down proceeds, novel MCU modes are being reported as charge sharing among memory storage nodes in the vicinity or bipolar effects in p-well (Ibe 2011).

The paper by Gill et al. (2009) addresses the importance of combinational SER in modern 32 nm CMOS circuits and products in terrestrial radiation environments. To quantitatively estimate the contribution of combinational SER relative to sequential SER, a test chip has been designed and exposed to alpha particle and neutron radiation. The test chip implements four different types of data paths feeding into nonhardened sequentials. Balanced (nominal) paths reflect P/N ratios that are close to those implemented in products, whereas skewed paths maximize single event transients (SETs) generation and propagation to test the accuracy of our simulation methodology against measured data. The design strategy chosen ensures a sequential SER that is independent of clock speed and minimal contributions due to clock nodes strikes. The obtained results are consistent with combinational SER increasing linearly with clock frequency. For the test chip analyzed, the neutron combinational

SER is about 2× higher than the alpha particle–induced one. Further, the combinational SER contribution per sensitive static logic gate is less than 1% of the nominal latch SER at 1 GHz at 0.75 V. Sequential soft error simulation results based on calibrated SER models are within ~2× of the measured error rates, whereas simulated combinational SER projections are within 2–3× of our test chip data, which is surprisingly accurate. The latter result indicates that applying voltage-independent current sources to mimic particle strikes yields sufficiently accurate estimates for terrestrial radiation environments. Based on measured and simulated SER, chip-level conservative estimates are derived by factoring in product P/N and logic depth distributions and typical use conditions. The results indicate that the chip-level SER contribution of combinational logic is well below 30% of the chip-level nominal latch SER in the investigated 32 nm process technology (Gill et al. 2009).

SEUs have been observed in implantable cardiac defibrillators by Bradley and Normand (1998). The incidence of SEUs is well modeled by upset rate calculations attributable to the secondary cosmic-ray neutron flux. The effect of recent interpretations of the shape of the heavy ion cross-section curve on neutron burst generation rate calculations is discussed. The model correlates well with clinical experience and is consistent with the expected geographical variation of the secondary cosmic-ray neutron flux. The observed SER was 9.3×10^{-12} upsets/bit-h from 22 upsets collected over a total of 284,672 device days. This is the first clinical data set obtained indicating the effects of cosmic radiation on implantable devices. Importantly, it may be used to predict the susceptibility of future implantable device designs to cosmic radiation. The significance of cosmic radiation effects relative to other radiation sources applicable to implantable devices is discussed. Implantable devices may be implanted for periods of between 3 and 8 years before battery depletion requires explant. Improving the longevity of the device drives the design toward low supply voltages since the current consumed by the integrated circuits is the dominant contributor to power consumption (followed by shock power; devices have the capacity to deliver up to several hundred charges; however, under clinical conditions a typical patient requires around 10 shocks per year). The low voltage operation greatly increases susceptibility to soft-error upsets. Microprocessor-based systems in which critical controlling software is in RAM, as opposed to ROM, are especially prone to SEUs. Clearly, an understanding of the SER is vitally important given the high reliability requirements and life-supporting nature of the application.

Interest in 14 MeV neutron tests is interesting over the years because of its possible advantages: (1) tabletop irradiation configuration, (2) isotropic source and experiments can be put all around, (3) no need of device opening, and (4) not as expensive as usual accelerator tests. In general, 14 MeV neutron generator is flexible and cost-efficient radiation tool. Analysis of nuclear databases showed that

14 MeV neutrons produce ions with characteristics able to trigger events; potential issue for representativeness of MCUs (especially multiplicity) due to the limited range of secondary ions; and capability to extrapolate from experimental 14 MeV neutron test (high confidence to neutron or protons within a ×2 margin whatever the considered heavy-ion LET for a nonhardened technology (LET<5 MeV cm^2/mg); see descriptions in the literature (Flament et al. 2004; Dyer et al. 2006; Baggio et al. 2007; Irom et al. 2007; Normand and Dominik 2010; Miller et al., 2013).

REFERENCES

Allred, J. C., Armstrong, A. H., and Rosen, L. 1953. The interaction of 14-MeV neutrons with protons and deuterons. *Phys. Rev.* 91(1): 90–99.

Anderson, J. D., Hansen, L. F., and Wong, C. 1974. Measurements of neutron emission spectra from 14-MeV neutrons on thick targets, in *Lawrence Livermore Laboratory Report UCRL-76033 Presented at International Symposium on Radiation Physics*, 1974, Bose Institute, Calcutta, India.

Antolkovic, B., Paic, G., Tomas, P., and Rendic, D. 1967. Study of neutron-induced reactions on He-3 at E_n = 14.4 MeV. *Phys. Rev.* 159: 777–781.

Arita, Y., Takai, M., Ogawa, I., and Kishimoto, T. 2004. Influence of elastic scattering on the neutron-induced single-event upsets in a static random access memory. *Jpn. J. Appl. Phys.* 43(9A/B): L1193–L1195.

ATSB. 2011. ATSB transport safety report aviation occurrence investigation AO-2008-070 final, in-flight upset 154 km west of Learmonth, WA, October 7, 2008, VH-QPA Airbus A330-303, 313p.

Baggio, J., Lambert, D., Ferlet-Cavrois, V., and Paillet, P. 2007. Single event upsets induced by 1–10 MeV neutrons in static-RAMs using mono-energetic neutron sources. *IEEE Trans. Nucl. Sci.* 54: 2149–2155.

Barschall, H. H. and Taschek, R. F. 1949. Angular distribution of 14-MeV neutrons scattered by protons. *Phys. Rev.* 75(12): 1819–1822.

Bassel, R. H., Drisco, D. H., and Satchler, G. R. 1962. The distorted-wave theory of direct nuclear reactions: “Zero-range” formalism without spin-orbit coupling, and the code SALLY. Oak Ridge National Laboratory Report ORNL-3240.

Battat, M. E. and Graves, E. R. 1955. Gamma rays from 14-MeV neutron bombardment of ^{12}C. *Phys. Rev.* 97: 1266–1267.

Bauer, R., Anderson, J. D., and Christensen, L. J. 1963. Scattering of 14 MeV neutrons from nitrogen and oxygen. *Nucl. Phys.* 47: 241–250.

Baumann, R. 2002. The impact of technology scaling on soft error rate performance and limits to the efficacy of error correction, in *International Electron Devices Meeting (IEDM) Tech. Dig.*, pp. 329–332. San Francisco, CA, December 8-11, 2002.

Begum, A. and Galloway, R. B. 1979. Polarization of 14 MeV neutrons in forward angle scattering by Cu and Pb. *Phys. Rev. C* 20(5): 1711–1715.

Berick, A. C., Riddle, R. A. J., and York, C. M. 1968. Elastic scattering of 14-MeV neutrons by deuterons. *Phys. Rev.* 174(4): 1105–1111.

Bjorklund, F. and Fernbach, S. 1958. Optical-model analyses of scattering of 4.1-, 7-, and 14-MeV neutrons by complex nuclei. *Phys. Rev.* 109: 1295–1298.

Bonnel, C. and Lévy, P. 1961. Study of the stripping reaction of the deuteron by 14.1-MeV neutrons. *Compt. Rend.* 253: 635–637.

Bouchez, R., Duclos, J., and Perrin, P. 1963. Scattering of 14 MeV neutrons from ^{12}C and excitation of the 7.65 MeV level. *Nucl. Phys.* 43: 628–635.

Bradley, P. D. and Normand, E. 1998. Single event upsets in implantable cardioverter defibrillators. *IEEE Trans. Nucl. Sci.* 45(6): 2929–2940.

Bramlitt, E. T. and Fink, R. W. 1963. Rare nuclear reactions induced by 14.7 MeV neutrons. *Phys. Rev.* 131: 2649–2663.

Broadhead, K. G., Shanks, D. E., and Heady, H. H. 1965. 14-MeV-neutron production of isomeric states for several rare-earth elements. *Phys. Rev.* 139(6B): 1525–1528.

Bychkov, V. M., Plyaskin, V. I., and Toshinskaya, Eh. F. 1982. Evaluation of the (n,2n) and (n,3n) cross-sections for heavy nuclei with allowance for non-equilibrium processes. INDC(CCP)-184/L.

Cerineo, M., Ilakovac, K., Šlaus, I., Tomaš, P., and Valkovic, V. 1964. Charge dependence of nuclear forces and the break-up of deutrons and tritons. *Phys. Rev. B* 133: 948–955.

Chatterjee, S. and Chatterjee, A. 1969. Single-particle behaviour in fast neutron (n,2n) reactions. *Nucl. Phys. A* 125: 593–612.

Clemens, M. A., Sierawski, B. D., Warren, K. M. et al. 2011. The effects of neutron energy and high-Z materials on single event upsets and multiple cell upsets. *IEEE Trans. Nucl. Sci.* 58: 2591–2598.

Coon, J. H., Bockelman, C. K., and Barschall, H. H. 1951. Angular distribution of 14-MeV neutrons scattered by tritons. *Phys. Rev.* 81: 33–35.

Coon, J. H., Graves, E. R., and Barschall, H. H. 1952. Total cross sections for 14-MeV neutrons. *Phys. Rev.* 88(3): 562–564.

Csikai, J. 1987. *CRC Handbook of Fast Neutron Generators.* CRC Press, Boca Raton, FL.

Demetriou, P., Kanjanarat, P., and Hodgson, P. E. 1994. FKK analysis of 14 MeV neutron-induced reactions. *J. Phys. G: Nucl. Part. Phys.* 20: 1779–1788.

Doczi, R., Semkova, V., Majdeddin, A. D., Buczko, Cs. M., and Csikai, J. 1997. Investigations on (n,p) cross sections in the 14 MeV region. IAEA Nuclear data Section Report INDC(HUN)-032, Distr. L. Vienna, Austria.

Dyer, C., Hands, A., Ford, K., Frydland, A., and Truscott, P. 2006. Neutron induced single event effects testing across a wide range of energies and facilities and implications for standards. *IEEE Trans. Nucl. Sci.* 53: 3596–3601.

Fang, Y.-P., Oates, A. S., Belhaddad, K., Correas, V., Nofal, I., and Lauzeral, O. 2013. Neutron and alpha soft error rate simulations for memory and logic devices at advanced technologies, in *Proceedings of IEEE Workshop on Silicon Errors in Logic—System Effects (SELSE).* Stanford University, Stanford, CA, March 26–27, 2013.

Fernbach, S., Serber, R., and Taylor, T. B. 1949. The scattering of high energy neutrons by nuclei. *Phys. Rev.* 75: 1352–1355.

Feshbach, H. and Weisskopf, V. F. 1949. Schematic theory of the nuclear cross-section. *Phys. Rev.* 76: 1550–1560.

Feshbach, H., Kerman, A., and Koonin, S. 1980. The statistical theory of multi-step compound and direct reactions. *Ann. Phys.* 125(2): 429–476.

Feshbach, S., Porter, C. E., and Weisskopf, V. F. 1954. Model for nuclear reactions with neutrons. *Phys. Rev.* 96(2): 448–464.

Flament, O., Baggio, J., D'hose, C., Gasiot, G., and Leray, J. L. 2004. 14 MeV neutron-induced SEU on SRAM devices. *IEEE Trans. Nucl. Sci.* 51: 2908–2911.

Forbes, S. G. 1952. Activation cross sections for 14 MeV neutrons. *Phys. Rev.* 88(6): 1309–1311.

Furić, M., Valković, V., Miljanić, Đ., Tomaš, P., and Antolković, B. 1970. Neutron-proton bremsstrahlung at 14.4 MeV. *Nucl. Phys. A* 156: 105–112.

Gadioli, E. and Hodgson, P. E. 1992. *Pre-equilibrium Nuclear Reactions*. Clarendon Press, Oxford, U.K.

Galloway, R. B. and Waheed, A. 1979. Polarization of 16-MeV neutrons due to elastic scattering. *Phys. Rev. C* 19: 268–271.

Ge, Z. and Ruan, X. 2013. Progress of nuclear data measurement in China during 2011–2012, in *25th Meeting of the NEA WPEC*, Paris, France, May 23–24, 2013.

Gill, B., Seifert, N., and Zia, V. 2009. Comparison of alpha-particle and neutron-induced combinational and sequential logic error rates, in *The 32 nm Technology, Proceedings of the 2009 IEEE Workshop on Silicon Errors in Logic—System Effects* (*SELSE 5*), 199–205. http://publications.lib.chalmers.se/publication/106944-evaluation-of-low-cost-detection-and-reco.

Glendenning, N. K. 1962. The two-nucleon stripping reaction. *Nucl. Phys.* 29: 109–119.

Glendenning, N. K. 1965. Nuclear spectroscopy with two-nucleon transfer reactions. *Phys. Rev. B* 137: 102–103.

Hands, A., Morris, P., Dyer, C., Ryden, K., and Truscott, P. 2011. Single event effects on power MOSFETs and SRAMs due to 3 MeV, 14 MeV and fission neutrons. *IEEE Trans. Nucl. Sci.* 58: 952–959.

Hansen, L. F., Anderson, J. D., Brown, P. S. et al. 1973. Measurements and Calculations of the Neutron Spectra from Iron Bombarded with 14 MeV Neutrons. *Nucl. Sci. Eng.* 51: 278–295.

Hansen, L. F., Anderson, J. D., Goldberg, E., Plechaty, E. F., Stelts, M. L., and Wong, C. 1969. Time Spectra from Spheres Pulsed with 14 MeV Neutrons. *Nucl. Sci. Eng.* 35: 227–239.

Hansen, L. F., Anderson, J. D., Goldberg, E., Kammerdiener, J. L., Plechaty, E. F., and Wong, C. 1970. Predictions for Neutron Transport in Air, Based on Integral Measurements in Nitrogen and Oxygen at 14 MeV Neutrons. *Nucl. Sci. Eng.* 40: 262–282.

Hayakawa T., Matsumoto T., Yahagi Y., Saito A., Hidaka M., Ibe E., and Itoh M. 2010. Study of neutron-induced soft error rate on advanced DRAM. CYRIC Annual Report. Tohoku University, Sendai, Japan.

Hlavač, S., Oblozinsky, P., Turzo, I., Dostal, L., and Kliman, J. 1994. Cross sections of the $^{16}O(n,\alpha\gamma)$ reaction at 14.7 MeV. IAEA Nuclear Data Section Report INDC(SLK)-002, IAEA, Vienna, Austria.

Holub, E. and Cindro, N. 1976. Study of some systematic trends and non-equilibrium effects in (n, 2n) reactions for nuclei far from the symmetry line. *J. Phys. G: Nucl. Phys.* 2(6): 405–419.

Ibe, E. 2011. Terrestrial neutron-induced failures in semiconductor devices and relevant systems and their mitigation techniques. Chapter 2, in Kanekawa, N., Ibe, E. H., Suga, T., and Uematsu, Y. (Eds.), *Dependability in Electronic Systems, Mitigation of Hardware Failures, Soft Errors, and Electro-Magnetic Disturbances*. Springer Science+Business Media, LLC, New York.

Ibe, E., Taniguchi, H., Yahagi, Y., Shimbo, K., and Toba, T. 2010. Impact of scaling on neutron-induced soft error in SRAMs from a 250 nm to a 22 nm. Design rule, electron devices. *IEEE Trans.* 57(7): 1527–1538.

Ibe, E., Yahagi, Y., Kataoka, F., Saito, Y., Eto, A., Sato, M., Kameyama, H., and Hidaka M. 2002. *IEEE First International Conference on Information Technology & Applications*, Paper No. 273-21.

Ilakovac, K., Kuo, L. G., Petravic, M., Slaus, I., and Tomas, P. 1961. Proton spectra from the D(n,p)2n reaction at 14.4 MeV. *Phys. Rev. Lett.* 6: 356–358.

Irom, F., Miyahira, T. F., Nguyen, D. N., Jun, I., and Normand, E. 2007. The results of recent 14 MeV neutron single event effects measurements conducted by jet propulsion laboratory, in *Proceedings of IEEE Radiation Effects Data Workshop*. Honolulu, Hawaii, 23–27 July, 2007.

Iwamoto, O. 2007. Development of a comprehensive code for nuclear data evaluation, CCONE, and validation using neutron-induced cross sections for uranium isotopes. *J. Nucl. Sci. Technol.* 44(5): 687–697.

Jahinuzzaman, S., Gill, B., Ambrose, V., and Seifert, N. 2013. Correlating low energy neutron SER with broad beam neutron and 200 MeV proton SER for 22 nm CMOS tri-gate devices, in *International Reliability Physics Symposium (IRPS), 2013 IEEE*, 3D.1.1–3.D.1.6.

Konobeyev, A. Yu. and Korovin, Yu. A. 1999. Semi-empirical systematics of (n, 2n) reaction cross-section at the energy of 14.5 MeV. *Nuovo Cimento A* 112(9): 1001–1013.

Lambert, D., Baggio, J., Hubert, G. et al. 2005. Neutron-induced SEU in SRAMs: Simulations with n-Si and n-O interactions. *IEEE Trans Nucl. Sci.* 52(6): 2332–2339.

LANL. 2014. http://wnr.lanl.gov/_assets/flight_paths/4FP30L_about.php. Accessed October 25, 2014.

Levkovskii, V. N. 1957. Žurn. Eksp. Teor. Fiz. 33: 1520. *Soviet Phys. JETP* 6, 1958: 1174–1175.

Levkovskii, V. N. 1974. Cross-sections of (n,p) and (n,alpha) reactions at a neutron energy 14–15 MeV. *Soviet J. Nucl. Phys. USSR* 18(4): 361–363.

Lu, W.-D. and Fink R. W. 1971. Applicability of the constant nuclear temperature approximation in the statistical model analysis of neutron cross sections at 14.4 MeV for medium-Z nuclei. *Phys. Rev. C* 4: 1173–1181.

Marshak, H., Richardson, A. C. B., and Tamura, T. 1966. Total cross section for 14 MeV neutrons using aligned ^{165}Ho nuclei. *Phys. Rev. Lett.* 16(5): 194–197.

Melkanoff, M. A., Nodvik, J. S., Saxon, D. S., and Woods, R. D. 1957. Diffuse-surface optical model analysis of elastic scattering of 17 and 31.5 MeV Protons. *Phys. Rev.* 106: 793–801.

Miljanić, D., Furić, M., and Valković, V. 1968. About pick-up reactions on ^{11}B. *Nucl. Phys. A* 119: 379–388.

Miljanic, D., Paic, G., Antolkovic, B., and Tomas, P. 1967. (n, d) reactions on ^{14}N, ^{35}Cl, ^{39}K, ^{40}Ca and ^{75}As at 14.4 MeV. *Nucl. Phys. A* 106: 401–416.

Miljanić, D. 1970. PhD thesis, University of Zagreb, Zagreb, Croatia.

Miljanić, D., Valković, V., Rendić, D., and Furić, M. 1970a. Angular distribution of tritons from the ^{11}B(n,t)^{9}Be reaction at 14.4 MeV. *Nucl. Phys. A* 156: 193–198.

Miljanić, D., Furić, M., and Valković, V. 1970b. A study of (n,d) and (n,t) reactions on ^{7}Li. *Nucl. Phys. A* 148: 312–324.

Miljanić, D. and Valković, V. 1971. The (n,d) reactions on light nuclei. *Nucl. Phys. A* 176: 110–128.

Miller, F., Weulersse, C., Carriere, T., Guibbaud, N., Morand, S., and Gaillard, R. 2013. Investigation of 14 MeV neutron capabilities for SEU hardness evaluation. *IEEE Trans. Nucl. Sci.* 60(4): 2789–2796.

Nakamura, T., Ibe, R., Baba, M., Yahagi, Y., and Kameyama, H. 2008. *Terrestrial Neutron-Induced Soft Errors in Advanced Memory Devices*. World Scientific, Singapore.

Normand, E. and Dominik, L. 2010. Cross comparison guide for results of neutron SEE testing of microelectronics applicable to avionics, in *Proceedings of IEEE Radiation Effects Data Workshop*, pp. 50–57. Denver, Colorado, 19-23 July, 2010.

Normand, E. 2002. Single-event effects in avionics. *IEEE Trans. Nucl. Sci.* 43(2): 461–474.

Ohlsson, M., Dyreklev, P., Johansson, K., and Alfke, P. 1998. Neutron single event upsets in SRAM-based FPGAs. *Radiation Effects Data Workshop, 1998*. IEEE, pp. 177–180. Newport Beach, CA, July 24, 1998.

Paic, G., Slaus, I., and Tomas, P. 1964. O^{16}(n,d)N^{15}gs and O^{16}(n,p)N^{16} reactions at 14.4 MeV. *Phys. Lett.* 94: 147–149.

Paić, G., Šlaus, I., Valković, V., and Cerineo, M. 1964. Direct effects in (n,p) reactions, Gugenberger, P. (Ed.), *C.R. Congres Internat. de Physique Nucleaire*, Paris, France, Vol. II., pp. 934–955.

Palau, J.-M., Hubert, G., Coulie, K., Sagnes, B., Calvet, M. C. and Fourtine, S. 2001. Device simulation study of the SEU sensitivity of SRAM's to internal ion tracks generated by nuclear reactions. *IEEE Trans. Nucl. Sci.* 48: 225 –231.

Palau, J.-M., Wrobel, F., Castellani-Coulie, K., Calvet, M.-C., Dodd, P. E., and Sexton, F. W. 2002. Monte Carlo exploration of neutron-induced SEU-sensitive volumes in SRAMs. *IEEE Trans. Nucl. Sci.* 49(6): 3075–3081.

Pashchenko, A. B. 1990. Cross-sections for reaction induced by 14.5 MeV neutrons and by neutrons of Cf-252 and U-235 fission spectra. Report of the Institute of Physics and Power Engineering, No. 0236, Moscow, TSNII atominform.

Phillips, D. D., Davis, R. W., and Graves, E. R. 1952. Inelastic collision cross sections for 14 MeV neutrons. *Phys. Rev.* 88(3): 600–602.

Phillips, G. C., Griffy, T. A., and Biedenharn, L. C. 1960. The significance of the generalized density of states function for nuclear spectra. *Nucl. Phys.* 21: 327–339.

Pierre, C. St., Machwe, M. K., and Lorrain, P. 1959. Elastic scattering of 14-MeV neutrons by Al, S, Ti, and Co. *Phys. Rev.* 215(4): 999–1003.

Poss, H. L., Salant, E. O., Snow, G. A., and Yuan, L. C. L. 1952. Total cross sections for 14-MeV neutrons. *Phys. Rev.* 87(1): 11–20.

Qaim, S. M. 1974. Total (n,2n) cross sections and isomeric cross-section ratios at 14.7 MeV in the region of rare earths. *Nucl. Phys. A* 224: 319–330.

Rayburn, L. A. 1959. Differential elastic scattering of 14-Mev neutrons by Zn, Sn, Sb, Pb, and Bi. *Phys. Rev.* 116(6): 1571–1574.

Reimer, P. 2002. Fast neutron induced reactions leading to activation products: Selected cases relevant to development of low activation materials, transmutation and hazard assessment of nuclear wastes, PhD Thesis, University of Köln, Druckerei des Forschungszentrums Jülich, Germany.

Ribe, F. L. and Seagrave, J. D. 1954. The pickup reactions ^{10}B(n,d)^{9}Be for 14-Mev neutrons. *Phys. Rev.* 94: 934–940.

Rosen, L. and Stewart, L. 1957. Neutron emission probabilities from the interaction of 14-Mev neutrons with Be, Ta, and Bi. Phys. Rev. 107(3): 824–829.

Serber, R. 1947. Nuclear reactions at high energies. *Phys. Rev.* 72: 1114–1115.

Schectman, R. M. and Anderson, J. D. 1966. Inelastic scattering of 14 MeV neutrons. *Nucl. Phys.* 77(2): 241–253.

Scherrer, V. E., Theus, R. B., and Faust, W. R. 1953. Gamma rays from interaction of 14-MeV neutrons with various materials. *Phys. Rev.* 91(6): 1476–1478.

Shibata, K. 1999. Figures of cross sections in JENDL dosimetry file 99 (JENDL/D-99). JAERI Nuclear Data Center, Japan Atomic Energy Research Institute.

Shin, K., Nishimura, S., Yamamoto, J., and Takahashi, A. 1988. Measurements of back-scattered neutrons and secondary gamma rays from stainless steel and limestone concrete for 14 MeV source neutrons. *J. Nucl. Sci. Technol.* 25(4): 333–340.

Singletary, J. B. and Wood, D. E. 1959. Scattering of 14-Mev neutrons by carbon. *Phys. Rev.* 114(6): 1595–1599.

Šlaus, I., Tudorić, J., Valković, V., Rendić, D., Tomaš, P., and Cerineo, M. 1964. Few nucleon problem, in Gugenberger, P. (Ed.), *C.R. Congres Internat. de Physique Nucleaire*, Paris, France, Vol. II, pp. 244–246.

Spaargaren, D. and Jonker, C. C. 1971. Angular correlations in inelastic neutron scattering by carbon at 15.0 MeV. *Nucl. Phys. A* 161: 354–374.

Tel, E., Okuducu, Ş., Bölükdemir, M. H., and Tanir, G. 2008. Semi-empirical systematics of (n, 2n), (n, α) reactions cross sections at 14–15 MeV neutron energy. *Int. J. Mod. Phys. E* 17, 567. DOI: 10.1142/S0218301308009914.

Thompson, L. C. and J. R. Risser, J. R. 1954. Gamma rays from the inelastic scattering of 14-MeV neutrons in ^{12}C and ^{16}O. Phys. Rev. 94(4): 941–943.

Tsao, C. H., Silberberg, R., and Letaw, J. R. 1988. A comparison of neutron-induced SEU rates in Si and GaAs devices. *IEEE Trans. Nucl. Sci.* 35(6): 1634–1637.

Valković, V. 1964a. Triton spectrum from n+7Li reaction. *Nucl. Phys.* 60: 581–587.

Valković, V. 1964b. Angular distribution of tritons from the reaction ^{10}B+n at 14.4 MeV. *Nucl. Phys.* 54: 465–471.

Valković, V. and Tomaš, P. 1964. Study of (n,t) reactions on some light nuclei, in Gugenberger, P. (Ed.), *C.R. Congres Internat. de Physique Nucleaire*, Paris, France, Vol. II., p. 936–938.

Valković, V., Tomaš, P., Šlaus, I., and Cerineo, M. 1964. Study of (n,p) and (n,d) reactions on ^{10}B at 14.4 MeV. *Glasnik Mat. Fiz. i Astr.* 19: 284–300.

Valković, V., Paić, G., Šlaus, I., Tomaš, P., Cerineo, M., and Satchler, G. R. 1965. Reactions ^{48}Ti(n,d)^{47}Sc, ^{16}O(n,d)^{15}N, ^{10}B(n,d)^{9}Be, ^{6}Li(n,d)^{5}He at 14.4 MeV. *Phys. Rev.* 139: B331–B339.

Valković, V. and Tomaš, P. 1971. A position sensitive counter telescope for the study of nuclear reactions induced by 14 MeV neutrons. *Nucl. Instr. Meth.* 92: 559–562.

Valković, V., Šlaus, I., Tomaš, P., and Cerineo, M. 1967. The reactions ^{10}B(n,t)α,α, ^{7}Li(n,t)α,n and ^{6}Li(n,d)α,n at 14.4 MeV. *Nucl. Phys. A* 98: 305–322.

Valković, V., Miljanić, D., Tomaš, P., Antolković, B., and Furić, M. 1969. Neutron-charged particle coincidence measurements from 14.4 MeV induced reactions. *Nucl. Instr. Meth.* 76: 29–34.

Valković, V., Furić, M., Miljanić, D., and Tomaš, P. 1970. Neutron–proton coincidence measurement from the neutron induced break-up of deuteron. *Phys. Rev. C* 1: 1221–1225.

Valković, V., Furić, M., Miljanić Đ., and Tomaš, P. 1970. Neutron-induced non-elastic processes on light nuclei, in *Proceedings of International Conference on Birmingham* (*North-Holland*), pp. 436–445.

Veljukov, G. E. and Prokofjev, A. T. 1962. Scattering of neutron on neutron (in Russian) *Izv. Akad. Nauk SSSR, Ser. Fiz.* 26: 1113–1116.

Wahl, A. C. 1955. Fission of U by 14-MeV neutrons: Nuclear charge distribution and yield fine structure. *Phys. Rev.* 99(3): 730–739.

Watanabe, Y. 2007. Nuclear data relevant to single event upsets in semiconductor memories induced by cosmic-ray neutrons and protons, in *Proceeding of 2006 Symposium on Nuclear Data, Jan. 25–26, 2007*, SND2006-III.03-1-SND2006-III.03–7.

Wong, D. Y. and Noyes, H. P. 1962. Electrostatic corrections to nucleon–nucleon dispersion relations. *Phys. Rev.* 126: 1866–1872.

Yahagi Y., Ibe R., Saito Y. et al. 2002. Final Report of IEEE 2002 International Integrated Reliability Workshop, 143.

Chapter 6

Fast Neutron Activation Analysis

An Analytical Method

6.1 FNAA: FUNDAMENTALS AND APPLICATIONS

Neutron activation analysis (NAA) and prompt gamma neutron activation analysis (PGNAA) are proven methods of elemental analyses traditionally performed at nuclear research reactors. With NAA, samples are irradiated with highly thermalized neutrons inside the reactor and then taken to a low background counting area where the induced radioactive decay products are analyzed. The intensities of the obtained specific gamma rays provide information about the number of atoms in the sample. Hence, the information on its chemical composition can be extracted from the measured gamma ray spectrum. NAA can be used to measure the concentrations of many elements, often with very high (<ng/g) sensitivity, but it is not sensitive for about one third of the elements where no radioactive product can be produced. PGNAA utilizes a neutron beam, external to the reactor, to irradiate a sample, which then emits prompt γ-rays with energies and cross sections characteristic of all elements. With PGNAA, sensitivities up to <0.1 mg of any element, except helium, can be achieved; see some of the numerous textbooks on the subject (Koch 1960; Bowen and Gibbons 1963; Taylor 1964; Lernihan and Thomson 1965; Kruger 1971; De Soete et al. 1972; Nargolwalla and Przybylowicz 1973). Both NAA and PGNAA offer the distinct advantage that many elements in a small sample can be simultaneously and accurately analyzed. Traditionally, NAA and PGNAA have been limited to research reactor laboratories that are not always conveniently accessible to scientists around the world, although, there are currently about 380 research reactors worldwide.

Activation analysis with 14 MeV neutron generators using short-lived radionuclides is described in many papers, see, for example, Salma and Zemplen-Papp (1993) and references therein. Application of the 14-MeV neutron activation analysis is usually restricted to analytical problems where the detection limit of approximately 0.1–10 mg for elements of interest is low enough. It is the relatively low flux of neutron generators and the medium activation cross section for nuclear reactions induced by fast neutrons that does not make it possible to improve detection limits and sensitivities. Various experimental arrangements and methodological developments have been devised to enable more accurate determinations.

Higher sensitivity (i.e., more counts) can be achieved by utilizing the special advantage of short-lived radionuclides. Theoretically, they can be activated to saturation in every repeated irradiation, while the background activity is increasing slowly. The method implies a purely instrumental character of analysis. The short-lived radionuclides can only find wide application after the introduction of high resolution semiconductor detectors for γ-ray spectrometry. These detectors are able to resolve complex spectra without the need for chemical separations.

To minimize the delay between irradiation and measurement, samples have to be transported quickly from the irradiation position to the detector. Since the radioactivity of a short-lived radionuclide soon decays, prolonged counting will not produce a larger net peak area. In cyclic, pseudocyclic, or cumulative activation analysis, the sample is reirradiated and measured several times and the spectra from these successive measurements are added together to give one final spectrum. Since the background will increase through successive activations, the signal-to-background ratio will become progressively smaller after each irradiation. Care must, therefore, be taken to choose the number of cycles and the cycle time.

Let us mention some of the numerous reports. For example, in the investigations reported by Salma and Zemplen-Papp (1993) 12 elements producing short-lived radionuclides by 14 MeV neutron irradiation were selected on the basis of their nuclear characteristics (F, Na, Se, Br, Y, Ce, Ti, Zr, Hf, Au, Pb, Ba). The effect of the fast transport system is significant for some elements and, according to the theoretical expectations, the relative improvement of sensitivity covers 5%–50%, depending on the half-life of the radionuclides. The cyclic activation method is absolutely necessary for successfully investigating short-lived radionuclides and for accumulating sufficient counts for good statistics. The effect of both the fast pneumatic transport system and cyclic activation was very efficient. For the radionuclides with a half-life above 1 s, the improvement of sensitivity is about 60%–90%. Sensitivity for the elements determined via the radionuclides with a very short half-life (subsecond range) can be enhanced by about one order of magnitude using the combined technique of fast transport and cyclic activation. Cyclic activation method for Pb and Zr utilizing an irradiation, decay, measuring, and recycling time of 2, 0.16, 2, and 0.16 s, respectively, and a cycle number of 10, improves the sensitivity by about 7 times, while the same cycle schedule repeated 30 times enhances the sensitivity for both elements by a factor of 20.

PGNAA performed by neutron sources other than the reactor has been used for prospecting of minerals, for process control in the cement industry and in coal-fired power plants, for the detection of explosives at airports and for demining of land, for in vivo measurement of various elements in animal and human bodies, and for the determination of hydrogen in metals and various elements in biological and environmental materials. The equipment needed for PGNAA is virtually identical to conventional neutron activation analysis. Thus, all research facilities where Instrumental NAA is carried out could potentially install PGNAA if a beam line from a research reactor or any other neutron source is made available for the irradiation of samples. New generations of neutron generators are available and offer increased possibilities for any NAA applications especially toward in situ measurement.

The renewal of interest in neutron activation analysis is a consequence of increased terrorism threat to the civilization. The vulnerability of societies to terrorist attacks results in part from the proliferation of chemical, biological, and nuclear weapons of mass destruction but is also consequence of the highly efficient and interconnected systems that we rely on for key services such as transportation, information, energy, and health care.

Nuclear techniques have been applied in the detection of hidden explosives for a number of years. Basically, they work on the principle that the nuclei of the chemical elements in the investigated material can be bombarded by penetrating nuclear radiation (mainly neutrons). As results of the bombardment, nuclear reactions occur and a variety of nuclear particles and gamma and x-ray radiation is emitted, specific for each element in the bombarded material.

The problem of material (explosive, drugs, chemicals, etc.) identification can be reduced to the problem of measuring elemental concentrations and/or ratios. Nuclear reactions induced by neutrons can be used for the detection of chemical elements, their concentrations, and concentration ratios or multielemental maps, within the explosives.

Use of slow/thermal/neutrons is convenient for elements H, N, Cl, Na, Al, Fe, and Pb. The radioactive capture of slow neutrons results in the characteristic gamma rays as follows: H (2.2 MeV), N (10.8 MeV), Cl (1.95 and 6.11 MeV), Na (2.98 and 7.39 MeV), Al (2.96 and 7.72 MeV), Fe (7.63 and 7.64 MeV), and Pb (6.73 and 7.36 MeV).

On the other hand, the use of fast (14 MeV) neutrons is convenient for the detection of the presence of elements: C (4.4391 MeV), N (2.3129 and 5.1059 MeV), O (6.1304 MeV), Al (2.211 and 2.981 MeV), and Fe (1.2383, 1.8107, and 2.5984 MeV) (see McKlveen 1981; Csikai 1987; IAEA 2012).

Neutron scanning technology offers capabilities far beyond those of conventional inspection systems. This highly sophisticated equipment could be deployed as part of a country-wide system of deterrence. The unique automatic, material-specific detection of terrorist threats can significantly increase the security at ports, border-crossing stations, airports, and even within the domestic transportation infrastructure of potential urban targets as well as protecting armed forces and infrastructure wherever they are located.

There are several distinguished methodologies of using neutrons as an analytical tool best summarized in an article by Gozani (1997). They are listed as

TNA: Thermal neutron analysis
FNA: Fast neutron analysis
PFNA: Pulsed fast (nanosecond) neutron analysis
PFNTS: Pulsed fast neutron transmission spectroscopy
API: Associated particle imaging
PFTNA: Pulsed fast and thermal neutron analysis
FNSA: Fast neutron scattering analysis

Table 6.1 shows selected features of these techniques for the nonintrusive interrogation of bulk samples. Neutron-based technique is an excellent choice to rapidly determine elemental content of the sample in situ in nonintrusive manner. It is a great choice for in situ applications that involve samples that are hard to reach or unsafe to handle and that require the analysis to be performed rapidly, in real time.

It is of interest to describe the components of pulse neutron–based analysis systems. A pulse neutron–based material analysis system consists of a neutron source, gamma and particle radiation detector(s), a shadow radiation shielding to cover detectors from direct source neutrons, and associated hardware and software for system control, data acquisition, and processing. Figure 6.1 shows the scheme of a typical system. The system operates as follows. Emitted by a source, neutrons induce nuclear reactions in the irradiated object and excite nuclei. Excited nuclei emit photons due to various deexcitation processes that are measured by a gamma ray detector. The gamma ray spectrum is analyzed providing information on the chemical composition of the irradiated sample.

There are several reported neutron sensors based on gamma ray spectroscopy described in the literature:

1. Pulsed fast neutron–time of flight; PFNA-TOF with 8.2 MeV pulsed neutrons from an accelerator is described in the report by Office of National Drug Control Policy, Washington, DC, September 1996, see also report by Hurwitz et al. (1992).

TABLE 6.1 SELECTED FEATURES OF THE MAIN NUCLEAR PHYSICS-BASED TECHNIQUES FOR THE NONINTRUSIVE INTERROGATION OF BULK SAMPLES

Technique	Radiation Source	Probing Radiation	Main Reaction	Detected Radiation	Signatures
TNA	^{252}Cf, d–D or d–T STNG	Thermalized neutrons	(n,γ)	Prompt γ-rays from neutron capture	H, N, Cl (others)
FNA	d–D or d–T STNG	Fast neutrons	(n,n′γ)	γ-rays from inelastic neutron scattering	C, O, Cl (N, others)
PFNA	ns-pulsed accelerator	Fast neutrons	(n,n′γ)	γ-rays from inelastic neutron scattering	C, O, Cl (N, others)
PFNTS	ns-pulsed accelerator	White spectrum of fast neutrons	All available	Source neutrons that are transmitted	H, C, N, O (others)
API	Associated particle d–T STNG	14 MeV neutrons with associated α-particles	(n,n′γ)	γ-rays in coincidence with α-particles	C, N, O (others)
PFTNA	µs-pulsed d–T STNG	Fast neutrons during pulse, and then thermalized	(n,n′γ) + (n,γ)	γ-rays from inelastic neutron scattering, capture and activation	H, C, N, O (others)
FNSA	ns-pulsed or DC accel., STNG	Monoenergetic fast neutrons	(n,n) + (n,n′)	Elastically and inelastically scattered neutrons	H, C, N, O (others)

Source: Adapted from Gozani, T. et al., Nuclear assay of coal, Volumes 1–8, EPRI Report FP989. RP983-1, 1979.
Note: STNG, sealed tube neutron generator.

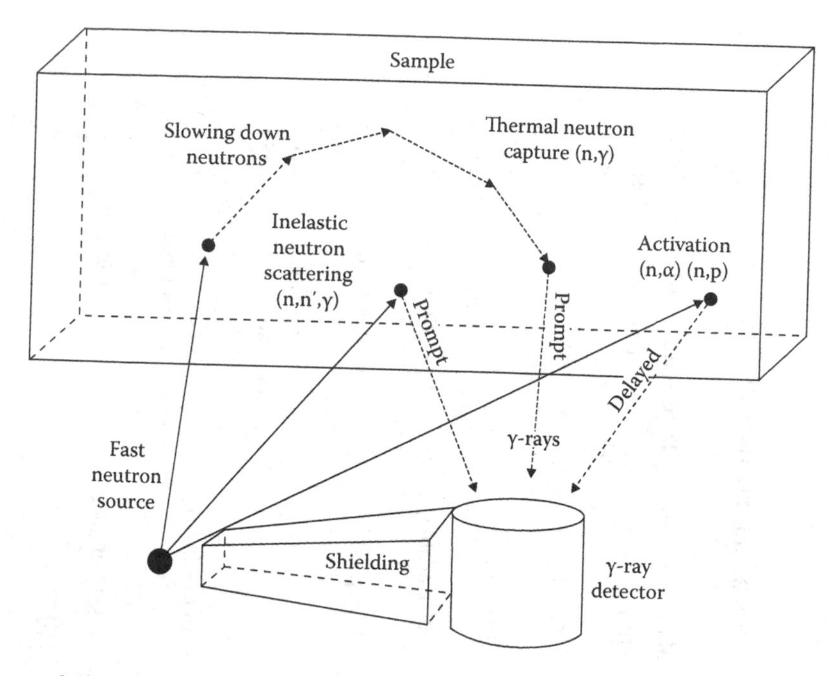

Figure 6.1 Pulse neutron–based elemental analysis scheme.

2. Coded aperture fast neutron analysis (CAFNA) as reported by Lamza (1997).
3. A system called pulsed elemental analysis with neutrons (PELAN) developed by Vourvopoulos et al. (1997). PELAN is a transportable neutron-based UXO identification system using pulsed beam of fast neutrons.

Material analysis using the detection of characteristic gamma rays induced by neutrons is discussed in detail in the paper by Barzilov et al. (2012). This technique is an excellent choice to rapidly determine elemental content of the sample in situ in a nonintrusive manner. It is also a great choice for in situ applications that involve samples that are hard to reach or unsafe to handle, and that require the analysis to be performed rapidly, in real time.

Accelerator-based neutron sources, neutron generators, can also provide the electronic control of neutron emission including its time structure. The pulse mode of neutron generator allows the use of coincidence methods to segregate prompt and delayed gamma ray signatures emitted from neutron-induced nuclear reactions. The kinematics of fusion reactions allows "tagging" of outgoing neutrons using the associated particles; this is discussed in Chapter 7 in details.

The pulse neutron systems are used in industry for analysis of coal (Dep et al. 1998; Sowerby 2009), cement (Womble et al. 2005), metal alloys (James and Fuerst 2000), in geological and soil analysis (Haskin et al. 1968; Bibby and Sellschop 1974; Wielopolski et al. 2008), and oil well logging (Czubek 1972; Csikai et al. 1982; see also Chapter 8). Security applications of neutron-based systems are for chemical and explosive threats detection (Vourvopoulos and Womble 2001; Aleksandrov et al. 2005; Lanza 2007), including the search for threats in cargo containers (Barzilov and Womble 2003; see also Chapter 7) and vehicles (Reber et al. 2005; Koltick et al. 2007), humanitarian demining, and confirmation of unexploded ordinance (Womble et al. 2002; Holslin et al. 2006). Such technologies are also considered in astrochemistry applications for in situ analysis of planetary samples (Parsons et al. 2011).

The 14 MeV NAA is capable of determining many light elements that cannot be analyzed using classical NAA. One of the most important applications of this technique is the determination of oxygen using the reaction $^{16}O(n,p)^{16}N$. ^{16}N decays with a half-life of 7.13 s and emits exceptionally high energy gamma photons with the energies of 6.13 and 7.12 MeV (Van Grieken and Hoste 1972). The following elements are routinely determined using 14 MeV NAA: F, Mg, Al, Si, Cu, Fe, P, and Zn. The most important reactions are listed in Table 6.2. Sensitivity data are taken from Csikai (1987) in the unit of decay per second per gram, that is, it gives the activity per unit target mass after 1 h of irradiation assuming a fast neutron flux of $10^9/cm^2/s$. The detection limits are calculated to be 0.1–1 mg for the given elements, when using a generator with an output of 10^{11} s^{-1}. As it can be seen from Table 6.2, there are interference problems even in this short list of elements in case the spectroscopic analysis is performed with scintillator detectors. Close-lying energies (e.g., Br and Ag from the given cases) can be resolved using HPGe detectors (see also Wood 1971; Krivan and Krivan 1976; Williams 1981).

Applications of fast neutron activation analysis (FNAA) technique to element determination at trace level have been described by Ene and Frontasyeva (2013). They present the results for analysis of samples from the technological process of an integrated iron and steel plant and variety of geological materials.

Using fast neutrons obtained from industrial neutron generators in geochemical research and mineral exploration, a sensitive determination of gold in auriferous sands, rocks, and concentrate (Nat et al. 2004; Ene 2011) and of alkali metals (Nat and Ene 2006) in iron ores used in metallurgical industry is possible. In geochemical exploration, where a large number of auriferous samples have to be analyzed at a low cost, the use of the neutron generators, short irradiation, and NaI(Tl) spectrometry is justified for gold analysis at trace level (Nat et al. 2004; Ene 2011); care must be taken in this case to small corrections for the interferences due to Rb, V, and Ti. In the case of $^{197}Au(n,n')^{197m}Au$ reaction, if available, a Ge(Li) detector can be used for the detection of 279 keV gamma rays of the gold short-lived radionuclide without any interference.

TABLE 6.2 MOST IMPORTANT REACTIONS USED FOR ANALYSIS BY 14 MEV NEUTRONS

Element	Reaction	Half-Life	Gamma Energy (MeV)	Sensitivity 10^6 decay/s/g
B	$^{11}B(n,p)^{11}Be$	13.8 s	2.125	15
O	$^{16}O(n,p)^{16}N$	7.13 s	6.13, 7.12	1.3
N	$^{14}N(n,2n)^{13}N$	9.97 min	0.511 (annih.)	15
Al	$^{27}Al(n,p)^{27}Mg$	9.46 min	0.844	1.6
Si	$^{28}Si(n,p)^{28}Al$	2.25 min	1.78	4.5
P	$^{31}P(n,\alpha)^{28}Al$	2.25 min	1.78	2.3
Cu	$^{63}Cu(n,2n)^{62}Cu$	9.74 min	0.511 (annih.)	3.6
Br	$^{79}Br(n,2n)^{78}Br$	6.5 min	0.614	3.7
Ag	$^{109}Ag(n,2n)^{108}Ba$ ^{137}Ba	2.37 min	0.633	2.3
Ba	$^{138}Ba(n,2n)^{137}Ba$	2.55 min	0.662	3.7

Source: After Csikai, J., *Handbook of Fast Neutron Generators*, CRC Press, Inc., Boca Raton, FL, 1987; Ehmann, W.D. and Vance, D.E., *Radiochemistry and Nuclear Methods of Analysis*, Wiley, New York, 1991.

When only a few number of samples are to be evaluated or gold must be determined more accurate in a sample, the long irradiation and Ge(Li) spectrometry of the 355.7 keV, most intense γ-ray of ^{196}Au from the reaction $^{197}Au(n,2n)^{196}Au$ is clearly the method of choice, taking into account only the small contribution of Se to the gold concentration. The contribution of the nuclear interfering elements, Hg and Pt, to the concentration of gold in the samples has been calculated and from the obtained data it can be deduced that the nuclear reactions $^{197}Au(n,2n)^{196}Au$, $^{197}Au(n,2n)^{196m}Au$, and $^{197}Au(n,n')^{197m}Au$ can be used for gold determination, with minimal errors. In conclusion, two methods for rapid determination of gold in auriferous geological materials in the range 20–2500 ppm were proposed by Nat et al. (2004) using optimum calculated experimental times, so that the systematic errors of analysis due to the gold-accompanying elements should be considerably diminished: a method using short irradiation (25 s) and a NaI(Tl) detector for measuring the induced gamma radioactivity in the samples and a method using long irradiation (3000 s) and a Ge(Li) detector. From this work it results that by applying INAA technique, a very good overall picture of the elemental composition of a multielemental sample may be obtained. For light element determinution, INAA could be coupled with other nuclear and ion beam analysis techniques; see, for example, Ene (2004).

Let us describe an interesting problem. Reports of the trafficking of red mercury (claimed to have the composition $Hg_2Sb_2O_7$) have been circulating for many years. Red mercury was touted as a mediator in nuclear weapons design, particularly as an essential ingredient in pure-fusion weapons, a view expressed by Barnaby (1994), a former director of the Stockholm International Peace Research Institute. What is known about red mercury is that it was the Russian code name for the production of ^{6}LiD—a legitimate component of thermonuclear weapons, but not some mystical or magical ingredient for other purposes. In recent years, red mercury has been widely discredited, and the "market" for it appears to be diminishing. Some authors claim that "red mercury" was a label on tritium containers sent from Israel to South Africa.

Here we shall describe one such event. A metallic cylinder with rounded ends has been brought to the laboratory with the assumption that it contains *red mercury*. The object had the following dimensions: length: 54.3 mm, diameter of the base: 19.8 mm, weight: 0.1106 kg, and the average density of the object: 6 615 kg/m^3. By shaking the ampoule one could conclude that there is a liquid inside, but from the average density of the object one could conclude that the liquid has high density. Since the unknown material was incapsulated inside the metallic cylinder the analysis has been done by using two analytical methods: activation analysis with 14 MeV neutrons and x-ray fluorescence (Obhodas et al. 2007). By using these two methods one can separate the elemental composition of the ampoule from that of its content.

For the FNAA analysis the sample was irradiated for 40 min with 14 MeV neutrons generated by SODERN GENIE16C sealed tube. The neutron beam intensity was ~7×10^7 n/s in 4π. The gamma rays were detected by the Ge(Li) detector, 5.5 cm thick and 5 cm in diameter. The activation analysis with 14.1 MeV neutrons has shown that the container and its content were characterized by the following chemical elements: Hg, Fe, Cr, and Ni, as it is seen in Figure 6.2. The last three elements are contained in the metallic ampoule while Hg is a component of the unknown material inside it. Antimony as a hypothetical component of the red mercury (hypothetical chemical formula of red mercury is $Hg_2Sb_2O_7$) has not been observed.

By using the EDXRF analysis it was shown that the elements Fe, Cr, and Ni were constituents of the capsule. Therefore, it was concluded that these three elements were present in the capsule only, while the content of the unknown material inside was Hg. Antimony as a hypothetical component of red mercury has not been detected. The analytical results are in an agreement with the reported results of another laboratory analyzing substances claimed to be RM (Grant et al. 1998). Their experience with reported RM samples has been that those specimens were nothing more but the elemental Hg, HgO, or HgI_2.

Red mud characterization using nuclear analytical techniques has been described by Obhodas et al. (2012). Red mud is a toxic waste left as a byproduct in aluminum production, Bayer process. Since it contains significant concentrations of other chemical elements interesting for industry, including REE, it

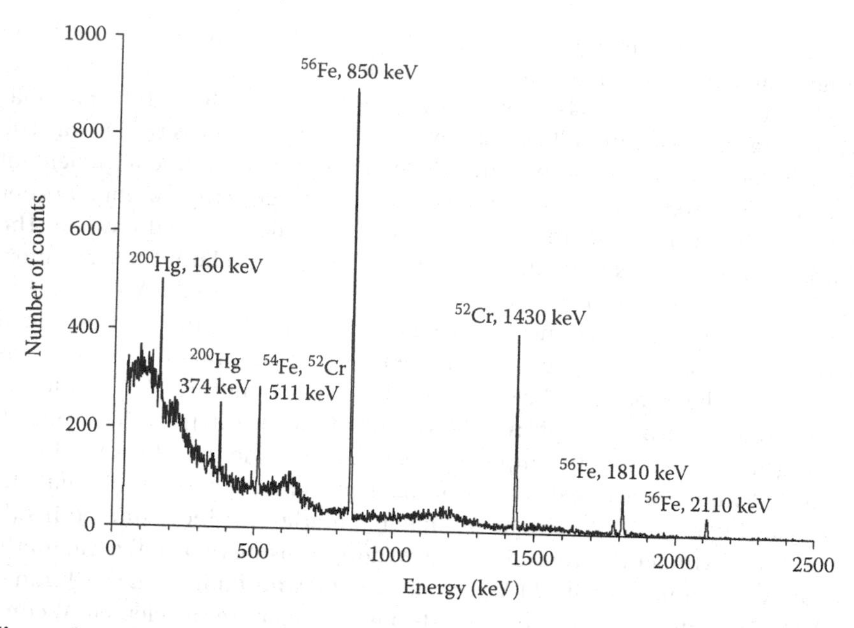

Figure 6.2 Gamma ray spectrum after bombardment with 14 MeV neutrons.

is also potential secondary ore source. The authors have used energy dispersive x-ray fluorescence analysis (both tube and radioactive source excitation), FNAA, and passive gamma spectrometry to identify a number of elements present in the red mud as well as to determine their concentration levels.

FNAA was used for Si and Al analysis. Samples of 50 g were placed in plastic holders and were irradiated 1 h by 14 MeV neutron beam of 10^7 n/s in 4π intensity, produced by the portable tube generator Genie 16C, Sodern. Gamma rays were collected with a NaI(Tl) gamma detector (3 in. × 3 in.), Ortec 855 dual spectroscopic amplifier, and ADLINK Technology NuDAQ PCI-9812710 analog input card. ROOT program package was used for spectrum analysis. Nuclear reactions data (gamma energy lines, decay half-life calculated and measured) used for Al and Si analysis are shown in Table 6.3. Spectrum of Si and Al in RM obtained by FNAA is shown in Figure 6.3. Inovision 451B-DE-S was used for passive gamma spectroscopy measurements.

Work on chemical composition identification of interrogated material by using fast neutrons has been done in several laboratories; see, for example, Sudac et al. (2004). The goal is determination of empirical nonhydrogen part of chemical formula of explosives, drugs, threat materials: $C_xN_yO_z$ from fast neutron induced gamma emission.

A pulsed fast/thermal neutron system for the detection of hidden explosives is described in Vourvopoulos and Schultz (1993). It is known that for all known

TABLE 6.3 NUCLEAR DATA FOR 14 MEV NEUTRONS REACTIONS USED FOR ANALYSIS OF AL AND SI IN RM

Nuclear Reaction	Gamma Energy (MeV)	Half-Life (s)	Half-Life Measured (s)
$^{27}Al(n,p)^{27}Mg$	0.84	567.5	583 ± 3
$^{27}Al(n,p)^{27}Mg$	1.01	567.5	588 ± 7
$^{28}Si(n,p)^{28}Al$	1.78	134.5	130 ± 1
$^{29}Si(n,p)^{29}Al$	1.27	393.6	301 ± 52

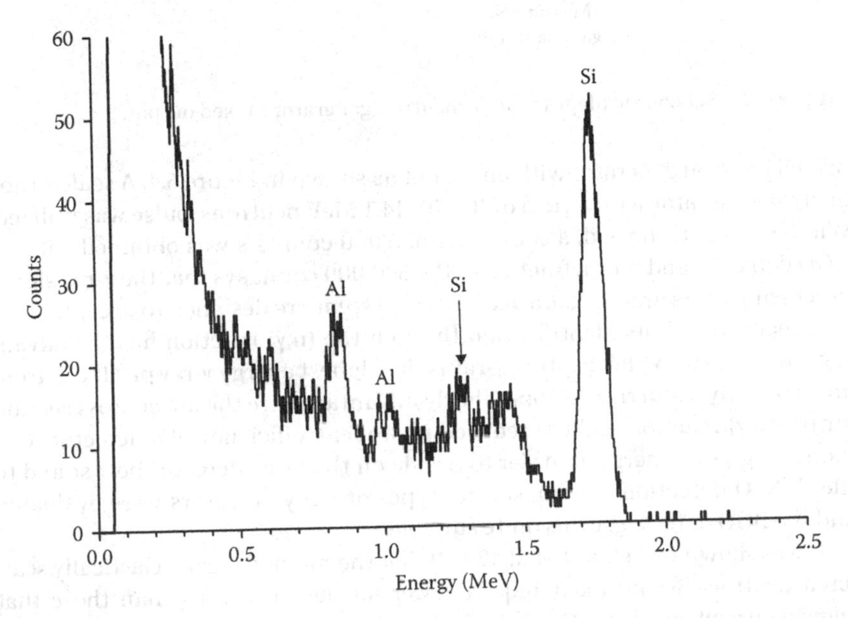

Figure 6.3 Spectrum of RM obtained by the FNAA. Ordinate presents number of counts.

explosives a strong correlation exists between their O and N concentrations. In addition to H, which is not seen readily with 14 MeV neutrons, C is another element that is common to all explosives. A detection system that could quantitatively determine the C, N, O content of an interrogated material could assist in identifying a hidden explosive. Among the various options for the C, N, O identification, a pulsed fast/thermal neutron technique was chosen. The three elements C, N, and O are identified through the characteristic γ-rays emitted from the interaction of neutrons with the corresponding nuclei. C and O are identified through the (n,n′γ) reaction, while N is identified through the (n,γ) reaction. The combination of fast and thermal neutrons is provided from a

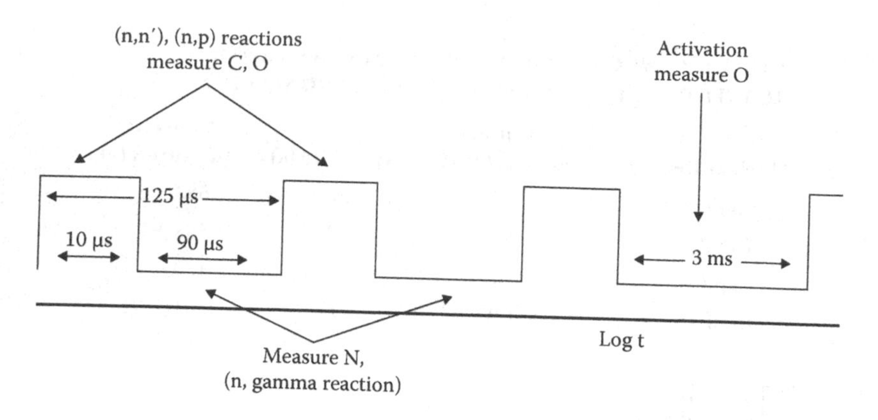

Figure 6.4 Schematic diagram of the neutron generator pulsed output.

pulsed neutron generator with an output as shown in Figure 6.4. A sealed-tube neutron generator with a yield of 10^3–10^4 14.1 MeV neutrons/pulse was utilized. With such neutron yield, a count rate of 5,000 counts/s was obtained with an HGO detector, and only a fraction of the 500,000 count sys that the γ-ray signal processing units and the data acquisition system are designed to accept.

Concerning N, its identification through the (n,γ) reaction has the advantage that the 10.83 MeV capture γ-ray is the highest energy γ-ray produced from any naturally occurring isotope. The disadvantages are the lower cross section for the (n,γ) reaction and the reduced photopeak efficiency of a detector with increasing γ-ray energy. In order to decide on the γ-ray detector best suited to the C, N, O detection system, several types of γ-ray detectors were evaluated and the BGO detector seems quite suitable.

It was shown by Lehnert et al. (2007) that the methods using elastically scattered neutrons would likely require fewer incident neutrons than those that depend on neutron absorption or inelastic scatter due to an increased interaction probability. This reduction would dramatically reduce dose rates and induced target radioactivity. In their study, a series of Monte Carlo (MCNP5) simulations were completed in order to investigate the use of elastically scattered neutrons in explosives detection, specifically explosives hidden inside standard-sized cargo containers. Fast neutrons were simulated with energies of 2.4, 14.1 MeV, or equivalent to that of a ^{252}Cf neutron source. Neutrons and gamma rays were tallied in seven idealized detectors arranged around a cargo container. Initial simulations characterized neutron-scattering behavior in several different materials including water, oil, steel, air, soil, and the explosive, RDX ($C_2H_6N_6O_6$). Further simulations dealt with a theoretical scenario in which a 1.87 m diameter solid sphere of RDX was placed inside a cargo container, shielded with water, oil, or steel and probed using 14.1 neutrons. Next, the presence of the ground was incorporated into this scenario. Several strategies were then used to determine possible flags that could determine the presence of the RDX. One strategy consisted of searching for changes in the backscattered neutron tallies due to changes in hydrogen concentration when RDX was present. A second strategy involved investigating the changes in scattered neutron flux at specific energies as a function of angle. Other potential flags were found by comparing the ratios of elastic and inelastic peaks in the neutron spectra at seven different angles. Preliminary analysis using this method showed several promising possibilities for detecting hidden explosives such as an almost 20-fold decrease in the ratio of 9.8 and 9.2 MeV neutrons at 60° when RDX is present in an isolated oil-filled cargo container.

This approach has been further elaborated in Lehnert and Kearfott (2011) and Lehnert (2012). In the work by Lehnert and Kearfott (2011), a method that involves the usage of flags relating the secondary radiation to the presence of explosives is discussed. These flags would consist of mathematical relationships among scattered neutrons at specific energies and angles, possibly combined with data concerning emitted photons from neutron inelastic scatter off light elements. This study involves the preliminary identification of potential flags by simulating a large RDX ($C_3H_6N_6O_6$) mass in an idealized cargo container using Monte Carlo radiation transport simulations. An analysis of the effect of changing hidden explosive mass and position on the detectability of the explosive was completed. Preliminary analysis revealed several promising algorithmic flags for the new method, although more realistic detection scenarios and experiment are needed to fully assess the approach's viability.

In the work by Lehnert (2012), an algorithm that uses flags has been further developed, calculated from specific measurements of the reflected neutrons and photons produced during active neutron interrogation, to discern explosives hidden in cargo containers. Steps in algorithm development included Monte Carlo simulations for scatter characterization, identification of flags in idealized scenarios, refinement of flags in realistic scenarios, combining the flags into a detection algorithm, and evaluation of the algorithm and associated detection system. Simulations compared favorably with small-scale neutron-scatter measurements using the explosives surrogate, melamine. The detection algorithm included corrections for different types of cargo contents and cargo inhomogeneity, surrounding environment, and realistic neutron sources and radiation detectors. The proposed algorithm has two variations, one of which can be easily implemented with today's technology. The proposed scanning system utilizes a shielded 14.1 MeV neutron generator, 11 large liquid scintillator neutron detectors, and several inorganic scintillators for photon spectroscopy. Dose estimates for this system fall well within acceptable levels for both operators and smuggled persons. Algorithm performance has been quantified with various explosive sizes and positions, as well as heterogeneous cargo configurations, with typical minimum detectable amounts not exceeding 200 kg.

While the commonly used military explosives are characterized by the presence of only four chemical elements (C, H, N, O), chemical warfare agents usually have in addition one or more chemical elements (P, S, Cl, F, As, or Br); see Table 6.4. Numerous nuclear reactions may be used for their detection. Some useful prompt gamma lines of CW detection are shown in Table 6.5.

The full size experiment performed by Bach et al. (1993) has shown the following first results for cyclic activation:

- CW shells are detectable within 5 min, from the identification of chlorine-based agents (yperites and lewisite) and phosphorus-based agents (tabun, sarin, VX, etc.).
- W contents are identified within 30 min, with the exception of sulfur-based agents.

TABLE 6.4 ELEMENTARY COMPOSITION OF SOME CW AGENTS AND EXPLOSIVES

Materials of Interest	Molar Mass (g)	Elementary Composition (Atom Number/Molecule)								
		Cl	P	As	S	F	O	N	C	H
CW agents										
Mustard gas (HD) (Yperite sulfur)	159	2			1				4	8
Nitrogen Yperite (HN3)	203	3						1	6	12
Lewisite (L1)	207	3		1					2	2
Tabun (GA)	162		1				2	2	5	11
Sarin (GB)	140		1			1	2		4	10
Amiton (VX)	267		1		1		2	1	11	26
Explosives										
Pentrite	316						12	4	5	8
Hexogen	296						8	8	4	8
Ammonium nitrate	80						3	2		4

TABLE 6.5 PROMPT GAMMA LINES FROM ELEMENTS FOUND IN CW AFTER 14 MeV BOMBARDMENT

Elements	Cl	P	F	S	As	N
γ-lines/keV	1763, 2645	1266, 2230	197	2230	279	5106, 2313

Another approach described in Sudac et al. (2004) is based on using the value of C/O ratio; there was no need for the questionable determination of nitrogen concentration. However, it was important that the experimental parameters are chosen in such a way so that the value of C/O is representative of the material inside the interrogated object. The measured γ-spectra of elements expected to be signature of CW presence are discussed in more detail in Chapter 7.

A pulsed neutron generator–based system for measuring fissile material mass in waste containers, presenting a direct control of ^{235}U and/or ^{239}Pu has been described in report by Batyaev et al. (2013).

The status of the application of activation analysis to meteoritic and lunar studies has been discussed by Showalter and Schmitt (1972) and Laul (1979). The aim of the investigations was to determine the abundance and distribution of chemical elements in meteorites and lunar materials. Chemical methods are complemented well by 14-MeV NAA, especially in cases where rapid, nondestructive, and subsequent investigation of small quantities of meteorites or lunar materials is necessary.

The major chemical elements in earth or earth-related materials (O, Na, Mg, Al, Si, K, Ca, Ti, Fe, Sr, Ni, Y, Zr, Nb, and Ce) except Ca can be determined advantageously by fast neutron activation. Ehmann and Morgan (1970a) have developed a method for the rapid, precise, and accurate determination of O, Al, Si, and Fe in small lunar samples. Because of the small sample size, they used a single transfer system with sequential irradiation and counting to increase the sensitivity and accuracy. The gamma ray from ^{16}N and the neutron flux during the irradiation were recorded by a multiscaler. Results from the determination of O, Al, Si, and Fe are in good agreement with those obtained by different methods (atomic absorption, classic gravimetry, spectrophotometry, and x-ray fluorescence); this proves the applicability of simple accelerators for the analysis of small samples.

From an analysis of Apollo 11 material, Ehmann and Morgan (1970b) concluded that the crystalline rocks fall into two distinct chemical groups based on their Al abundance: the breccias and fines are enriched in O, Si, and Al, as compared to the crystalline rocks; Si and O are strongly correlated in the Apollo 11 rocks. The O and Si abundances for Apollo 12 rocks are generally higher than for the corresponding Apollo 11 rocks. The O–Si correlation for Apollo 11 and 12 crystalline rocks is similar, and both differ significantly from that observed in terrestrial igneous rocks. For additional reading on NAA applications to lunar, meteorite and other space material see (Wing 1964; Vogt and Ehmann 1965; Morgan and Ehmann 1970; Ehmann and Morgan 1971; Janghorbani et al. 1973; Hamrin et al. 1979; Laul and Wogman 1981).

6.2 QA/QC MEASURES, REFERENCE MATERIALS

The importance of achieving laboratory accreditation under ISO 17025 (ISO 1999), which is the implementation of quality assurance (QA) in a laboratory, is growing in time. Quality assurance is defined (Taylor 1981, 1987; Taylor and Oppermann 1986; ISO 1987) as a system of activities whose purpose is to provide the producer or user of a product or service with the assurance that it meets defined standards of quality with a stated level of confidence. QA deals with the approach of the whole organization and it spreads over the details of personnel, documents, and facilities involved in the service. The small part of the QA that deals with the process itself is called quality control (QC). The major aspects of QC are related with the calibration, reference materials or standards, control samples, control charts, and statistical analysis. QC in gamma-ray spectrometry requires monitoring of the spectrometer, the methodology, and the consistency of the results. Calibration, a very important part of QC, provides a way to convert instrumental signals from a less usable to more usable form. In gamma-ray spectrometry, mostly three types of calibrations are performed, namely, energy-channel, energy-resolution or full width half maximum (FWHM), and energy-efficiency calibration. Reference materials, standards, and control samples are used to check the validity of a method (Guzzi et al. 1978). Different parameters in a method and the performance of the method itself are monitored by control charts developed by Shewhart (1931). Control charts provide a mean to investigate whether a method is under statistical control or not. They indicate trends, sudden shifts, or changes in a process. They also provide variability in the method. Control charts are characterized by a centerline, which is the desired value of a process; confidence limits are then calculated about the centerline as warning and action limits on the process.

The QC also includes replicate measurements and the analyses of reference materials, participation in the proficiency tests, and intercomparison exercises (Dybczinski 1980; Bleise and Smodis 1999, 2001). In a modern laboratory, all these processes generate a huge data and the best way to store and retrieve data for traceability and QA purposes is by electronic databases (Wasim 2007). A database is defined as a collection of related data (Elmasri and Navathe 2000). A relational database model stores data in tables and relate one table with the other using one or more common fields and data in each table is in nonrepeated form. Presently, several relational database management software are available on the market, such as Oracle (Oracle Corp.), SQL Server (Microsoft Inc.), PARADOX (Corel Inc.), ACCESS (Microsoft Inc.), and others. There are several examples of databases created to store nuclear data, such as GAMCAT (Tepel and Muller 1990) for alpha-particles, NESSY (Boboshin and Varlamov 1996) for nuclear spectroscopy, NUBASE (Audi et al. 1996) for nuclear and decay properties, a database (Be et al. 1996) containing decay constants, BANDRRI

(Arcos et al. 2000) for ionizing radiations, NUCData (Wasim and Zaidi 2002) for neutron activation analysis (NAA) and gamma-ray spectrometry, and a compilation of k_0 and related data (Kolotov and De Corte 2003).

6.2.1 Explosive and CW Simulants

In order to check the capability of neutron-based analytical techniques to detect explosives and CW agents, a very convenient approach is the use of simulants. A simulant is a material with approximately the same ratio of chemical elements as the genuine material of interest. A simulant is usually a harmful mixture of easily available compounds. It might have some additional chemical elements, especially those not seen in neutron activation analysis. Simulants are often used in the execution of intercomparison exercises involving different laboratories. Such a TNT simulant has been incorporated in the dummy land mine DLM2 (Brooks et al. 2004), which was designed in the framework of a Coordinated Research Project (CRP) sponsored by IAEA and 13 replicas of this design were constructed and distributed among CRP participants. TNT simulant was sealed in a polymethylmethacrylate container (IAEA 2003) and the characteristics of DLM2 were container diameters, 80 mm (outer) and 70 mm (inner); container lengths, 34 mm (outer) and 22 mm (inner); and the mass of container, 100 g. Mass of the TNT simulant was 100 g. Composition of TNT simulant was 17.3 g graphite + 23.9 g oxalic acid crystals + 58.8 g cyanuric acid. Elemental composition of container was [H:C:N:O] = [4:2:0:1].

A list of simulants available from A.&C.T., Zagreb, Croatia, is presented in Table 6.6. It is certified that simulants for the dangerous materials are prepared from commercially available benign materials as indicated. The materials, components of simulants, were grinded, mixed, and homogenized in such a way that the resulting mixture has a relative concentration of chemical elements as shown in the column "simulant chemical formula."

The purpose of this product is to imitate dangerous substances by mixing safe ordinary materials containing the same chemical elements. Only ordinary and widely utilized safe materials, which do not chemically react with each other when mixed, are used for production of simulants. These simulants are used for performance check, adjustment, and calibration of special equipment for the detection of dangerous materials by measuring elemental composition. Application of this product for some other purpose is not provided.

Materials with chemical formula containing elements present in imitated explosive substances are chosen as primary products. Imitation of mass ratios of the elements, which are also present in imitated substances with the same ratio, is achieved by mixing selected components in required proportions. Thus, a homogeneous mixture of safe components with individual chemical formulae is produced. In the aggregate, these components provide the same

TABLE 6.6 LIST OF SIMULANTS IN 1 L CONTAINERS

No.	Item	Chemical Formula	Density (g/cm³)	1 kg Simulant Components	Gross Weight (g)	Simulant Net Weight (g)	Simulant Chemical Formula
S-1	TNT	$C_7H_5N_3O_6$	1.654	Quartz sand 583 g, melamine 204 g +graphite 214 g	1.440	860	$Si_3C_7H_3N_3O_6$
S-2 A	TNT	$C_7H_5N_3O_6$	1.654	Graphite 158 g, oxalic acid cyanuric acid ($C_3H_3N_3O_3$) 566 g			$C_7H_6N_3O_6$
S-2/B	TNT	$C_7H_5N_3O_6$	1.654	Same as above	1490	910	$C_7H_6N_3O_6$
S-2/C	TNT	$C_7H_5N_3O_6$	1.654	Same as above	1200	700	$C_7H_6N_3O_6$
S-2/D	TNT	$C_7H_5N_3O_6$	1.654	Same as above	1000	500	$C_7H_6N_3O_6$
S-3	Hexogen, RDX	$C_3H_6N_6O_6$	1.82	Melamine 412 g + quartz sand 588 g	1.560	980	$Si_3C_3H_6N_6O_6$
S-4 A	TATP Non-nitrogen	$C_9H_{18}O_6$	1.0–1.2	Sucrose 844 g + paraffin 156 g	1.380	800	$C_9H_{16.9}O_6$
S-4 B	TATP Non-nitrogen	$C_9H_{18}O_6$	1.0–1.2	Sucrose 844 g + paraffin 156 g	1.440	860	$C_9H_{16.9}O_6$
S-5 A	Iprit, sulfur mustard	$C_4H_8Cl_2S$	1.27	Paraffin 274 g, sodium chloride 570 g + sulfur 156 g	1.480	900	$C_4H_{8.2-8.4}Cl_2SNa_2$
S-5 B	Iprit, sulfur mustard	$C_4H_8Cl_2S$	1.27	Paraffin 274 g, sodium chloride 570 g + sulfur 156 g	1.530	970	$C_4H_{8.2-8.4}Cl_2SNa_2$

(*Continued*)

TABLE 6.6 (*Continued*) LIST OF SIMULANTS IN 1 L CONTAINERS

No.	Item	Chemical Formula	Density (g/cm³)	1 kg Simulant Components	Gross Weight (g)	Simulant Net Weight (g)	Simulant Chemical Formula
S-6	Iprit, sulfur mustard	$C_4H_8Cl_2S$	1.27	Paraffin 250 g, zinc chloride 607 g + sulfur 143 g	—	—	$C_4H_{8.2-8.4}Cl_2SZn$
S-7	Sarin	$C_4H_{10}FO_2P$	1.09 at 25°C	Teflon 176 g, paraffin 345 g, KH_2PO_4 479 g	1400	820	$C_4H_{7.6}FO_2P$
S-8	Sarin	$C_4H_{10}FO_2P$	1.09 at 25°C	Teflon 183 g, paraffin 135 g, black phosphor 227 g + sucrose 455 g	—	—	$C_4H_{6.6}FO_2P$
S-9 A	Cocaine	$C_{17}H_{21}NO_4$	1.216	Melamine 69 g, graphite 243 g, sucrose 410 g + paraffin 277 g	1.370	580	$C_{17}H_{21}NO_4$
S-9 B	Cocaine	$C_{17}H_{21}NO_4$	1.216	Melamine 69 g, graphite 243 g, sucrose 410 g + paraffin 277 g	1.400	820	$C_{17}H_{21}NO_4$
S-10 A	Heroine	$C_{21}H_{23}NO_5$	1.35	Melamine 57 g, graphite 294 g, sucrose 421 g + paraffin 227 g	1.440	860	$C_{21}H_{23}NO_5$
S-10 B	Heroine	$C_{21}H_{23}NO_5$	1.35	Melamine 57 g, graphite 294 g, sucrose 421 g + paraffin 227 g	1.400	820	$C_{21}H_{23}NO_5$

Note: Empty container weight = 580 g.

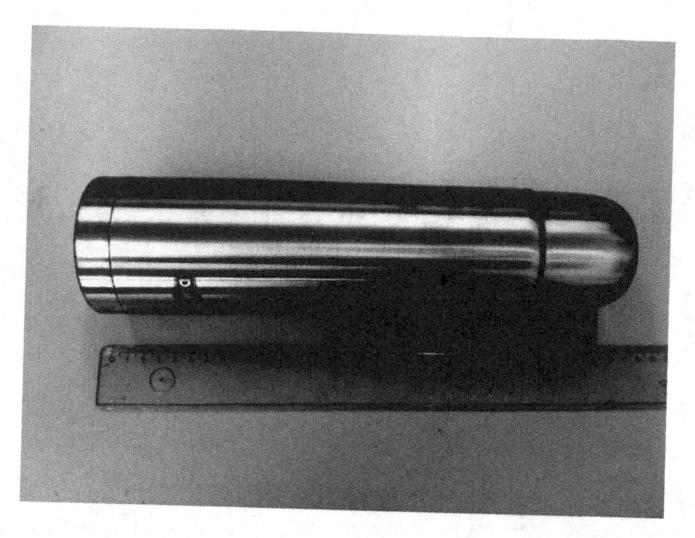

Figure 6.5 The 1 L waterproof, stainless steel flask is used as simulant container.

ratio of elements as in the chemical formula of the imitated material, while these simulant components do not mix with each other on the elemental level. Such approach allows a complete exclusion of production of chemically dangerous simulants with a joint chemical formula and all these simulants represent just a mixture of safe materials with individual chemical content.

Simulants may be packed and transported in any convenient package. Nevertheless, to avoid penetration of water as moisture and thus misrepresentation of the imitated chemical formula, it is recommended to use hermetically sealed package in the form of metallic and plastic containers with tightly closing lids for transportation purpose. No packaging material containing the same chemical elements as the imitated material should be utilized while using the simulants at the equipment for the detection of explosive and narcotic materials. Therefore, it is recommended to utilize waterproof metallic containers with the equipment. The best option for the recommended package, also used for safe transportation of simulant samples in small quantities, is a thermal waterproof stainless steel flask of 580 g in weight as shown in Figure 6.5.

6.3 PROCESS CONTROL OF INDUSTRIAL PRODUCTION PROCESSES

On-stream elemental analysis is a further development of methods for sample analysis. Such methods are important for continuous industrial processes because they enable the operator to act quickly to achieve the optimum yield and quality. In addition to other physical and chemical methods, NAA—being independent of chemical bond—can also be used for continuous analysis. The most frequent fields of application are the on-stream analysis of solutions and the analysis of solid substances on conveyer belts. Continuous neutron activation analysis (CNAA) has been reviewed by Kliment and Tölgyessy (2007). The activity of the material measured by the detector depends on the nuclear properties of the elements of the sample, the neutron flux, the flow rate, the layer thickness, the geometry of the arrangement, etc.

Industrial applications of neutron methods include characterization of coal (Belbot et al. 1999; Sowerby 2009), cement (Womble et al. 2005), and metal alloys (James and Fuerst 2000). The flux of small neutron generator is quite sufficient for CNAA because there is seldom any need to determine trace element in the material investigated.

Online nuclear analysis of coal based on prompt neutron activation represents a technological breakthrough for real-time process control necessary for optimum efficiency in the use of coal. These analyzers are presently being installed to solve a variety of current problems in coal usage. In the paper Brown et al. (1982), the general features of these instruments are described and a detailed discussion of a high counting rate spectroscopy system used with an analyzer based on a germanium detector is given. A brief discussion of various applications of these analyzers include control of coal blending, control of coal burning efficiency, and quality control in coal beneficiation and synfuel processes (for more reading see: Kliment and Tolgyessy 1972; Al-Shahristani and Jervis 1973; James et al. 1976; Elias et al. 1979; Khalil et al.1980; Ehman et al. 1982; Wilde and Herzog 1982).

The need for real-time analysis of the composition of bulk quantities of coal and the applicability of prompt neutron activation analysis (PNAA) to accomplish this has long been recognized (Stewart 1967; Gozani et al. 1977). Since 1975, Science Applications, Inc., under sponsorship by the Electric Power Research Institute (EPRI) has developed the prompt neutron analysis to a mature technology (Elias et al. 1976; Gozani 1978; Gozani et al. 1978, 1979). Two online nuclear analyzers of coal called nucoalyzers have been fabricated. These instruments that are based on lower resolution NaI detectors perform a limited compositional analysis of coal, mainly for its sulfur content.

A germanium detector–based nuclear coal analyzer that performs complete elemental analysis of coal in the process stream is described in Brown et al. (1982). The paper describes some general features of this instrument and the details are given on the high counting rate spectroscopy system developed for the high resolution online analyzer.

French company SODERN has developed a line of products for the online analysis of raw materials. For over 15 years, SODERN has been supplying online systems for the real-time elemental analysis of bulk raw materials moving on a conveyor belt (Le Tourneur et al. 2009); see Figures 6.6 and 6.7.

Figure 6.6 Controlled neutron analyzer (CNA) for cement, coal, and mining industry. (From Le Tourneur, P. et al., ULIS: A portable device for chemical and explosive detection, paper SM/EN-17 in *Proceedings of the International Topical Meeting on Nuclear Research Applications and Utilization of Accelerators*, IAEA, Vienna, Austria, 2009. With permission.)

Figure 6.7 Raw material analyzer. (From Le Tourneur, P. et al., ULIS: A portable device for chemical and explosive detection, paper SM/EN-17 in *Proceedings of the International Topical Meeting on Nuclear Research Applications and Utilization of Accelerators*, IAEA, Vienna, Austria, 2009. With permission.)

Unlike conventional analyzers that use a continuous-emission isotopic source, the online analyzer made by SODERN—the CNA (controlled neutron analyzer)—uses an electrical neutron source with stabilized emission and can be switched off at any time. The CNA is composed of a measuring chamber, to which is attached an electrical box that uses digital signal processing technology and an operator interface. The CNA's measuring chamber is installed around the conveyor belt and analyzes the composition of the material. It comprises a neutron emission unit and a gamma detection unit installed either in the upper or lower part of the chamber, depending on the characteristics of the material to be analyzed. The CNA is interfaced with a high-performance automatic system that makes it possible to both measure and control the chemical composition of the raw materials and adapt the manufacturing process. It is widely used to control and optimize various manufacturing processes, in particular those used by cement manufacturers, the mining industry, thermal power plants, and waste treatment plants.

The cement CNA is installed at source between the first crusher and the prehomogenization stockpile or before the raw mill for the blending of raw mix. Controlling the raw materials makes it possible to optimize the composition of the mix before the kiln operations. So far, over 70 SODERN online analyzers have been installed and approved by leading cement manufacturers in South Africa, one of them shown in Figure 6.8.

Figure 6.8 Cement plant in South Africa. (From Le Tourneur, P. et al., ULIS: A portable device for chemical and explosive detection, paper SM/EN-17 in *Proceedings of the International Topical Meeting on Nuclear Research Applications and Utilization of Accelerators*, IAEA, Vienna, Austria, 2009. With permission.)

Figure 6.9 Coal analysis system, China. (From Le Tourneur, P. et al., ULIS: A portable device for chemical and explosive detection, paper SM/EN-17 in *Proceedings of the International Topical Meeting on Nuclear Research Applications and Utilization of Accelerators*, IAEA, Vienna, Austria, 2009. With permission.)

Online analyzers are also widely used for coal, notably in thermal power plants, mines, gasification plants, and sometimes in coke plants where measuring the chemical composition of the coal is indispensable for the process. The CNA is measuring online and in real time all the major elements of coal, including carbon and oxygen. Through this comprehensive elemental analysis, the coal CNA can define the main characteristics of the coal including total ash, calorific value (BTU), ash viscosity, volatile elements, etc. Figure 6.9 shows such a system in China.

The robust minerals CNA produced by SODERN and featuring PFTNA technology is making its way into the mining industry for the analysis of nickel, copper, and iron, in the aim of improving processing efficiency and profitability. See Figure 6.10 showing the CNA system analyzing nickel ore in New Caledonia.

Online neutron activation cement element analyzer, DF-5701, manufactured by Dandong Dongfang Measurement & Control Technology Co., DFMC, China, is used for every related method of measuring material composition in the process of cement production, especially for raw ingredients and preblending stockpile. It not only directly measures bulk materials through the belt, providing composition results without the need to sample, but also does so through automatic control that improves the product pass rate. Some other interesting industrial applications of 14 MeV neutrons are also described in (Van Grieken et al. 1970; Hayward et al. 1971).

Figure 6.10 CNA system analyzing nickel ore in New Caledonia. (From Le Tourneur, P. et al., ULIS: A portable device for chemical and explosive detection, paper SM/EN-17 in *Proceedings of the International Topical Meeting on Nuclear Research Applications and Utilization of Accelerators*, IAEA, Vienna, Austria, 2009. With permission.)

6.4 THREAT MATERIAL DETECTION WITH PFNA

6.4.1 Explosives

The list of explosive compounds has more than 100 items including some improvised primary explosives like acetone peroxide, ddnp/dinol, double salts, hmtd, leadazide, lead picrate, mekap, mercury fulminate, "milk booster," nitromannite, sodium azide, tacc, etc. Instructions regarding how to prepare them can be found in the open literature and on the Internet.

However, the most often used explosives are trinitrotoluene (TNT), pentaerythritoltetranitrate (PETN), cyclotrimetilentrinitramin (RDX), trinitrophenylmethylnitramine (tetryl), tetrytol, and hexatol.

Trinitrotoluene, commonly known as TNT, is a constituent of many explosives, such as amatol, pentolite, tetrytol, hexatol, torpex, tritonal, picratol, ednatol, and composition B. It has been used under such names as triton, trotyl, trilite, trinol, and tritolo. In a refined form, TNT is one of the most stable of high explosives and can be stored over long periods of time. It is relatively insensitive to blows or friction. TNT is used in pressed and cast form. Pressed TNT can be used as a booster or as a bursting charge for high-explosive shells and bombs.

Pentaerythritoltetranitrate (PETN) is one of the strongest known high explosives. It has also been used under such names as pentrit, nitropenta, niperyt, and TEN. It is more sensitive to shock or friction than TNT or tetryl, thus it is never used alone as a booster. It is primarily used in booster and bursting charges of small caliber ammunition, in upper charges of detonators in some land mines and shells, in shaped charges, and as the explosive core of primacord. It is also used as an explosive compound in plastic explosives (PEP).

Cyclotrimetilentrinitramin (RDX) is one of the strongest known high explosives. It has also been used under such names as hexogene, cyclonit, SDX, and T-4. It is primarily used in booster and bursting charges of small caliber ammunition, in upper charges of detonators in some land mines and shells, and in shaped charges.

Tetryl can be initiated from flame, friction, shock, or sparks; it burns readily and is quite likely to detonate if burned in large quantities. Tetryl is the standard booster explosive and is sufficiently insensitive in pressed form to be used safely as a booster explosive.

Tetrytol is a cast mixture of tetryl and TNT and is designed to obtain a tetryl mixture that may be used in burster tubes for chemical bombs, in demolition blocks, and in cast shaped charges. Hexatol is a cast mixture of RDX and TNT. It is used as the main explosive charge of some landmines. Table 6.7 shows molecular weights and chemical composition of the most often used explosives.

Ratios of elemental concentrations for some explosives are shown in Table 6.8. Ratios of elemental concentrations C/N and C/O for explosives are different from the values of these ratios for other types of material.

List of materials that can be used as an explosive is rather large. It is rather disturbing that one can find on the web detailed instructions how to make some of them; see, for example, Goldmann (1998). List of improvised explosives

TABLE 6.7 NITROCOMPOUND EXPLOSIVES

Name	Molecular Weight	C	H	N	O	Density (g/cm³)
TNT	227.13	7	5	3	6	1.65
RDX	222.26	3	6	6	6	1.83
HMX	296.16	4	8	8	8	1.96
Tetryl	287.15	7	5	5	8	1.73
PETN	316.2	5	S	4	12	1.78
Nitroglycerin	227.09	3	5	3	9	1.59
EGDN	152.1	2	4	2	6	1.49
DNB	168.11	6	4	2	4	1.58
Picric acid	229.12	6	3	3	7	1.76
AN	80.05	—	4	2	3	1.59

TABLE 6.8 C/N AND C/O CONCENTRATIONS RATIOS

Name	C/O	H/N	C/N	O/N
NG	0.33	1.67	1	3
TNT	1.17	1.67	2.33	2
RDX	0.5	1	0.5	1
PETN	0.42	2	1.25	3
AN	0	2	0	1.5

includes primary explosives and exotic and friction primers. Primary explosives being acetone peroxide, ddnp/dinol, double salts, hmtd, leadazide, lead picrate, mekap, mercury fulminate, "milk booster," nitromannite, sodium azide, and tacc. Exotic and friction primers: leadnitroanilate, nitrogen sulfide, nitrosoguanidine, tetracene, chlorate-friction primers, chlorate-trimercury-acetylide, and trihydrazine-zinc (ii) nitrate.

Presence of any amount of explosives is of interest. However, of special interest is the detection of large quantities being transported in vehicles. According to the U.S. Department of Treasury and Bureau of Alcohol, Tobacco, Firearms, and Explosives (ATF), the vehicle bomb explosion hazard and evacuation distances are shown in Table 6.9.

How to detect the presence of explosives? Explosives and chemical agents' detection systems can be based on the fact that the problem of explosive detection and identification can be reduced to the problem of the measurement of elemental concentrations and ratios of elemental concentrations.

TABLE 6.9 VEHICLE BOMB EXPLOSION HAZARD AND EVACUATION DISTANCES

Vehicle Description	Max. Explosives Capacity	Lethal Air Blast Range (m)	Min. Evacuation Distance (m)	Falling Glass Hazard (m)
Compact sedan	227 kg (in trank)	30	457	381
Full size sedan	455 kg (in trank)	38	534	534
Passenger or cargo van	1,818 kg	61	838	838
Small box van (14 ft box)	4,545 kg	91	1,143	1,143
Box van or fuel truck	13,636 kg	137	1,982	1,982
Semitrailer	27,273 kg	183	2,134	2,134

Different nuclear analytical techniques could be used for this purpose; however, the use of nuclear analytical techniques has some specific advantages. There have been a number of excellent general reviews of neutron techniques for nonintrusive inspection; see Gozani (2002), Buffler (2004), and Runkle et al. (2009). An early review of neutron-based nonintrusive inspection technique is presented by Gozani (1999). His company Ancore Corporation, Santa Clara, CA, has for some time operated the Ancore cargo inspector (ACITM). The system was based on PFNA technology. ACI's neutron production system consists of a pulsed injector, accelerator, high energy beam transport, hydraulically actuated scan arm with neutron production target and vertical neutron collimator, and the horizontal neutron collimator. These subsystems, working together, produce pulses of neutrons that scan the cargo. During operation, the neutron production system is controlled by the ACI system operator station. The system is also equipped with an independent control station that provides diagnostic as well as manual control of the system during maintenance. The detector arrays and signal processors detect signals from the cargo contents and process them into ACI images. The detector arrays consist of two sets of gamma ray detectors with signal amplification and drive electronics. One set of detectors, inside the beam, obtains neutron and gamma ray radiography signals of the object. The second set of detectors, outside the beam, detects the material specific signals produced from the pulsed fast neutron analysis (PFNA) process. Model ACI-2001 requires a facility that will allow maximum cargo container throughput. Dimensions for the inspection tunnel depend on the dimensions of the inspected object; a smaller opening is required for cargo containers than for large trucks. In addition to the structure housing the ACI equipment, consideration should also be given to staging areas and stations. Additionally, allowances need to be made for an inspection processing room, utilities building, and operations control/screening room. A minimal facility requires approximately 400–500 m^2.

Neutron systems are employed for the detection and analysis of chemical compounds and explosive threats (Vourvopoulos and Womble 2001; Gozani 2004; Strellis et al. 2009; Eleon et al. 2010). They are also utilized in vehicle- and cargo-scanning scenarios (Barzilov and Womble 2003; Reber et al. 2005; Perret et al. 2006; Koltick et al. 2007), humanitarian demining, and confirmation of unexploded ordinance (Aleksandrov et al. 2005; Lanza 2007).

Detection of explosives buried in soil is generally much more difficult than detection of explosives in baggage or parcels. Even though the basic nuclear reactions that can be used are the same, for buried explosives certain of these reactions may not be applicable because of the interference from competing reactions due to the soil constituents surrounding the IED or landmine or the explosive signal may be too weak with respect to the high background generated by the soil under neutron interrogation. For example, hydrogen, although a useful indicator of explosives in baggage and useful for process monitoring in coal slurry analysis, may be of limited use for landmine detection because of the typically wide variability of water content in the soil (Obhodas et al. 2004). Silicon dioxide significantly interferes with buried explosives detection but it is not a factor in the baggage scenario. The gamma-ray spectrum from explosives in baggage would likely be significantly different from that of a buried landmine when irradiated with fast neutrons. This is due to the enormous effect of neutron moderation in the soil. The monoenergetic distribution of incident fast neutrons from the source is transformed into a complex energy spectrum by the time the neutrons reach the explosive. This leads to significantly different ratios of nuclear reaction rates within the explosive itself, compared to the ratios that would be obtained in the absence of soil. The resultant neutron energy spectrum irradiating the explosive will be a function of soil depth and soil composition. Further, the attenuation by the soil of the gamma rays from the landmine may affect the relative importance and indeed the applicability of a particular reaction. Additionally, a fieldable system that must operate robustly under various adverse environmental conditions gives rise to additional design constraints not found in a system situated in an air-conditioned building, such as an airport. This difference in operational environments and scenarios affects the choice of neutron source, the sensor head design, the choice of radiation detector, and data processing electronics (Faust et al. 2011). In the report by Coleman et al. (1974) nuclear reactions induced with 14 MeV neutrons were judged not to be applicable for landmine detection.

Faust et al. (2011) considered FNA-augmented TNA and came to the conclusions that pulsed beam FNA approach would not improve the detection performance of TNA system for landmine detection in a confirmation rôle, and could not be made into a practical stand-alone detection system for buried landmines. Detection of buried landmines through FNA may, however, be possible using the tagged-neutron (API) approach. Both, simulations and experiments have shown that target signal to background ratios can be significantly increased by using associated particle neutron generator and time gating to define a depth slice containing the target (Faust et al. 2011).

The role of soil in landmine detection has been discussed in detail in Obhodas et al. (2003). The physics of the use of 14 MeV neutrons in the explosive detection is described in Sudac et al. (2003). Of interest is also the discussion on the detection of landmines by neutron backscattering with exploring the limits of the technique as presented in Viesti et al. (2006).

Here we shall present the systems explosive detection using the 14 MeV neutron generators.

PELAN is a device that has been developed to identify a landmine through the elemental constituents of its explosive (Vourvopoulos and Womble 2001). The International Atomic Energy Agency (IAEA) evaluated PELAN as a mine-confirmation sensor in November 2002 near Zagreb, Croatia, at a test minefield of the Croatian Mine Action Center. The evaluation included one type of anti-tank (AT) mine and three types of antipersonnel (AP) mines. These are typical mines, found in many mine fields in Croatia. The AP mines were buried at depths up to 15 cm and the AT mine up to a depth of 20 cm. The soil moisture varied between 25% and 29% (weight). PELAN was operated with its laptop computer through a hard-wired connection from a distance of at least 20 m. PELAN was asked to check 127 "flagged" positions that could contain one of the earlier mentioned mines. The analysis of the 2002 Croatian blind tests gave

$$\text{Probability of detection: } P_d = 0.85$$

$$\text{Probability of false alarm: } P_{fa} = 0.23$$

As stated in the report to IAEA by the test supervising team, although the time was inadequate for systematic testing, it could be said that with control of system drifts, PELAN should have no problems with recognizing antitank mines buried under 15 cm of soil. In that case also recognition of small antipersonnel mines such as PMA-3 under 5–10 cm of soil will be quite probable. Multielemental analysis is a strong point of PFTNA analysis as was demonstrated in the case of TMM-1 and PMA-1 mines. Although the authors stress the need to improve the reproducibility of data, they believe that this test has demonstrated great potential of PELAN for humanitarian demining.

The authors concluded that the use of pulsed neutrons in PELAN has been shown to be an effective method for the detection of explosives. PELAN's strength lies in its quantitative multielemental analysis, which allows it to distinguish explosives from other substances that contain mostly the same chemical elements. For landmines, its current role is that of a confirmation sensor. For recovered ordnance, PELAN can differentiate between an explosive or inert fill, and in many cases can identify the fill. The mine casing does not present a hindrance to PELAN's ability; indeed, PELAN can tell with a very high probability of detection whether the detected mine is plastic or metal encased. The current PELAN configuration allows its operation from a distance as large as 45 m. It can be mounted to a robot and guided to within a few inches/centimeters from a "flagged" position. PELAN did not make it! Today it is more or less forgotten.

Let us turn to the thinking described in the paper by Maglich (2005). His presentation concentrated on military explosives that consist of four elements: H, C, N, and O, for example, stoichiometry of TNT is $C_7N_3O_6H_5$. For RDX, used in plastic bombs, it is $C_6N_6O_6H_6$. Fermi argued that the presence of nitrogen is the "explosive signature" and should be detected *via* 10.8 MeV γ's emitted by thermal neutron capture in ^{14}N. Westinghouse researchers followed his idea by a development program that was abandoned 10 years and $100 million later with the admission: "We developed a nitrogen detector, not an explosive detector." One cubic meter of air contains nearly a kilogram of N_2. Qualitatively detecting the mere presence of one or more elements of the explosive does not make an explosive detector.

Since there are no excited states of H, the task of atometry is to obtain, in a shortest time possible, quantitative atomic ratio of the three elements, that is, the subscripts a, b, and c in $C_aN_bO_c$, to an accuracy sufficient to discriminate explosives from 1000 odd innocuous substances also containing C, N, and O. Atometry is defined as stoichiometry by means of elementary particles. Atometry is achieved by quantitative high-resolution analysis of γ spectra from the inelastic scattering of fast neutrons. Neutrons of E > 5 MeV, colliding with C, N, and O produce characteristic γ's from each of the three elements, γ energies being 4.4, 5.1, and 6.1 MeV, respectively.

It was Maglich (2005) who concluded that scintillation detectors were the menace. Solid state gamma detector *is sine qua non.* This point is illustrated in Figure 6.11.

Figures 6.12 and 6.13 show a robot (Andros Mark V-A1) operating atometer (SIEGMA™ 3E3) inspecting a suspicious package, while Figure 6.14 shows the components of the atometer: accelerator (neutron source), germanium detector, and electronics. A picture of the atometer system display for the operator is given in Figure 6.15. High false alarms. SIEGMA should be considered as a "prior art."

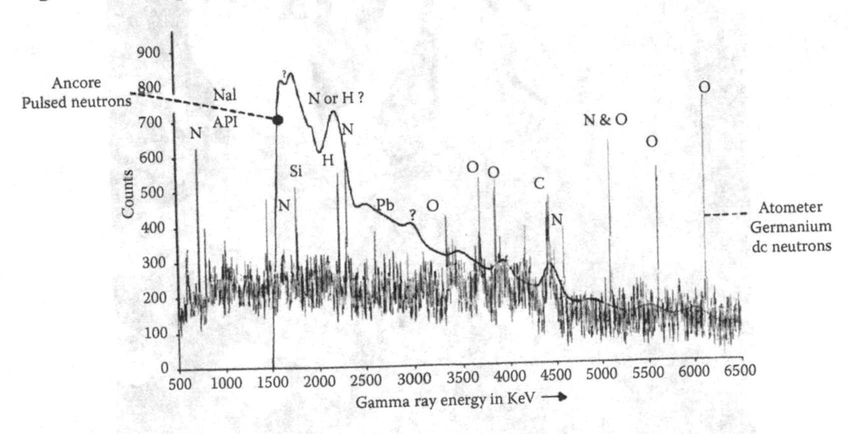

Figure 6.11 The γ-spectra of the same material, solid line NaI detector, and histogram Ge detector.

Figure 6.12 Ready for inspection.

Figure 6.13 Robot-borne atometer "suitcase" model SIEGMA™ 3E3 inspects a briefcase.

Figure 6.14 The atometer system including a particle accelerator for the neutron production, the germanium detector for the γ detection, and the required electronics.

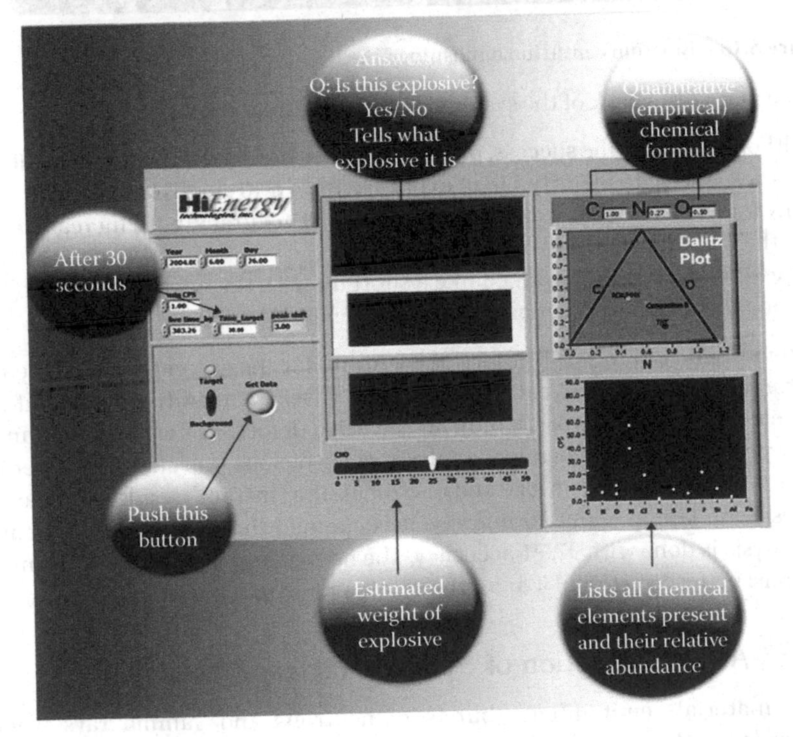

Figure 6.15 Display for the operator of the atometer system.

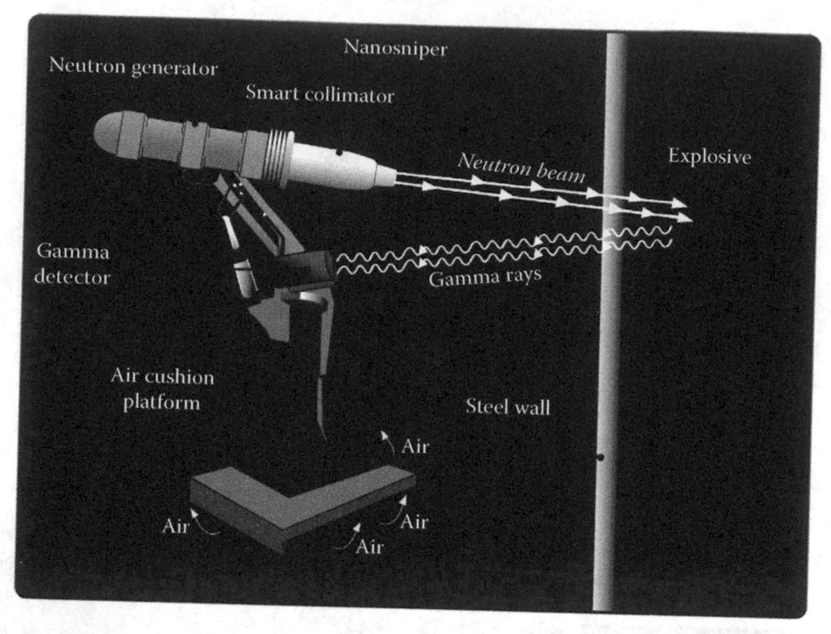

Figure 6.16 Bioatom centurion nanosniper™.

The main drawback of the system was its nondirectionality.

Although a scientific success, prior art devices never became practical. It took 10 min to identify 1 lb of explosive from a 10 cm distance.

Any increase in speed would result in a loss of accuracy, both increasing the false alarm rate.

The SIEGMA machine worked like a shotgun with the neutrons irradiating the target with a most unfavorable energy.

Bioatom Centurion Inc., subsidiary of BioAtom, Inc. has recently announced a neutron-based instrument called nanosniper™ (see Figure 6.16). The manufacturer claims this to be world's only noninvasive detector of explosives. Nanosniper performs online chemical analysis of suspicious objects in a matter of 16 s to 3 min (depending on size and distance). The "instant" result is displayed wirelessly on laptop about 90 m away. It is claimed that the real time chemical analysis is done with 97.5% accuracy. The nanosniper uses directed beam of neutrons; being directional it analyzes gamma rays from target only.

6.4.2 Active Detection of Special Nuclear Material

Fissile materials emit intense bursts of neutrons and gamma rays when bombarded with neutrons. These particles are easily detected and signal the presence of radioactive material. It is necessary to have a very high-flux neutron source for this detection method to be practical. When applied to ship container control the detection scheme must be implemented in a manner that allows for high throughput and does not disrupt port productivity. Passive detection technologies do not work well because fissile material is weakly radioactive and easy to shield. Active interrogation has been effectively demonstrated, but previous neutron sources have led to excessively long detection times.

Phoenix Nuclear Labs (PNL 2012) has reported the development of a neutron generator with steady state neutron production as high as 10^{14} neutrons per second allowing for high throughput screening with a single container scanned in seconds. PNL scientists have demonstrated the detection concept (shown in Figure 6.17) in a laboratory setting.

The low natural background rates and the penetrating nature of neutron radiation make neutron detection (particularly time-correlated neutrons) a good method for quantifying and accounting for large amounts of special nuclear material (SNM) capable of neutron-induced fission and fission chains. Fission is one of the few natural processes that produce time-correlated neutrons—the others are spallation-type processes, like (n,xn) and cosmic-induced background—that have low but measurable rates in common terrestrial

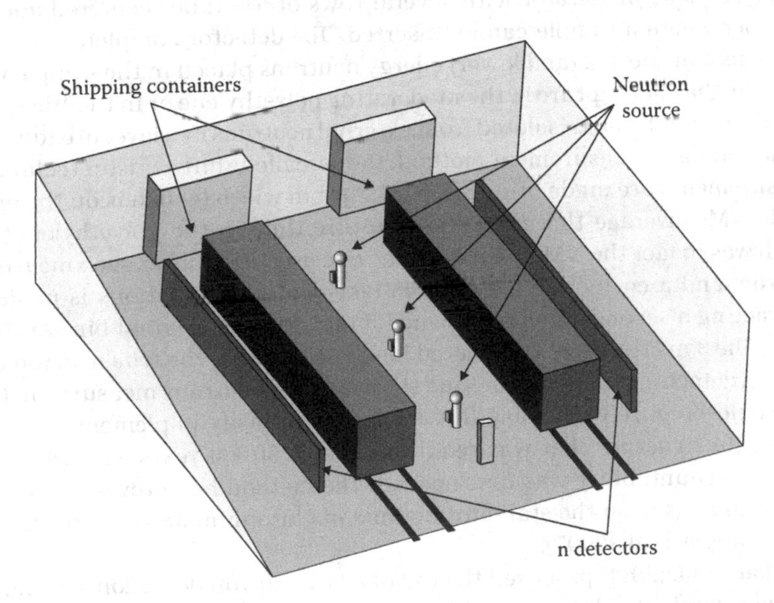

Figure 6.17 Concept of container screening for SNM presence with high intensity neutron generators, as reported by Phoenix Nuclear Labs. (From PNL, Phoenix Nuclear Labs, Madison, WI, Fact Sheet, 2012.)

material. The high rates of most transuranic spontaneous fission sources (like Pu) of even a gram or less usually swamp any cosmic-induced background rate (in comparison, kg quantities of natural uranium produce neutrons only on the same order as that of the cosmic background). All other sources of neutrons are generally from alpha-n production (energetic alpha particles fusing with light elements) that generate single neutrons that will be uncorrelated in time. The exploitation of the time-correlated neutron flux from the spontaneous fission of ^{240}Pu is what is used for materials accounting (Nakae et al. 2014).

Historically, the standard method employed for the material accounting of plutonium masses is to measure the time-correlated neutron flux from a given sample of material. In principle, one can pick any number of ways to determine the level of time correlation. In practice, almost all methods employ the measurement of neutron counts over a fixed period of time, that time chosen to be appropriate for the instrument employed. The bin can either be randomly triggered or triggered by a measured neutron. The statistics of these measured bins then can be summed and compared to what would have occurred from a random distribution and from the difference and level of correlation, a quantity determined.

The standard instrument roundly employed around the world is a high efficiency neutron multiplicity counter (NMC). This is essentially a large block of polyethylene moderator with several rows of ^{3}He tubes centered about a chamber where a sample can be inserted. The detector completely absorbs all the fission spectra (and lower) energy neutrons placed in the sample well, either by thermal capture in the moderating polyethylene or in the ^{3}He where they are counted and is isolated from external neutrons by a layer of cadmium. In the standard measurement method, the so-called shift-register technique, measurements are made by opening a 32 μs bin width (which is on the order of the NMC average thermalization/capture time in the polyethylene/^{3}He, which was in fact the NMC ~26 μs for the one used here) after each measured neutron, and a correction for random (accidental) correlations is made by subtracting a second 32 μs bin opened 3 ms after the original bin was triggered. The 3 ms time was considered long compared to the true neutron correlation detector constant time, and therefore any neutrons measured in that bin would be purely random. This method was easily implemented with a serial gate structure that was readily available 50–60 years ago when plutonium accountability was first needed. The method has proven to be very robust and has been the standard method of plutonium accountability ever since Dougan et al. (2007).

Nakae et al. (2014) proposed the use of a fast neutron detection system, for example, employing liquid scintillator (LS) with pulse shape discrimination (PSD) as a possible alternative to a ^{3}He thermal-based system. Even though LS detectors relying on neutron detection through proton recoil have intrinsically low efficiency, the inherent faster timing of the signal allows for much shorter measurement binning and therefore a much lower rate of random correlations. This makes it an attractive method for assaying nuclear material, especially material with relatively high neutron fluxes.

6.5 SAFETY ISSUES

The recommended exposure limits for radiation workers is limited to 5 R/year (50 mSv/year) by the International Commission on Radiological Protection (ICRP). Most laboratories, however, set lower limits based on the *as low as reasonably achievable* (ALARA) principle. A lethal dose is defined as a whole body dose that results in 50% mortality in 30 days. This is usually identified as 500 Rads (5 Gy).

The dose from an isotropic source reduces by a factor of four as one moves twice the distance from the source following a $1/r^2$ falloff. The safety requirement for a flux of 2×10^7 neutrons/s is shown in Figure 6.18. At a distance of 4 m, the dose rate falls below the criteria set by the Nuclear Regulatory Commission (NRC) for safety of the general public, a maximum of 2 mREM/h, as shown with dashed horizontal line in Figure 6.18. The NRC rules also stipulate that the maximum dose a member of the general public can receive in a year is 100 mR/year. This number is well below the typical level of background radiation of 360 mREM/year accumulated by the general public from natural and

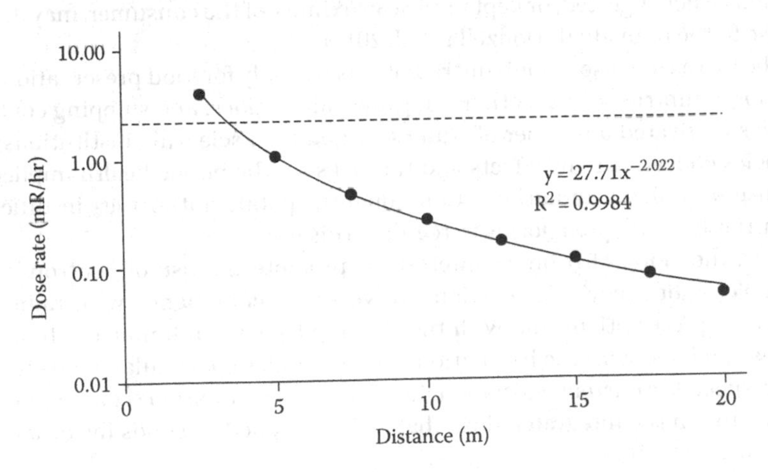

Figure 6.18 Dose rate as a function of distance from the NG running at a typical 2×10^7 neutrons per second. (After Litz, M. et al., Neutron-activated gamma-emission: Technology review, Army Research Laboratory, Adelphi, MD, Report ARL-TR-5871, 2012.)

manmade sources. Following the ALARA policy convention is performed if 12.5 m radius is enforced for compact electronic generators of neutrons, where the dose rate is 10 times below the legal limit.

6.5.1 Radiological Risks from Irradiation of Food in Neutron Inspection Systems

6.5.1.1 Introduction

One important aspect of the neutron inspection portal to be clarified concerns the possible effects of irradiation on the goods inside the inspected container. Particular attention must be devoted to the irradiation of foodstuff, pharmaceuticals and medical devices.

The possible effect of irradiation with ionizing radiations of transported goods on the public health is twofold; the first effect concerns possible modifications of the properties of the irradiated materials, induced by the ionizing radiation. This problem is particularly relevant for foodstuff, whose nutritional and organoleptic properties must not be altered or modified by any treatment. The second effect is connected to the possibility that the ionizing radiation might induce activation of nuclides of the irradiated products. This problem could be particularly relevant when the irradiation system makes use of neutrons, since it is unavoidable that some nuclei are converted to radioactive nuclides. This fact is particularly significant when the irradiated goods are foodstuff, pharmaceuticals, or medical devices. Indeed, these particular type of goods when ingested, or kept in close proximity of the consumer, may deliver a dose to the individual (Donzella et al. 2010).

The increasing use of ionizing radiations not only for food preservation, but also for nonintrusive inspection of luggage, air cargoes, and shipping containers has motivated a number of studies by qualified scientific institutions and agencies concerning the effects and the risks for the public health implied by the use of ionizing radiations. Consequently, public authorities in different countries have adopted norms to regulate this use.

It is, therefore, of primary interest, to promote the use of neutron-based technology for security inspections, to verify its compliance with radiological safety prescriptions and with the existing legislation limits and to assess the risks related with the irradiation of goods and, in particular, food stuff. In this respect, the current legislation limits for food irradiation set an important constraint on the integrated dose that can be applied to goods inside a cargo container.

In evaluating the amount of activation of foodstuff and other materials, in particular, pharmaceuticals and medical devices, significant comparison can be made with the levels of natural radioactivity present in those materials, the effective doses received annually by any individual due to natural sources of exposure, as well as the effective dose limits for the public set by international and European institutions.

The exposure of human beings to ionizing radiation from natural sources is a continuing and inescapable feature of life on earth. For most individuals, this exposure exceeds that from all man-made sources combined. There are two main contributors to natural radiation exposures: high-energy cosmic ray particles incident on the earth's atmosphere and radioactive nuclides that originated in the earth's crust. Those radionuclides are present everywhere in the environment, including the human body itself. Both external and internal exposures to humans arise from these sources (UNSCEAR 2000).

Internal exposure derives from both inhalation and ingestion of natural radionuclides. The major source of dose to humans due to natural radionuclide inhalation is the radon (Rn) gas, present in the atmosphere of indoor environments. The resulting worldwide average of the annual effective dose from Rn gas absorbed by an adult is 0.48 mSv, whereas the amount for specific countries stays generally within the range of 0.3–0.6 mSv. For children and infants, the values are about 10% and 30% higher.

Ingestion intake of natural radionuclides depends on the type of natural radionuclides involved, on the consumption rates of food and water and on the radionuclide concentrations in the environment. Food and water consumption varies a lot all over the world. Also, the concentrations of naturally occurring radionuclides (i.e., ^{3}H, ^{40}K, U, and Th series) in foods vary widely because of the different natural radioactivity background levels, climate, and agricultural conditions. There are also differences in the types of local foods that may be included in the categories such as vegetables, fruits, or fish.

The total average effective dose from inhalation and ingestion of terrestrial radionuclides is 0.31 mSv, of which 0.17 mSv is from ^{40}K and 0.14 mSv is from the long-lived radionuclides in the uranium and thorium series. In Table 6.10, the most important natural radionuclides with their average annual intake (activity introduced in body because of air inhalation and ingestion of food contaminated with natural radionuclides), averaged all over the world, are shown (UNSCEAR 2000). The total activity contained in the body of any adult person (averaged all over the world) is 104 Bq. In Table 6.11, the average activity concentration of natural radioactive potassium (^{40}K) inside some food is also reported.

TABLE 6.10 AVERAGE ANNUAL INTAKE (BY ADULT) OF URANIUM (UM) AND THORIUM SERIES RADIONUCLIDES IN DIET (BQ/YEAR)

^{235}U	^{238}U	^{230}Th	^{226}Ra	^{210}Pb	^{210}Po	^{232}Th	^{228}Ra	^{228}Th
0.2	5.7	3.0	22	30	58	1.7	15	3.0

TABLE 6.11 ^{40}K AVERAGE ACTIVITY CONCENTRATION INSIDE SOME TYPE OF FOODS

Food Product	Specific Activity (Bq/kg)
Banana	110
Lean beef	104
Dry cocoa powder	481
Milk	48
Average of many green vegetables	89

TABLE 6.12 AVERAGE WORLDWIDE EXPOSURE TO NATURAL RADIATION SOURCES

Source of Exposure	Effective Dose (mSv/year)	
	Average	Typical Range
Cosmic radiation (directly ionizing and photon component, neutron component) and cosmogenic radionuclides	0.39	0.3–1.0
External terrestrial radiation (outdoor and indoor)	0.48	0.3–0.6
Inhalation exposure (U and Th series, ^{222}Rn, ^{220}Rn)	1.26	0.2–10
Ingestion exposure (^{40}K and U and Th series)	0.31	0.2–0.8
Total	2.4	1–10

In Table 6.12, the average and typical ranges of doses received by the world population due to the natural radiation background (cosmic radiation and terrestrial radionuclides) are listed. The natural radiation doses are larger than the dose limits stated by international organizations (and local legislations) for the maximum admitted exposure of members of the population due to human activities, which amounts to 1 mSv of effective dose per year. This means that mentioned limits imposed by the international legislations are significantly lower than the doses normally and unavoidably received by all human population in the natural environment.

In the previous discussion, the doses received by the patients in the health institutions during medical practices (diagnostic and therapy) are not mentioned. These should be added to the values reported in Table 6.12, to obtain the total doses received yearly by the population. As an example, in European countries, the average doses received during medical practices by a member of the population are around 1.5–2 mSv per year.

In the following, expected effects of irradiation of goods, and in particular foodstuff, with neutron beams utilized for cargo inspection are described.

Possible radioactivity induced by neutron irradiation should be compared with natural radioactivity already present in the same food; the consequent possible internal or external radiological exposures of members of the population should be compared with the annual dose limits of maximum admitted exposure due to human activity stated by international organizations and in particular by European community rules (1 mSv of effective dose for any individual).

The use of ionization, as a method of food preservation, is harmonized in the European Union (EU) on the basis of the establishment of specific treatment conditions and of a list of foodstuffs that can be treated. The labeling of ionized foodstuffs must state that they have been treated by this method. The following norms have been adopted by the European Union:

1. Directive 1999/2/EC of the European Parliament and of the Council of February 22, 1999, on the approximation of the laws of the Member States concerning foods and food ingredients treated with ionizing radiation (framework Directive) (*Official Journal of the European Communities*, L 66 of 13.3.1999).
2. Directive 1999/3/EC of the European Parliament and of the Council of February 22, 1999 on the establishment of a community list of foods and food ingredients treated with ionizing radiation (implementing Directive) (*Official Journal of the European Communities*, L 66 of 13.3.1999). The list of foodstuffs that may be treated with ionizing radiation in different Member States, to the exclusion of all others, and the maximum radiation doses authorized is defined in *Official Journal of the European Communities*, C 112 of 12.5.2006. Maximum authorized absorbed doses ranges from 0.06 kGy for onions in Poland to 10 kGy for flakes and germs of cereals for milk product in Belgium and France, with average values of the order of some kGy.

These Directives apply to the manufacture, marketing, and importation of foods and food ingredients treated with ionizing radiation. They do not apply to the following:

1. Foodstuffs exposed to ionizing radiation, provided that the absorbed dose is not greater than 0.01 Gy for measuring or inspection devices that utilize neutrons and 0.5 Gy in other cases, at maximum radiation energy level of 10 MeV in the case of x-rays, 14 MeV in the case of neutrons, and 5 MeV in other cases.
2. Foodstuffs that are prepared for patients requiring sterile diets under medical supervision. Ionizing radiation is used to reduce the number of pathogenic microorganisms in food ingredients in order to increase the storage life of the end product.

In the United States, the legislation limits and risks associated with the use of ionizing radiation in the inspection of goods that might contain foodstuff was worked out in the past for the commonly employed x-ray scanners. The reference norm is the U.S.-FDA report 21 CFR Part 179, "Irradiation in the Production, Processing and Handling of Food," that defines radiation and radiation sources that can be used for food irradiation, general provisions for food irradiation, limitations in food and foodstuff that can be irradiated, maximum admitted doses (which range from 0.3 to 30 kGy) and admitted packaging.

Under 21 CFR 170.3(i), safety of a food ingredient means that "there is a reasonable certainty in the minds of competent scientists that the substance is not harmful under the intended conditions of use." FDA's regulations reflect the congressional judgment that the additive must be properly tested and such tests carefully evaluated, but that the additive need not, indeed cannot, be shown to be safe to an absolute certainty. The House Report on the Food Additives Amendment of 1958 stated: "Safety requires proof of a reasonable certainty that no harm will result from the proposed use of the additive. It does not—and cannot—require proof beyond any possible doubt that no harm will result under any conceivable circumstance" (House Report 2284, 85th Cong., 2nd sess. 1958).

In the cargo inspection final rule, FDA's evaluations of the safety of radiation for the inspection of cargo containers that may contain food (66 FR 18537) are discussed in detail. Under that regulation, machine sources producing x-rays at energies not greater than 10 MeV may be used to inspect containers of food, provided that the absorbed dose does not exceed 0.5 Gy. Among the reports submitted in the petition or that FDA identified in scientific publications, the agency explicitly cited three in its final rule. These reports, which were among the most recent studies or reviews, assessed the potential for induced radioactivity in food by experimental measurement and theoretical calculation and provided the primary basis for FDA's conclusions regarding safety of the petitioned use of 10 MeV x-rays at a dose not exceeding 0.5 Gy. The first report is from the World Health Organization (WHO 1990); according to it, no detectable radioactivity will be induced in foodstuffs by x-rays with a maximum energy level of 10 MeV, when a radiation dose of 0.5 Gy is not exceeded.

The second report (Wakeford and Blackburn 1991) discussed a study investigating the radioactivity induced in codfish, rice, and a macerated meat product irradiated with high energy bremsstrahlung x-rays produced by an electron linear accelerator that generated electrons at energies up to 12 MeV and predominantly at 8 MeV. The authors reported that the bremsstrahlung x-rays used to irradiate the food had a maximum energy in the region of 10 MeV. These foods received radiation doses ranging from 8.8 to 14 kGy, which is 1.8×10^4 to 2.8×10^4 times higher than the 0.5 Gy maximum dose permitted by the final rule. Induced activities were reported to be extremely small and of the same order of magnitude as natural background levels and any induced activities dropped quickly.

The third report (Findley et al. 1992a) summarized a study that investigated the induced radioactivity in chicken, prawns, cheeses, and spices irradiated with electron beams at two energies, 10 and 20 MeV and at different doses up to 10 kGy. The authors noted that any induced radioactivity was due to photonuclear reactions resulting from bremsstrahlung x-rays and electronuclear reactions induced by the electron beams. The authors found that even when the food was irradiated with electrons at 20 MeV and doses at 10 kGy, the highest energy and dose tested, any induced activity was negligible after 1 day. The authors reported that the measured values agreed well with calculated values.

Based on the totality of the data and other relevant material evaluated by FDA, the agency concluded that no detectable radioactivity will be induced in food when an x-ray energy of 10 MeV and a dose of 0.5 Gy are not exceeded, and that the use of x-rays, produced by a machine source at energies of 10 MeV or lower, to inspect food, is safe.

6.5.1.2 Irradiation with Neutron Beams

In a 2004 document, Food and Drug Administration 21 CFR Part 179, "Irradiation in the Production, Processing, and Handling of Food," the situation about the use of fast neutrons in inspection systems is summarized; it is shortly recalled in the following. In a notice published in the Federal Register of November 18, 1993 (58 FR 60860), FDA announced that a food additive petition (FAP 3M4399) had been filed by Science Applications International Corp. (SAIC), 2950 Patrick Henry Dr., Santa Clara, CA 95054 (the petition in the following). The petition proposed that the food additive regulations in § 179.21 "Sources of radiation used for the inspection of food, for inspection of packaged food, and for controlling food processing" (21 CFR 179.21) be amended to provide for the safe use of a source of fast (high energy) neutrons to inspect cargo containers that may contain food (Gonzani 2004). In a second time (1998), the rights to the petition FAP 3M4399 had been transferred from SAIC to Ancore Corporation (the petitioner in the following).

When the petition was filed on November 18, 1993, it contained an environmental assessment (EA). In the notice of filing for this petition, the agency announced that it was placing the submitted EA on display at the Division of Dockets Management for public review and comment. No comments on the EA were received. Based on the original EA, FDA prepared a finding of no significant impact to the environment dated May 31, 1994.

On July 29, 1997, FDA published revised regulations under part 25 (21 CFR part 25), which became effective on August 28, 1997. On May 12, 2003, the petitioner submitted a claim of categorical exclusion under the new § 25.32(j), in accordance with the procedures in § 25.15(a) and (d). Because the environmental record for the FAP was outdated, the agency reviewed the claim of categorical exclusion under § 25.32(j) for this final rule and found it to be warranted.

6.5.1.3 Definition of Safety for Food Irradiation with Neutrons

A source of radiation used for the purpose of inspection of foods meets the definition of a food additive under section 201 (s) of the Federal Food, Drug, and Cosmetic Act (the act) (21 U.S.C. 321 (s)). Under section 409(c)(3)(A) of the act (21 U.S.C. 348(c)(3)(A)), a food additive cannot be approved for a particular use unless a fair evaluation of the data available to FDA establishes that the additive is "safe" for that use. FDA's food additive regulations in 21 CFR 170.3(i) define safe as "a reasonable certainty in the minds of competent scientists that the substance is not harmful under the intended conditions of use."

The mentioned petition proposes the use of a pulsed fast neutron analysis system employing a beam of high energy neutrons at energies up to 9 MeV to inspect large cargo containers and trucks that may contain food, provided that the maximum dose absorbed by the food does not exceed 0.01 Gy. One of the safety issues considered by FDA is the potential for the formation of products generated in the food by radiation-induced chemical reactions (radiolysis products). The types and amounts of these products depend on the chemical constituents of the food and on the conditions of irradiation. Radiation chemistry of the components of food has been discussed previously in detail in the agency's final rule permitting the irradiation of meat (62 FR 64107). As stated in the meat irradiation final rule, most of the radiolysis products that are generated from food irradiation are also found in foods that have not been irradiated. Some of these compounds are also produced by heating foods; in that case they are produced in amounts far greater than the trace amounts that result from irradiating foods.

The amount of radiolysis products generated in food increases with increasing absorbed dose of radiation. FDA has previously established that gamma rays from cobalt-60 or cesium-137, high-energy electrons up to 10 MeV, and x-rays up to 5 MeV are safe for the treatment of different types of food at doses ranging from 0.3 to 30 kGy, depending on the type of food. Because the petition proposes to limit the maximum absorbed dose to 0.01 Gy (a dose at least 3×10^4 times less than these approved uses), the amounts of radiolysis products generated in food from the petitioned source of radiation will be less than from these approved sources. Accordingly, FDA has concluded that the proposed use is safe in terms of exposure to potential radiolysis products.

Neutrons have a greater propensity to induce radioactivity in scanned materials than x-rays and gamma rays of the same energy. To assess the induction of radioactivity in food by neutron irradiation from a cargo surveillance system, the petitioner submitted a 1992 report (the Harwell Report) that was prepared by Harwell Laboratory of the United Kingdom's Atomic Energy Authority (Findlay 1992b) and a study that was performed by the petitioner itself (Ryge et al. 1992). FDA contracted for an independent evaluation of the data in the petition by the U.S. Department of Energy, Oak Ridge National Laboratory (ORNL) (Easterly et al. 2003). The references provide the primary basis for FDA's conclusion regarding the safety of the petitioned use of neutron radiation.

The Harwell Report (Findlay 1992b) assessed the radiological implications of the use of neutron-based cargo surveillance techniques on cargoes of food. Three cargo scenarios were investigated: semi-infinite slabs (representing inspection of a large transport container of food), 1 kg of food in a 20 kg suitcase (representing airport inspection of a piece of luggage containing a small quantity of food), and 2 m high pallets of food. Calculations were made for 17 different types of food simulating exposure to 0.5 Gy of neutrons (50 times higher than the maximum petitioned dose level of 0.01 Gy) with energies of 1, 2, 5, 8, and 14 MeV. Calculations included induced activities and the resultant doses to consumers after ingesting foods 5 min to 1 month after inspection. Three types of food were considered for the calculations based on the chemical elements of the foods (e.g., calcium, iron, magnesium, sodium, and potassium): (1) a single distribution representing the maximum credible concentrations of the elements in any food, (2) a single "reference" distribution of 47 elements obtained from studies of dietary intake, and (3) distributions corresponding to elemental concentrations in 17 common food types. Of these three distributions, the one considered the most realistic was the single "reference" distribution, because it is based on the daily elemental requirements for "reference" man. For such distribution, the report provided calculations of radiation dose per unit activity intake into the body for induced activities of the neutron irradiated "reference" food at a consumption rate of 2.88 kg of food per day and the resultant dose to reference man after ingesting the foods immediately after inspection and up to 1 month after inspection. Prior to irradiation, the ingestion dose of "reference" food is reported to be 1.823 $\times 10^{-10}$ Sv/g, which is a typical value corresponding to natural radioactivity present in food. The authors calculated that, depending on the energy of the neutron beam, for an absorbed dose in the reference food of 0.5 Gy, the ingestion doses from consuming the "reference" food 1 hour, 8 h, and 1 day after irradiation would range from 9.2×10^{-10} to 3.2×10^{-9} Sv/g, from 5.3×10^{-10} to 1.7×10^{-9} Sv/g, and from 3.7×10^{-10} to 9.2×10^{-10} Sv/g, respectively. As this example and others that follow illustrate, any induced radioactivity is small and dissipates rapidly. Therefore, within 1 day, the ingestion dose from inspected foods would be essentially the same as the one generated by natural radioactivity in the same food.

FDA notes that the Harwell Report addresses a neutron dose 50 times higher than that proposed in the petition and reports radioactivity in the food within 24 h of inspection. Because food subject to this regulation would be inspected at a far lower dose, and would unlikely be consumed within 24 h of inspection considering the logistics of food transportation, any residual induced radioactivity would be well below what occurs naturally.

The calculations provided by the petitioner SAIC-ANCORE (Ryge et al. 1992) were based on computer modeling and estimated the committed effective dose equivalents to adults, children, and infants due to the ingestion of neutron-radiation inspected foods 12 h after exposure to an 8 MeV neutron fluence rate of 5×10^5 n/cm^2/s, for a period of 1 s, corresponding to a dose of 0.021 mGy. The petitioner identified representative foods, the elemental composition of each food, and typical values for the annual amount of each food ingested. The calculated annual effective doses from the consumption of foods that have been irradiated ranged from 3.4×10^{-11} to 2.1×10^{-8} Sv, which is significantly below the annual effective dose from natural radioactivity in food that is reported to be 3.9×10^{-4} Sv per year.

Although the absorbed dose in this study is approximately 500 times less than the maximum petitioned dose level of 0.01 Gy (10 mGy), the calculated annual effective dose from foods inspected with high energy neutrons is 2×10^4 to 1.1×10^7 times less than the annual effective dose from naturally occurring radioactivity in food.

ORNL (Easterly et al. 2003) performed an independent assessment for a subset of foods considered by the petitioner and ingestion doses per unit of food were found to be in general agreement with those presented in the petitioner's supportive documentation. In addition, ORNL designed three extreme-case scenarios regarding consumption of food inspected with pulsed fast neutrons. One scenario assumed the entire diet has been irradiated for 1 s and then consumed 12 h later. This scenario, although highly conservative, is considered to be the most reasonable of the three. The second scenario assumed infrequent (equivalent to 10 full days of dietary needs per year) consumption of food 5 min after it had been irradiated for 1 s. The third scenario assumes infrequent tasting of food immediately after it had been irradiated for 5 min. The calculated annual effective doses for each scenario is 4.8×10^{-8}, 3.4×10^{-7}, and 1.0×10^{-5} Sv, respectively, which are approximately 40 to 8×10^3 times less than the annual effective dose from the consumption of foods due to naturally occurring radioactivity.

Moreover, the petitioner proposed a range of up to 9 MeV, with no lower limit, for the source's average neutron energy. Fast neutrons with energy greater than 1 MeV are necessary to penetrate large cargo containers, whereas lower energy neutrons, including thermal neutrons, have less penetrating power and are more likely to induce radioactivity in food. Therefore, FDA considered whether the data in the petition demonstrate that a source of high energy neutrons would require a lower energy limit to ensure safe use. Although the petitioner originally proposed a neutron energy range up to 9 MeV, the Harwell Report is based on neutron energy levels ranging from 1 to 14 MeV, supporting the safety of neutron energies within that range.

Because the data do not adequately address the issue of induced radioactivity from neutrons of energy below 1 MeV and because neutrons with such energy levels are not explicitly intended to be used, FDA concludes that a minimum energy level requirement of 1 MeV is appropriate. In addition, FDA has also concluded that it is necessary to restrict the neutron source to one that produces monoenergetic neutrons.

FDA concludes that consumption of food inspected by a source of monoenergetic neutrons between 1 and 14 MeV is safe, and that the conditions listed in § 179.21 should be amended as set forth in the following. In accordance with § 171.1(h) (21 CFR 171.1 (h)), the petition and the documents that FDA considered and relied upon in reaching its decision to approve the petition are available for inspection at the Centre for Food Safety and Applied Nutrition.

In 1995, the U.S. petitioner (SAIC-ANCORE) submitted a request to the European Scientific Committee for Food (EU-SCF) to consider whether the use of neutron scanning devices, and pulsed fast neutron analysis (PFNA) in particular, would cause foodstuffs exposed to this type of interrogative irradiation to become a hazard to the health of the consumer (Report 1997). The petitioner provided the documentation relative to references Findlay et al. (1992b) and Ryge et al. (1992).

The opinion of the SCF was as follows: "On the basis of the above information, the SCF has concluded that surveillance devices using neutron scanning, in particular, PFNA systems operating at up to 14 MeV, which during interrogation do not impart to foodstuff radiation doses greater that 0.01 Gy, raise no safety concerns with regard to the negligible induced radioactivity, radiolytic products, their toxicity, nutritional and organoleptic properties of the neutron interrogated foodstuffs. The estimated maximum effective equivalent radiation dose to the individual consumer would amount to less than the 5 pSv due to natural background radiation. However, in order to allay any consumer concerns, it would be desirable to confirm the theoretically calculated induced radioactivity levels by testing food from a container which has been subject to neutron scanning by PFNA system for any induced radioactivity. A procedure should also be developed, which ensures that the same food items are not exposed repeatedly to neutron scanning, as this opinion of the committee applies only to foods which have been subject to a single exposure of PFNA scanning."

Two additional reports were produced, in the past years, devoted to the use of PFNA systems for cargo interrogation as proposed by SAIC-ANCORE

Corporation. In the first one, produced by ORNL (Slater et al. 2000) at the request of Safety and Ecology Corporation of Tennessee, fissile cargo were considered, dose rates at various locations within and just outside the facility building were calculated, and estimates of the activation of structures and typical cargoes were performed.

A second report was produced by NCRP (Tenforde 2002), in the form of letter report, at the request of Sensor Concepts and Applications, Inc., (SCA) of Phoenix, Maryland, working with the U.S. Department of Defense (DoD) and other federal agencies, with the responsibility of the control of the commerce between the United States, Mexico, and Canada. Specific questions addressed in the report were (1) what is the appropriate dose limits for persons inadvertently irradiated by the PFNA system? (2) What are the proper methods to determine the dose received? (3) Can the use of the PFNA system result in levels of activation products in pharmaceuticals and medical devices that might be of concern to public health?

In the former report by ORNL (Slater et al. 2000), fissile cargoes were considered to determine if a significant neutron signal would be observable during interrogation. Results indicate that ample multiplication would be seen for near critical bare targets. On the contrary, if the fissile targets are too subcritical, they may escape detection. A water-reflected near critical sphere showed relatively little multiplication. By implication, a fissile target shielded by hydrogenous cargo might not be detectable by neutron interrogation, particularly if reliance is placed on the neutron signal. The cargo may be detectable if the ample increase in the photon signal can be put to use.

Dose rates were calculated at various locations within and just outside the facility building. These results showed that some dose rates might be higher than the target dose rate of 0.5 μSv/h and, at some peripheral locations, may be at levels of concern for long periods of operation, since the accumulated dose may exceed regulatory limits. However, with limited exposure time, the total dose may be well below the allowed total dose.

Lastly, estimates were made for the activation of system structures and typical cargo. Indeed, the presence of certain elements (such as cobalt, iron, nitrogen, nickel, etc.) in materials undergoing neutron irradiation, can be of concern due to the possible activation of certain isotopes and the nontrivial residual activity still occurring minutes, hours, and even days after irradiation. Therefore, an attempt was made to quantify the activity levels in selected stream of commerce cargoes passing through the main beam in less than 8 s and in structural materials that, on the contrary, are irradiated for hours, day after day. MCNP-4B (Briemeister 1997) was used to calculate neutron fluxes in the regions of interest (at the center of the truck lane, in the collimator, and in front of the concrete beam stop) in a broad three-energy group structure, and the ORIGEN (Herman and Westfall 1984) isotope generation and depletion code was used to calculate residual activity.

Different cargo compositions were tested: salted beef, ball bearings, surgical implants, and fertilizers, whereas it was assumed that structural materials were irradiated continuously for 8 h each day. ORIGEN calculations were performed for 10 days in cycles consisting of 8 h irradiation followed by 16 h decay. Results show that most cargo will not be exposed long enough to be activated to levels of concern. On the other hand, portions of the structure may become activated because of many irradiation/decay cycles experienced over long periods of time and buildup of some radionuclide may reach levels of concern. In particular, portions of the structures lying within the path swept by the vertical collimator will experience the highest activation levels.

In the latter report by CRP (Tenforde 2002), CRP answered to the following three questions:

1. *What is the appropriate dose limits for persons inadvertently irradiated by the PFNA system?* CRP recommended that the PFNA system be designed and operated in a manner that ensures that an inadvertently exposed person will receive an affective dose (E) of less than 1 mSv. CRP further recommended that this limit could be raised to 5 mSv, if necessary, to achieve national security objectives. A limit of 5 mSv is allowed for infrequent annual exposures in CRP's current guidance for exposures to members of the public. In all cases, the PFNA system should be designed and operated in accordance with the principles of keeping exposures "as low as reasonably achievable" (ALARA). In forming this recommendation, CRP considered that an inadvertently exposed person would be exposed only once, at most only a few times, to the PFNA system; that the E limit should be consistent with previous NCRP recommendations (NCRP Report 1993) and provide a level of protection consistent with that accorded to members of the public; that the limit should consider the requirements for protecting individuals of all ages.
2. *What are the proper methods to determine the dose received?* The quantity that must be used to express the radiation level received by an exposed individual is the effective dose (E) (Tenforde 2002). Indeed, the use of fast neutrons for cargo inspection results in a complex distribution of photons and neutrons inside the cargo container. The radiation spectrum that reaches an individual inside the cargo is dependent on the container structural material, the contents of the container, and the position of the individual inside the container volume. In less complicated situations, when only X and gamma rays

interact with the human body, the mean absorbed dose in an organ or tissue D_T (the total amount of the energy deposited in the organ or tissue divided by its mass) is the basic quantity in radiation protection. In this more complicated system, D_T must be modified to reflect both the greater biological effect of neutrons, compared to X and gamma rays, and to the variation of the radiation risk among different organs or tissues in the body.

The values for effective dose (E) delivered by the PFNA system under various irradiation scenarios should be determined by mathematical simulation using radiation transport calculations and confirmed by experimental measurement. NCRP stated that it should be possible to evaluate, prior to routine use of a PFNA system, the potential unintended values of E to individuals associated with a range of irradiation conditions likely to be encountered during the implementation of the PFNA system. Moreover, NCRP stated that it should be possible to monitor appropriate performance characteristics during routine use of the PFNA system that would enable an adequate estimate of E (to an exposed individual) to be made, using the data obtained for the range of irradiation conditions. A following NCRP report (Tenforde 2003b) presented in detail the specific methods and instruments recommended for the measurement and the determination of the effective dose that an individual would receive by inadvertent exposure to radiation from the PFNA system.

3. *Can the use of the PFNA system result in levels of activation products in pharmaceuticals and medical devices that might be of concern to public health?* Since the application of the PFNA system is based on the interaction of neutrons with the nuclei of the atoms, it is unavoidable that some nuclei will be converted to radioactive nuclide. When ingested, or kept in close proximity to the consumer, these activated products deliver a low value of absorbed dose (D) to the individual. A detailed evaluation of D delivered to the maximally exposed individual through consumption or use of activated products would be complex and time consuming, requiring extensive neutron transport calculations as well as detailed evaluation of intake and retention of specific elements in pharmaceuticals and medical devices. However, an informed opinion of the significance of the D values received from activation products can be developed by considering those component elements of pharmaceuticals and medical devices that are likely to produce the highest values for D to consumers, and evaluating the conservative safe estimate of D to organs and tissues from those elements against the established E limit for the general public.

To this aim, NCRP considered a list of thermal and fast neutron activation products of elements with atomic number between 1 and 60, plus gold, platinum, iridium, and bismuth, high-Z elements that occur in pharmaceuticals and medical devices. For the fast neutron activation, a narrow energy distribution, with a peak at 8.5 MeV and a total fluence of 6.4×10^5 n/cm^2 was assumed. This is the specified neutron fluence at the surface of a container scanned by the PFNA system. This approximation neglects the decrease of the total fluence with neutron beam penetration inside the container as well as the lower energy neutrons generated by scattering. However, the structure of the fast neutron activation cross sections that produce the largest values of D justifies this assumption. For thermal neutron activation, a Maxwellian distribution of the neutron energies with 1/E tail (E being the energy of the neutron) extending up to about 0.1 MeV was assumed. Again, a neutron fluence of 6.4×10^5 n/cm^2 was assumed, equivalent to adopting the hypothesis that every fast neutron was reduced to thermal energy inside the container or was scattered back into it after being thermalized in the surrounding shielding. The assumed fluence is significantly overestimated since many of the low energy neutrons will not interact within the container and not contribute to activation. The generated activity of each product was calculated per unit mass of the naturally occurring target using a computer algorithm.

In case activation products might become uniformly distributed throughout the body of the consumer and considering that the quality factor of the radiation generated by radioactive decays is one (gamma rays and beta particles) the equivalent dose for each organ or tissue is numerically equal to the absorbed dose (D). When the data are used to calculate D absorbed by an individual, all the radionuclide remaining at 24 h post irradiation was assumed to contribute to the value of D; that is, the material is ingested and remains in the body permanently. For these calculations, a conservative value of 50 kg for the human body mass was assumed.

In the case of pharmaceuticals, the ^{24}Mg(n,p)^{24}Na reaction is likely to produce the highest absorbed dose due to a fast neutron reaction, because relatively large amount of magnesium, as MgOH (milk of magnesia, recommended dosage 2.7 g/day of magnesium), may be consumed. Assuming 10 g of magnesium, the amount ingested would be 200 weight parts per million (wppm) of the body mass. The calculated absorbed dose rate is 1.58×10^{-11} mGy/h per wppm after 24 h, and the ^{24}Na mean lifetime is 21.64 h. Thus, the total absorbed dose in case of ingestion after 24 h would be 6.84×10^{-8} mGy. The highest absorbed dose from a thermal neutron–activated product may result from ^{23}Na(n,γ)^{24}Na, because of the relatively high amount of sodium that might be consumed. For example, isotonic saline administered intravenously contains 3.45 g/L of sodium. Assuming treatment with 3 L, this would contribute approximately 10 g of sodium, or 200 wppm. The calculated absorbed dose rate is 8.94×10^{-11} mGy/h per wppm after 24 h, so the total absorbed dose, if the sodium is retained in the body, is 3.87×10^{-7} mGy.

In some situations, the irradiated element and the resulting radionuclide may not be uniformly distributed throughout the body. For examples, significant quantities of bismuth may be ingested as bismuth subsalicylate, an ingredient in an over-the-counter digestive remedy. Bismuth probably is retained in the contents of the intestinal tract. The recommended maximum dosage is equal to 2.5 g of bismuth in 48 h, and the weight of the colon is about 1 kg, resulting in a concentration of 2500 wppm. The resulting absorbed dose is 1.32×10^{-8} mGy if the bismuth is retained until it has all decayed. In this ease, the contribution to the effective dose (E) is lower because the tissue weighting factor for the colon is 0.12. All the absorbed doses calculated for activated soluble products are below 1.0×10^{-6} mGy. Since the gamma and the beta radiations from these radionuclides have a radiation weighting factor of one, the results is 10^{-6} times the recommended annual limit for E of 1.0 mSv. Therefore, these absorbed doses would contribute minimally to the value of E compared to the limit for E for the general public.

The significance of activation of implanted medical devices can be evaluated based on the absorbed dose rate at a distance from the device. For example, 1 g of cobalt, perhaps in the form of a stainless steel part used in orthopedic reconstruction, would deliver an absorbed dose of 1.25×10^{-10} mGy/h at 5 cm. In all cases the absorbed doses to organs and tissues would result in very low effective doses, and are not of further concern. Thus, the opinion of NCRP was that the activation of pharmaceuticals and medical devices by PFNA system would not result in effective doses of concern for public health.

In a following NCRP report (Tenforde 2003a), the effects of the activation of foodstuff were considered. These effects are similar to those of pharmaceuticals and medical devices. Daily intake of most elements, except hydrogen, oxygen, carbon, and sodium, in the form of food is typically less that the recommended intake in the form of dietary supplements. Thus the doses to these elements in food will be less than those found for pharmaceuticals. The highest dose from activated food would generally be from the ^{24}Na produced by capture reaction with natural sodium in salty processed food such as potato chips. Assuming an individual ingests 10 g sodium from a salty food immediately after PFNA inspection, the resulting dose would be approximately 1.0×10^{-6} mGy. All other elements commonly found in foods would produce much lower doses.

The effects of irradiation of goods and foodstuff both with x-rays and electron beams and with neutrons have been exhaustively investigated in the studies mentioned in the previous paragraphs. Therefore, it appears not necessary, at least in a first approximation, to repeat the same studies for any future neutron inspection systems. Conclusions on the future installation can be easily inferred from the comparison of its irradiation conditions with the limits set for x-ray irradiation and with the conditions in which the results for the PFNA irradiation system, mentioned earlier, have been obtained.

6.5.1.4 Eritr@C System

Biological shielding assessment and dose rate calculation for a neutron inspection portal EURITRACK is described in a paper by Donzella et al. (2012) and report (Donzella et al. 2010). They have confidently concluded that the activation of foodstuff, pharmaceuticals, or medical devices by Eritr@C systems would not result in effective doses of concern for public health.

In their report, possible risks connected with the irradiation of goods and in particular foodstuff, pharmaceuticals and medical devices, by EURITRACK/Eritr@C installations are evaluated and assessed. Two aspects of the problem were investigated in particular. The first one concerns possible modifications of the properties of the irradiated goods or materials. This problem is particularly relevant for foodstuff, whose nutritional and organoleptic properties must not be altered or modified by any treatment. The second one concerns the possibility that activation induced by neutron collisions in goods and in particular in foodstuff and pharmaceuticals may result in contamination and committed doses to individuals after ingestion. This second problem may be relevant when using neutron beams due to the propensity of fast and thermal neutrons to generate activated nuclides in different elements by inelastic and capture reactions, respectively.

Modifications of the Properties of the Irradiated Goods or Materials, in Particular, Foodstuff

Since the use of ionizing radiation, not only for food preservation but also for nonintrusive inspection of luggage, air cargoes, and shipping containers, has greatly increased recently, several studies have been performed to assess the effects and risks for the public health and public authorities, in different countries, have adopted norms to regulate this use.

As already mentioned, the following norms have been adopted by the European Union: Directive 1999/2/EC of February 22, 1999, concerning foods and food ingredients treated with ionizing radiation (framework Directive) and Directive 1999/3/EC of February 22, 1999, on the establishment of a community list of foods and food ingredients treated with ionizing radiation, implementing Directive (*Official Journal of the European Communities*, L 66 Of 13.3.1999). The list of foodstuffs that may be treated with ionizing radiation in different Member States, to the exclusion of all others, and the maximum radiation doses authorized is also defined (*Official Journal of the European Communities*, C 112 of 12.5.2006). Average values of the admitted doses are of order of some kGy. The Directives do not apply to foodstuffs exposed to ionizing radiation, provided that the absorbed dose is not greater than 0.01 Gy for measuring or inspection devices that utilize neutrons and 0.5 Gy in other cases, at maximum radiation energy level of 10 MeV in the case of x-rays, 14 MeV in the case of neutrons, and 5 MeV in other cases.

Therefore, irradiation with neutron beams of 14.1 MeV that cause an absorbed dose of less than 0.01 Gy can be considered safe for the preservation of nutritional and organoleptic properties of foodstuff. Calculations of the absorbed doses after neutron irradiation in typical conditions for Eritr@C installations have been performed by means of the MCNP modeling of Eritr@C installations (Zenoni et al. 2009). The absorbed doses by food (red delicious apples) placed in different positions inside the container after a 10 min irradiation amount, in the worst case, to some 10^{-6} Gy, several orders of magnitude lower than the limit of 0.01 Gy established for the application of the norms.

Induced Radioactivity by Neutron Beams

Several reports by reference public institutions and agencies have been devoted to the study of effects of neutron irradiation of goods and in particular foodstuff and pharmaceuticals with reference to the PFNA project (Findlay et al. 1992a,b; Slater et al. 2000; Tenforde 2002, 2003b; Easterly et al. 2003). Concerning irradiation of foodstuff, a common result of these studies is that the ingestion dose from inspected foods, in the worst case, would be well below the one generated by natural radioactivity in the same food. The opinion of NCRP for the irradiation of pharmaceuticals and medical devices is that activation by PFNA systems would not result in effective doses of concern for public health.

On the basis of some of these studies, FDA concluded that the consumption of food inspected by a source of monoenergetic neutrons between 1 and 14 MeV in the PFNA conditions is safe (U.S.-FDA Report 21 CFR Part 179—Docket No. 93F-0357). The European Scientific Committee for Food (EU-SCF) concluded, basing on the same studies, that surveillance devices using neutron scanning, in particular PFNA systems operating at up 14 MeV, which during interrogation do not impart to foodstuff radiation doses greater that 0.01 Gy, raise no safety concerns with regard to the negligible induced radioactivity, radiolysis products, toxicity, and nutritional and organoleptic properties of the neutron-interrogated foodstuffs (Reports—European Commission 1997).

PFNA and EURITRACK/Eritr@C systems are based on the same principle of fast neutron interrogation of the elemental composition of a material, even if they differ in several important aspects concerning the neutron source and the interrogation procedure. However, in spite of these differences, the total neutron fluences needed to obtain adequate answers from the interrogation are essentially the same. Reference total neutron fluence assumed for PFNA is 6.4×10^5 n/cm^2 separately for fast and thermal neutron components; calculations for an Eritr@C installation are reported and showed that the total fluence (including fast and thermal neutron components) ranges from 6.4×10^5 n/cm^2 in. on axis positions to 7.8×10^2 n/cm^2 in. off axis positions. For this reason, the already comfortable results obtained by the studies devoted to PFNA systems can be confidently scaled to the Eritr@C conditions.

Nevertheless, activation calculations in typical Eritr@C conditions have been directly performed for two reactions that are likely to produce the highest absorbed dose for fast and thermal neutron components, respectively, $^{24}Mg(n,p)^{24}Na$ and $^{23}Na(n,\gamma)^{24}Na$, because of the large amount of magnesium and sodium that might be consumed in pharmaceuticals and in foodstuff (several grams per day). Obtained results in Eritr@C conditions for residual specific activity of ^{24}Na after neutron interrogation are shown to be similar or lower than those obtained in the PFNA case (Tenforde 2002). Specific activities generated by an Eritr@C 10 min irradiation for the most relevant activation reactions on the pure elements are orders of magnitude less than the specific activities present in some common type of food products due to natural radioactivity. Consequent committed doses in the case of ingestion of foods activated after neutron interrogation are orders of magnitude lower than doses normally received by any individual of the public due to natural radioactivity.

Therefore, it has been concluded that cargo containers can be inspected with Eritr@C installation without any limitation due to the type of transported goods. The only concern raised in the ORNL report (Slater et al. 2000) for the PFNA system regards possible activation of fixed structures of the installation, containing elements such as cobalt, iron, nitrogen, and nickel, which may become activated because of many irradiation/decay cycles experienced over long periods of time. Specific calculations should be addressed to evaluate and assess this particular radiological risk for the Eritr@C installations.

Irradiation of Inadvertently Exposed Individuals (Stowaways)

Finally, it is worth recalling the problem of possible irradiation of inadvertently exposed individuals (stowaways), which was specifically addressed for the Eritr@C installations in the Eritr@C report (Zenoni et al. 2009). Calculations performed in that report show that, in the extreme case of a stowaway hidden in the centre of a void container, the resulting value for the neutron ambient dose equivalent H* (10) rate is around 750 μSv/h, whereas the contribution from photons is at least a factor 100 lower and can be neglected. Close to the container walls, on the axis of the neutron beam, the dose rate increases by a factor two.

Therefore, the total ambient dose equivalent H* (10) accumulated in one single Eritr@C irradiation, lasting 10 min, is not larger than 0.25 mSv. In the case of uniform irradiation, H* (10) represent quite well the effective dose E, therefore, the dose is less than the annual effective dose limit foreseen by European

Legislation for the public. The dose limit foreseen by NCRP for PFNA inspection systems in one irradiation is 1.0 mSv and may be raised to 5.0 mSv, if necessary, to achieve national security objectives.

It is worth remarking that the probability of inadvertently irradiating a stowaway hidden in the cargo container, with an Eritr@C system, should be almost negligible, given the availability of the prior x-ray image of the container content.

6.5.1.5 Conclusions

When considering possible alteration of foodstuff nutritional and organoleptic properties following is of importance. For what concerns irradiation of foodstuff with x-ray for food preservation, the Directive 1999/2/EC of the European Parliament and of the Council of February 22, 1999, on the approximation of the laws of the Member States concerning foods and food ingredients treated with ionizing radiation, constitutes the reference norm in Europe framework Directive (*Official Journal of the European Communities*, L 66 of 13.3.1999).

As previously mentioned, this Directive defines its own limits of application and shall not apply to foodstuffs exposed to ionizing radiation generated by measuring or inspection devices provided that the dose absorbed is not greater than 0.01 Gy for inspection devices that utilize neutrons and 0.5 Gy in other cases, at a maximum radiation energy level of 10 MeV in the case of x-rays, 14 MeV in the case of neutrons, and 5 MeV in other cases. Member States shall take all measures necessary to ensure that irradiated foodstuffs can be placed on the market only if they comply with the provisions of this Directive.

Therefore, irradiation with neutron beams of 14.1 MeV that cause an absorbed dose of less than 0.01 Gy can be considered safe for the preservation of nutritional and organoleptic properties of foodstuff. It is easy to estimate, by Monte Carlo calculation using modeling of particular installations by MCNP5 transport code (MNCP 2003), the dose absorbed by different kinds of food and foodstuff in typical irradiation conditions.

Therefore, in conclusion, systems described in this and the next chapter would absolutely not result either in any modifications of the properties of the irradiated material or in effective doses of concern for public health.

6.6 FUSION WITH OTHER METHODS

When the report by Yang et al. (2013), the authors have discussed the fusion of x-ray imaging and photoneutron-induced gamma analysis for contraband detection. In their work, a 7 MV LINAC-based photoneutron interrogating system has been setup to fulfill the demand of contrabands detection in homeland security. Both x-ray imaging and photoneutron-induced γ-ray analysis are used to extract the information of inspected materials. Four hundred and eighty CsI detectors of 5 mm (height) × 10 mm (width) × 20 mm (length) are used to form the detector array to measure the attenuation information of penetrating x-rays. Sixteen NaI(Tl) detectors of 127 mm (height) × 127 mm (diameter) are used to register the photoneutron-induced γ-ray spectra of inspected materials. Two-dimensional elemental distributions of H, N, Fe, and Cl are extracted by calculating the area of 2.223, 10.829, 7.64/9.298 MeV, and 1.165/1.951/1.959 MeV γ-ray peaks in the spectra measured by NaI(Tl) arrays, respectively. Mixed materials like iron, salt water, melamine, and sugar are scanned to test the contraband detention capability. The images of x-ray attenuation and four elemental two-dimensional distributions are fused together to separate suspicious materials. Areas with high concentrations of nitrogen and chlorine are easily identified and can indicate the potential existence of illicit substances.

As reported by Sowerby et al. (2009), the Commonwealth Science and Industrial Research Organisation (CSIRO) has developed a scanner that combines fast (14 MeV) neutron and gamma-ray (or X-ray) radiography (FNGR). This direct, dual-beam radiography technique is much more efficient than methods that measure secondary radiation and has much better material discrimination sensitivity than the dual-energy x-ray technique. A full scale prototype FNGR scanner was trialed by Australian Customs Service to screen incoming air cargo at Brisbane International Airport in 2005/2006. The trial of the scanner demonstrated the material discrimination capability of the technology and its ability to make concealed organic materials more obvious; CSIRO and Nuctech Company Limited have developed a new version of the scanner suitable for commercial deployment that combines a 14 MeV fast neutron radiography system with dual-energy x-ray radiography. The x-ray system uses a 6 MeV LINAC x-ray source and binocular stereoscopic imaging with much better spatial resolution than the scanner trialed at Brisbane airport. The improved resolution, combined with binocular stereoscopic imaging, allows complex cargo images to be separated into multiple layers, making it easier to identify threat items. The first unit of the commercial scanner was recently commissioned in Beijing and results are presented in Sowerby et al. (2009). The system implemented in their scanner utilizes a single x-ray source and twofolded x-ray detector arrays (see Figure 6.19). This system allows complex cargo images to be separated into multiple layers, making it easier to identify threat items.

The AC6015XN scanner has been evaluated on a range of test objects and cargos. Both fast neutrons and high energy x-rays have sufficient penetration for air cargo screening.

However, the penetration of fast neutrons depends much more on the atomic number than it does for high-energy x-rays, and the most significant obstacle

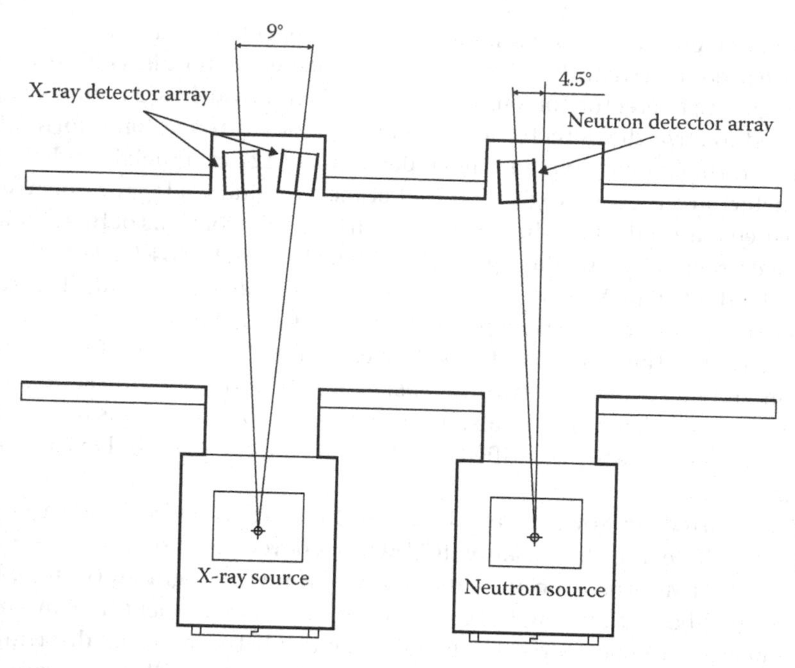

Figure 6.19 Schematic plan view of the layout of the x-ray and neutron beams in the scanner as described by Sowerby et al. (2009). (From Sowerby, B.D. et al., Recent developments in fast neutron radiography for the interrogation of air cargo containers, in *IAEA Conference*, Vienna, Austria, May 4–8, 2009, Paper SM/EN-01, 2009.)

to the application of the FNGR technique to the scanning of sea freight is the neutron penetration required to image through thick, organic cargos.

In addition to the report by Runkle et al. (2009), there is a number of interesting papers discussing the subject of using fast neutrons for the inspection of air cargo containers; see Ederhardt et al. (2005), Liu et al. (2008), and Eberhardt et al. (2007).

An interesting development is so-called Xn tube, its principle being shown in Figure 6.20.

The proposed novel techniques for explosive and fissile material detection make use of the peculiar capability of producing a tagged neutron beam to confine the inspection to a predetermined volume element. A straightforward application of these techniques would imply coupling the inspection by tagged neutron beams to a commercial imaging device based on either x-ray or gamma-ray radiography that performs a fast scan of the container, identifies a "suspect" region and provides its coordinates to the neutron-based device for the final "confirmatory" inspection. This is to be discussed in great detail in the next chapter.

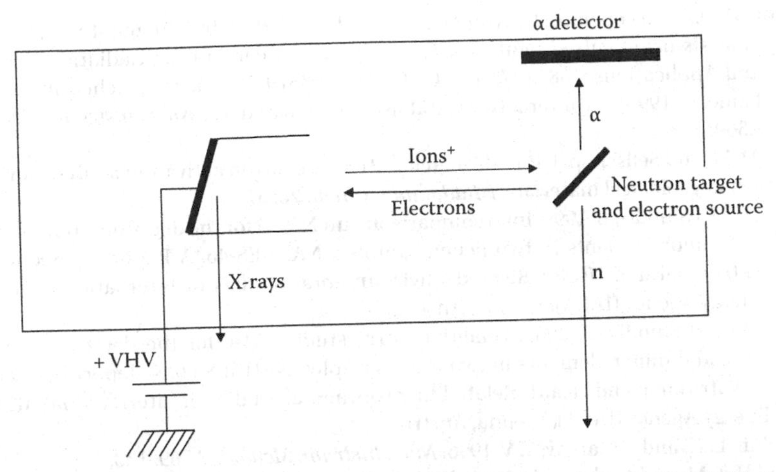

Figure 6.20 The principle of Xn tube. (After Le Tourneur, P. et al., ULIS: A portable device for chemical and explosive detection, paper SM/EN-17 in *Proceedings of the International Topical Meeting on Nuclear Research Applications and Utilization of Accelerators*, IAEA, Vienna, Austria, 2009.)

REFERENCES

Aleksandrov, V. D., Bogolubov, E. P., Bochkarev, O. V. et al. 2005. Application of neutron generators for high explosives, toxic agents and fissile material detection. *Appl. Radiat. Isot.* 63: 537–543.

Al-Shahristani, H. and Jervis, R. E. 1973. On-stream nuclear activation of trace uranium and thorium using a 14 MeV neutron generator. *J. Radioanal. Chem.* 13: 459–468.

Arcos, J. M. L., Bailador, A., Gonzalez, A. et al. 2000. The Spanish National Reference Database for Ionizing Radiations (BANDRRI). *Appl. Radiat. Isot.* 52: 335–340.

Audi, G., Bersillon, O., Blachot, J., and Wapstra, A. H. 1996. The NUBASE evaluation of nuclear and decay properties. *Nucl. Instrum. Methods* A369: 511–515. See also: *Nucl. Phys.* A729: 3–128, 2003.

Bach, P., Ma, J. L., Froment, D., and Jaureguy, J. C. 1993. Chemical weapons detection by fast neutron activation analysis techniques. *Nucl. Instrum. Methods Phys. Res.* B79: 605–610.

Barnaby, F. 1994. Red mercury: Is there a pure-fusion bomb for sale?. International Defence Review 6: 94.

Barzilov, A. and Womble, P. C. 2003. NELIS - A neutron inspection system for detection of illicit drugs. *AIP Conf. Proc.* 680: 939–942, ISSN 1551-7616.

Barzilov, A., Novikov, I. S., and Womble, P. C. 2012. Material analysis using characteristic gamma rays induced by neutrons, in Adrovic, F. (Ed.) *Gamma Radiation.* Intech Europe, Rijeka, Croatia. pp 17–40.

Batyaev, V. F., Bochkarev, O. V., Larionov, P. V., Sklyarov, S. V., Belevitin, A. G., Romodanov, V. L., and Chernikova, D. N. 2013. A pulsed neutron generator-based system for measuring fissile material mass in waste containers, in *Proceedings of the International Conference on Portable Neutron Generators and Technologies on Their Basis (in Russian)*, VNIIA, Moscow, October 22–26, 2012, pp. 460–467.

Belbot, M. D., Vourvopoulos, G, Womble, P. C. and Paschal, J. 1999. Elemental online coal analysis using pulsed neutrons. *Proc. SPIE* 3769, Penetrating Radiation Systems and Applications, 168–177; doi:10.1117/12.363679. Be, M. M., Duchemin, B., and Lame, J. 1996. An interactive database for decay data. *Nucl. Instrum. Methods* A369: 523.

Bibby, D. M. and Sellschop, J. P. F. 1974. Accurate fast neutron activation analysis for oxygen in geological materials. *J. Radioanal. Chem.* 20: 677.

Bleise, A. and Smodis, B. 1999. Intercomparison run NAT-5 for the determination of trace and minor elements in two lichen samples. NAHRES-46, A Report by Section of Nutritional and Health Related Environmental Studies of International Atomic Energy Agency (IAEA), Vienna, Austria.

Bleise, A. and Smodis, B. 2001. Quality control study NAT-6 for the determination of trace and minor elements in two moss samples. NAHRES-66, A Report by Section of Nutritional and Health Related Environmental Studies of International Atomic Energy Agency (IAEA), Vienna, Austria.

Boboshln, L. N. and Varlamov, V. V. 1996. *Nucl. Instrum. Methods* A369: 133.

Bowen, H. J. M. and Gibbons, O. 1963. *Radioactivation Analysis.* Oxford University Press, London, U.K.

Briesmeister, J. F. 1997. MCNP - A general Monte Carlo N-particle transport code. Version 4B, LA- 2625-M.

Brooks, F. D., Drosg, M., Buffler A., and Allie, M. S. 2004. Detection of anti-personnel landmines by neutron scattering and attenuation. *Appl. Radiat. Isot.* 61: 27–34.

Brown, D. R., Bozorgmanesh, H., Gozani, T., and McQuaid, J. 1982. On-line nuclear analysis of coal and its uses, in Filby, R. H., Carpenter, B. S., and Ragaini, R. C. (Eds.) *Atomic & Nuclear Methods in Fossil Energy Research.* Springer, New York, pp. 155–162.

Buffler, A. 2004. Contraband detection with fast neutrons. *Radiat. Phys. Chem.* 71: 853–861.

Coleman, W., Ginaven, R., and Reynolds, G. 1974. Nuclear methods for mine detections. AD A167 968, U.S. Army Mobility Equipment Research and Development Center, Ft. Belvoir, Virginia, final report for contract DAAK02–73-C-0139.

Csikai, J. 1987. *Handbook of Fast Neutron Generators.* CRC Press, Inc., Boca Raton, FL.

Csikai, J., Al-Jobori, S. M., Buczkč, M., and Szegedi, S. 1982. Determination of major and trace elements in crude oils by neutron activation and reflection methods. *J. Radioanal. Chem.* 71(1–2): 215–223.

Czubek, J. A. 1972. Pulsed neutron method for uranium well logging. *Geophysics* 37(1): 160–173.

De Soete, O., Gilbels, R., and Hoste, J. 1972. Chemical analysis: A series of monographs on analytical chemistry and its applications, in Elving, P. J. and Kolthoff. I. M. (Eds.) *Neutron Activation Analysis,* Vol. 34. John Wiley & Sons. New York.

Dep, L., Belbot, M., Vourvopoulos, G., and Sudar, S. 1998. Pulsed neutron-based on-line coal analysis. *J. Radioanal. Nucl. Chem.* 234(1–2): 107–112.

Donzella, A., Bonomi, G., Giroletti, E. et al. 2010. Irradiation of container cargoes with neutron inspection devices, WP5 D5.3 of the EU project ERITR@C.

Donzella, A., Bonomi, G., Giroletti, E., and Zenoni, A. 2012. Biological shielding assessment and dose rate calculation for a neutron inspection portal. *Radiat. Phys. Chem.* 81: 414–420.

Dougan, A. D., Snyderman, N. J., Nakae, L. F. et al. 2007. New and novel nondestructive neutron and gamma-ray technologies applied to safeguards, presented at *2007 JAEA-IAEA Workshop on Advanced Safeguards Technology for the Future Nuclear Fuel Cycle*, #4A.2, November 13–16, 2007, Tokai-mura, Japan.

Dybczynski, R. 1980. NAA as viewed from the perspective of International Atomic Energy Agency intercomparaison tests. *J. Radioanal. Chem.* 60: 45–54.

Easterly, C. E., Eckerman, K. F., Ross, R. H., and Opresko, D. M. 2003. Assessment of petition to use pulsed fast neutron analysis (PFNA) in inspection of shipping containers containing foods. Oak Ridge National Laboratory, Oak Ridge, TN.

Eberhardt, J., Liu, Y., Roach, G., Sowerby, B., and Tickner, J. 2007. Feasibility study of the detection of special nuclear materials in air cargo using fast-neutron/gamma-ray radiography, presented at *2007 IEEE Nuclear Science Symposium (NSS) and Medical Imaging Conference*, Honolulu, Hawaii, October 28–November 3, 2007.

Eberhardt, J. E., Rainey, S., Stevens, R. J., Sowerby, B. D., and Tickner, J. R. 2005. Fast neutron radiography scanner for the detection of contraband in air cargo containers. *Appl. Radiat. Isot.* 63: 179–188.

Ehmann, W. D. and Morgan, J. W. 1970a. Oxygen, silicon and aluminium in Apollo 11 rocks and fines by 14 MeV neutron activation, in *Proceedings of Apollo 11 Lunar Science Conference*, Vol. 2, Houston, TX, January 5–8, 1970. Pergamon Press, New York; *Geochim. Et Cosmochim. Acta* Supp. 1: 1071–1079.

Ehmann, W. D. and Morgan, J. W. 1970b. Oxygen, silicon and aluminum in lunar samples by 14 MeV neutron activation, in *Proceedings of Apollo 11 Lunar Science Conference*, Houston, TX, January 5–8, 1970, p. 528.

Ehmann, W. D. and Morgan, J. W. 1971. Major element abundances in Apollo 12 rocks and fines by 14 MeV neutron activation, in *Proceedings of the 2nd Lunar Science Conference*, January 11–14, 1971, p. 1237, Houston, Texas.

Ehmann, W. D. and Vance, D. E. 1991. *Radiochemistry and Nuclear Methods of Analysis*. Wiley, New York.

Ehmann, W. D., Koppenaal, D. W., and Khalil, S. R. 1982. Fast neutron activation analysis of fossil fuels and liquid-fraction products, in Filby, R. H. (Ed.) *Atomic and Nuclear Methods in Fossil Energy Research*. Plenum Press, New York, pp 69–82.

Eleon, C., Perot, B., Carasco, C., Sudac, D., Obhodas, J., and Valkovic, V. 2010. Experimental and MCNP simulated gamma-ray spectra for the UNCOSS neutron-based explosive detector. *Nucl. Instrum. Methods A* 629: 220–229.

Elias, E., Segaland, Y., Notea, A. 1976. Dual Gaugeing Utilizing Penetrating and Scattered Photon Fluxes. *Nucl.Tech.* 33: 305–313.

Elias, E., Gozani, T., Orphan, V., Reed, J., and Shreeve, D. 1979. Prompt neutron activation analysis - Applications to coal analysis. *Trans. Am. Nucl. Soc.* 26: 160–161.

Elmasri, R. and Navathe, S. B. 2000. *Fundamentals of Database Systems*, 3rd edn. Pearson Education Inc., Delhi, India.

Ene, A. 2004. Improvement of sensitivity in PIGE analysis of steels by neutron-gamma coincidences measurement. *Nucl. Instrum. Methods Phys. Res.* B222: 228–234.

Ene, A. 2011. Analytical applications of thermal and 14 MeV neutron activation analysis in rnetallurgical industry, in *The 3rd Joint Seminar JINR-Romania on Neutron Physics for Investigations of Nuclei, Condensed Matter and Life Sciences*, Targoviste, Romania, July 24–30, 2011, pp. 34–35. Book of Abstracts. Bibliotheca Publishing House, Targoviste, Romania.

Ene, A. and Frontasyeva, M. V. 2013. Applications on neutron activation analysis technique in element determination at trace level, in TEHNOMUS J., *17th International Conference on New Technologies and Products in Machine Manufacturing Technologies*, May 17–18, 2013, Suceava, Romania, pp. 165–171.

Faust, A. A., McFee, J. E., Bowman, C. L. et al. 2011. Feasibility of fast neutron analysis for the detection of explosives buried in soil. *Nucl. Instrum. Methods Phys. Res. A* 659: 591–601.

Findley, D. J. S., Parson, T. V., and Sene, M. R. 1992a. Experimental electron beam irradiation of food and the induction of radioactivity. *Appl. Radiat. Isot.* 43: 567–575.

Findlay, D. J. S., Forrest, R. A., and Smith, G. M. 1992b. Neutron-induced activation of food (Harwell Report). UK Atomic Energy Authority Industrial Technology *Report* AEA-InTec-1051.

Goldmann, D. 1998. Improvised primary explosives. http://www.libertyreferences.com/Improvised_Primary_Explosives.pdf.

Gozani, T., Evans, E., Orphan, V., Reed, J. and Shreve, D. 1977. Prompt Neutron Activation Analysis—Applications to Coal Analysis *Trans. Am. Nucl. Soc.* 26: 160–161.

Gozani, T. 1997. Neutron based non-intrusive inspection techniques. *Proc. Int. Soc. Opt. Eng.* 2867: 174–181.

Gozani, T. 1978. The development of continuous nuclear analysis of coal - A review. *Trans. Am. Nucl. Soc.* 28: 97.

Gozani, T. 1999. A review of neutron based non-intrusive inspection techniques. http://media.hoover.org/documents/gozani_TechConf.pdf, downloaded March 19, 2014.

Gozani, T. 2002. A review of neutron based non-intrusive inspection technologies, in *Conference on Technology for Preventing Terrorism*, Stanford, CA, March 12–13, 2002. pp 1–12.

Gozani, T. 2004. The role of neutron based inspection techniques in the post 9/11/01 era. *Nucl. Instrum. Methods B* 213: 460–463.

Gozani, T., Bozorgmanesh, H., Brown, D., Elias, E., Maung, T., and Reynolds, G. 1978. Coal elemental analysis by prompt neutron activation analysis. *Trans. Am. Nucl. Soc.* 28: 88.

Gozani, T., Bozorgmanesh, H., Brown, D. et al. 1979. Nuclear assay of coal, Volumes 1–8, EPRI Report FP989. RP983-1.

Grant, P. M., Moody, K. J., Hutcheon, I. D. et al. 1998. Nuclear forensics in law enforcement applications. *J. Radioanal. Nucl. Chem.* 235(1–2): 129–132.

Guzzi, G., Pietra, R., and Sabbioni, E. 1976. Determination of 25 elements in biological standard reference materials by neutron activation analysis. *J. Radioanal. Chem.* 34: 35–57.

Hamrin, C. E., Jr., Johannes, A. H., James, W. D., Jr., Sun, G. H., and Ehmann, W. D. 1979. Determination of oxygen and nitrogen in coal by instrumental neutron activation analysis. *Fuel* 58: 48–54.

Haskin, L. A., Wildeman, T. R., and Haskin, M. A. 1968. An accurate procedure for the determination of the rare earths by neutron activation. *J. Radioanal. Chem.* 1: 337–348.

Hayward, C. C., Oldham, G., and Ware, A. R. 1971. 14 MeV neutron activation analysis applied to a nonaqueous flowing system. *J. Radioanal. Chem.* 7: 341–346.

Hermann, O. W. and Westfall, R. M. 1984. ORIGEN-S: SCALE system module to calculate fuel depletion, actinide transmutation, fission product buildup and decay, and associated radiation source terms. Section F7 of *SCALE: A Modular Code System for Performing Standardized Computer Analyses for Licensing Evaluation*, Book III, Vol. 2, Oak Ridge National Laboratory Report ORNLINUREG/CSD-2 (NUREG/CR-0200).

Holslin, D. T., Shyu, C. M., Sullivan, R. A., and Vourvopoulos, G. 2006. PELAN for non-intrusive inspection of ordnance, containers, and vehicles. *Proc. SPIE* 6213: 621307.

Hurwitz, M. J., Smith, R. C., Noronha, W. P., and Tran, K.-C. 1992. Detection of Illicit Drugs in Cargo Containers. *Proceedings of the Contraband and Cargo Inspection Technology International Symposium*, Office of National Drug Control policy, The White House, Washington, DC, p. 29.

IAEA. 2003. *Proceedings of the Final RCM on the Application of Nuclear Techniques to Anti-Personnel Landmines Identification*, September 15–18, 2013, IAEA, Vienna, Austria.

IAEA. 2012. Neutron generators for analytical purposes. IAEA Radiation Technology Reports No. 1, IAEA, Vienna, Austria.

ISO. 1987. International Standards Organizations, International vocabulary of basic and general terms in metrology, Geneva, Switzerland.

ISO 17025. 1999. General requirements for the competence of testing and calibration laboratories. International Standards Organization, Geneva, Switzerland.

James, W. D. and Fuerst, C. D. 2000. Overcoming matrix effects in the 14-MeV fast neutron activation analysis of metals. *J. Radioanal. Nucl. Chem.* 244: 429–434.

James, W. D., Ehmann, W. D., Hamrin, C. E., and Chyi, L. L. 1976. Oxygen and nitrogen in coal by instrumental neutron activation analysis. *J. Radioanal. Chem.* 32: 195–205.

Janghorbani, M., Miller, M. O., Ma, M. S., Chyi, L. L., and Ehmann, W. O. 1973. Oxygen and other elemental abundance data for Apollo 14, 15, 16, and 17 samples, in *Proceedings of the 4th Lunar Science Conference*, March 5–8, 1973, p. 1115 , Houston, Texas.

Khalil, S. R., Koppenaal, D. W., and Ehmann, W. D. 1980. Oxygen concentrations in coal and fly ash standards. *Anal. Lett.* 13(A12): 1063–1071.

Kliment, V. and Tolgyessy, J. 1972. Continuous neutron activation analysis. *J. Radioanal. Chem.* 10: 273–297.

Kliment, V. and Tölgyessy, J. 2007. Continuous neutron activation analysis. *J. Radioanal. Nucl. Chem.* 10: 273–297.

Koch, R. C. 1960. *Activation Analysis Handbook*. Academic Press, New York.

Kolotov, V. P. and De Corte, F. 2003. An electronic database with a compilation of k(0) and related data for NAA. *J. Radioanal. Nucl. Chem.* 25: 501–508.

Koltick, D. S., Kim, Y., McConchie, S., Novikov, I., Belbot, M., and Gardner, G. 2007. A neutron based vehicle-borne improvised explosive device detection system. *Nucl. Instrum. Methods Phys. Res. B* 261: 277–280.

Krivan, V. and Krivan, K. 1976. Tabulation of calculated data on primary reaction interferences in 14 MeV neutron activation analysis. *J. Radioanal. Chem.* 29: 145–173.

Kruger, P. 1971. *Principles of Activation Analysis*. John Wiley & Sons, New York.

Lamza, R. C. 1997. Explosive detection with application to landmines. Report of an Advisory Group Meeting, IAEA, Vienna, Austria, 9-12.12.

Lanza, R. C. 2007. Nuclear techniques for explosive detection: Current prospects and requirements for future development, in *Combined Devices for Humanitarian Demining and Explosive Detection*. International Atomic Energy Agency, Vienna, Austria, ISBN 978-92-0-157007-9.

Laul, J. C. 1979. Neutron activation analysis of geologic materials. *At. Energy Rev.* 17: 603–695.

Laul, J. C. and Wogman, N. A. 1981. 14 MeV neutron activation analysis of geological and lunar samples. *IEEE Trans. Nucl. Sci.* 28: 1703–1705.

Le Tourneur, P., Dumont, J. L., Groiselle, C. et al. 2009. ULIS: A portable device for chemical and explosive detection, paper SM/EN-17 in *Proceedings of the International Topical Meeting on Nuclear Research Applications and Utilization of Accelerators*, IAEA, Vienna, Austria, pp 1–29.

Lehnertr, A. L. 2012. A flag-based algorithm for explosive detection in sea-cargo containers using active neutron interrogation. PhD thesis, University of Michigan, Ann Arbor, MI, http://hdl.handle.net/2027.42/91602.

Lehnert, A. L. and Kearfott, K. J. 2011. Preliminary identification of flags for a novel algorithm-based approach for explosives detection using neutron interrogation for simulated idealized cargo container scenario. *Nucl. Instrum. Methods Phys. Res. A* 638: 201–205.

Lehnert, A. L., Whetstone, Z. D., Zak, T., and Kearfott, K. J. 2007. Preliminary simulations in the use of fast neutrons to detect explosives hidden in cargo containers, in *IEEE Nuclear Science Symposium Conference Record, NSS'07*, Honolulu, HI, October 26–November 3, 2007, Vol. 2, pp. 1134–1137.

Lernihan, J. M. A. and Thomson, S. J. (Eds.). 1965. *Activation Analysis Principles and Applications*. Academic Press, New York.

Litz, M., Waits, C., and Mullins, J. 2012. Neutron-activated gamma-emission: Technology review. Army Research Laboratory, Adelphi, MD, Report ARL-TR-5871.

Liu, Y., Sowerby, B. D., and Tickner, J. R. 2008. Comparison of neutron and high energy X-ray dual-beam radiography for air cargo inspection. *Appl. Radiat. Isot.* 66: 463–473.

Maglich, B. C. 2005. Birth of "Atometry" - Particle physics applied to saving human lives. *AIP Conf. Proc.* 796: 431–438.

McKlveen, J. W. 1981. *Fast Neutron Activation Analysis, Elemental Data Base*. Ann Arbor Science, Ann Arbor, MI.

MCNP. 2003. A general Monte Carlo N-particle transport code, Version 5, X-5 Monte Carlo Team.

Morgan, J. W. and Ehmann, W. O. 1970. Precise determination of oxygen and silicon in chondritic meteorites by 14 MeV neutron activation with a single transfer system. *Anal. Chem. Acta* 49: 287–299.

Nakae, L. F., Chapline, G. F., Glenn, A. M. et al. 2014. The use of fast neutron detection for materials accountability. Applications of Nuclear Techniques (CRETE13). *Int. J. Mod. Phys.: Conf. Ser.* 27: 1460140 (8 pp.).

Nargolwalla, S. S. and Przybylowicz, E. P. 1973. *Activation Analysis with Neutron Generators*. John Wiley & Sons, New York.

Nat, A. and Ene, A. 2006. 14 MeV neutron activation analysis of alkali in ores used in iron and steel industry. *The Annals of "Dunarea de Jos" University of Galati*, Fascicle II, year XXIV (XXIX): 55–62.

Nat, A., Ene, A., and Lupu, R. 2004. Rapid determination of gold in Romanian auriferous alluvial sands, concentrates and rocks by 14 MeV NAA. *J. Radioanal. Nucl. Chem.* 261(1): 179–188.

NCRP Report. 1993. National Council on Radiation Protection and Measurement. Limitation of exposures to ionizing radiation, NCRP Report No 116. National Council on Radiation Protection and Measurement, Bethesda, MD.

Obhodaš, J., Sudac, D., and Valković, V. 2003. The role of soil in landmine detection, in Sahli, H., Bottoms, A. M., and Cornelis, J. (Eds.) *Proceedings of EUDEM2-SCOT 2003, International Conference on Requirements and Technologies for the Detection, Removal and Neutralization of Landmines and UXO*, Brussels, Belgium, September 15–18, 2003, pp. 101–106.

Obhođaš, J., Sudac, D., Nad, K., Valković, V., Nebbia, G., and Viesti, G. 2004. The soil moisture and its relevance to the landmine detection by neutron backscattering technique. *Nucl. Instrum. Methods B* 213: 445–451.

Obhodas, J., Sudac, D., Blagus, S., and Valkovic, V. 2007. Analysis of an object assumed to contain "Red Mercury." *Nucl. Instrum. Methods Phys. Res. B* 261: 922–924.

Obhodas, J., Matjacic, L., Sudac, D., Nad, K., and Valkovic, V. 2012. Red mud characterization using nuclear analytical techniques. *IEEE Trans. Nucl. Sci.* 59: 1453–1457.

Parsons, A., Bodnarik, J., Evans, L. et al. 2011. Active neutron and gamma-ray instrumentation for in-situ planetary science applications. *Nucl. Instrum. Methods Phys. Res. A* 652: 674–679.

Perret, G., Perot, B., Artaud, J.-L., and Mariani, A. 2006. EURITRACK tagged neutron inspection system. *J. Phys.: Conf. Series* 41: 375–383.

PNL. 2012. Phoenix Nuclear Labs, Madison, WI. Fact Sheet. -http://phoenixnuclearlabs.com/.

Reber, E. L., Blackwood, L. G., Edwards, A. J., Jewell, J. K., Rohde, K. W., Seabury, E. H., and Klinger, J. B. 2005. Idaho explosives detection system. *Nucl. Instrum. Methods Phys. Res. B* 241(1–4): 738–742.

Reports of the Scientific Committee for Food (Thirty-sixth series) Directorate-General Industry, ISBN 92-827-9580-2, European Commission, 1997.

Runkle, R. C., White, T. A., Miller, E. A., Caggiano, J. A., and Collins, B. A. 2009. Photon and neutron interrogation techniques for chemical explosives detection in air cargo; A critical review. *Nucl. Instrum. Methods A* 603(3): 510–528.

Ryge, P., Bar-Nir, P. I., and Simic, M. 1992. Food safety effects of inspection by SAIC pulsed fast neutron analysis explosive detection system, SAIC. Report submitted to Ministry of Agriculture, Fisheries and Food (MAFF), UK by SAIC, USA on June 12, 1992.

Salma, I. and Zemplen-Papp, E. 1993. Activation analysis with neutron generators using short-lived radionuclides. *Nucl. Instrum. Methods Phys. Res. B* 79: 564–567.

Shewart, W. 1931. *The Economic Control of Quality of Manufactured Products*. D. Van Nostrand, New York.

Showalter, S. and Schmitt, K. 1972. Applications of activation analysis to geochemical, meteoritic and lunar studies, in *Advances in Activation Analysis*, Vol. 2. Academic Press, London and New York, pp. 185–219.

Slater, C. O., Pace III, J. V., and Santoro, R. T. 2000. *Irradiation Effects for the Pulsed Fast Neutron Analysis (PFNA) Cargo Interrogation System*. Oak Ridge National Laboratory, Oak Ridge, TN, ORNLITM-2000/352.

Sowerby, B. D. 2009. Nuclear techniques for the on-line bulk analysis of carbon in cola-fired power stations. *Appl. Radiat. Isot.* 67: 1638–1643.
Sowerby, B. D., Cutmore, N. G., Liu, Y., Peng, H., Tickner, J. R., Xie, Y., and Zong, C. 2009. Recent developments in fast neutron radiography for the interrogation of air cargo containers, in *IAEA Conference*, Vienna, Austria, May 4–8, 2009, Paper SM/EN-01.
Stewart, R. F. 1967. Nuclear measurements of carbon bulk materials. *ISA Trans.* 6 (3): 200–208.
Strellis, D. A., Gozani, T., and Stevenson, J. 2009. Air cargo inspection using pulsed fast neutron analysis, in *IAEA Proceedings Series: Topical Meeting on Nuclear Research Applications and Utilization of Accelerators*, International Atomic Energy Agency, Vienna, Austria, Paper SM/E-05.
Sudac, D., Blagus, S., Matika, D., Kollar, R., Grivičić, T., and Valković, V. 2003. The use of 14 MeV neutrons in the explosive detection, in Sahli, H., Bottoms, A. M., and Cornelis, J. (Eds.) *Proceedings of EUDEM2-SCOT 2003, International Conference on Requirements and Technologies for the Detection, Removal and Neutralization of Landmines and UXO*, Brussels, Belgium, September 15–18, 2003, pp. 749–754.
Sudac, D., Blagus, S., and Valković, V. 2004. Chemical composition identification using fast neutrons. *Appl. Radiat. Isot.* 61/1: 73–79.
Taylor, J. K. 1981. Quality assurance of chemical measurements. *Anal. Chem.* 53: 1588A–1596A.
Taylor, J. K. 1987. *Quality Assurance of Chemical Measurements*. CRC Press, Boca Raton, FL.
Taylor, J. K. and Oppermann, H. V. 1986. *Handbook for the Quality Assurance of Meteorological Measurements*. National Bureau of Standards, Handbook 145, Gaithersburg, MD.
Taylor, O. 1964. Neutron irradiation and activation analysis, George Newnes, London, U.K.
Tenforde, T. S. 2002. Letter report on radiation protection advice for pulsed fast neutron analysis system used in security surveillance. NCRP September 20.
Tenforde, T. S. 2003a. Part II. The ALARA principle and related issues. NCRP February 21.
Tenforde, T. S. 2003b. Part III. Methods for the determination of effective dose to inadvertently exposed individuals. NCRP July 31.
Tepel, J. W. and Muller, H.-W. 1990. GAMCAT – A personal computer database on alpha particles and gamma rays from radioactive decay. *Nucl. Instrum. Methods A* 286: 443.
UNSCEAR, 2000. Report of the United Nations Scientific Committee on the Effects of Atomic Radiation to the General Assembly, Washington DC.
Van Grieken, R. and Hoste, J. 1972. Annotated bibliography on 14 MeV neutron activation analysis. CEC, Eurisotop Office Information Booklet 65–8, Saclay.
Van Grieken, R., Specke, A., and Hoste, J. 1970. Simultaneous determination of silicon and phosphorus in cast iron by 14 MeV neutron activation analysis. *J. Radioanal. Chem.* 6: 385.
Viesti, G., Lunardon, M., Nebbia, G. et al. 2006. The detection of landmines by neutron backscattering: Exploring the limits of the technique. *Appl. Radiat. Isot.* 64: 706–716.
Vogt, J. R. and Ehmann, W. O. 1965. An automated procedure for the determination of oxygen using fast neutron activation analysis: Oxygen in stony meteorites. *Radiochim. Acta* 4: 24–28.
Vourvopoulos, G. and Schultz, F. J. 1993. A pulsed fast-thermal neutron system for detection of hidden explosives. *Nucl. Instrum. Methods Phys. Res. B* 79: 585–588.

Vourvopoulos, G. and Womble, P. 2001. Pulsed fast thermal neutron analysis: A technique for explosives detection. *Talanta* 54: 459–468.

Vourvopoulos, G., Dep, L., Paschal, J., and Spichiger, G. 1997. PELAN- A transportable neutron based UXO identification technique, in *Proceedings of UXO Forum'97*, Nashville, TN, May 28–30, 1997, pp. 342–349. (See also: Applied Physics Institute, Department of Physics and Astronomy, Western Kentucky University, Bowling Green, KY 42101, http://www.wku.edu/API/.)

Wakeford, C. A. and Blackburn, R. 1991. Induction and detection of radioactivity in foodstuffs irradiated with 10 MeV electrons and x-rays. *Radiat. Phys. Chem.* 38(1): 29–38.

Wasim, M. 2007. A database for QA/QC in neutron activation analysis and gamma-ray spectrometry. *J. Radioanal. Nucl. Chem.* 272: 61–67.

Wasim, M. and Zaidi, J. H. 2002. NUCDATA: A useful database for NAA lab. *Nucl. Instrum. Methods A* 481:760.

WHO. 1990. Food safety aspects relating to the application of X-ray surveillance equipment: Memorandum from a WHO meeting. *Bull. World Health Org.* 31: 297–301.

Wielopolski, L., Hendrey, G., Johnsen, K. et al. 2008. Nondestructive system for analyzing carbon in the soil. *Soil Sci. Soc. Am. J.* 72(5): 1269–1277.

Wilde, H. R. and Herzog, W. 1982. On-line analysis of coal by neutron induced gamma spectrometry. *J. Radioanal. Chem.* 71(1–2): 253.

Williams, R. E. 1981. Gamma-ray spectroscopy following high flux 14 MeV neutron activation. UCRL-53208. PhD thesis, Lawrence Livermore Laboratory, Livermore, CA.

Wing, J. 1964. Simultaneous determination of oxygen and silicon in meteorites and rocks by nondestructive activation analysis with fast neutrons. *Anal. Chem.* 36: 559.

Womble, P. C., Vourvopoulos, G., Paschal, J., Novikov, I., and Barzilov, A. 2002. Results of field trials for the PELAN system. *Proc. SPIE* 4786: 52–57.

Womble, P. C., Paschal, J., and Moore, R. 2005. Cement analysis using d + D neutrons. *Nucl. Instrum. Methods Phys. Res. B* 241(1–4): 765–769.

Wood, D. E. 1971. Industrial applications of activation analysis with 14 MeV neutrons. *Nucl. Instrum. Methods* 92: 511–515.

Yang, Y., Yang, J., and Li, Y. 2013. Fusion of x-ray imaging and photoneutron induced gamma analysis for contraband detection. *IEEE Trans. Nucl. Sci.* 01/2013; 60(2). DOI: 10.1109/TNS.2013.2248095.

Zenoni, A., Bonomi, G., Donzella, A., Giroletti, E., Nechi, M. M., and Perot, B. 2009. Radiological risk assessment and shielding study for Eritr@C installations. Eritr@C Report ID 5.4, unpublished.

Chapter 7

Applications of Tagged Neutron Beams

7.1 USE OF TAGGED NEUTRON BEAMS

The physics of tagged neutron beam applications to sample characterization and identification is shown in Figure 7.1. The $E_n = 14$ MeV neutrons (mass – m_n) are produced by a small neutron generator inside which an alpha particle detector is mounted. By the detection of associated alpha particles, an electronically collimated neutron beam is produced, which upon the bombardment of the investigated voxel produces gamma rays characteristic of the material present within the voxel. The measurement of time-of-flight spectrum, ΔT, (alpha particle: start, gamma ray: stop) determines the distance, s, of the investigated volume following relation:

$$s = Vx\Delta T = \left(\frac{2E_n}{m_n}\right)^{1/2} x\Delta T \text{ s}$$

The basic components of such a laboratory system are shown in Figure 7.2a where the source of tagged neutron beam is API-120 neutron generator with the cone of shielding material protecting gamma detector from direct irradiation and with an oil-in-water target being measured. This experimental arrangement (for different target) is also shown in Figure 7.2b. Sealed tube neutron generator with α-particle detector is shown on the right side of the photo, and γ-ray detector with photomultiplier is shown on the left side of the photo; between them is a cone preventing detector to see directly the neutron source, while the white box is being investigated for its content.

As an example, the measured α–γ time spectrum is shown in Figure 7.3, resulting from the neutron bombardment of oil-in-water target. The time spectrum was obtained by signals from YAP(Ce) alpha detector inside API-120 neutron generator and from LaBr3(Ce) gamma detector. Windows on the regions of the spectrum marked "a" and "b" were applied when analyzing the gamma

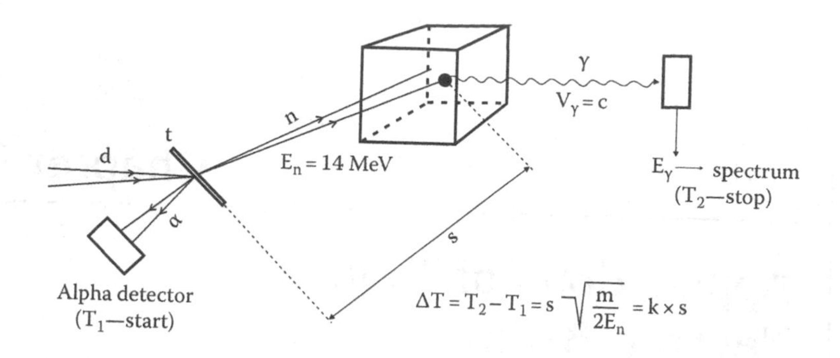

Figure 7.1 Schematic presentation of the use of tagged neutrons.

ray energy spectrum. This procedure results in two quite different energy spectra as shown in Figure 7.4.

In order to scan the interrogated object one can rotate the neutron generator around its axis, which results in the movement of electronically collimated neutron beam across the object. In this case only a single alpha detector is required. The neutron generator can rotate around its axis, which results in the movement of electronically collimated neutron beam across the object to

(a)

Figure 7.2 (a) Experimental setup with API-120 neutron generator. *(Continued)*

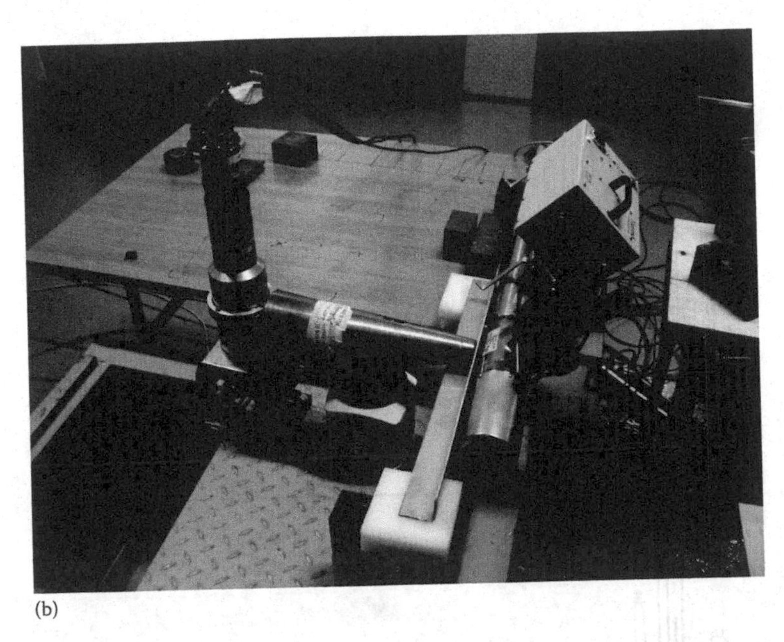
(b)

Figure 7.2 (*Continued*) (b) Essential components of the associated particle neutron probe.

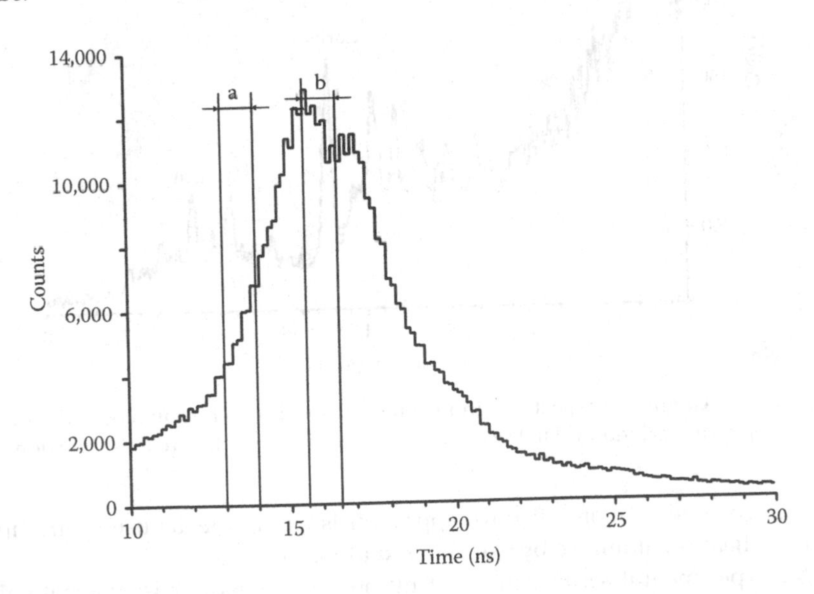

Figure 7.3 Time spectrum from bombardment of oil-in-water target using API-120 neutron generator and LaBr3(Ce) detector.

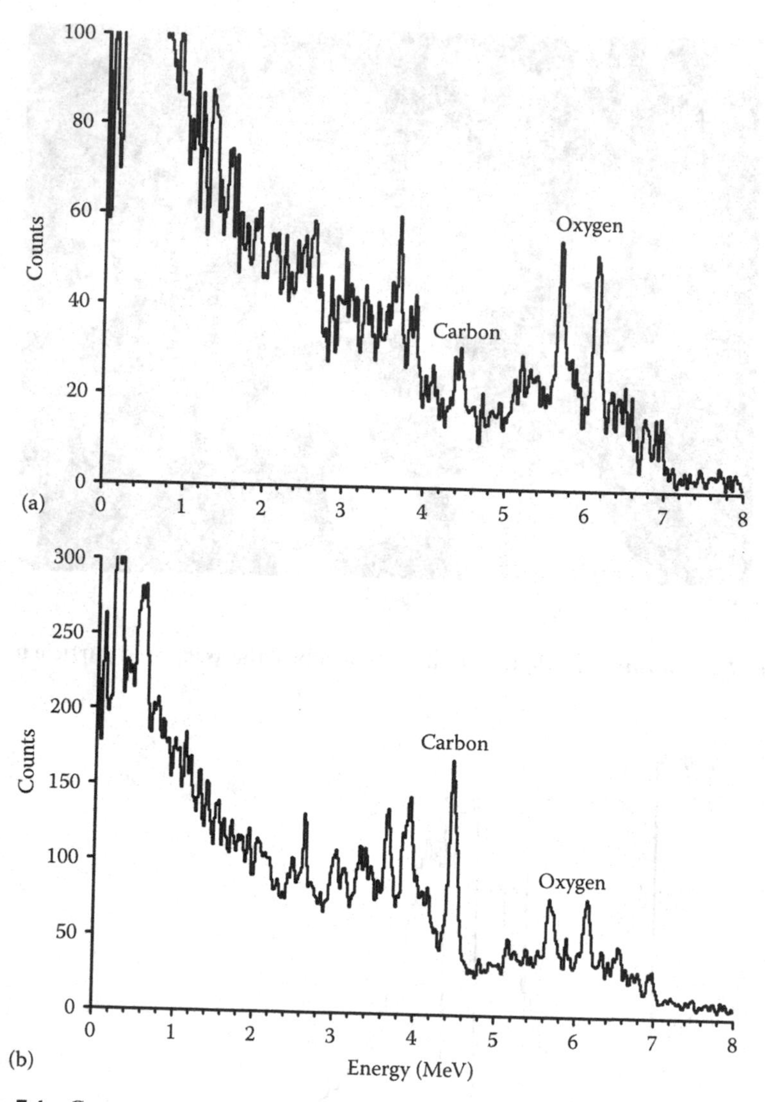

Figure 7.4 Gamma ray spectrum from time window (a) corresponding to the water portion and time window (b) in the time spectrum corresponding to diesel portion.

be interrogated. A more effective approach is simultaneous use of multiple neutron beams collimated by a segmented alpha detector.

An experimental setup with rotating neutron generator is schematically presented in Figure 7.5. The graphite block positioned 10 cm from the 3″ × 3″ NaI (Tl) gamma detector was scanned by electronically collimated neutron

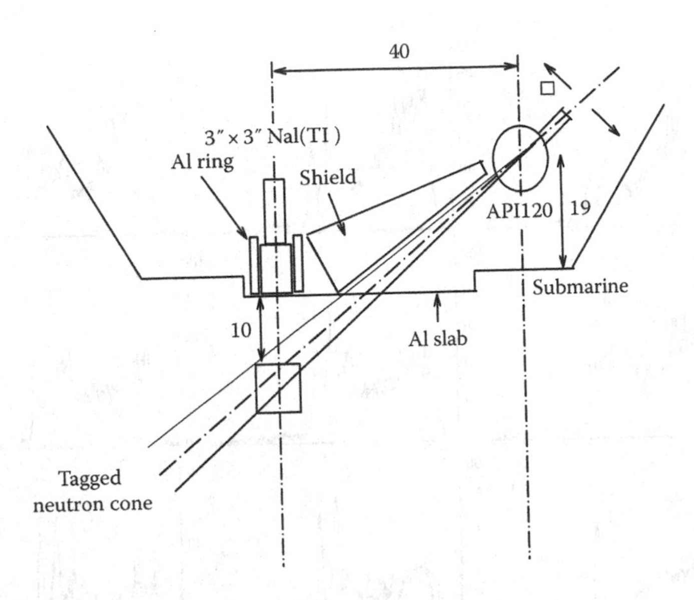

Figure 7.5 Experimental setup with rotating neutron generator.

beam by the rotation of API-120 neutron generator in steps of 2°. At each position resulting gamma spectrum was detected and recorded as shown in Figure 7.6. The measured gamma spectra are dominated by 4.43 MeV peak from the first exited state of carbon and its escape peaks. From Figure 7.6 one can see that the intensity of carbon peaks varies with the angle of the neutron generator; the variation of carbon gamma ray intensity, shown in Figure 7.7, defines the size of the carbon block. This approach can be used in the determination of the size of the interrogated hidden suspicious threat material, for example, explosive hidden in the automobile.

The early work on the subject of the use of associated particle sealed-tube neutron probe for nonintrusive inspection at the Argonne National Laboratory (ANL), Illinois, has been summarized in Rhodes and Dickerman (1997). The neutron probe used in ANL was developed primarily by nuclear diagnostic systems (Gordon and Peters 1990). Similar associated particle systems were later developed by Beyerle et al. (1990) and Ussery and Hollas (1994).

Since 2004, Oak Ridge National Laboratory (ORNL) has been developing fast-neutron imaging with an associated particle imaging d-t neutron generator. The pixilated alpha detector embedded in the generator time and directionally tags some (5%) of the neutrons. These neutrons have been very useful for a wide variety of fast-neutron imaging measurements with various configurations of fissile and nonfissile materials. For example, these neutron imaging systems make tomographic images, and in addition, by associating the spatial

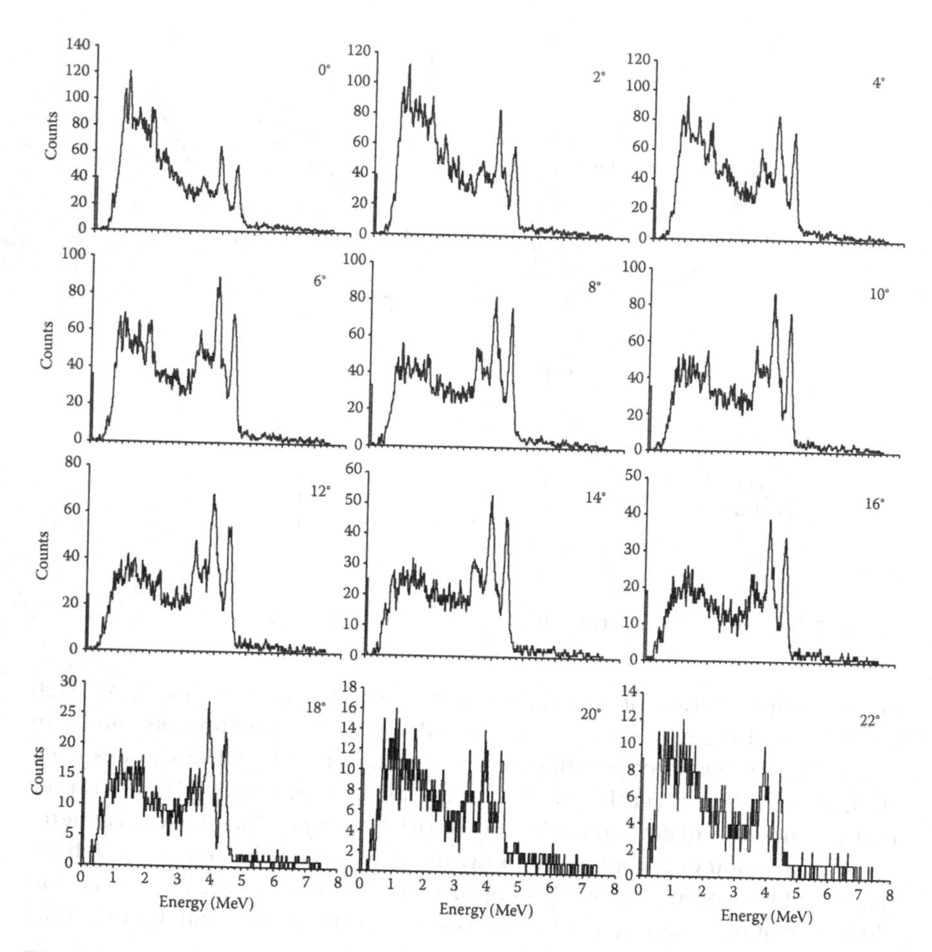

Figure 7.6 Carbon gamma ray spectra for different positions of neutron generator, 0°–22° in steps of 2°. Measurement time for each angle is ~250 s with the total number of neutrons, TOTN = 1.2×10^7.

distribution of doubles with the various pixels of the alpha detector, an image of the spatial distribution of induced-fission sites is obtained (Mullens et al. 2011). Furthermore, neutron scattering (both elastic and nonelastic) from light elements can be used to image and identify materials (Grogan and Mihalczo 2012). The paper (Mihalczo et al. 2013) reviews the development of the alpha detectors, imaging measurements, and recent results of fast-neutron measurements done at ORNL.

Literature on the problem of hidden substances identification by detection of fast neutron–induced γ-rays using associated α-particle technique

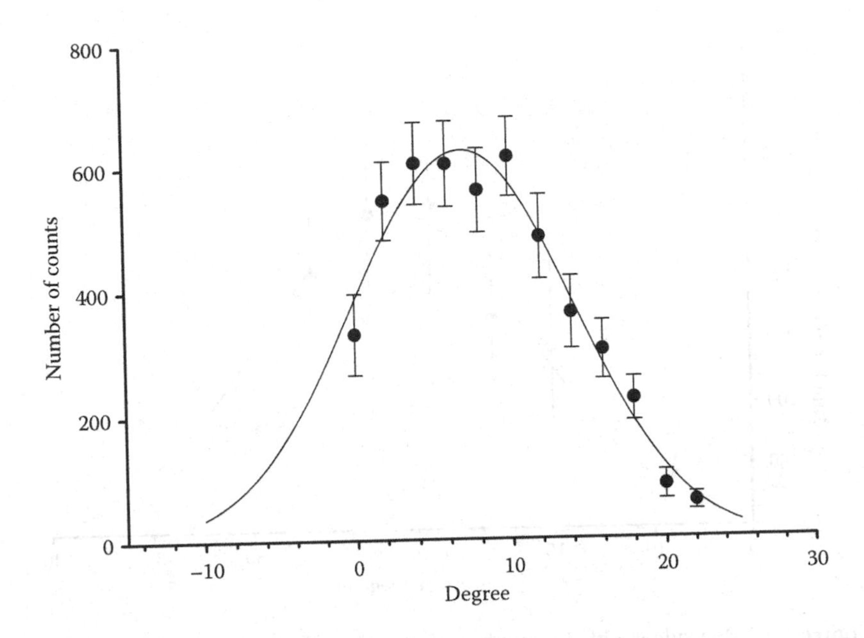

Figure 7.7 Scan of the graphite block. Distribution of carbon gamma ray intensity defines the size of the object.

contains many published reports. For the first time it has been discussed in the papers by Blagus et al. (2004) and Sudac et al. (2004). Other reports include detection of hidden explosives by using tagged neutron beams with subnanosecond time resolution by Pesente et al. (2004), detection of hidden explosives in different scenarios with the use of nuclear probes by Nebbia et al. (2005), tagged neutron inspection system (TNIS) based on portable sealed generators by Pesente et al. (2005), applications of nuclear techniques relevant for civil security by Valkovic (2006), and inspecting the minefield and residual explosives by fast neutron activation method by Sudac et al. (2012).

Determination of inspection parameters to be used in the investigation of the objects in the water for the presence of threat materials has been elaborated in detail by Valkovic et al. (2012). Measurements reported have been done with the target being the simulant of the explosive RDX with a mass of 5 kg, fixed under the submarine. Measurements were performed for target–detector distances of 0, 4, 5, and 9 cm. For each distance a number of measurements were done. As an example, Figure 7.8 shows the number of detected events in the carbon 4.44 MeV peak as a function of neutron generator rotation angle in the case of distance between sensor and stimulant being 5 cm. The angle of rotation

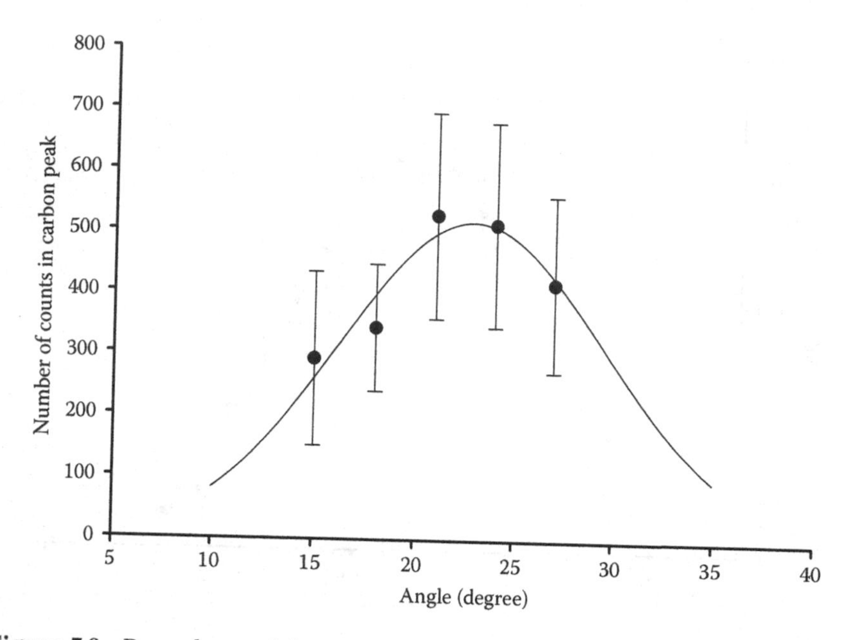

Figure 7.8 Dependence of the number of events in the carbon 4.44 MeV peak on the rotational angle of the neutron generator. Distance between neutron sensor and investigated object was 5 cm.

is measured with respect to arbitrary chosen initial position. Experimental values were fitted with Gaussian distribution:

$$N = a\mathrm{Exp}\left\{-0.5\left[\frac{(x - x_0)}{b}\right]^2\right\}$$

Parameters of the fit were

$$a = 516 \pm 36$$

$$(7 \pm 1)^\circ$$

$$x_0 = (22.8 \pm 0.8)^\circ$$

$$\mathrm{FWHM} = 2.34b = (16.4 \pm 2.3)^\circ$$

FWHM on the distance of half meter from the source corresponds to the diameter of 14.4 cm. The actual target diameter was 16 cm. Similar measurements have been performed for a distance of 4 cm. Parameters of the fit are

$$a = 478 \pm 47$$

$$(4.4 \pm 1.1)^\circ$$

$$x_o = (26.5 \pm 0.7)^\circ$$

$$\text{FWHM} = 2.34b = (10.3 \pm 2.6)^\circ$$

For the distance of 0 cm, it has been chosen for parameter b value b = 7,

$$a = 870 \pm 193$$

$$x_o = (33 \pm 2)^\circ$$

For the distance of 9 cm, only for the rotation angle of 22° the number of events in the carbon 4.44 MeV peak was bigger than critical limit, therefore, this angle was taken as x_o.

The same type of data could be collected by the use of a generator with associated particle detector having five or more sectors in a single measurement. Figure 7.9 shows the calibration line that defines the rotation angle of the neutron generator for a given distance of the object.

In order to satisfy the required performances, the system had to be housed in the water leak-proofed container that would house all the required power supplies and electronics modules and variety of other sensors. These were ingredients needed for realization of device called Surveyor (Valkovic et al. 2007). This is shown in Figure 7.10. The outside view of the device is shown in Figure 7.11. Although the device is optimized for underwater

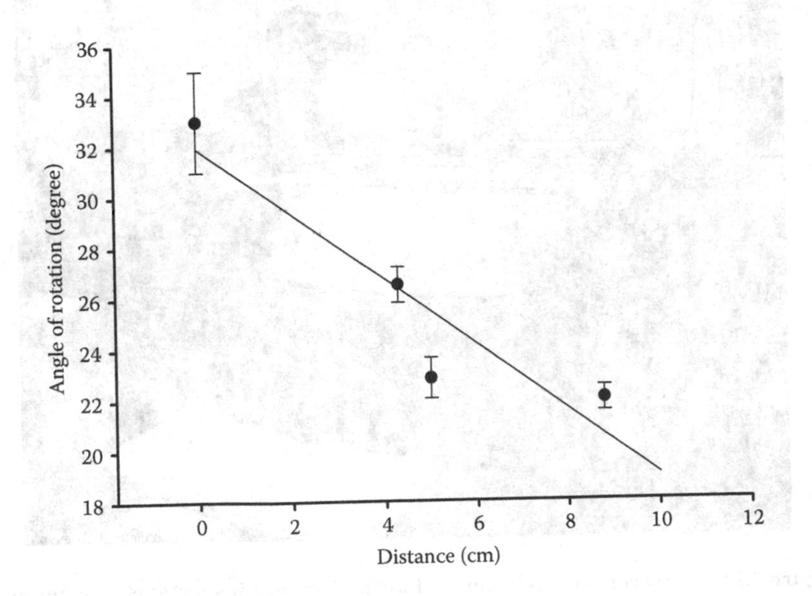

Figure 7.9 Rotational angle—distance calibration.

Figure 7.10 Surveyor inside: testing of the components.

Figure 7.11 Surveyor body is optimized for underwater inspections. The author of this book at left, Mr. Robert Kollar, electronics expert at right.

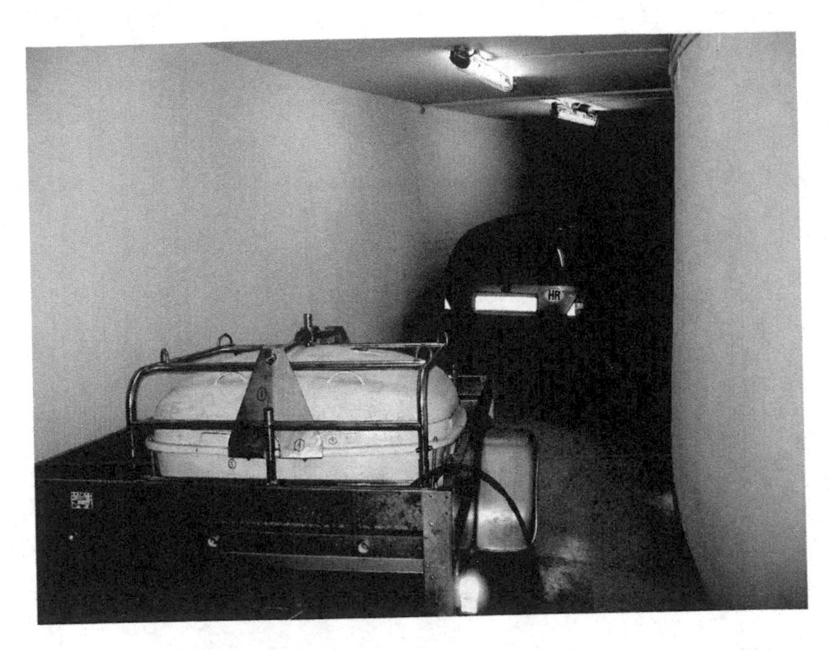

Figure 7.12 A suspicious car and Surveyor on a trailer.

applications (to be discussed later), it can be used for the inspection of objects on the ground as illustrated in the example of car inspection for the presence of the bomb.

The scenario for this test is indicated in the series of photos that follows. A car was found parked in the garage with a suspicious material in its trunk. The Surveyor was towed to this area and positioned against the rear of the car as shown in Figures 7.12 through 7.15.

In the first measurement, the explosive simulant, mass 48.8 kg, volume 57 × 37 × 42 cm^3, was found placed inside the car. Plastic box in the trunk contained four sacs (5 kg) of melamine and four sacs (7.2 kg) of sand (SiO_2) as shown in Figure 7.16. This is approximately 59% SiO_2 and 41% of $C_3H_6N_6$, which makes a bomb simulant (RDX)—$C_3H_6N_6O_6$ and Si.

Measurements were done by rotation of neutron generator inside Surveyor, by remote control. In such a way not only the presence of simulant was confirmed but also the size of package was estimated. This is of importance to first responders for the determination of the radius of eventual evacuation zone. Since the material was not mixed the system would observe better one or other component stimulant depending on the rotation angle. The obtained gamma ray spectra are shown in Figures 7.17 through 7.19 for different neutron generator angles.

Figure 7.13 Surveyor neutron (interrogation) beam is now pointing toward the car trunk.

Figure 7.14 Surveyor is positioned for measurement—back side.

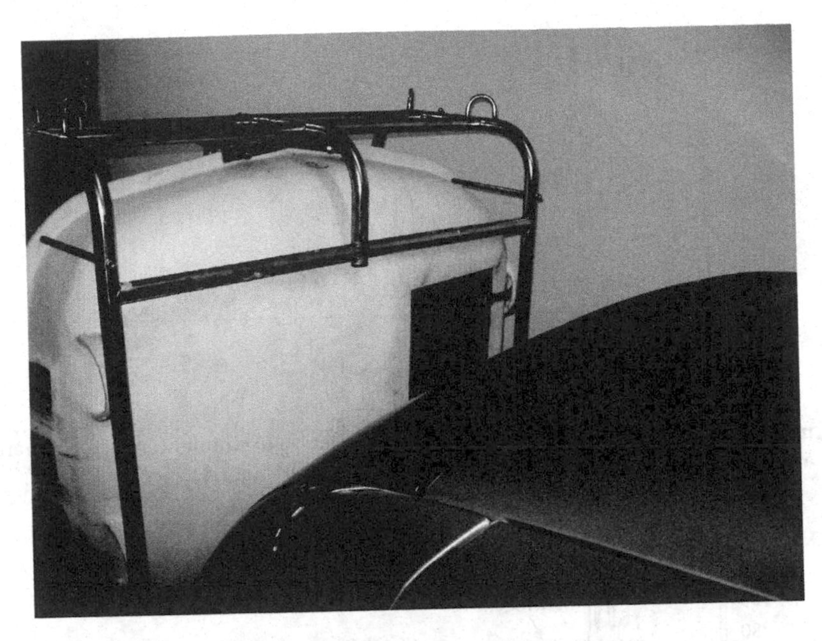

Figure 7.15 Surveyor is positioned for measurement—front side.

Figure 7.16 The mass of 48.8 kg of RDX simulant in the trunk of the car.

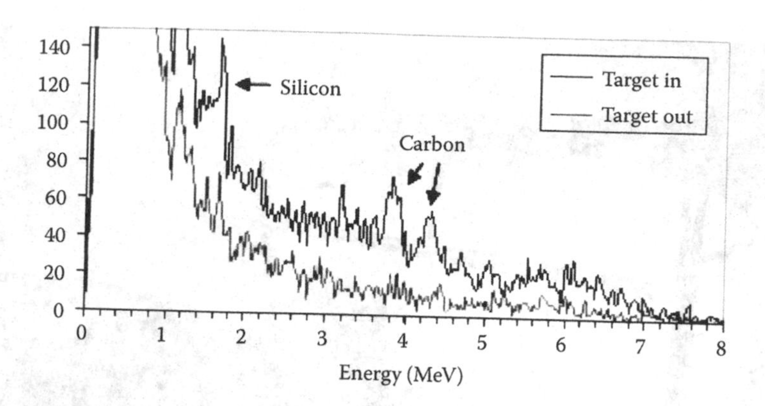

Figure 7.17 Gamma ray spectrum from the irradiation of simulant: neutron generator angle 0°, measurement time 4531 s, and the number of tagged neutrons 12×10^7.

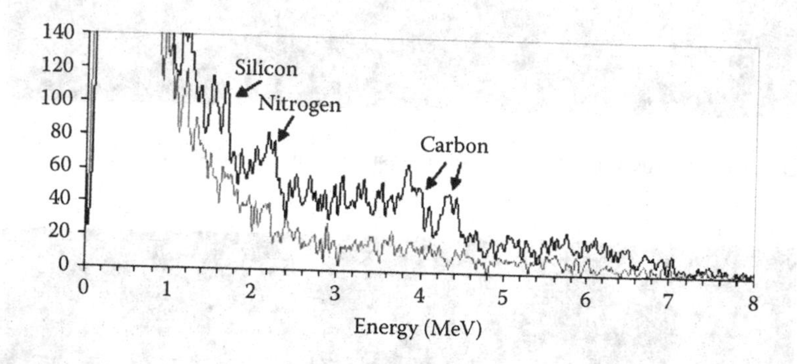

Figure 7.18 Gamma ray spectrum from the irradiation of simulant: neutron generator angle 4° (upper) and 28° (lower) and the number of tagged neutrons 12×10^7.

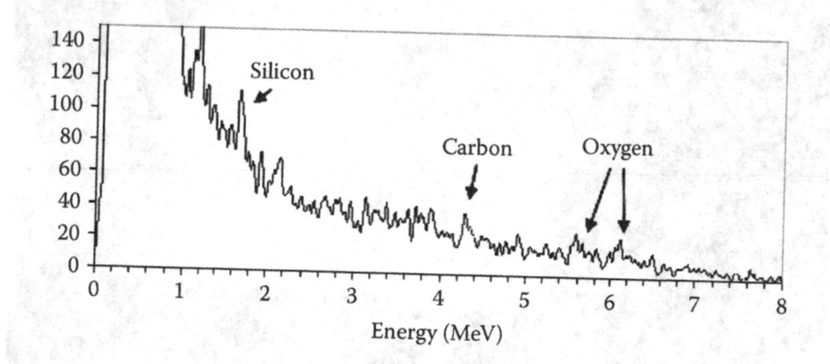

Figure 7.19 Gamma ray spectrum from the irradiation of simulant: neutron generator angle 8°, measurement time 4625 s, and the number of tagged neutrons 12×10^7.

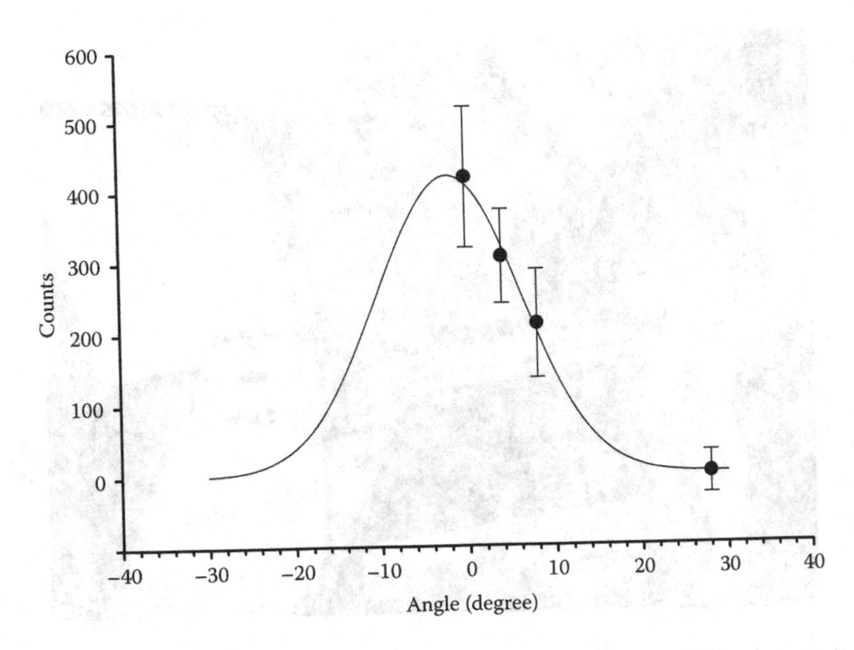

Figure 7.20 Number of detected events in 4.44 MeV carbon peak from the explosive simulant as a function of neutron generator rotation angle.

The presence of C, N, O peaks indicates that the material hidden in the car trunk could be an explosive. Figure 7.20 shows a number of detected events in 4.44 MeV carbon peak from the simulant as a function of neutron generator angle. The experimental points were fitted with Gaussian curve having full width at half maximum, FWHM = $(19.3 \pm 4)°$. Since the distance from the neutron source to the investigated object is 85 cm, this value corresponds to the object height of 29 cm. The real object size being 32 cm, this could be considered as a fair agreement.

In the second measurement a package containing real TNT explosive, mass 5 kg, volume $31 \times 12 \times 15$ cm^3, was placed into the car trunk as shown in Figure 7.21. The center of explosive package was about 50 cm from the gamma detector. The intensity of neutron beam is approximately 10^7 neutrons per second in 4π. Opening angle of tagged neutron beam was $(12.6 \pm 1.6)°$. Only three measurements were performed: with neutron generator positioned at angles of 0°, 8°, and 16°. The measured gamma spectra show peaks corresponding to O, C, and N at angles of 0° and 8° while at 16° no explosive is detected. The measured gamma ray spectra at neutron generator angle 8° is shown in Figure 7.22, and the intensity of measured peaks counted in the number of events per corresponding gamma peak is presented in Table 7.1.

Figure 7.21 Package of 5 kg of TNT in the car.

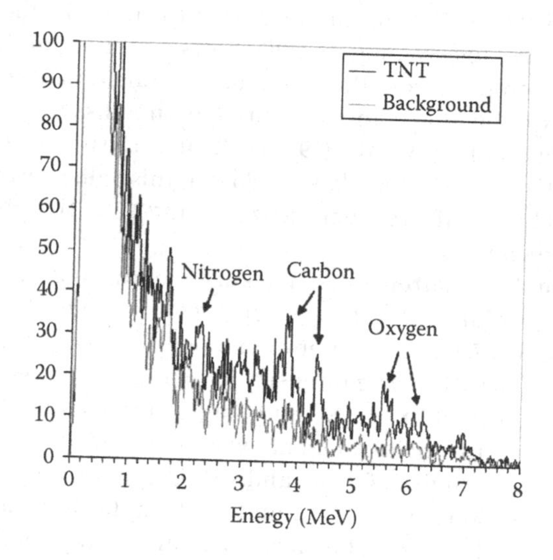

Figure 7.22 Gamma ray spectra from irradiation of explosive TNT mass 5 kg, volume $31 \times 12 \times 15$ cm^3 (target in), and background (target out): measurement time 2651 s, total number of tagged neutrons 6×10^7, and neutron generator angle 8°.

TABLE 7.1 NUMBER OF DETECTED PULSES FOR THE ABOVE GEOMETRY

Carbon 4.44 MeV Peak	Oxygen 5.62 Peak	Nitrogen 2.31 MeV Peak
153 ± 45	118 ± 33	130 ± 68

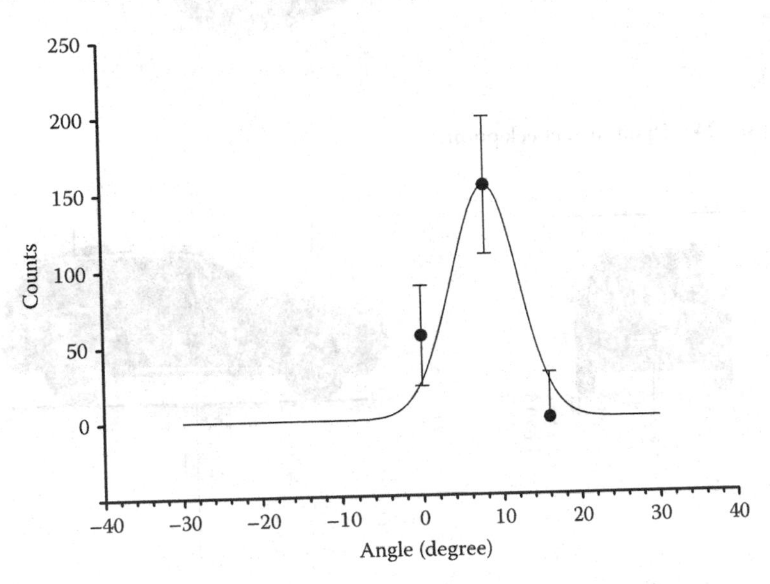

Figure 7.23 Number of detected events in the carbon 4.44 MeV peak when TNT package is investigated. FWHM = (10 ± 2.35)°; for a distance of 85 cm from neutron source it corresponds to the object height of 15 cm (the real dimension is 12 cm).

The number of detected events in the carbon 4.44 MeV peak plotted as a function of neutron generator angle when TNT package was investigated is shown in Figure 7.23. The Gaussian curve best fitting these three points has a FWHM = (10 ± 2.35)°; for a distance of 85 cm from neutron source this corresponds to the object height of 15 cm (the real dimension is 12 cm). Again, it is a fair agreement!

Based on these experimental results a concept of "open road check point" has been developed (see Figure 7.24). The realization of the check point is foreseen in three possible geometries which are proposed as follows:

1. Underground, under the pavement, placement of both neutron generator and detector battery with remote operational control unit (see Figure 7.25).
2. Under the pavement, placement of detector battery with (1) fixed position of x–y–z movable neutron generator; (2) no outside indication for the presence of neutron generator, placed on robotic vehicle that can

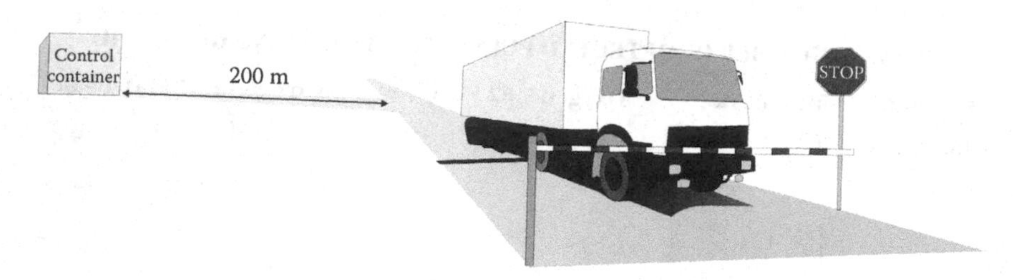

Figure 7.24 Open road check point.

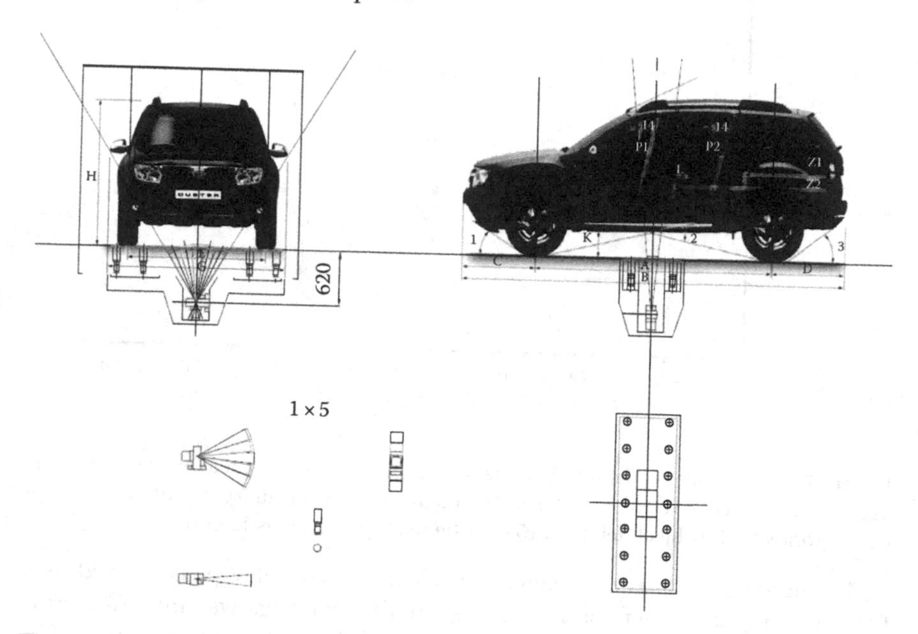

Figure 7.25 Check point with detectors and neutron generator in a container located under the road pavement.

position the neutron generator head against any part of the inspected vehicle (remotely controlled); and (3) use of dummies (poles, etc.) with generator either in position (1) or (2); see Figure 7.26.

The system can detect the explosive also while the automobile is in transit over it. It can be calculated that it is possible to detect 100 kg of explosive hidden in the car while it is moving with the speed of <5 km/h with the driver in it.

The major issues to be resolved include the determination of the minimum detectable limit (MDL) within predetermined measurement time and the

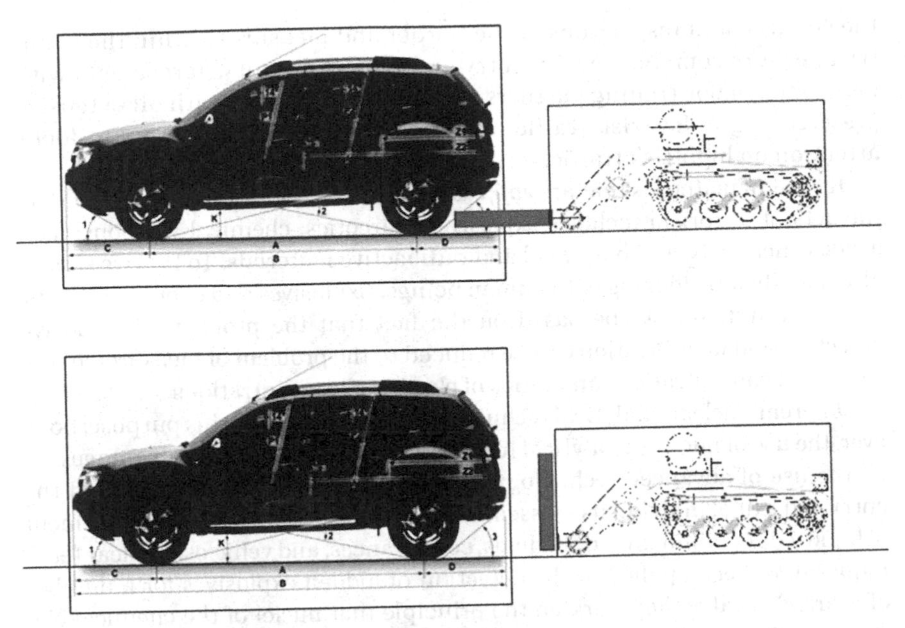

Figure 7.26 Gamma detectors in the container placed under the road pavement; generator brought in when and where needed.

reduction of the false alarm rate. This could be resolved with the optimization of detectors geometry and improvements of software that will be user friendly, indicating the presence and type of threat material.

7.2 HOMELAND SECURITY APPLICATIONS

Terrorism is a major threat to twenty-first century civilization and an enduring challenge to human ingenuity. The vulnerability of societies to terrorist attacks results in part from the proliferation of chemical, biological, and nuclear weapons of mass destruction but is also consequence of the highly efficient and interconnected systems that we rely on for key services such as transportation, information, energy, and health care. In today's society acts of terrorism must involve in some stages the illicit trafficking either of explosives, chemical agents, nuclear materials, and/or humans. Therefore, the society must rely on the antitrafficking infrastructure that encompasses responsible authorities: their personnel and adequate instrumental base.

Border management system should keep pace with expanding trade while protecting from the threats of terrorist attack, illegal immigration, illegal drugs, and other contraband. The border of the future must integrate actions abroad to screen goods and people prior to their arrival in the country, and inspections at the border and measures within the country to ensure compliance with entry and import permits, agreements with neighbors, major trading partners, and private industry with all extensive prescreening of low-risk traffic, thereby allowing limited assets to focus attention on high-risk traffic.

The list of materials that are subject to inspection with the aim of reducing the acts of terrorism includes explosives, narcotics, chemical weapons, hazardous chemicals, and biological and radioactive materials. To this we should also add illicit trafficking with human beings. Explosives and chemical agents' detection systems can be based on the fact that the problem of explosive detection and identification can be reduced to the problem of measurement of elemental concentrations and ratios of elemental concentrations.

Different nuclear analytical techniques could be used for this purpose; however, the use of nuclear analytical techniques has some specific advantages.

The use of advanced technology to track the movement of cargo and the entry and exit of individuals is essential to the task of managing the movement of hundreds of millions of individuals, conveyances, and vehicles. Nuclear techniques have been applied in the detection of hidden explosives for a number of years. Basically, they work on the principle that nuclei of the chemical elements in the investigated material can be bombarded by penetrating nuclear radiation (mainly neutrons). As results of the bombardment, nuclear reactions occur and a variety of nuclear particles and gamma and x-ray radiation is emitted, specific for each element in the bombarded material.

The risk of nuclear terrorism carried out by subnational groups should be also considered not only in construction and/or use of nuclear device, but also in possible radioactive contamination of large urban areas. The threats to security from nuclear and radiological terrorism could be grouped into the following three categories:

1. *Stolen state-owned nuclear weapons or weapons components*, modified as necessary to permit terrorist use.
2. *Improvised nuclear devices* (INDs) fabricated from stolen or diverted special nuclear material (SNM)—plutonium and especially highly enriched uranium (HEU).
3. Attacks on *nuclear reactors* or *spent nuclear fuel* or attacks involving *radiological devices*.

INDs are nuclear weapons fabricated by terrorists, with or without state assistance, using stolen or diverted SNM. It should be noted that the basic technical information needed to construct a workable nuclear device is readily available in the open literature. Therefore, the primary impediment that prevents countries or technically competent terrorist groups from developing nuclear weapons is the availability of SNM, especially HEU.

Much easier to produce is the so-called dirty bomb that might result in a threat involving the dispersal of radiological material in an effort to contaminate a target population or distinct geographical area. The material could be spread by radiological dispersal devices (RDDs), that is, "dirty bombs" designed to spread radioactive material through passive (aerosol) or active (explosive) means. Alternatively, the material could be used to contaminate food or water. There are a number of possible sources of material that could be used to fashion such a device, including nuclear waste stored at a power plant (even though such waste is not highly radioactive) or radiological medical isotopes found in many hospitals or research laboratories. Radioactive materials are often sintered in ceramic or metallic pellets. Terrorists could then crush the pellets into a powder and put the powder into an RDD. The RDD could then be placed in or near a target facility and detonated, spreading the radiological material through the force of the explosion and in the smoke of any resulting fires.

Schematic of nuclear device is simple: the center of the hypothetical device is fissile core made from weapon-grade uranium or weapon-grade plutonium. The isotopic composition of these two materials is shown in Table 7.2. The dimensions of the core of such hypothetical device are as follows: for weapon-grade uranium—approximately ring with R = 7 cm, d = 1.2 cm, and mass 12 kg; for weapon grade plutonium—ring with R = 5 cm, d = 0.75 cm, and mass 4 kg. Neutron emission rate on the surface of such device would be 1400 neutrons/s in the case of uranium and 400,000 neutrons/second in the case of plutonium with temper material being depleted uranium in both cases. Significant gamma ray emission rate at the surface of such a device would exist only if the temper material is depleted uranium (60,000–100,000 gammas/s), while in the case of tungsten being temper material this rate would be significantly lower (30–1.000 gammas/s).

Number of neutrons per second per kilogram from spontaneous fission and (α,n) reactions in weapon-grade uranium is relatively small: the highest being

TABLE 7.2 WEAPON-GRADE MATERIAL

Weapon-Grade Uranium		Weapon-Grade Plutonium	
Isotope	%	Isotope	%
Uranium-234	1.0	Plutonium-238	0.005
Uranium-235	93.3	Plutonium-239	93.3
Uranium-238	5.5	Plutonium-240	6.0
Other	0.2	Plutonium-241	0.44
		Plutonium-242	0.015
		Other	0.2

for spontaneous fission of ^{238}U being 14 s^{-1}/kg and for (α,n) reaction on ^{234}U being 50 s^{-1}/kg. This is contrary to weapon-grade plutonium where the number of neutrons per second per kilogram for spontaneous fission of ^{238}Pu is 2.6×10^6, ^{240}Pu is 0.9×10^6, and ^{242}Pu is 1.7×10^6 s^{-1}/kg. Taking into the account the contributions of (α,n) reactions and isotopic compositions of weapon-grade materials, the total number of neutrons per second per kilogram is 1.6 for U and 56,000 for Pu.

It should be noted that the U.S. and Soviet authorities are believed to have built several hundred portable atomic bombs. The small atomic demolition munition (SADM) might weight around 45 kg and be carried in two parts. Such a device has a warhead consisting of a tube with two pieces of uranium, which, when rammed together would cause the atomic blast. The bomb container also includes a firing unit and possibly a device that would have to be decoded for detonation. The SADM power is equivalent to 1 kiloton or less of TNT (Hiroshima bomb was 13 kilotons).

IAEA, Vienna, has a mandate to document worldwide events of illicit trafficking of weapons-usable nuclear material. Until now, only 25 highly credible cases of illicit trafficking in weapons-usable nuclear material have become known since the recording of such incidents was started in 1991. By comparison, there have been over 800 cases involving illicit trafficking in other nuclear and radioactive material, such as low-enriched uranium yellowcake, and medical and industrial radiation sources, during the same period of time. The inherent uncertainties in current knowledge on nuclear smuggling make it difficult to judge whether trafficking in weapons-usable nuclear material is really such a relatively rare phenomenon, or whether it was and still is carried out in such a clandestine, professional (in criminal terms) manner that it remains largely undetected (Zaitseva and Steinhausler 2004).

There have been at least 150 incidents of nuclear smuggling in the last decade of 20th century, half involving SNM. As little as 16 kg of highly enriched uranium (HEU) or 6 kg of plutonium can be used to produce a 20 kiloton weapon, even with low technology levels. Developing technology for the detection of HEU has become a priority for the U.S. armed forces. Passive detection technologies do not work well because fissile material is weakly radioactive and easy to shield. Active interrogation has been effectively demonstrated, but previous neutron sources have led to excessively long detection times.

Detection Strategy is based on the fact that fissile materials emit intense bursts of neutrons and gamma rays when bombarded with neutrons. These particles are easily detected and signal the presence of radioactive material. It is necessary to have a very high-flux neutron source for this detection method to be practical. Furthermore, the detection scheme must be implemented in a manner that allows for high throughput and does not disrupt port productivity. Phoenix Nuclear Labs, PNL, (phoenixnuclearlabs.com) has developed a neutron generator that is 1000 times stronger than any commercially available product other than a nuclear reactor. PNL's source makes it possible to detect SNM with greatly increased sensitivity. Steady state neutron production as high as 10^{14} neutrons per second allows for high throughput screening with a single container scanned in seconds.

In the report by Valkovic et al. (2007) the possibility of the detection of "dirty bomb" presence inside sea containers was evaluated. The method proposed for explosive and fissile material detection makes use of two sensors (x-rays and neutrons). A commercial imaging device based on the x-ray radiography performs a fast scan of the container, identifies a "suspect" region, and provides its coordinates to the neutron-based device for the final "confirmatory" inspection. In this two-sensor system, a 14 MeV neutron beam defined by the detection of the associated alpha particles is used for the interrogation of only volume elements marked by x-ray sensor. The object's nature is determined from passive, and neutron induced, gamma energy spectra measurements. Experimental results (time-of-flight and gamma energy spectra) obtained for the irradiation of 30 kg of TNT, depleted uranium, and other materials hidden inside the container are presented. The technique proposed by Valkovic et al. (2007) for explosive and fissile material detection makes use of the capability of a tagged neutron beam to confine the inspection to a predetermined volume element. The aim of the work described was to develop such a technique and to demonstrate the possibility of identifying explosives and fissile materials hidden in cargo containers inaccessible to present nonintrusive inspection technologies.

A straightforward application of the proposed approach is the coupling of the inspection by tagged neutron beams to a commercial imaging device based on either x-ray or gamma-ray radiography that performs a fast scan of the container, identifies a "suspect" region, and provides coordinates of the suspicious object to the neutron-based device for the final "confirmatory" inspection.

The x- and y-coordinates of a suspicious object are obtained from the x-ray image and define the position to which the neutron generator is moved. The z-coordinate is obtained by neutron-alpha particle time-of-flight measurements. The object's nature is determined from passive, and neutron induced, gamma energy spectra measurements.

In order to investigate different scenarios of illicit trafficking of explosive and radioactive materials, the experimental setup with a 3 m long section of the real container has been installed in the neutron laboratory (Valkovic et al. 2007). Either the container or detectors' portal could be moved in x-direction, while the support of investigated object inside the container can be moved in y and z directions. All movements should be done by the computer-controlled motors.

During preliminary measurements, tests were performed in two different geometries corresponding to two different ways of realization of the proposed neutron sensor. In geometry #1 the array of gamma detectors was placed on the opposite side of the container with respect to the position of neutron generator, outside the neutron beam cone at d = 87 cm from the investigated sample.

Experimental results obtained with 30 kg of TNT inside the container have shown that TAC information can be used to localize the interrogated object, while the information of the object's nature was contained in the measured gamma ray spectrum.

In geometry #2, the array of gamma detectors was placed perpendicularly to the neutron beam axis. The interrogated target was made of carbon blocks with a total mass 15.9 kg, in the volume $10.2 \times 30.6 \times 30.6$ cm^3. They were placed at the positions (1) 37 cm, (2) 65 cm, and (3) 150 cm from the right side of the container.

The information on the position of the interrogated object was contained in the TAC spectrum, while the object's composition was giving rise to the characteristic gamma ray spectrum. The summary of the experimental results presented in Table 7.3 indicates that the geometry #2 results in a much shorter measuring time and a better signal to the background ratio.

Two types of measurements were performed with depleted uranium: passive and active. In the passive measurements, approximately 27.3 kg of depleted uranium was placed into the iron case having 3 mm thick walls. Iron case was estimated to weight about 10.1 kg. Depleted uranium assembly was placed inside the shipping container masked by the lead shield.

Gamma spectrum of the depleted uranium was measured with the handheld spectrometer, the so-called identiFINDER (FLIR, 2011), and with the 4″ × 4″ NaI(Tl).

TABLE 7.3 SUMMARY OF EXPERIMENTAL RESULTS

Geometry	#1	#2	#2	#2
Position	—	(a)	(b)	(c)
Distance from neutron source (cm)	212	37	65	150
Neutron flux (n/s)	2.4×10^7	1.17×10^7	1.63×10^7	1.32×10^7
Exposure (s)	11,253	8,530	6,127	7,573
Number of tagged neutrons	8.3×10^8	2.5×10^8	2.5×10^8	2.5×10^8
Distance of gamma detector from the object	87 cm	42 cm	31 cm	90 cm
Number of counts in object gamma peaks	3,899	38,681	48,914	4,372
$N_{TAC,min} \leq 3\sqrt{N_B}$	224	346	606	275
Time required to obtain $N_{TAC,min}$ for 10^8 n/s source (s)	155	9	13	63

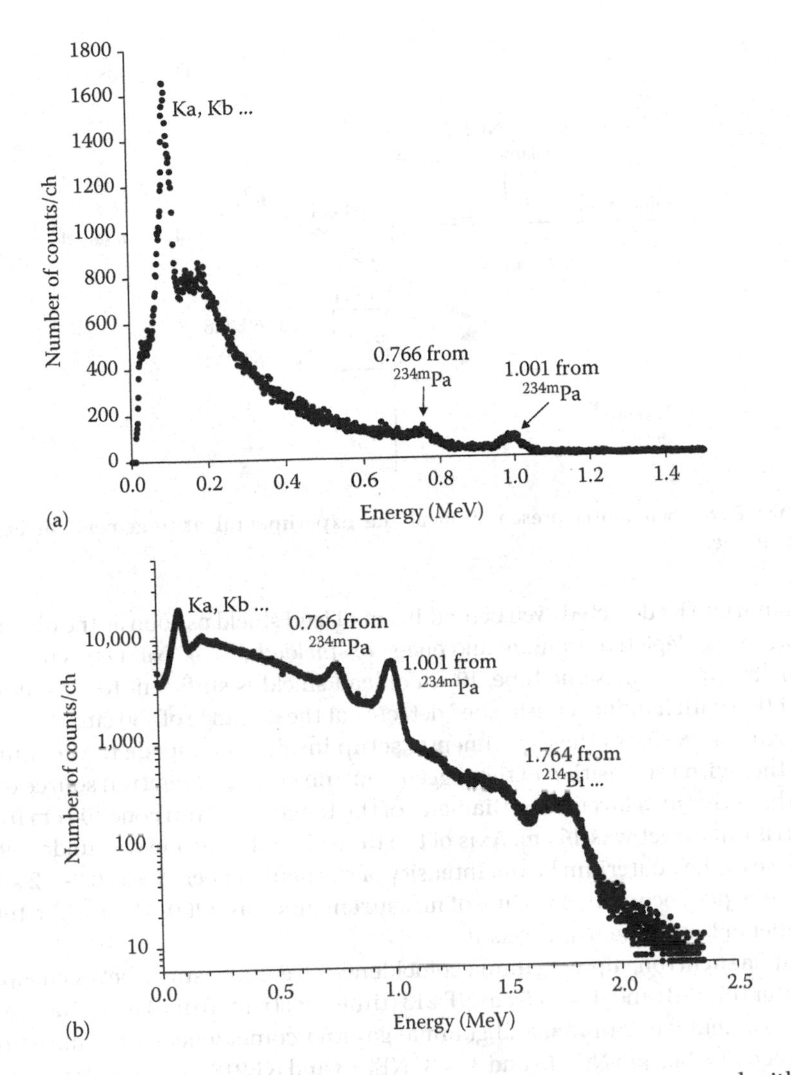

Figure 7.27 (a) Gamma ray spectrum from depleted uranium measured with the handheld detector. (b) Gamma ray spectrum from depleted uranium as measured with 4″ × 4″ NaI (Tl).

Figure 7.27 shows the gamma ray spectra of the depleted uranium measured with the "identiFINDER" (a) and with the 4″ × 4″ NaI (Tl) (b). The experimental results from the passive measurements are the following: total number of counts from 27.3 kg of depleted uranium placed behind 10 cm of lead shield as measured by 4″ × 4″ NaI (Tl) detector in 570.8 s was 203,487 ± 1,831 counts, compared with the background level 173,873 ± 1,565 counts. This means that 27.3 kg of depleted

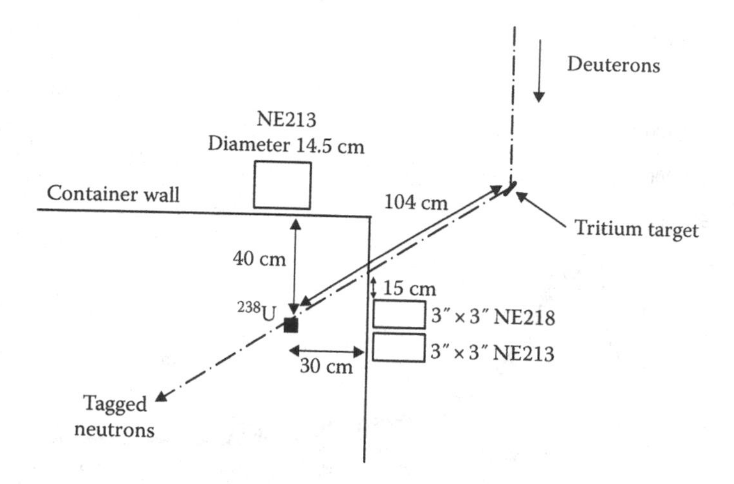

Figure 7.28 Schematic presentation of the experimental arrangement for active measurements.

uranium can be detected even behind 10 cm of lead shield as soon as the distance between the depleted uranium and one-side shielded 4″ × 4″ NaI (Tl) is no more than ~30 cm. At the same time, 10 cm of lead shield is sufficient to prevent the detection with handheld unshielded detector at the distance of ~30 cm.

Figure 7.28 shows the experimental setup inside and outside the container together with the position of the tagged neutron cone and neutron source during the active measurements. Diameter of the tagged neutron cone 104 cm from the tritium target was 16 cm. Axis of the tagged neutron cone was under 39.2° relative to the deuterium beam. Intensity of the neutron beam was $0.9–1.2 \times 10^7$ neutrons per second in 4π. Time of measurements was 6200–7900 s. The total number of tagged neutrons was 10^8.

Alpha–neutron, alpha–gamma coincidences were measured between alpha counter (Pilot B) and $\Phi = 14.5$ cm NE 213 (time spectrum from TAC1). Neutron–neutron, neutron–gamma, and gamma–gamma coincidences were measured between $\Phi = 14.5$ cm NE 213 and 3″ × 3″ NE213 and NE218 (time spectrum from TAC2).

The time spectra from TAC1, alpha–gamma, and alpha-n coincidences for different materials are shown in Figure 7.29. The spectra are shown for actual masses of materials as used in the measurements. Even in this case, an excess of α–n coincidences for uranium is observed. The shift of iron α–γ peak with respect to the background (container) α–γ peak is related to the iron target position inside the container.

Figure 7.30 shows the time spectra from TAC2 for various materials with an adequate time window on TAC1 that corresponds to alpha–neutron

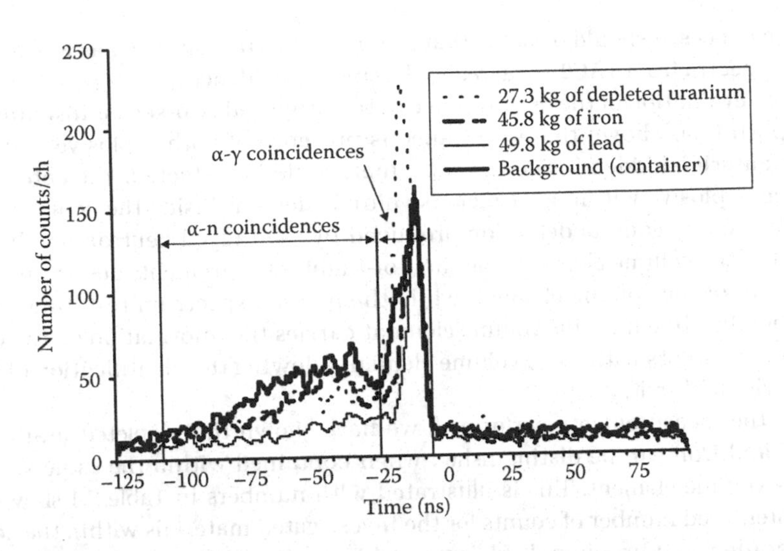

Figure 7.29 The time spectra from TAC1: alpha-gamma and alpha-n coincidences for depleted uranium, iron, lead, and background.

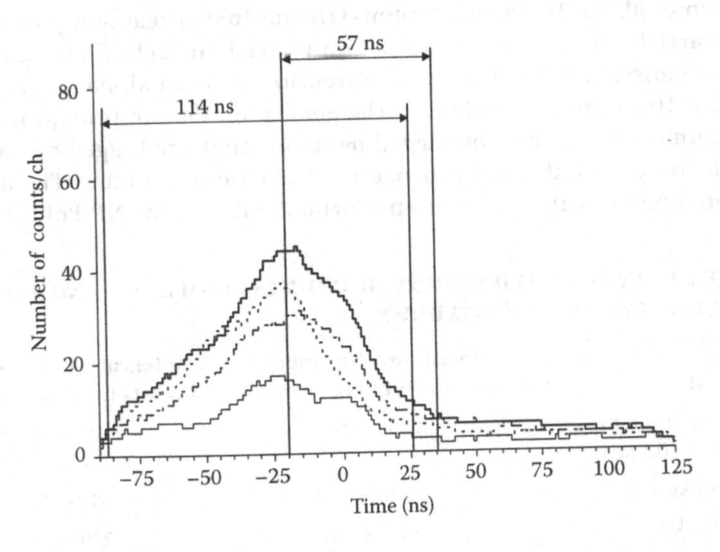

Figure 7.30 The time spectra from TAC2: n-n and n-gamma coincidences gated by α-n coincidence window on TAC1 for depleted uranium (top) and iron, lead, and background (bottom).

coincidences. It should be noted that the time spectra from TAC1 as well as the time spectra from TAC2 are gated with triple coincidences.

The evaluation of the performance of the proposed two-sensor instrumental portal has shown that simultaneous presence of both explosive and fissile material, hidden inside the container, could be detected. The detection of the explosive within a suspicious volume element inside the container is performed by gamma detection produced by the tagged neutron bombardment of the volume element. The time-of-flight measurements determine the position of the volume element, while the gamma spectrum resulting from the bombardment of this volume element carries the information on the elemental contents within the volume element allowing the identification of the material within it.

In the measurements performed, we have shown that depleted uranium, lead, and iron can be distinguished when contained within the same suspicious volume element. This is illustrated with numbers in Table 7.4 showing an integrated number of counts for the investigated materials within the container (depleted uranium, lead, iron, and background) for n–n–α and n–γ–α coincidences.

In the frame of the French transgovernmental R&D program against chemical, biological, radiological, nuclear, and explosives (CBRN-E) threats, CEA is studying the detection of SNMs by neutron interrogation with fast neutrons produced by an associated particle sealed-tube neutron generator (Carasco et al. 2012). The deuterium–tritium fusion reaction produces an alpha particle and a 14 MeV neutron almost back to back, allowing tagging neutron emission both in time and direction with an alpha particle position-sensitive sensor embedded in the generator. Fission prompt neutrons and gamma rays induced by tagged neutrons that are tagged by an alpha particle are detected in coincidence with plastic scintillator. The authors presented numerical simulations performed with the MCNP-PoliMi Monte

TABLE 7.4 INTEGRATED NUMBER OF COUNTS FROM n–n–α AND n–γ–α COINCIDENCES, FOR TAC WINDOWS

Material	Integrated Number of Counts in 114 ns Window	Integrated Number of Counts in 57 ns Window
Depleted uranium 27.3 kg + 10.1 kg of iron	2285 ± 183	1101 ± 88
Iron 45.8 kg	1780 ± 142	636 ± 51
Lead 49.8 kg	1479 ± 118	719 ± 58
Background	750 ± 60	331 ± 26

Note: See Figure 7.29.

Carlo computer code and with postprocessing software developed with the ROOT data analysis package. False coincidences due to neutron and photon scattering between adjacent detectors (cross talk) are filtered out to increase the selectivity between nuclear and benign materials. Accidental coincidences, which are not correlated to an alpha particle, are also taken into account in the numerical model, as well as counting statistics, and the time–energy resolution of the data acquisition system. Such realistic calculations show that relevant quantities of SNM (few kg) can be distinguished from cargo and shielding materials in 10 min acquisitions. First laboratory tests of the system under development in CEA laboratories are also presented.

Significant amount of the work on this problem has been performed at ORNL. The nuclear material identification system, NMIS, (Mihalczo et al. 2000) was first used at National Nuclear Security Administration's Y-12 National Security Complex (Y-12 NSC) in 1984 with a time-tagged californium neutron source (Mihalczo 1973) for time-of-flight transmission measurements through weapons components. The U.S. DOE Office of Nuclear Verification (ONV) in the mid-1990s supported the development of NMIS for potential application in the arms control treaties. In 2007 ONV decided to support the development of a fieldable, user-friendly version of NMIS designated as FNMIS (Radle et al. 2009, 2010), and this is scheduled to be completed sometime in 2015. ORNL has also developed a higher resolution fast-neutron imaging system designated the Advance Portable Neutron Imaging System, APNIS, (Blackston 2009). APNIS utilizes data acquisition and processing methods developed for positron emission tomography (PET) applications.

7.3 DETECTION OF THREAT MATERIALS (EXPLOSIVES, DRUGS, AND DANGEROUS CHEMICALS)

As already mentioned, the problem of material (explosive, drugs, chemicals, etc.) identification can be reduced to the problem of measuring elemental concentrations. Nuclear reactions induced by neutrons can be used for the detection of chemical elements, measurements of their concentrations, concentration ratios, and multielemental maps, within the explosives. Neutron-scanning technology offers capabilities far beyond those of conventional inspection systems. This highly sophisticated equipment is now ready to be deployed as part of a worldwide system of deterrence. The unique, automatic, material-specific detection of terrorist threats can significantly increase the security at ports, border-crossing stations, airports, and even within the domestic transportation infrastructure of potential urban targets as well as protecting forces and infrastructure wherever they are located.

Let us describe the experimental arrangement that can be used for the determination of the presence of hidden explosives. In these measurements, the neutron beam was obtained from $^3H(d,n)^4He$ bombarding water-cooled standard tritium target with 150 KeV deuterons beam from Texas Nuclear Corporation 150 KeV electrostatic accelerator (neutron generator), as seen in Figure 7.31. The active part of the target was 6 mm in diameter and consisted of titanium (0.95 mg/cm^2) deposited on copper backing (7 mm in diameter and 0.3 mm thickness). The activity was 0.8 Ci/cm^2. The associated alpha particle detector for fast coincidence measurement was 0.5 mm thin Pilot B fast organic scintillator with decay time 1.8 ns, coupled via Perspex light guide to an RCA-6342A photomultiplier. Its position (81.7° relative to deuteron beam) was fixed in such a way that the associated neutron beam was in vertical plane at 90° relative to the deuteron beam, as shown in Figure 7.32 (photo) and Figure 7.33 (schematic presentation). The 20 mm diameter slit in front of the scintillator was used to define geometrical dimensions of neutron beam on interrogated object. In front of the scintillator, a 9 μm thick Al foil was mounted to prevent the scattered deuterons, light, and secondary electrons from reaching the scintillator. Since the scintillator detector was placed at a distance of 110 mm from the tritium target, and the usually used slit diameter was 20 mm, the neutron spot on the place of interrogated object, at distance 650 mm from tritium target, was 50 mm. Position and the approximate dimension of the neutron spot was checked by the means of NE-213 neutron detector in the coincidence with associated alpha particle detector. Neutron

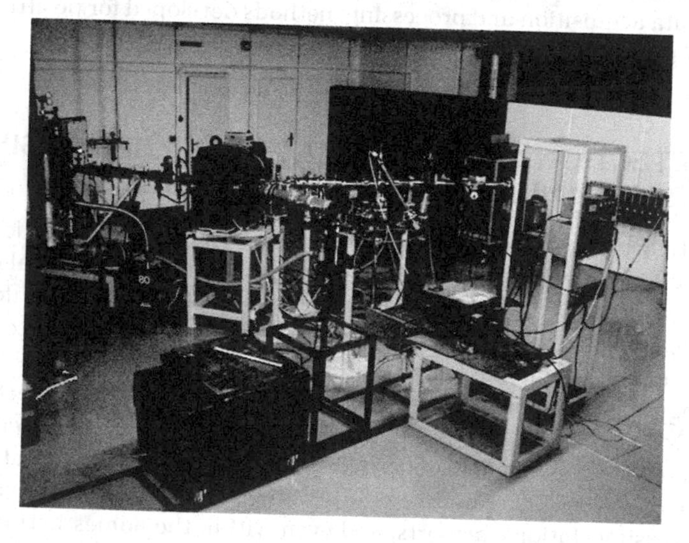

Figure 7.31 Texas Nuclear Corporation 300 KeV electrostatic accelerator at the Ruder Boskovic Institute, Zagreb, Croatia.

Figure 7.32 End of the accelerator. Tritium target assembly.

beam monitoring was accomplished by silicon photodiode maintained at 135° relative to the deuteron beam.

Let us describe an interesting application of such experimental setup: detection of explosive devices in passenger's luggage. A set of explosive devices, landmines as shown in Figure 7.34, has been acquired and used in this experiment. One of those explosive devices has been hidden in the suitcase containing the different clothing items in addition (as shown in Figure 7.35). The suitcase has then been positioned below a set of gamma detectors and irradiated by tagged neutrons produced in the chamber shown in the same figure. In addition to tritium target, the associated α-particle detector was also positioned at the appropriate angle inside the chamber. The α-particle energy spectrum, as measured by the α-particle detector, is shown in Figure 7.36a. Measurement of time-of-flight spectrum (time difference between the signals in α-detector and γ-detector) revealed several peaks corresponding to neutrons arrival at different positions. The position of the suitcase is marked as "sample" and the window on this peak is used for the evaluation of the measured gamma-ray spectrum. The obtained spectrum shows peaks corresponding to chemical elements N, C, and O, indicating the presence of a suspicious object (threat material) inside the suitcase.

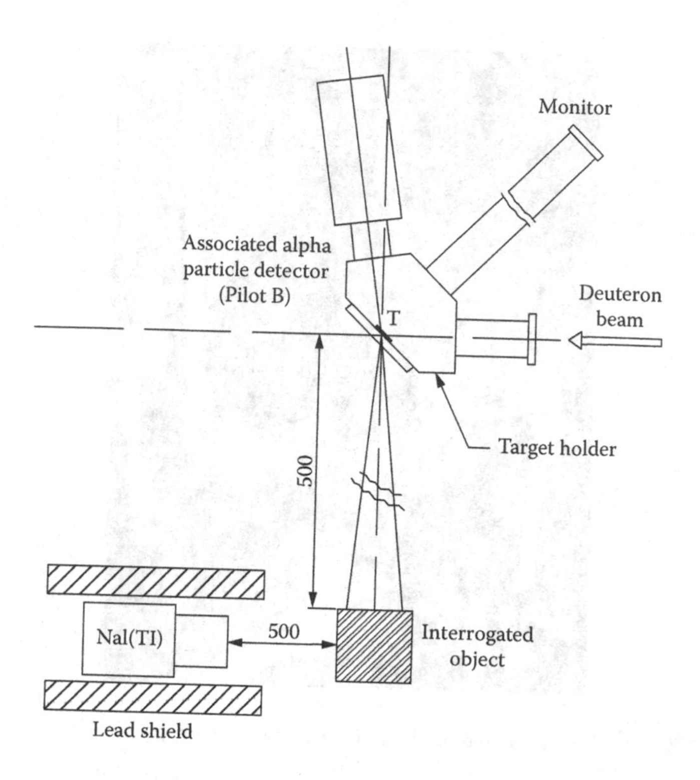

Figure 7.33 Scheme of the experimental setup.

Figure 7.34 Some investigated objects. Collection of landmines.

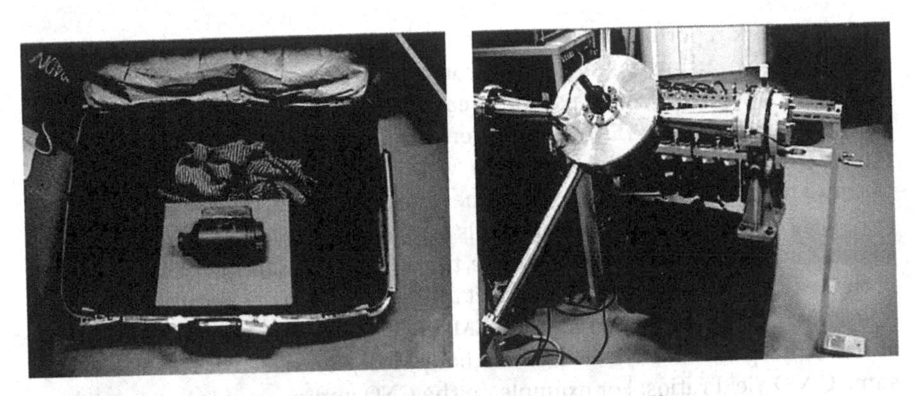

Figure 7.35 An explosive device hidden in the passenger coffer.

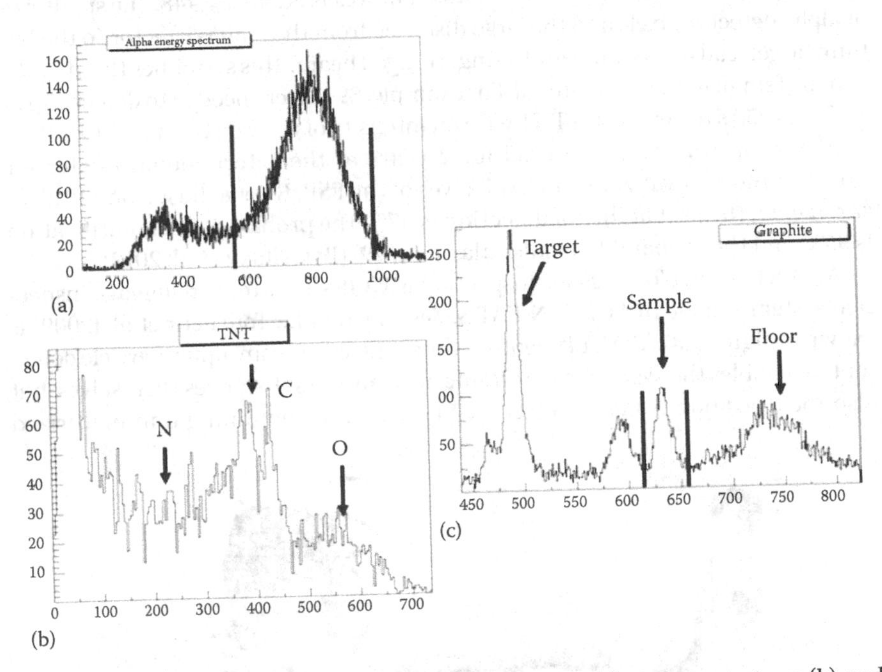

Figure 7.36 Associated alpha particle energy spectrum (a), gamma spectrum (b), and time spectrum (c). The position of the suitcase is marked as «sample» and the window on this peak is used for the evaluation of the measured gamma-ray spectrum. The obtained spectrum shows peaks corresponding to chemical elements N, C and O, indicating the presence of a suspicious object inside the suitcase.

In the work reported by Bystritsky et al. (2013) a stationary setup for identifying explosives using the tagged neutron method has also been described. The stationary setup has been in use for 2 years. A large number of tests with different explosive materials (EM) and conventional materials have been performed during this period. The photon spectra were analyzed in two modes: the detection and the identification of EMs. In the detection mode, the decision program (DP) has to determine if the material is dangerous or not. A typical detection time is 4 min. In the identification mode, the dangerous material is classified into groups of explosives. For identification it takes 16–20 min to collect the statistics. The standard approach is to identify EMs by determining the relative yields of carbon, nitrogen, and oxygen (CNO method). However, some EMs have the exact same C:N:O yield ratios. For example, in the CNO approach, it is not possible to distinguish between octogen ($C_4H_8O_8N_8$) and cyclonite ($C_3H_3O_6N_6$). Therefore, for identification, the EMs were grouped into nine classes with similar C:O:N ratios.

Another distinguishing feature of this stationary setup is the large granularity: the inspection zone is divided into many elementary volumes, or voxels, in which the analysis is performed. The total number of voxels is $64 \times 7 = 448$. The small size of alpha detector pixels and the large distance from the alpha detector to the tritium target lead to a small size of a single tagged beam. This simplifies the identification of small masses of material. For example, 393 s were needed to detect 25 g of TNT and 250 s to detect 50 g TNT with an intensity of neutron beam, $I = 4 \times 10^7\ s^{-1}$.

The setup was routinely used for 2 years at the Interregional Center for Analysis and Neutralization of Explosives of the FSB, Russia. Based on 147 measurements, the probability of detection is 97%, the probability of identification is 95%, and the probability of false alarms is 2% (Bystritsky et al. 2013).

Another system for baggage inspection is ULIS—unattended luggage inspection system—made by SODERN-EADS, described in Le Tourneur et al. (2009) is shown in Figure 7.37. The ULIS neutron tube is fitted with an alpha-particle detector that enables the system to determine not only the substances themselves, but also their position inside an object. ULIS is a case containing a miniaturized

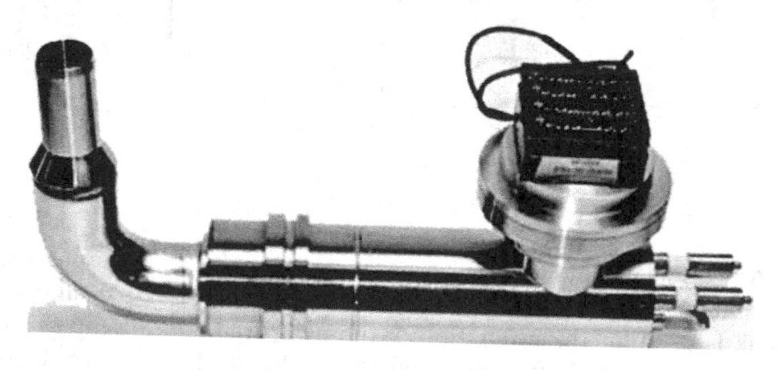

Figure 7.37 ULIS—unattended luggage inspection system—made by SODERN-EADS.

neutron tube, a gamma-ray detector, an electronics module, batteries and a high voltage power supply, and two small video cameras. The case is connected to a remote laptop computer, so that the system can be operated from a safe distance.

To investigate a suspicious item of luggage, the ULIS case is placed near the object and activated from the laptop. The tube emits neutrons, which penetrate the object under investigation. Some of the neutrons interact with the atomic nuclei of the various chemical elements inside and produce characteristic gamma rays, which are captured by the gamma-ray detector in the ULIS case. The ULIS software compares the gamma-ray spectra with a "library" of known signatures and thus unambiguously identifies the contents of the object under investigation.

The ULIS neutron tube incorporates an important additional component: the associated particle detector. This attachment makes use of the fact that every emitted neutron releases an alpha particle in exactly the opposite direction. High-speed processing by the ULIS electronics module takes this directional information into account and combines it with time-of-flight measurements in a triangulation algorithm, thereby producing an image showing the location of suspect material inside the object under investigation. The result is shown within a few minutes. If no suspect material is found, ULIS indicates "No threat detected."

Another device for explosive detection based on nanosecond neutron analysis, called SENNA, is described by Kuznetsov et al. (2006) and is shown in Figure 7.38. This is a portable device consisting of a single suitcase weighting 35 kg; it is remotely controlled from any PC computer. Inside is an APT neutron generator with a 3 × 3 matrix of semiconductor detectors of associated alpha particles, two BGO-based detectors for gamma-rays detection, fully digital data acquisition electronics, data analysis and decision-making software, and batteries. Detection technology is based on determining the chemical composition of the concealed substance by analyzing gamma rays resulting from interaction of tagged fast neutrons with its material. A combination of position–sensitive alpha detector and time-of-flight analysis allows one to determine the location of the detected

Figure 7.38 SENNA, the device for the detection of explosives.

material within the inspected volume and its approximate mass. Fully digital data acquisition electronics is capable of performing alpha–gamma coincidence analysis at very high counting rates, which leads to the reduction of the detection time down to dozens of seconds. SENNA's scenario-driven automatic decision-making algorithm based of "fuzzy logic" mechanism allows one to detect not only standard military or industrial explosives, but also improvised explosives (including those containing no nitrogen), even if their chemical composition differs from that of standard explosives. According to the authors SENNA can also be "trained" to detect other hazardous materials, such as chemical and toxic materials, if their chemical composition is in any way different from that of the surrounding materials.

In conclusion, the device for detection of concealed explosives, called SENNA, has characteristics shown in Table 7.5, and it is capable of finding small amounts of explosives (starting with hundreds of grams) concealed among large amount of material within about a minute. Components of the device can be used to build specialized devices for specific applications (inspection of luggage, examination of suspicious objects, analysis of contents of cargo containers, etc.), as well as to create universal portable explosive detectors (see also Vakhtin et al. 2006 and Evsenin et al. 2009). In addition to SENNA III, which is dedicated to explosive detection, the same group has developed SENNA IV for "dirty bombs" and SNM detection, SENNA V for luggage control, and a system for the detection of liquid explosives.

The batch production of such detectors is organized in the town of Dubna by the company, Neutron Technologies. The company has been established by Joint Institute for Nuclear Research (JINR Dubna) and Rusnano Company. Their most popular product is the portable detector DVIN-1 (see Figure 7.39). Inspection block

TABLE 7.5 CHARACTERISTICS OF THE SENNA PROTOTYPE

Detection/identification method	Nanosecond neutron analysis/associated particles technique (NNA/APT)
Decision-taking algorithm	Automatic
Simultaneously inspected volume	~30 × 30 × 50 cm^3
Spatial resolution	7–10 cm in-plane, 10 cm in-depth
Total mass of the device	35 kg
Dimensions of the device	90 × 50 × 30 cm^3
Life time, before the change of the vacuum tube (estimated)	Over 200,000 measurement cycles
Radiation safety	Safe when switched off; safe distance when switched on: 5m
Power consumption	<100 W

Figure 7.39 Portable explosive detector DVIN-1.

consists of a neutron generator with a built-in alpha detector, gamma-ray detectors, DAQ electronics, and power system, and weights 34 kg. Overall dimensions: 740 × 510 × 410 mm. Operator's position: notebook connected to inspection block with cable 50 m long. Gamma detectors are based on BGO, crystal size 76 × 76 mm. Maximal neutron flux: 5×10^7 n/s. Neutron energy: 14.1 MeV. Operation mode: continuous. Peak power consumption: 70 W. Number of tagged neutron beams: 9 or 64.

The tagged-neutron method(TNM) has been developed by the scientists of the Joint Institute of Nuclear Research. Neutron Technologies Company was established in 2010 to commoditize the researches by JINR. Rusnano's investment contributed to making a viable business of a promising technology. The established production infrastructure allowed delivery of over 90 DVIN-1 portable explosives detection complexes (as of February 2014) to the stations of the major Russian railroads, including Oktyabrskaya, Gorkovskaya, Kavkazskaya, and Severo-Kavkazskaya, as well as the subways of Moscow, Saint Petersburg, Samara, Yekaterinburg, Kazan, and Novosibirsk. Over 3 years, the company's total revenue has exceeded RUB 650 mln. Besides various explosives detector modifications, Neutron Technologies Company has successfully designed other solutions based on the tagged-neutron method. In particular, on request of Alrosa (Alrosa's share in global diamond production is nearly 27 percent), the company created a technology of nondestructive detection of diamonds in blue earth (Aleksakhin et al. 2013).

There are a number of chemical warfare (CW) agents that could be found in the old ammunition (from WWI) lying at the bottom of the sea, rivers, and elsewhere. In order to identify the CW agents one needs information about its chemical composition. Table 7.6 lists the most commonly used CW agents by giving their chemical formula, name that is used by chemical abstracts service (CAS), and their CAS registry number. By inspection of Table 7.6, we can conclude that the presence of some chemical elements inside the investigated object (mine, shell, grenade, and similar) can be related to the presence of CW agents. Elements of interest are phosphorus (P), sulfur (S), chlorine (Cl), fluorine (F), arsenic (As), and bromine (Br).

TABLE 7.6 CHEMICAL WARFARE AGENTS

Common Name	Chemical Abstracts Service (CAS) Name	Molecular Formula	CAS Registry Number
DA, diphenylchloroarsine, Clark I	Diphenylarsinouschloride	$C_{12}H_{10}AsCl$	712-48-1
DC, diphenylcyanoarsine, Clark II	Diphenylarsinouscyanide	$C_{13}H_{10}AsN$	23525-22-6
DM, adamsite	10-Chloro-5,10-dihydrophenarsazine	$C_{12}H_9AsClN$	578-94-9
GA, tabun	Dimethyl phosphoramidocyanidicacid, ethylester	$C_5H_{11}N_2O_2P$	77-81-6
GB, sarin	Methylphosphonofluoridicacid, (1-methylethyl)ester	$C_4H_{10}FO_2P$	107-44-8
H, mustard gas, yperite, sulfur mustard	1,1′-Thiobis(2-chloroethane)	$C_4H_8Cl_2S$	505-60-2
HN-1, nitrogen mustard	2-Chloro-N-(2-chloroethyl)-N-ethylethanamine	$C_6H_{13}Cl_2N$	538-07-8
HN-2, nitrogen mustard, mechlorethanamine	2-Chloro-N-(2-chloroethyl)-N-methylethanamine	$C_5H_{11}Cl_2N$	51-75-2
HN-3, nitrogen mustard, nitrogen Lost	2-Chloro-N,N-bis(2-chloroethyl)ethanamine	$C_6H_{12}Cl_3N$	555-77-1
L, Lewisite	(2-Chloroethenyl) arsonousdichloride	$C_2H_2AsCl_3$	541-25-3

While the commonly used military explosives are characterized by the presence of only four chemical elements (C, H, N, O), chemical warfare agents usually have in addition one or more chemical elements (P, S, Cl, F, As, or Br). The results from the laboratory tests for the detection of the presence of these chemical elements as reported by Valkovic et al. (2009) are presented next.

During these tests two sets of measurements were done: materials placed in air at the distance of 10 cm from the submarine window and with 10 cm water layer between the window and the investigated material.

The results of irradiation of sulfur, chlorine, arsenic, and fluorine targets in air are shown in Figure 7.40. The measured gamma ray spectra indicated in red are obtained with following experimental conditions:

1. Sulfur mass 1 kg. Measurement time 3008 s, 12×10^7 emitted tagged neutrons.
2. Sodium chloride mass 1 kg. Measurement time 6000 s, 24×10^7 emitted taggedneutrons. Chlorine gamma lines are marked.

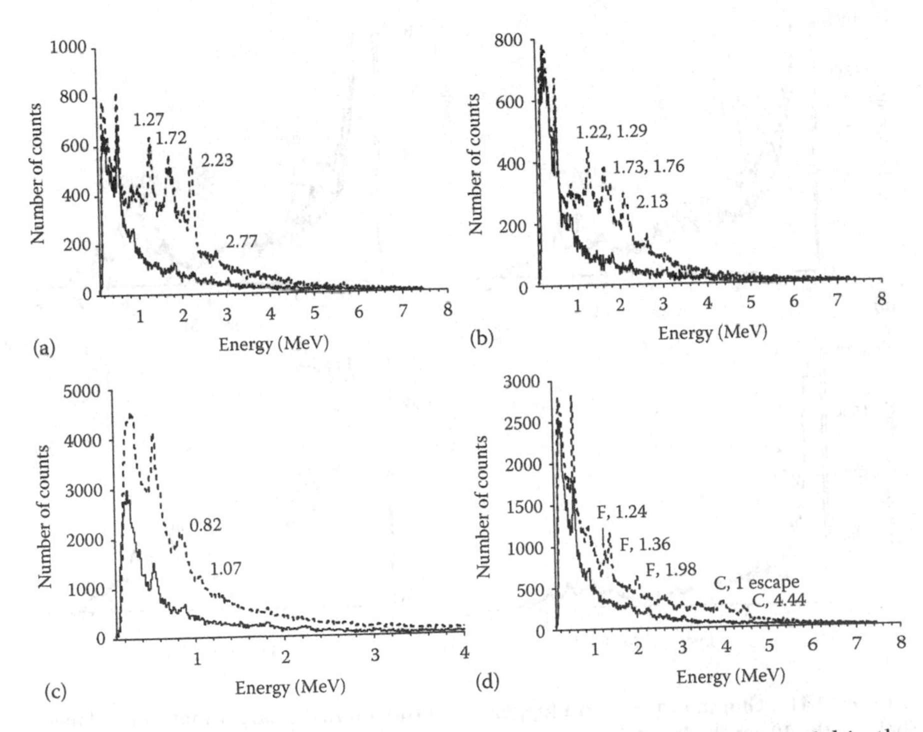

Figure 7.40 Gamma ray spectra (upper) from the following target material in the air: (a) sulfur, (b) sodium chloride, (c) arsenic trioxide, and (d) teflon; target out—lower spectra.

3. Arsenic trioxide mass 2 kg. Measurement time 9189 s, 36 × 10^7 emitted taggedneutrons. Arsenic gamma lines are marked.
4. Teflon mass 1 kg. Measurement time 6214 s, 24 × 10^7 emitted tagged neutrons.
5. The background spectra, lower curve, are obtained for the same experimental conditions with target out.

The results of the irradiation of sulfur, chlorine, arsenic and fluorine targets behind the 10 cm thick water layer are shown in Figure 7.41. The measured gamma ray spectra, upper curves, are obtained with the following experimental conditions:

1. Sulfur mass 1 kg. Measurement time 6351 s, 24 × 10^7 emitted tagged neutrons.
2. Sodium chloride of mass 1 kg. Measurement time 10,969 s, 36 × 10^7 emitted tagged Neutrons. Chlorine gamma lines are marked.

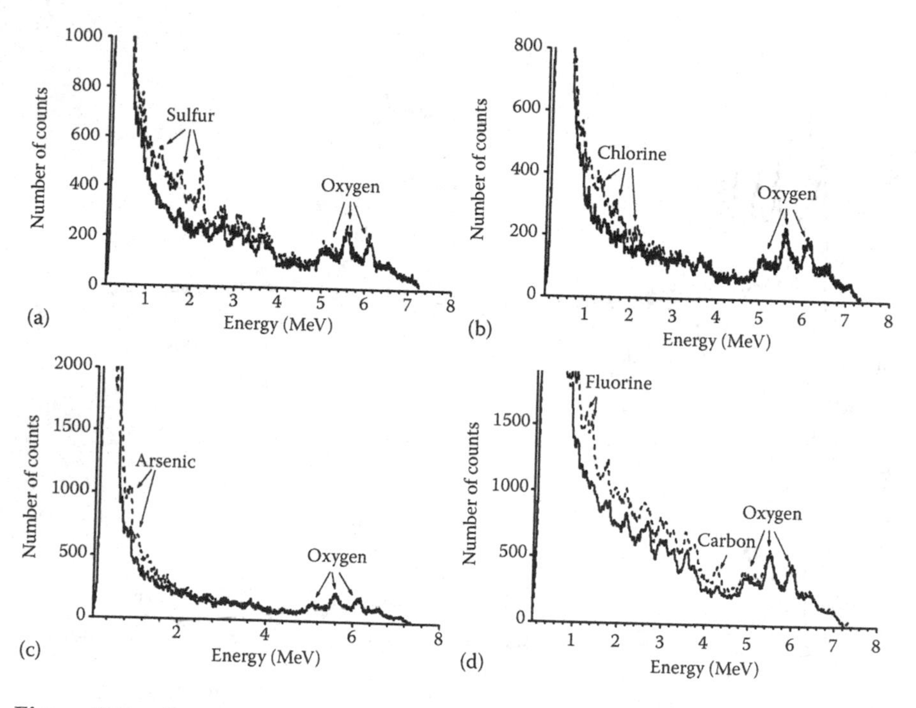

Figure 7.41 Gamma ray spectra (upper) from the following target materials placed behind the 10 cm thick water layer: (a) sulfur, (b) sodium chloride, (c) arsenic trioxide, and (d) teflon; target out—lower spectra.

3. Arsenic trioxide of mass 2 kg. Measurement time 7000 s, 24 × 10^7 emitted tagged Neutrons. Arsenic gamma lines are marked.
4. Teflon of mass 1 kg. Measurement time 10,026 s, 36 × 107 emitted tagged neutrons.

The background spectra indicated in black are obtained for the same experimental conditions with the target out.

The source of the background can be also the sediments on which the object is placed on the bottom of the sea or riverbed. They have shown that the gamma ray spectra obtained from the measurements of the sediments are dominated by lines of silicon, oxygen, and carbon. Different sediments have different spectra and in some cases silicon lines are absent; the spectra are then characterized by the presence of calcium, oxygen, and carbon lines.

These results demonstrate the possibility of detecting the presence of some chemical elements present in the CW agents in air and behind the 10 cm layer of water. The only element whose presence was difficult to quantify "in water" measurements was nitrogen. The usual materials for the CW agents' containers are iron and glass. Therefore the tests reported by Valkovic et al. (2008) should be repeated with the experimental geometry where the measured material is in the water and behind the assumed wall of the container.

Two detectors were used in these preliminary experiments: NaI (Tl) and BaF_2; their characteristics are shown in the Table 7.7. The ^{60}Co gamma ray spectra as measured by these two gamma ray detectors are shown in Figure 7.42. The energy resolution of BaF_2 detector is poor and as a result the two gamma ray peaks (1.173 and 1.332 MeV) are seen as a single peak centered at 1.2525 MeV. The resulting carbon signature as obtained by using these two gamma detectors is shown in Figure 7.43, indicating that NaI (Tl) detector as a more appropriate choice.

The capability of the methodology using the associated alpha particle technique is best illustrated by the following example. The raw data, gamma ray spectra with no window on TAC (100 ns), for different materials exposed to the

TABLE 7.7 PROPERTIES OF NA(TL) AND BAF2 GAMMA DETECTORS

Material	Density (g/cm³)	Emission Maximum (nm)	Decay Time	Refraction Index	Conversion Efficiency (%)	Hydroscopic
Na (Tl)	3.67	415	0.23 ms	1.85	100	Yes
BaF_2	4.88	315	0.63 ms	1.5	16	No
		220	0.8 ns	1.54	5	

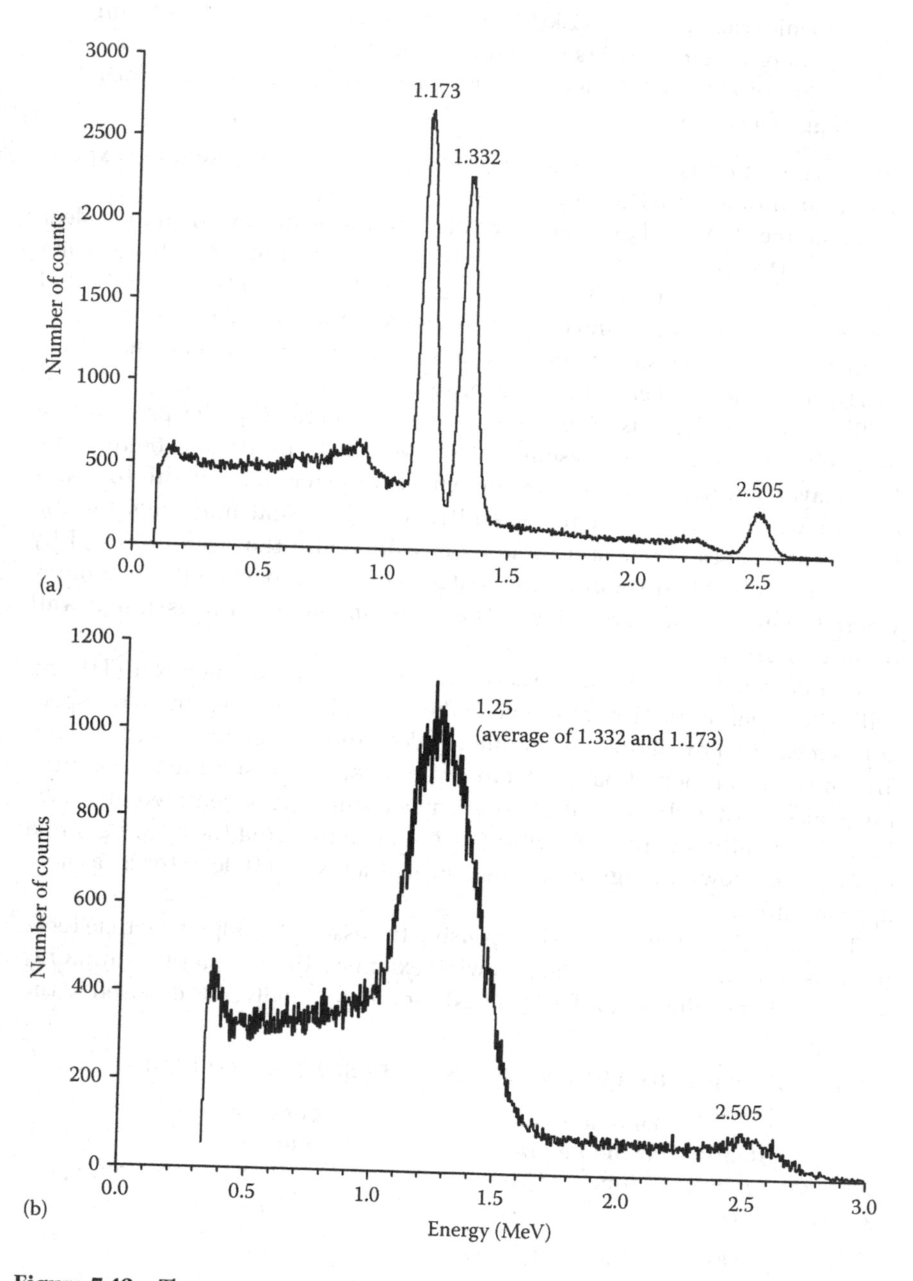

Figure 7.42 The gamma ray spectra of ^{60}Co as measured by (a) NaI (Tl) and (b) BaF_2 detectors.

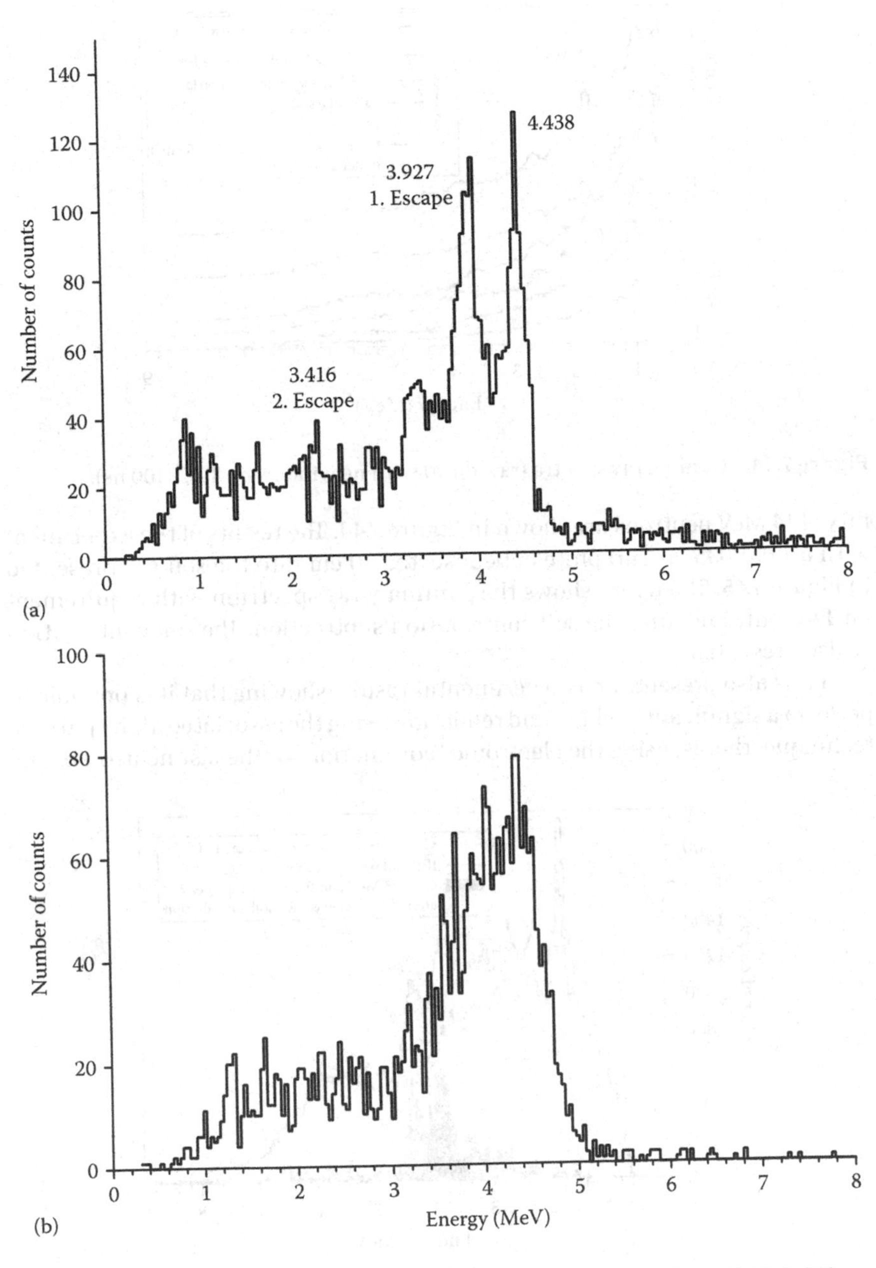

Figure 7.43 The gamma ray spectra of graphite as measured by (a) NaI (Tl) and (b) BaF_2 detectors.

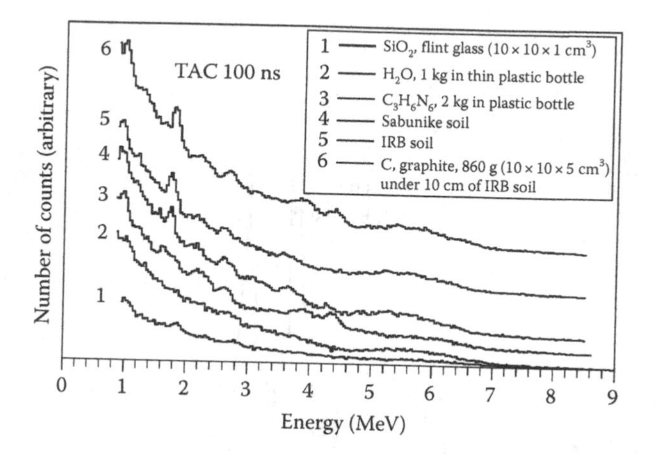

Figure 7.44 Gamma ray spectra (raw data) with no window on TAC (100 ns).

flux of 14 MeV neutrons are shown in Figure 7.44. The results of the experiment with a 1,031,035 cm^3 graphite cube inserted 10 cm into the soil are presented in Figure 7.45. The figure shows the gamma γ-ray spectrum with requirement on TAC only and after the soil contribution subtraction. The soil contribution is also presented.

Let us also present some experimental results showing that it is possible to perform a significant background reduction using the associated alpha particle technique, that is, using the electronic "collimation" of the fast neutron beam.

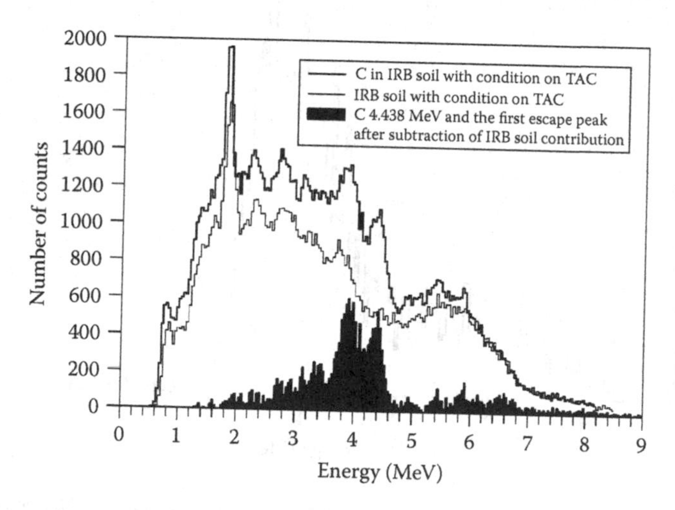

Figure 7.45 The results of the experiment with a 1,031,035 cm^3 graphite cube inserted 10 cm into the IRB soil.

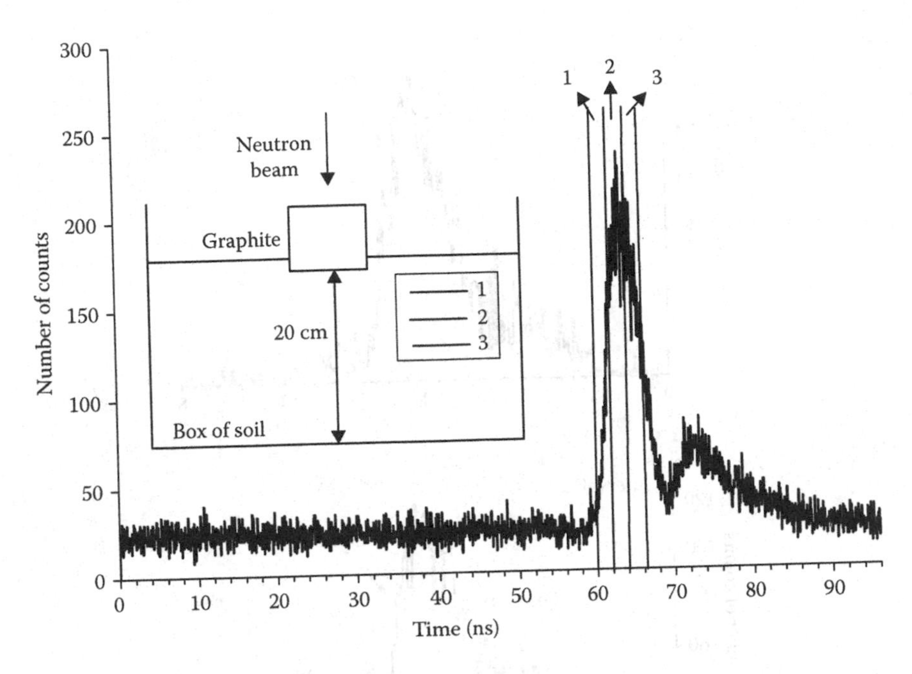

Figure 7.46 Experimental setup for the experiment with graphite on the soil and the corresponding measured TAC spectrum.

Although this method requires more electronics, the results show that it is superior in comparison with noncoincidence measurements. Figure 7.46 shows the experimental setup for the experiment #1: graphite on the soil together with the corresponding measured TAC spectrum. The resulting gamma spectra after the application of window on TAC spectrum are shown in Figure 7.47. In experiment #2 the carbon cube was burred under the layer of soil as shown in Figure 7.48; the same figure shows the measured TAC spectrum. The gamma ray energy spectra corresponding to different TAC intervals are shown in Figure 7.49.

From the presented experimental results it is obvious that by using the off-line analysis and by putting the appropriate window on the TAC spectrum, it is possible to identify the characteristic γ-rays generated by A(n,n′γ)A reaction in all interrogated samples. Even in the condition of considerably inferior time resolution, as with NaI (Tl) detector, it is still possible to reduce background radiation and to perform qualitative chemical analyses of interrogated objects. In this respect, the experiment with a sample of graphite under the layer of soil is very encouraging.

It is well known that mechanical demining, by using the special machines, provides significant savings over manual demining. Not only can more land be

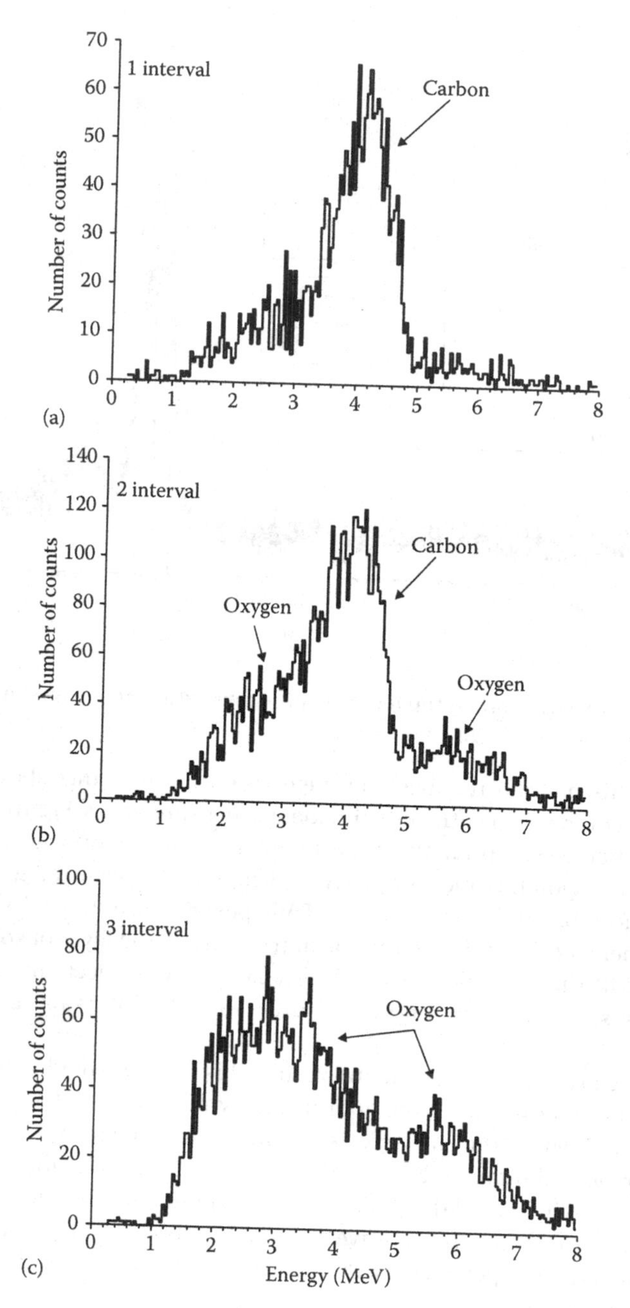

Figure 7.47 The resulting gamma spectra after applications of windows on time spectrum from Figure 7.46.

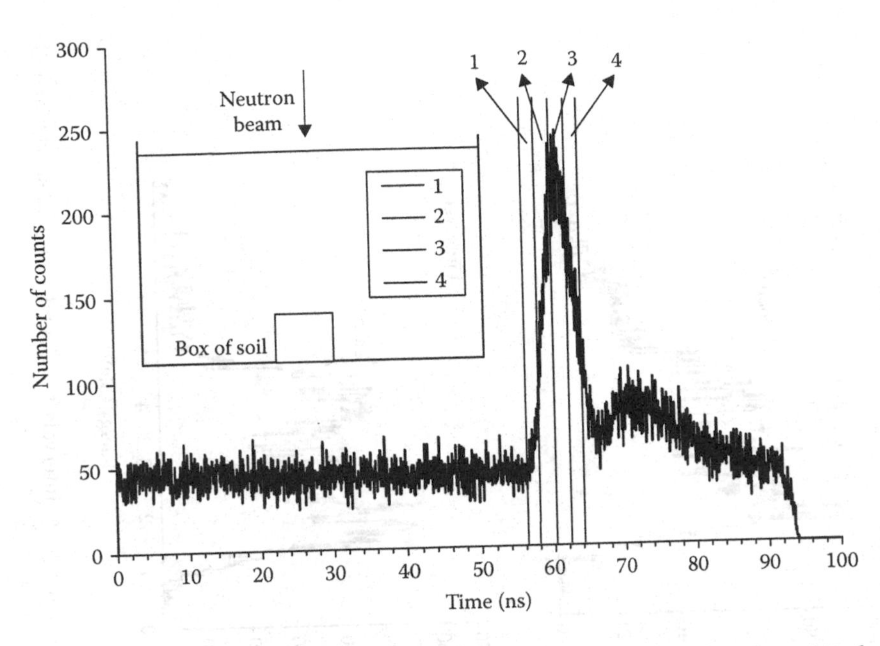

Figure 7.48 Experimental setup for the experiment with graphite cube buried in the soil with corresponding measured TAC spectrum. Intervals 1, 2, 3, and 4 correspond to depths 1, 2, 3, and 4, as shown in figure.

cleared, but human lives are saved as the process of manual demining is extremely dangerous. DOK-ING Ltd., Zagreb, Croatia (www.dok-ing.hr), developed MV-4 and MV-10 mine clearance systems designed to clear various types of terrain containing antipersonnel (AP) mines, antitank (AT) mines, and unexploded ordnance (UXO). AP and AT mines are destroyed by the force of impact of the flail tool attachment. The flail tool attachment is a hardened steel shaft with hammers attached at the end of the chains. During mine clearance activities, the shaft rotates and the hammers strike the ground and shatter or activate embedded mines. The force of the flail hammers are calculated to enable cutting through dense vegetation and digging into soil. After the treatment of a minefield with such a system it is necessary to check the minefield for the presence of explosive residues that are pieces of explosive fragments weighting 100 g or more.

In the article by Sudac et al. (2012) a neutron-based system was proposed for the detection of TNT fragments and AP mines as an upgrade of a robotic mobile system like MV-4 or MV-10. Proposed system contains 21 7.62 cm × 7.62 cm $LaBr_3$:Ce gamma ray detectors and 6 NGs. Such a system has a major drawback of being very expensive and producing significant amount of radiation during the operation. Rather than using only one sensor for the residual explosive detection, a multisensor approach is considered. A number of prototype

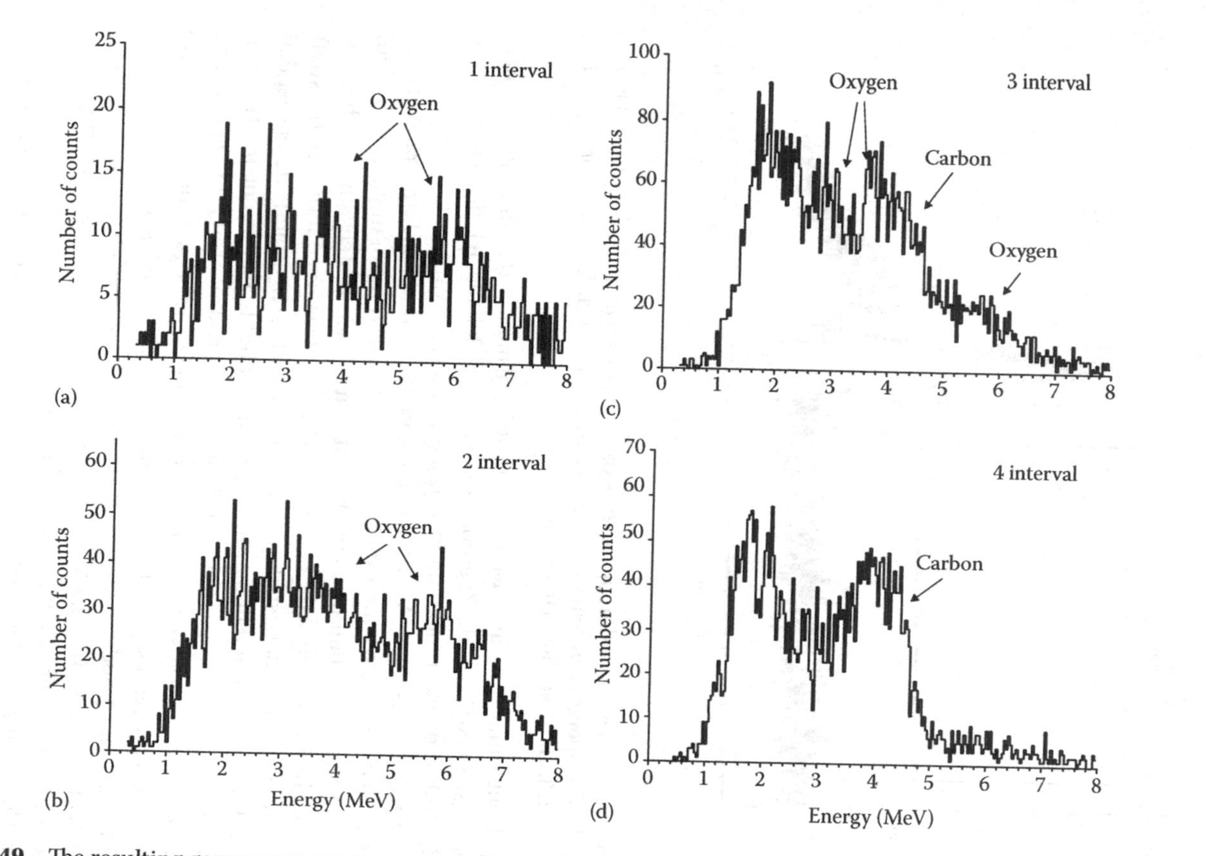

Figure 7.49 The resulting gamma ray energy spectra after applications of windows, intervals 1, 2, 3, and 4, on TAC spectrum indicated in Figure 7.48.

multisensor mine detector systems have been built. Canadian teleoperated landmine detection systems, for example, use five different technologies: electromagnetic induction, visible wavelength imaging, ground-penetrating radar for a quick scan of the minefield, and thermal neutron analysis as a confirmation tool (Faust et al. 2005). Combination of GPR and metal detector was discussed in Bruchini et al. (1998) and Lopera and Milisavljevic (2007). The above technologies depend on various parameters like the soil texture, bulk density, and soil moisture (Miller et al. 2004; Das et al. 2001; Kazunori et al. 2011). Especially the GPR strongly depend on the soil water content. It should be pointed out that the fast neutron (FN) and the associated alpha particle technique could be used not only as a confirmation tool, but it can measure the soil moisture content also.

In the experimental setup described by Sudac et al. (2012), a neutron generator API-120 manufactured by ThermoElectron was used as a source of 14 MeV neutron beam that was produced in $^{3}H(d,n)^{4}He$ nuclear reaction. The alpha detector incorporated inside the NG was made from the YAP:Ce scintillator fixed to the NG and from the removable photomultiplier tube (PMT). The collimator in front of the PMT defines the opening angle of the tagged neutron cone that is 12°. The gamma ray detector was 7.62 cm × 7.62 cm BGO. Between the gamma ray detector and the NG is a shield that protects the detector from the directly hitting neutrons. Below the gamma ray detector was put a target, for example, a box of loam soil a few centimeters apart from the detector. Figure 7.50 (adopted from Sudac et al. 2012) shows the gamma ray spectrum of TNT surrogate in a plastic container with a mass of 1.06 kg and the gamma ray spectrum of dry soil. The spectra are normalized in the sense that the integral over the gamma ray energy is equal to one. The TNT surrogate contained 0.158%wt of graphite, 0.276%wt of dihydrate oxalic acid, and 0.566%wt of cyanuric acid. The density was 1.06 g/cm^3. In this mixture, the carbon to oxygen and carbon to nitrogen ratio is the same like in the real TNT explosive, but carbon to hydrogen ratio is 7/6 instead of 7/5. Carbon, oxygen, and nitrogen peaks are visible in the TNT surrogate gamma ray spectrum. Carbon, oxygen, silicon, and aluminum peaks are visible in the gamma ray spectrum of dry soil. Presence of aluminum and silicon in the soil was confirmed also by the FN activation analysis. The presence of aluminum and silicon in the soil ruins the possibility of nitrogen detection at least in small AP mines field with the only TNT explosives.

In the measurements done on a simulant of AP mine, it was confirmed that nitrogen could not be detected by 14 MeV neutrons because the only clear difference between the mine in soil and the soil alone is in the carbon content. Similar results was obtained with the 7.62 cm × 7.62 cm $LaBr_3$:Ce gamma ray detector also. Contrary to the carbon content of soil, its oxygen content depends on the soil moisture as expected. Figure 7.51 shows the oxygen content in dependence

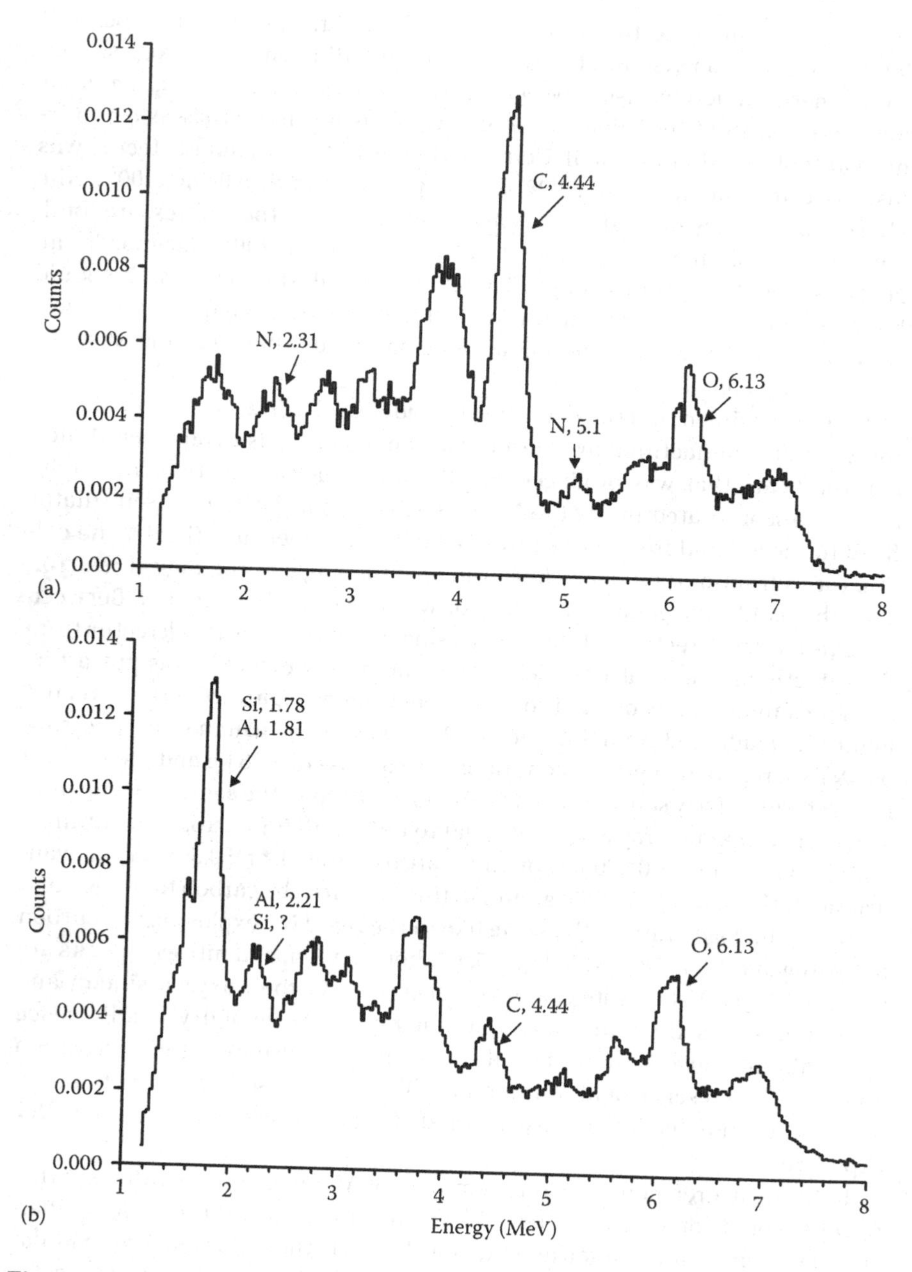

Figure 7.50 (a) The gamma ray spectrum of TNT surrogate $C_7N_3O_6H_6$ and (b) the gamma ray spectrum of dry soil.

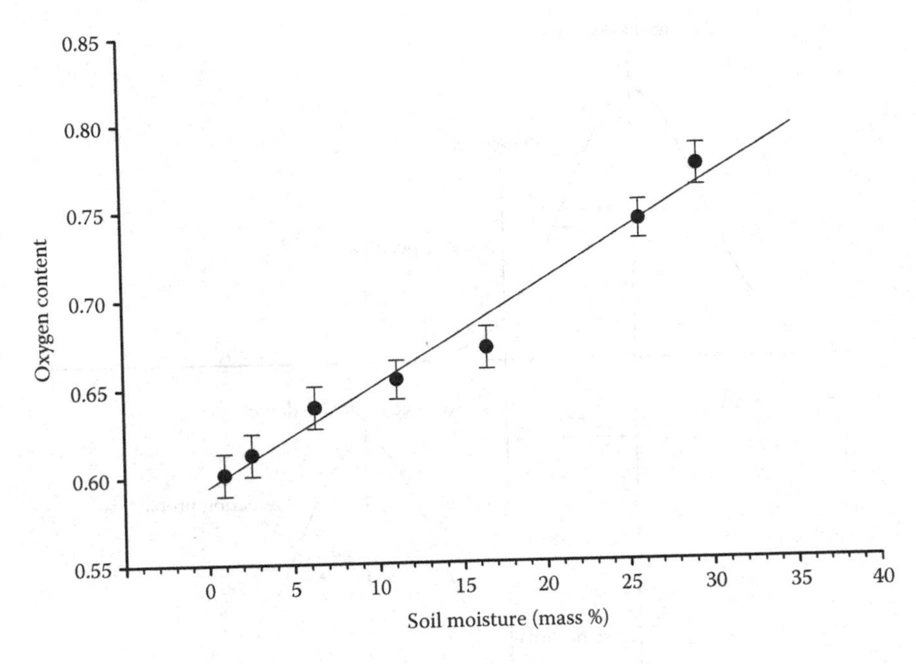

Figure 7.51 Oxygen content in dependence on the soil moisture. The gamma ray detector was 7.62 cm × 7.62 cm BGO.

of soil moisture as measured with the BGO detector. The soil moisture content may also be measured by measuring the hydrogen content. The hydrogen can be detected by its captured gamma ray line at 2.2 MeV, which usually is a part of a random background spectrum.

The difference in carbon content between the soil and the TNT explosive enables the possibility to detect the AP mine. Figure 7.52a shows the theoretical normal distributions of the carbon content in soil and from the mine buried into the soil (Figure 7.52b). It can be seen that the two distributions overlap partially. Overlapping decreases as the time of measurement extends. False positive is defined by an area over the threshold of the carbon distribution in the soil. Detection probability is defined by an area over the threshold of the carbon distribution from mine buried into the soil. Threshold is defined by the formula

$$\text{Threshold} = \text{Average background} + \kappa\sigma_b$$

where parameter κ depends on the accepted false positive alarm rate. Here, $\kappa = 1.282$, which corresponds to the 10% of the false positive. Table 7.8 shows the detection probability and carbon content for target in—target out

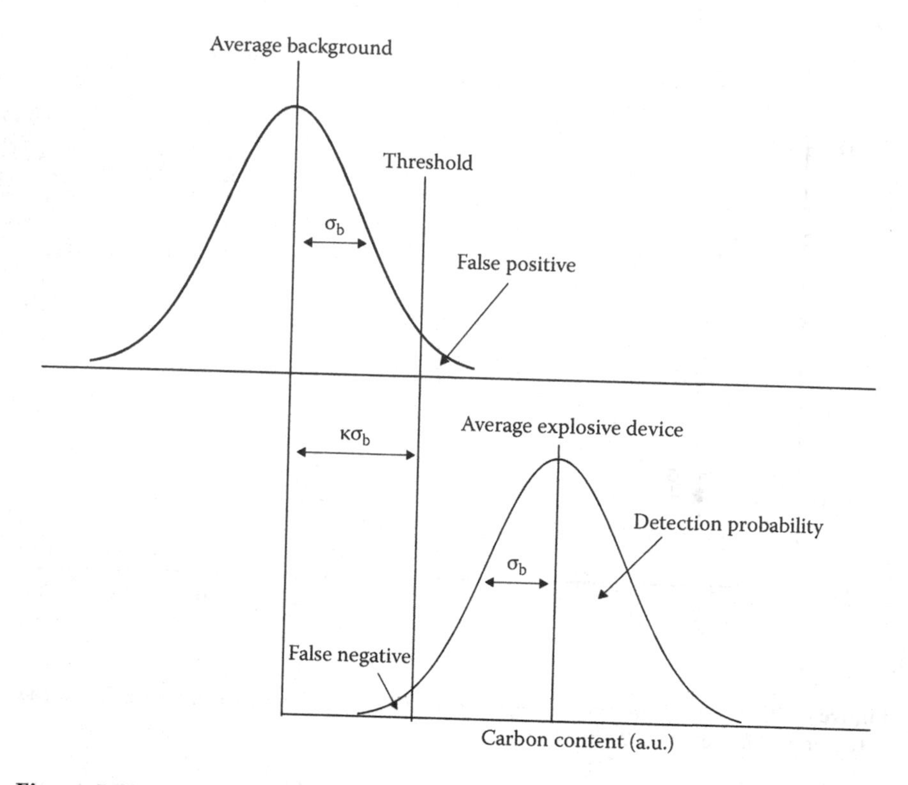

Figure 7.52 The normal distribution for the carbon content in the soil and for the carbon content in TNT fragment/explosive device buried into the soil.

configurations in the case of BGO gamma ray detector. Data were averaged over the soil moisture content; the errors shown are results of the fitting procedure. While the average carbon content was more or less the same, or otherwise did not depend on the time measurement, the error bar does. It became bigger as the measurement time decreases. It turns out that the BGO detectors show better results in comparison to the $LaBr_3$:Ce gamma ray detector. It can detect the AP mine earlier.

Although nitrogen could be used for the TNT explosive detection, it was found that the best way to detect the small quantity of it is by measuring the carbon content. The method can be used, regardless of the soil moisture content. Some problems may be found around the trees. As one could expect this method to work in combination with the others method and with MV-4/MV-10 (which does not work in woods), this problem was not studied by Sudac et al. (2012). BGO gamma ray detector is a better choice than the $LaBr_3$:Ce gamma ray detector. It is more stable on the energy calibration shifting, linear in energy calibration, and cheaper than the $LaBr_3$:Ce of the same dimensions. By using the

TABLE 7.8 DETECTION PROBABILITY AND CARBON CONTENT FOR TARGET IN/TARGET OUT CONFIGURATION

	BGO Detector			$LaBr_3$:Ce Detector		
Time (s)	Target in	Target out	Detection prob. (%)	Target in	Target out	Detection prob. (%)
1255	0.289 ± 0.011	0.166 ± 0.011	100	0.260 ± 0.013	0.156 ± 0.013	100
209	0.290 ± 0.028	0.166 ± 0.028	99.9	0.261 ± 0.032	0.154 ± 0.032	98
157	0.290 ± 0.033	0.166 ± 0.033	99.4	0.261 ± 0.037	0.155 ± 0.037	94.6
106	0.289 ± 0.04	0.167 ± 0.04	96.3	0.261 ± 0.045	0.154 ± 0.045	86.6
70	0.289 ± 0.049	0.166 ± 0.049	89.1	0.262 ± 0.055	0.157 ± 0.055	73.4
52	0.289 ± 0.056	0.166 ± 0.056	81.9	0.265 ± 0.063	0.155 ± 0.063	67.6

Note: The gamma ray detector is 7.62 cm × 7.62 cm BGO and LaBr3:Ce detectors.

BGO and the appropriate NG that can provide neutron beam intensity more than 2×10^7 n/s, the AP mine/TNT explosive residue can be found in less than a minute. However, the desired speed of 10 cm/s of the demining vehicle can hardly be obtained by using the FN/API method alone. However, by combination of the several methods like GPR, MD, IR, and FN/API the desired speed could be possible to achieve.

In the work by Batyaev (2013), capabilities of the tagged neutron method for detection and identification of explosives materials were explored using an idealized geometrical model that included a 14 MeV neutron generator with an integrated alpha detector, a gamma-ray detector based on BGO/LYSO crystals, and irradiated samples in the form of simulated explosive material (TNT, tetryl, RDX, etc.) or benign material such as cotton, paper, etc. Research was carried out under the framework of computational simulations of neutron physics processes by Monte Carlo methods as well as experimental measurements using an ING-27 neutron generator produced by VNIIA. The work resulted in a comparison between measured and simulated ROC (receiver operating characteristics) curves obtained via integration of analytically expressed functions of irradiation time, mass, and type of explosive materials and benign materials. Experimental results indicate that 0.3 kg of tetryl simulant located 45 cm from the neutron generator is detected in 97% of cases after a 1 min measurement, with the false alarm rate being highly dependent on the type of benign material present: from ~0% in the case of water to ~5% in the case of silk. Comparison of simulated and experimental data for these results showed they were in agreement in cases where the simulations account for neutron scattering from the object and background effects.

Material characterization in cemented radioactive waste with associated particle technique has been reported by Carasco et al. (2010). Test measurements were performed in order to study the capability of the associated particle technique to characterize the elemental composition of materials constituting radioactive waste enclosed in cemented packages. In order to simulate cemented radioactive waste, a mortar cube filled with various materials has been build and interrogated by a 14 MeV tagged neutron beam.

The tests have been performed with the EURITRACK tagged neutron inspection system. The detection of the alpha particle by a 64-element YAP:Ce scintillation array coupled to a multianode photomultiplier allows to define the direction of the neutron. The detection of the gamma ray by NaI (Tl) scintillators in coincidence with the YAP:Ce assembly with a dedicated electronics permits to build a time-of-flight (TOF) spectrum from which the flight path of the neutron is inferred, providing, in this way, in-depth information. Since the gamma-ray energy is specific to the element with which the neutron interacted, it is also possible to get information about the chemical composition of the inspected objects. The gamma-ray energy spectrum obtained from a given area of the TOF spectrum is a mixture of the signatures of the nuclei from the corresponding region of the inspected object. To get a chemical insight into the inspected area, the measured energy spectrum is unfolded between 1.35 and 8 MeV on the basis of the signature of elements from a database. Finally, by using appropriate correction factors, it is possible to recover the chemical proportions of the detected elements. The radioactive waste package has been simulated with a 6 cm thick mortar container. For practical reasons, the mortar container has been built inside a 3 mm iron cargo container on top of which a set 16 5″ × 5″ × 10″ NaI (Tl) shielded detectors have been placed. Put together, the mortar and the iron layers mimic iron containers in which radioactive waste is blocked by cement.

The energy spectrum associated to the mortar indicated that it is mainly composed of carbon and oxygen, with an oxygen-to-carbon ratio of 3.2 ± 0.3 that agrees with the chemical composition of calcium carbonate ($CaCO_3$). After determining its chemical composition, the mortar cubic package has been filled with materials of interest, which are prohibited or stringently controlled in cemented radioactive waste. Detection tests have been performed with samples of graphite, wood, grease, water, magnesium, aluminum, PVC, mercury, and boron in the mortar container.

In conclusions, measurements performed with the tagged neutron inspection system (TNIS) show the ability of the associated particle technique to detect materials that are forbidden or strictly controlled in cemented radioactive waste packages. The TNIS allows to identify and locate, via gamma-ray spectroscopy and neutron TOF, chemical elements inside the waste package. An element is identified by its characteristic gamma rays with an unfolding algorithm. The TNIS succeeded in detecting graphite, organic materials, PVC, aluminum, and magnesium inside a mortar container simulating a cemented radioactive waste package. However, the system showed difficulties to detect water and boron and it seems not to be able to identify mercury. Improvements can be brought to the TNIS in order to increase its performances. For instance, the database used to unfold the energy spectra can be modified to take advantage of low-energy gamma rays, which are strongly needed to identify metals and to increase the discrimination power of the unfolding algorithm. This last can also be improved. Finally, the geometry of the setup is not optimized. Indeed, the system has been designed for the inspection of cargo containers (project EURITRACK), whose shape and size are completely different to those of radioactive waste. A dedicated setup would greatly improve the accuracy of the system.

7.4 USE OF TAGGED NEUTRON BEAM FOR SEA CONTAINER INSPECTION, TRANSPORT INFRASTRUCTURE CONTROL

Invented during the Second World War as an efficient way of moving military equipment up to the front line without tying down too many soldiers for loading and unloading ships, the container has become indispensable to world commerce. Today, containers have helped to make the distribution of goods so efficient that manufacturers have been able to reduce inventories to a bare minimum. Containers also turned out to be handy ways to smuggle drugs, contraband, and illegal immigrants. A victim of its own success, the container offered criminals the same benefits as those enjoyed by ocean carriers and shippers: efficiency and security.

Let us mentioned some parameters on the size of container transport industry:

- Ninety percent of world cargo moves by container.
- In many nations such as the United Kingdom, Japan, and South Korea, over 90% of trade volume arrives or leaves by sea.
- In the United States, almost half of incoming trade (by value) arrives by ship.

Why is there risk to sea-going containers?

- Containerized shipping is a major vulnerability, and the global economy depends upon it.
- Over 200 million cargo containers move between major seaports each year.
- Some terrorist groups have stated that one of their goals is to destroy U.S. economic interests.

The movement of each container is part of a transaction that can involve up to 25 different parties: buyers, sellers, inland freighters and shipping lines, middlemen (customs and cargo brokers, for example), financiers, and governments. A single trade can generate 30–40 documents and each container can carry cargo for several customers, thus multiplying the number of documents still further.

A typical large container ship can carry up to 6,000 TEU and generate 40,000 documents. In 2001, around 9 m TEU arrived in America's container ports by sea, which translates into around 17,000 actual boxes a day. Many more arrive in America from Canada, mainly by train or truck.

The fact that containers are "intermodal"—that is, they can travel by sea and on land, by road or rail—means the system is difficult to regulate as it crosses jurisdictional boundaries. On a ship at sea, the container comes under the aegis of the International Maritime Organization (IMO), a United Nations body based in London. On land—in a port or on road or rail—it passes into the hands of national governments, which may have separate legislation for the different transport modes.

Modern personnel, parcel, vehicle, and cargo inspection systems are noninvasive imaging techniques based on the use of nuclear analytical techniques. The inspection systems are using penetrating radiations (neutrons, gamma, and x-rays) in the scanning geometry, with the detection of transmitted or radiation produced in investigated sample.

The research on the possibility of using tagged neutron beam for the inspection of shipping container for the presence of the threat materials was reported by Valkovic et al. (2004). A part of a real shipping container has been brought into the neutron laboratory to be used in variety of experiments. This is illustrated by a photo shown in Figure 7.53. Soon, the first experiments with the detection of real explosive were done and reported by Valkovic et al. (2004). The experimental setup with 30 kg of TNT is shown in Figure 7.54. The 30 kg of TNT (volume 35 × 30 × 30 cm^3) inside container were placed 2.12 m from the neutron source. Gamma ray detector was placed outside the container on the opposite side of the neutron source, d = 87 cm from the TNT sample; see Figure 7.55.

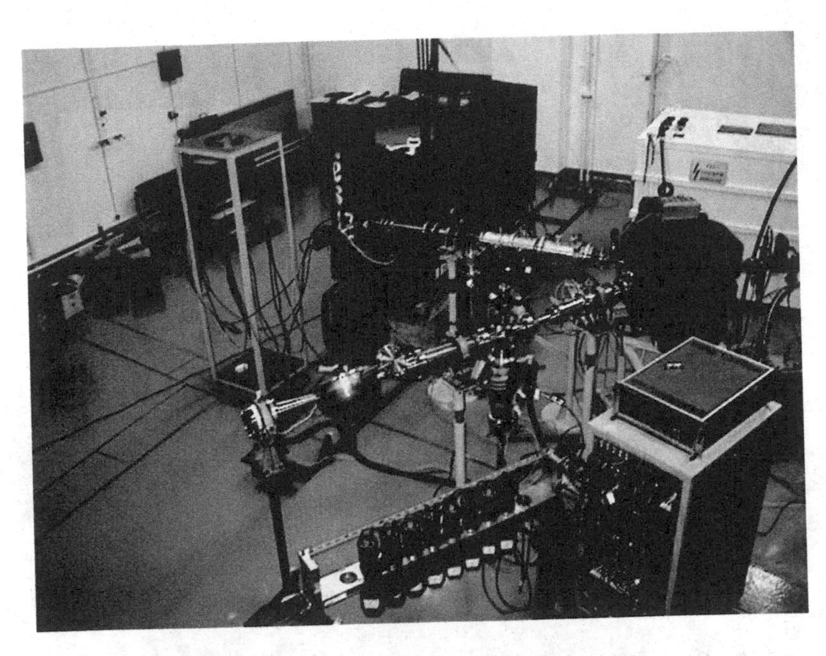

Figure 7.53 Biggest target ever in low energy nuclear laboratory.

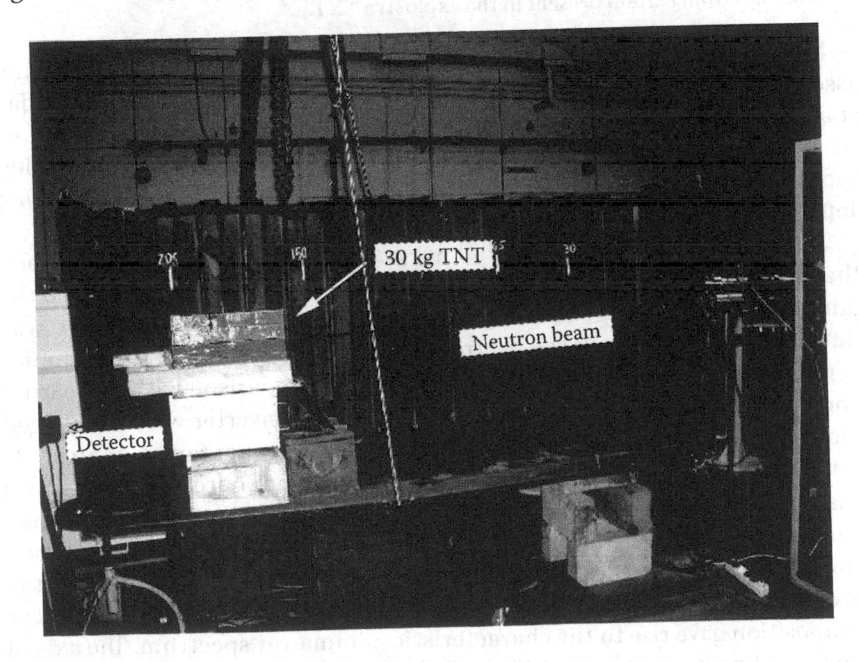

Figure 7.54 Experimental setup with 30 kg of TNT explosive.

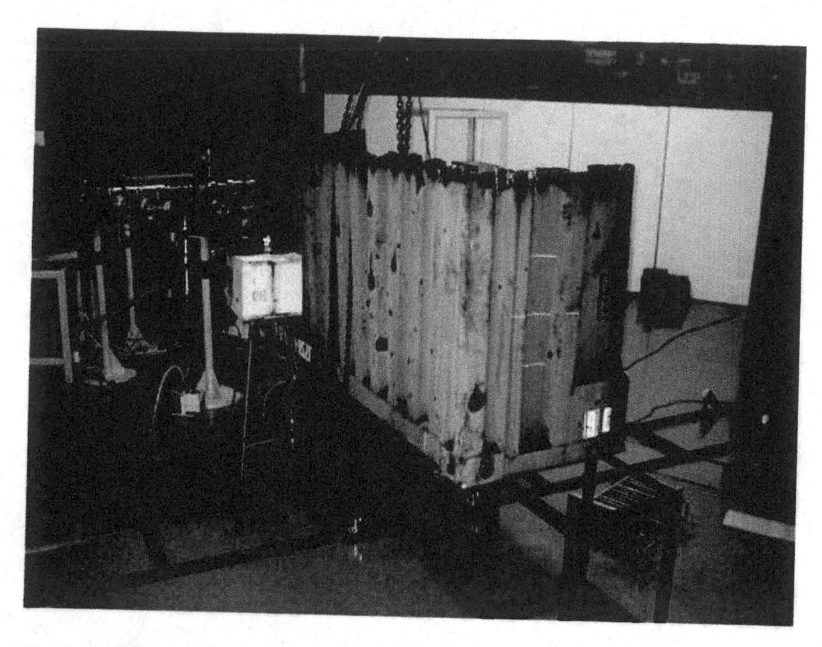

Figure 7.55 Gamma detector positioned outside the container was able to detect gamma rays from carbon present in the explosive TNT.

Inspections for contraband in a shipping container using fast neutron and associated alpha particle techniques: proof of principle is described by Sudac et al. (2005). The system was calibrated by inspection of 63.6 kg graphite target as shown in Figure 7.56. The gamma spectrum resulting from the 14 MeV neutron bombardment of this target is shown in Figure 7.57, spectrum being dominated by 4.43 MeV carbon peak.

The carbon target was placed at several positions inside the container. The neutron beam was entering the container from the right side, while the gamma detector (hexagonal BaF_2, area = 64.95 cm^2, thickness = 16 cm) was placed perpendicularly to the neutron beam axis. Standard NIM electronics were used as described in Sudac et al. (2004). In comparison with this earlier work, start and stop inputs in time to amplitude converter were exchanged and the alpha detector was placed 11 cm from the tritium target, at an angle of 135° from the deuterium beam. The 14 MeV neutron source (tritium target) was placed as close as possible to the right side of the container. The alpha–gamma coincidence spectra were measured by selection of alpha–gamma coincidences within 100 ns TAC range. The information on the position of the interrogated object was contained in the TAC spectrum, while the object's composition gave rise to the characteristic gamma-ray spectrum. The axis of the tagged neutron cone was directed through the container approximately

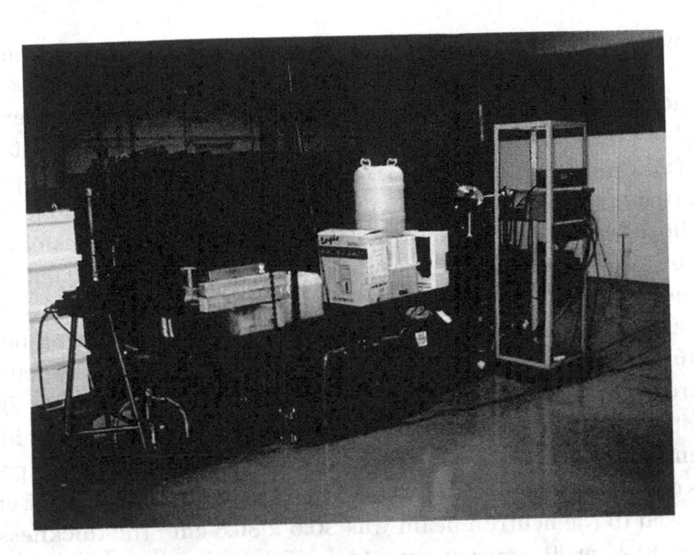

Figure 7.56 63.6 kg of graphite: d(target-detector) = 76 cm; d(target-neutron source) = 200 cm.

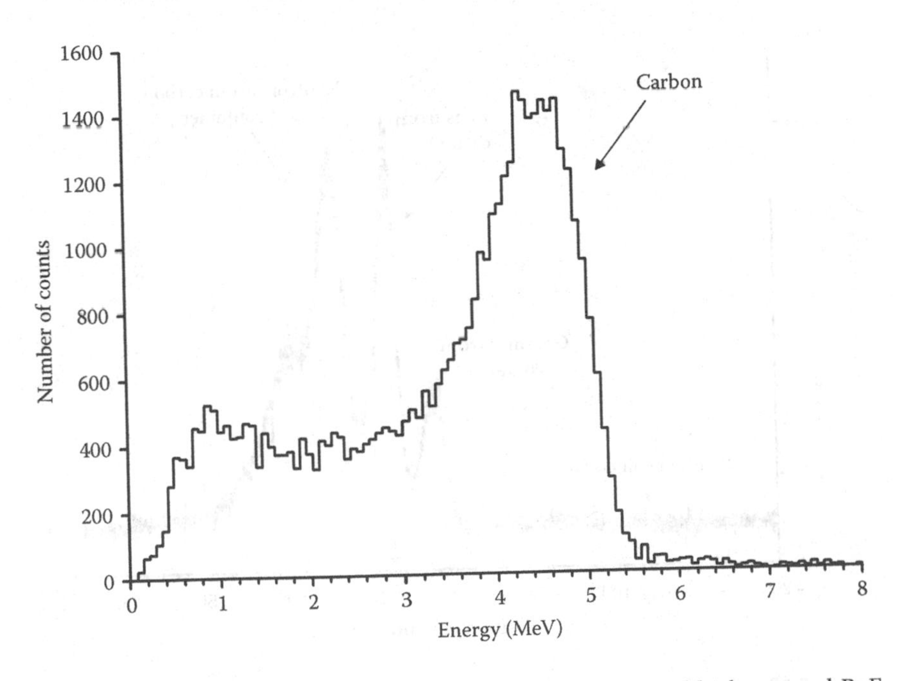

Figure 7.57 Carbon gamma ray energy spectrum as measured by hexagonal BaF_2 gamma detector.

parallel with the container side 25 cm from it. Vertical and horizontal profiles of the tagged neutron beam were measured with a 7.62 cm × 7.62 cm NE-213 neutron detector in coincidence with alpha counter. Experimental diameter of the neutron spot, 72 cm from the tritium target was (10.9 ± 0.3) cm and (11.5 ± 0.3) cm for the horizontal and vertical profile, respectively, in comparison with calculated 10.6 cm. Time calibration was done by introducing the different delays into the delay line (0.0513 ± 0.0003 ns/ch). Energy calibration was done by using the radioactive ^{60}Co source (0.065 MeV/ch, energy resolution 17.5% for the 1.1 MeV and 1.3 MeV lines). Spectra reported were measured for the same total number of the emitted tagged neutrons, NT = 1.46 × 10^8. Acquisition time varied from 4607 to 9230 s depending on the neutron intensity that varied from 0.935 × 10^7 to 2.5 × 10^7 n/s. The relative nonstatistical instrumental error was calculated from several identical measurements and its value varied between 2% and 4%. In this experiment the mass of carbon blocks was 15.9 kg in the volume 10.2 × 30.6 × 30.6 cm^3. The area exposed to the neutron beam was 30.6 × 30.6 cm^2; the thickness of the target was 10.2 cm. The measurements were done for the location of carbon blocks (1) 37 cm, (2) 47 cm, (3) 65 cm, and (4) 150 cm from the right side of container. Figure 7.58 shows the time spectrum for the carbon target placed

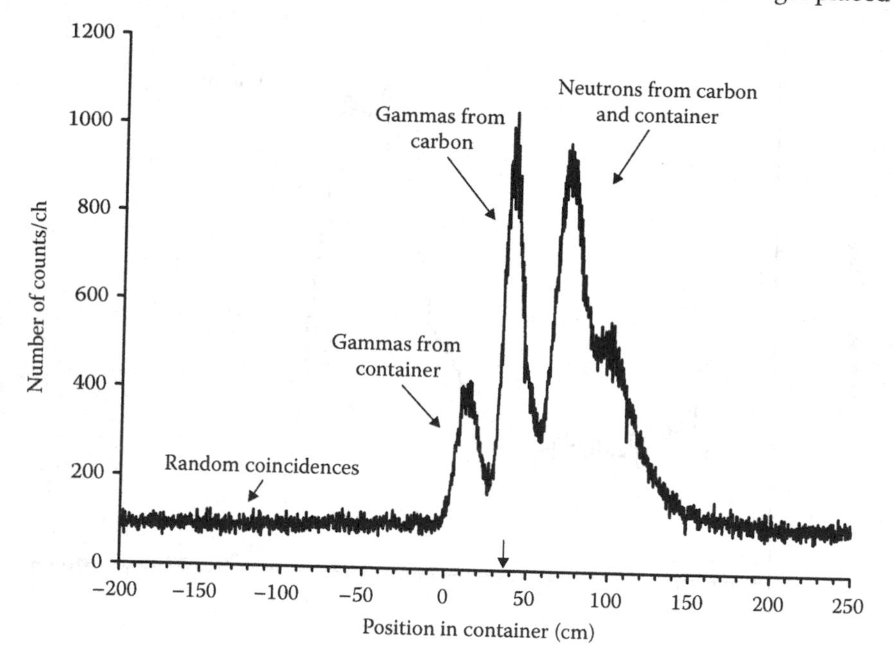

Figure 7.58 Time spectrum for the carbon target placed 37 cm from the container wall.

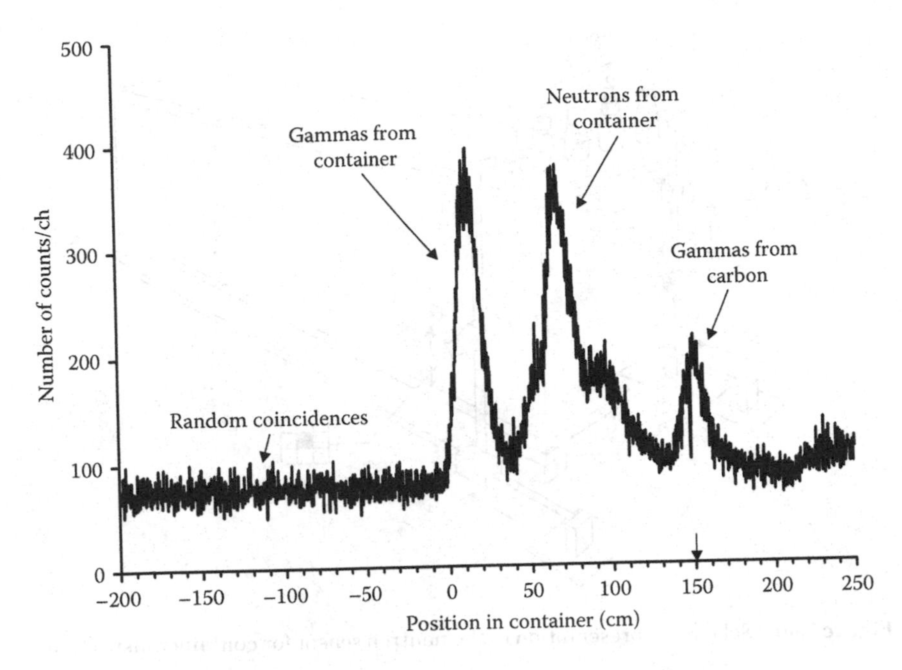

Figure 7.59 Time spectrum for the carbon target placed 150 cm from the container wall.

37 cm from the container wall while Figure 7.59 shows the time spectrum for the carbon target placed 150 cm from the container wall.

With this measurement it has been shown that it is possible to determine the location of the object inside the closed container (for carbon and TNT). This information is contained in the TAC spectrum (associated alpha particle being start impuls and the gamma ray produced in the object being stop impuls). Object identification could be done by the analysis of the measured coincidence gamma ray spectra. In such a way the "prove of principle" has been made: it is possible to construct multisensor system made of fast x-ray sensor (whole container) followed by detail elemental analysis of suspicious volume by using neutron sensor (Valkovic et al. 2004; Sudac et al. 2004; Blagus et al. 2004) as shown in Figure 7.60.

Significant improvement of this technique has been reported by Valkovic (2009) in the framework of NATO supported project SfP-980526. A number of experiments have been done using different materials including real explosive, 100 kg of Semtex 1a explosive, as shown in Figure 7.61. These results lead to an important conclusion: the material hidden in the container can be identified by measurements of its three characteristics, namely carbon and oxygen content and the neutron attenuation, which is determined mainly by its density.

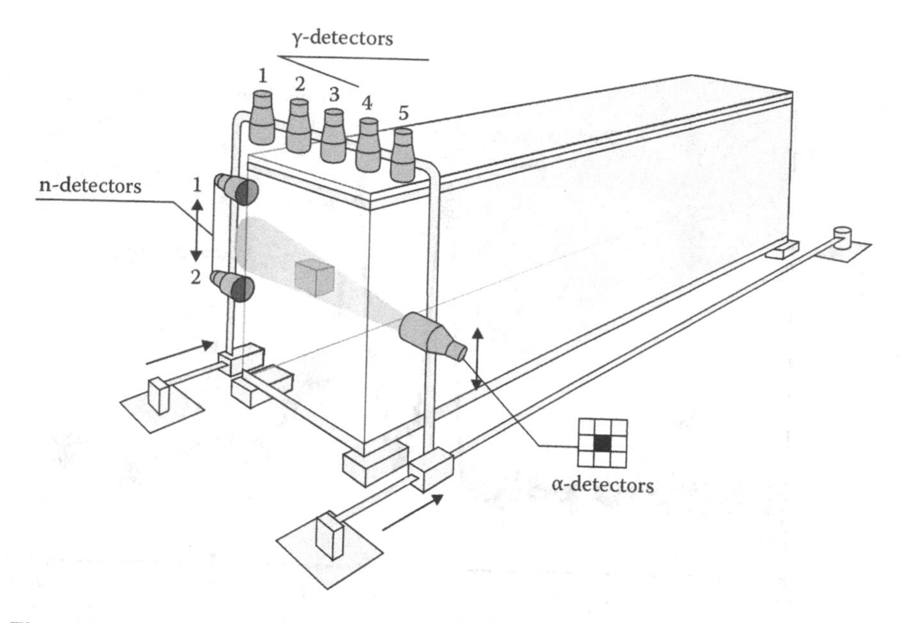

Figure 7.60 Schematic presentation of the neutron sensor for container inspection.

Figure 7.61 100 kg of SEMTEX 1A in the center of the container.

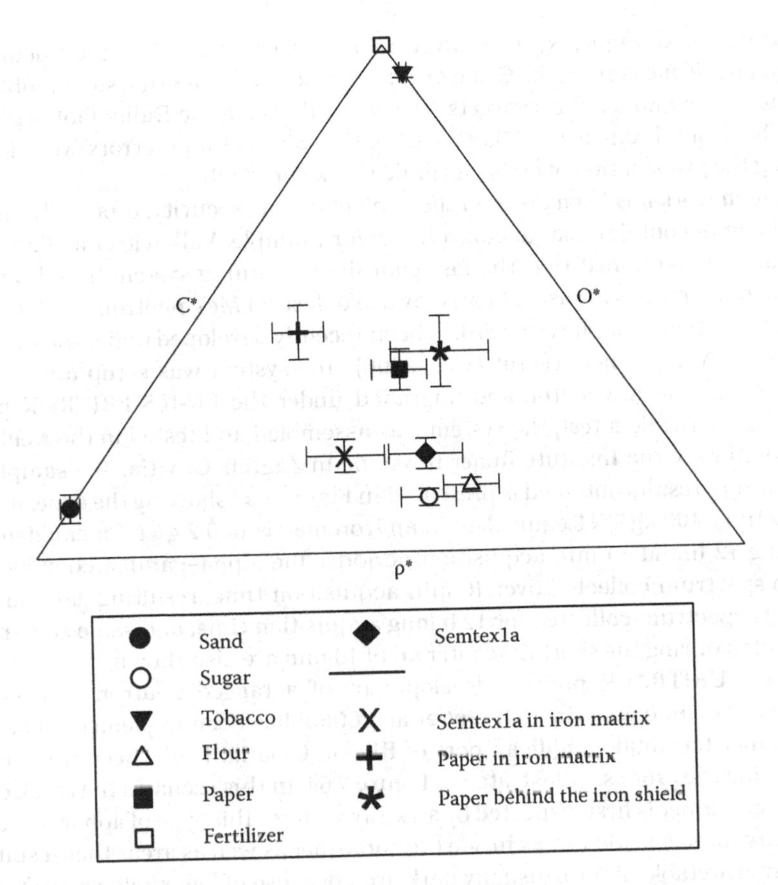

Figure 7.62 Triangular graph: distribution of measured data for different materials.

This is evident if the measured data are presented in a modified Dalitz plot; see Figure 7.62. In this triangular diagram the three axes are defined by

$$\rho^* = \frac{(N/N_{av})^2}{\left[(N/N_{av})^2 + (C/C_{av})^2 + (O/O_{av})^2\right]}$$

$$C^* = \frac{(C/C_{av})^2}{\left[(N/N_{av})^2 + (C/C_{av})^2 + (O/O_{av})^2\right]} \tag{7.2}$$

$$O^* = \frac{(O/O_{av})^2}{\left[(N/N_{av})^2 + (C/C_{av})^2 + (O/O_{av})^2\right]}$$

Here, for example, N_{av} is an average integral number of tagged neutrons taken while measuring the C and O gamma spectra from sand, sugar, tobacco, flour, paper, and fertilizer targets. The vertical axis in the Dalitz Plot is ρ^* and the horizontal axis is $0.57735\ (1 + C^* - O^*)$. Bidirectional errors were found using the propagation of errors formula (Valkovic 2009).

Much work has been done on the subject of port security, especially on the problem of container cargo control; see, for example, Valkovic et al. (2006). It should be mentioned that the fast container scanning system based on the elemental analysis of suspect cargo by use of fast, 14 MeV neutrons with detection of associated alpha particle has been recently developed under the EU-FP6 "EURITRACK" project (Perot et al. 2006). The system was setup and tested in port of Rijeka, Croatia, and upgraded under the EU-JLS ERITR@K project. Before the field test, the system was assembled and tested in the Neutron Laboratory of the Institute Ruder Boskovic in Zagreb, Croatia. The sample of laboratory results obtained is presented in Figure 7.63 showing the experimental setup a 100 kg TNT equivalent in an iron matrix of 0.2 g/cm^3 mean density during 12 h and 10 min acquisition periods. The alpha–gamma coincidence time spectrum collected over 10 min acquisition time, resulting gamma ray energy spectrum collected for 12 h long acquisition time, and same spectrum collected during the short time interval of 10 min are also shown.

The EURITRACK project, development of a tagged neutron inspection system for cargo containers (Perot et al. 2006b) has been implemented at the container terminal "Brajdica," port of Rijeka, Croatia. It has been organized following the proposal illustrated in Figure 7.64. In this scenario the truck carrying container is first inspected by an x-ray system. This type of apparatus can identify the suspicious areas inside the container as well as areas that result in noninterpretable picture (usually dark area because of low x-ray penetration). With this information in the hands of custom officers the truck is sent to the neutron system for the inspection of suspicious region. The experimental setup in the port of Rijeka is shown in Figure 7.65. There are other approaches to this problem especially related to detection of hidden explosives by using tagged neutron beams. For the description of the status and perspectives see Viesti et al. (2005).

In the EURITRACK setup tested in the container terminal Brajdica, port of Rijeka, the neutron generator used in the portal was SODERN with 8×8 α-detectors as shown in Figure 7.66. Before the system was completed, several studies have been performed including simulation of a tagged neutron inspection system prototype (Donzella et al. 2006); the Monte Carlo analysis of tagged neutron beams for cargo container inspection was performed by Pesente et al. (2007). Preliminary results were discussed and identification of materials hidden inside a container by using the 14 MeV tagged neutron beam was described (Sudac et al. 2007). Conversion factors from counts to chemical ratios for the

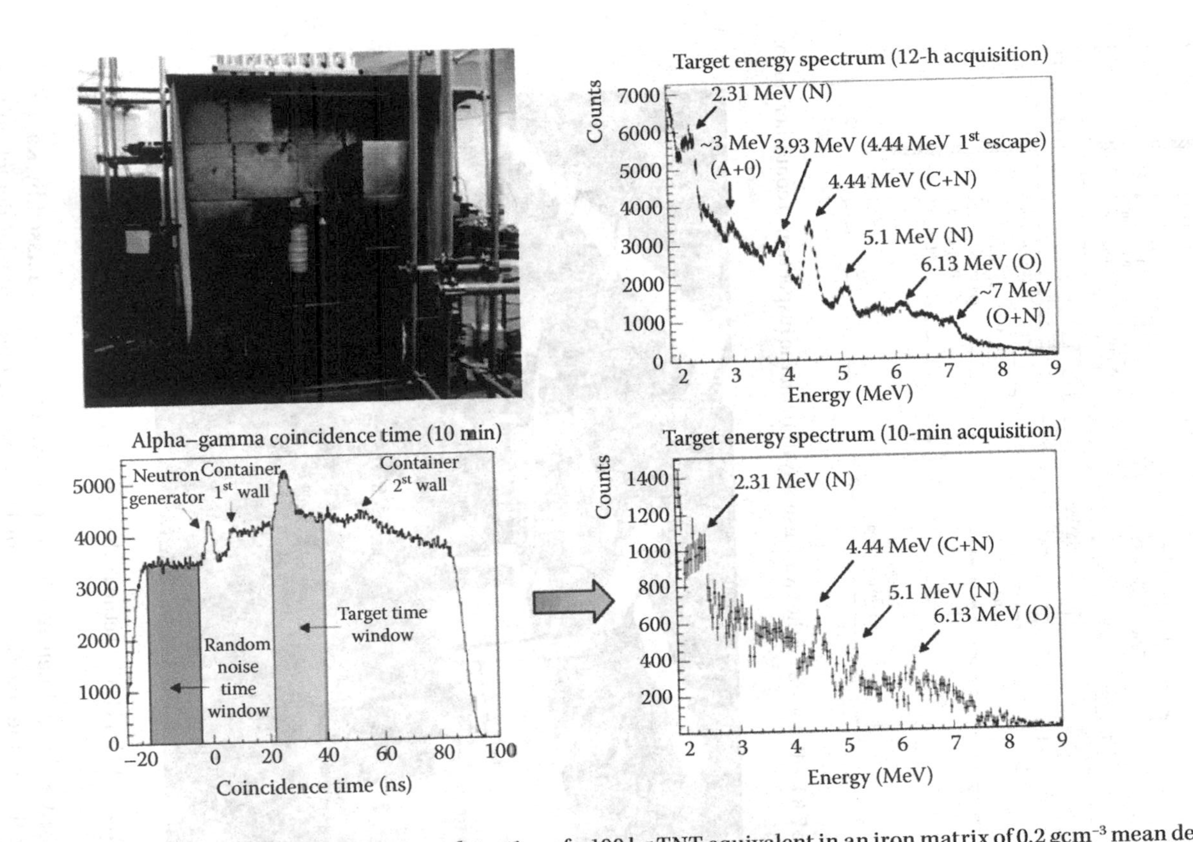

Figure 7.63 Laboratory test of EURITRACK system: detection of a 100 kg TNT equivalent in an iron matrix of 0.2 gcm^{-3} mean density during 12 h and 10 min acquisitions.

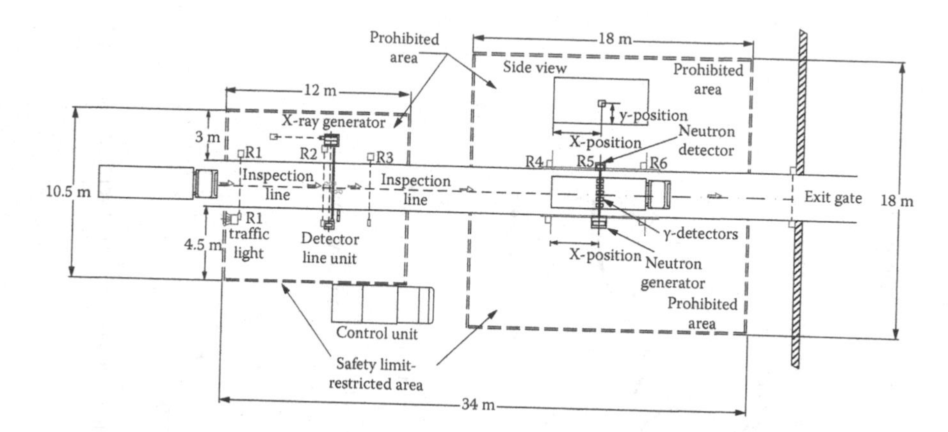

Figure 7.64 Proposal for the two sensors system for the inspection of container cargo.

Figure 7.65 EURITRACK setup in the container terminal "Brajdica," port of Rijeka.

EURITRACK tagged neutron inspection system were discussed by Kanawati et al. (2011), while matrix characterization in threat material detection processes was described in detail by Obhodas et al. (2009).

Although the calculated and measured doses from neutron and gamma ray inspection devices show that API-based systems deliver a small dose

Figure 7.66 Neutron generator, made by SODERN, with 8x8 α-detectors used in the EURITRACK portal.

(see Table 7.9), additional calculations of the neutron and photon dose rates were necessary to define a restricted area to public during operation of the neutron generator and the adequate radiological shields (Perret and Perot 2007). The restricted zone and shields are designed to ensure an annual dose below the legal limit of 1 mSv. Assuming an envelope operating duration of

TABLE 7.9 APPROXIMATE DOSES FROM NEUTRON AND GAMMA RAY EXPOSURE

No	Source	Effective Dose (mSv)
1	Computed tomography	35/exposure
2	Natural background	2.2/year
3	Gamma camera	2/exposure
4	X-ray tube	0.5/image
5	API	0.5/inspection
6	Medical	0.37/person–year
7	X-ray cargo inspection system	0.1/inspection
8	Cosmic rays	0.6/10 h flight
9	X-ray luggage inspection system	0.002/inspection

500 h, the limit dose rate outside the restricted area is 2 μSv/h and the maximum calculated dose rate for the defined restricted area is smaller, 1.3 μSv/h. Moreover, assuming that a public person would stay for 500 h at the border of the restricted area is very penalizing. Therefore, the maximum annual dose will be much lower than the legal limitation of 1 mSv.

The characteristics of the EURITRACK demonstrator on the site are assumed to be as follows: The isotropic neutron source emitting 10^8 n/s surrounded by a 1 m thick and 5 m high U-wall in concrete. The U-wall is extended on both sides by 4.5 m long concrete walls that are 20 cm thick and 5 m high. A container is centered on the neutron beam axis and an opposite 20 cm thick concrete wall is necessary to protect the adjacent area. Monte Carlo calculations were performed with the MCNP-4C2 code system, after Briesmeister (2000). The neutron generator was modeled by a punctual isotropic 14 MeV neutron source. Two configurations of interest were reported: without any container and with a 2.4 × 2.4 × 12 m^3 container filled with an iron matrix of density 0.5 g/cm^3. Preliminary calculations have shown that a container filled with an organic matrix of density 0.5 g/cm^3 leads to lower dose rates than the reported configurations.

Detector tallies (from 5 to 255) located on the boundaries of the restricted zone were used to calculate the neutron and photons ambient equivalent dose rates for a nominal 10^8 n/s source strength (Perret and Perot 2007). The neutron and photon ambient dose equivalent per unit fluence coefficients needed to convert the neutron and photon fluxes in ambient dose rates equivalent were extracted from the paper by Smith (1997). Both neutron and photon ambient dose equivalent rates and Monte Carlo statistical uncertainties were calculated. Tallies with uncertainties greater than 5% are unreliable but correspond to low dose rates and provide the order of magnitude of the result. The photon dose rate is much lower than the neutron dose rate and the total dose rate is lower than 1.3 μSv/h for any of the locations and configurations considered. The most penalizing configuration is without any container because the container closes the solid angle of the U-wall and acts as a biological shield. Neutrons reflected in the iron container are shielded by the 20 cm thick concrete lateral wall on each side of the U-wall. The opposite concrete wall length can be optimized to reduce the neutron reflection for the configuration without container. Other calculations with containers loaded with iron matrix of higher density should be investigated if required by the licensing study.

Inspection of cargo container by EURITRACK system in the port of Rijeka, Croatia, is shown in Figure 7.67; the truck carrying the ship container is driven inside the portal and the driver leaves the truck after positioning it following the operator's instructions. Analysis of containerized cargo in the ship container terminal is discussed in detail in the paper (Obhodas et al. 2010) where results of analysis performed on 152 screened containers selected by the Croatian

Figure 7.67 Inspection of cargo container by EURITRACK system in the port of Rijeka, Croatia. Person shown in the figure is Mr. Zvonko Valković in charge of site logistics.

Custom and detailed descriptive statistical analysis of their shipping manifests is presented. Statistics such as distribution functions of container types, type of packages, type of cargos, and average cargo densities were obtained in order to establish the properties of standard cargo traffic.

The analyses by using multivariate statistical approach that included principal component analysis (PCA) and between group analysis (BGA) have shown that cargo matrices can be classified as metallic (Fe, Al, Cu, Zn, Ni, etc.), ceramic glass (Si, O), and organic matrices (C, O, N).

In the paper by Obhodas et al. (2010), the organic matrices have been analyzed in more detail since they are identified as most difficult because of their potential interference with main elements contained in threat materials such as explosives or drugs. Density as a discriminator factor has been included for better recognition of different type of goods.

Several improvements of this system and applications different from the primary task of identifying explosive presence in the container have been discussed in several reports. For example, Kanawati et al. (2010) have discussed gamma-ray signatures improvement of the EURITRACK tagged neutron inspection system database. Some aspects of the identification of materials hidden inside a sea-going cargo container filled with an organic cargo by using the tagged neutron inspection system have been discussed by Sudac et al.

(2008). The limitations of associated alpha particle technique for contraband container inspections were presented in an earlier work by Sudac et al. (2007). Simulation of a tagged neutron inspection system prototype is reported by Donzella et al. (2006). Of special interest is a possibility of detecting the presence of a "dirty bomb" in the vehicles or containers (Sudac and Valkovic 2006; Valkovic et al. 2007).

Let us mention in some details the possibility of detection of illicit drugs with the EURITRACK System (Perot et al. 2009). This report presents tests performed with the EURITRACK neutron inspection system and the Silouhette 300 kV x-ray system on illicit drug samples in the seaport of Rijeka. Drugs and sugar, to allow comparison with a benign material, have been hidden in two different positions inside a metallic cargo made of iron boxes filled with iron wire balls. The average density of this metallic cargo is 0.2 g/cm^3. Two empty boxes have been used to hide the samples, as shown in Figure 7.68. The x-ray picture obtained from the inspection of the container shows only the presence of boxes not revealing their content; see Figure 7.69.

In the test with 20 kg of heroin in the iron cargo the interrogated sample has been measured only in the upper location. A 30 min acquisition has been performed. The distance spectrum is presented in Figure 7.70, with the marked area corresponding to the heroin position resulting in the gamma ray spectrum

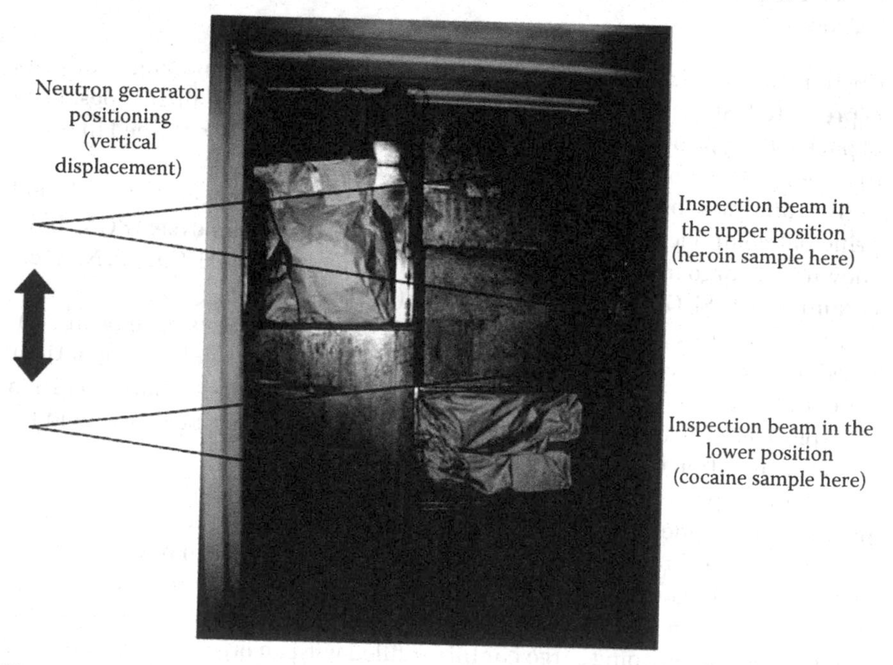

Figure 7.68 The location of heroin and cocaine samples inside metal boxes.

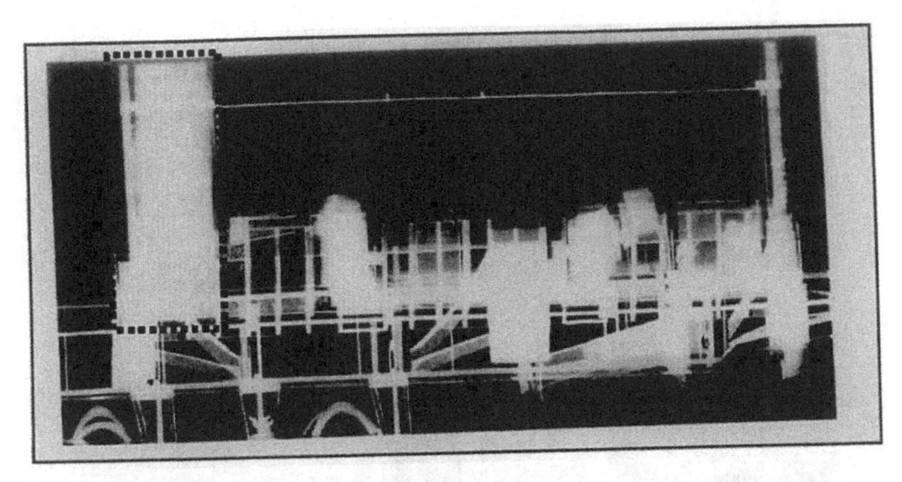

Figure 7.69 X-ray picture of the boxes within which the drugs were hidden during the experiments.

shown on the left side. An unfolding algorithm allows decomposing the measured energy spectrum into the sum of energy spectra and each spectrum in the sum corresponding to one element (C, N, O, Fe...) with a weight related to its chemical proportion. The obtained sum spectrum is plotted (black lines) in the previous figures to compare with the measured spectrum and thus verify the quality of the fit. The different energy spectra recorded in the same upper position for drugs and sugar are compared in Figure 7.71 confirming again that heroin and cocaine can be distinguished from sugar, but not marijuana.

The tests performed at the seaport of Rijeka with the EURITRACK neutron inspection on real drug samples demonstrate the possibility to detect

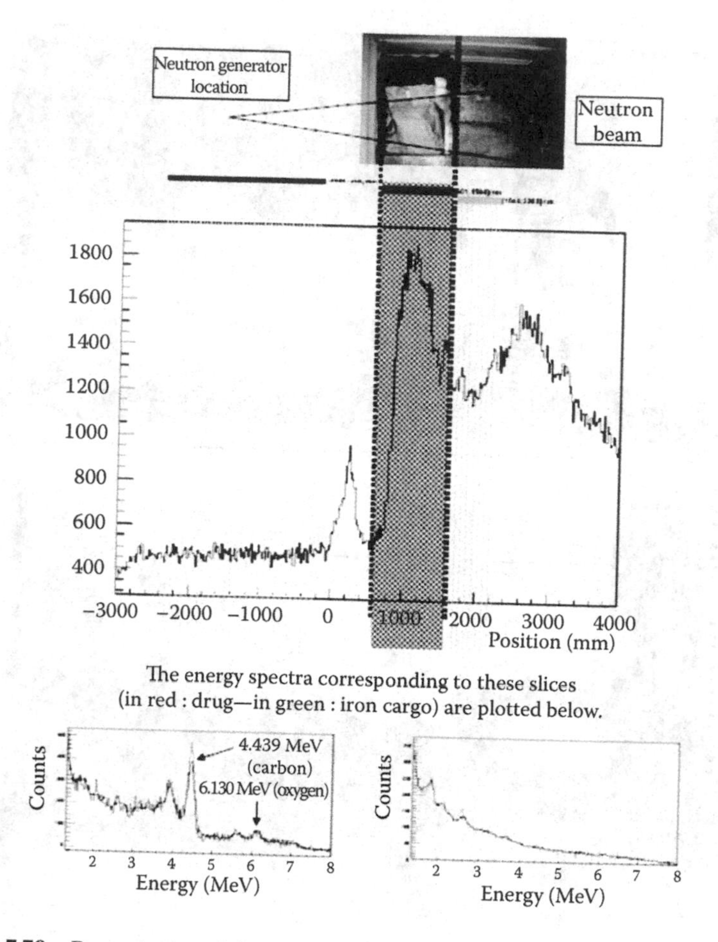

Figure 7.70 Presentation of the experiment setup and results of the measurements.

bulk cocaine and heroin (20–30 kg) in a metallic cargo. These illicit drugs have been distinguished from benign substances from their chemical composition. These experimental results confirm preliminary evaluations performed by numerical simulation. Marijuana, which chemical composition is similar to benign materials, cannot be distinguished from common organic materials, but the presence of an unexpected organic substance inside the metallic cargo can be used as an indication that a suspect object has been hidden in the container. Detection tests have also been performed with a small sample of cocaine (1 kg), showing the capability of the system to detect the presence of the carbon signature when the sample is hidden in a metallic structure. The low signal for oxygen does not allow the

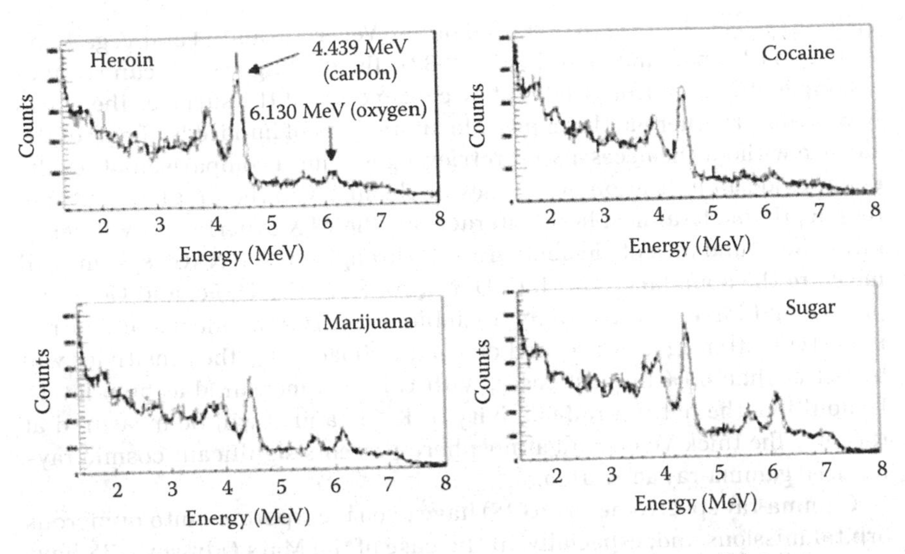

Figure 7.71 Comparison between different drugs and benign material—sugar.

identification of the organic material by O/C analysis, but the presence of carbon can be used as an indication that an organic material has been hidden in the metallic structure.

Systems for the inspection of cargo containers have been also described by several other groups, see, for example, Belichenko et al. (2009), Megahid et al. (2009), and Vakhtin et al. (2009).

7.5 NEUTRON GENERATORS IN EXTREME CONDITIONS

It has been illustrated by many research groups that neutron generators can be efficiently used to study problems in harsh environments: underwater, in soil wells, and in space. Here we should mention SAGE (surface atmospheric geochemistry experiment) mission (Bradley at al. 1994). The SAGE mission proposed that a select set of U.S.- and Russian-built instruments be flown on a Russian built Venus lander of the Vega design with a Russian-provided launch vehicle. The proposed instrument set includes a new atmospheric chemical analyzer for descent operation, a surface imaging spectrometer for mineralogical identification, an alpha proton x-ray analyzer for elemental analysis of a retrieved sample, and the new neutron activation gamma-ray spectrometer (NAGS) for analysis of the Venusian surface in situ. The APX and NAGS elemental analyzer systems are complementary. The APX system provides elemental analysis of a surface sample obtained by an external drilling system identical to that used on the Venera 13 and 14 and Vega 1 and 2 landers (Barman and Shevchenko 1983). The drilling system can retrieve a sample of 1–6 cm from the top few centimeters of the surface. The NAGS system on the other hand can provide an analysis of hundreds of cm^3 of the surface without the necessity for retrieving it. Thus a comparison of results can provide an indication of changes in chemistry in the first few centimeters depth due to atmospheric interaction. The APX experiment will determine the abundance of elements from C through Ni. The NAGS system will measure the abundances of H, C, O, Na, Al, Si, S, Ca, Ti, Fe, and Gd + Sm, plus Cl and Mg (if present in appreciable quantities) by means of neutron inelastic scattering, capture, and delayed radioactivity. The sensitivity will be better than 0.1% in most cases, with Gd + Sm measured at ppm levels. In addition, the natural radioactivity of K, Th, and U will be measured at the site. The thick Venusian's atmosphere prevents significant cosmic ray–induced gamma-ray activation.

Gamma-ray spectrometers (GRS) have been incorporated into numerous orbital missions and, especially, in the case of the Mars Odyssey GRS have contributed detailed maps of elemental composition over the entire surface of Mars (Boynton et al. 2007). While such remote-sensing measurements of planetary composition depend on cosmic rays as the material excitation source, the system described by Parsons et al. (2011) incorporates a much more powerful neutron excitation source that permits high sensitivity surface and subsurface measurements of bulk elemental compositions in only tens of minutes. The instrument described is capable of detailed in situ bulk geochemical analysis of the surface of planets, moons, asteroids, and comets. This instrument technology uses a pulsed neutron generator to excite the solid materials of a planet and measures the resulting neutron and gamma-ray emission with its detector system. These time-resolved neutron and gamma-ray data provide detailed information about the bulk elemental composition, chemical context, and density distribution of the soil within 50 cm of the surface. The goal is to apply active neutron scattering and neutron-induced gamma-ray techniques to surface instruments for use on any solid solar system body. The pulsed neutron generator-gamma ray and neutron detectors (PNG-GRAND) experiment is an innovative application of active (where neutrons are produced at specific times by the PNG) neutron-induced gamma-ray technology so successfully used in oil field well logging and mineral exploration on Earth. The objective of active neutron-induced gamma-ray technology program at NASA Goddard Space Flight Center (NASA/GSFC) was to develop the PNG-GRAND instrument to the point where it can be flown on a variety of surface lander or rover missions to the Moon, Mars, Venus, asteroids, comets, and satellites of outer planets.

PNG-GRAND combines a pulsed neutron generator (PNG) with gamma ray and neutron detectors to produce a landed instrument to determine subsurface elemental composition of a planet without the need to actually drill into the solid surface, a great advantage in mission design. The PNG-GRAND prototype was tested at a unique outdoor neutron instrumentation test facility constructed at NASA/GSFC that includes a 1.8 × 1.8 × 0.9 m^3 granite monument. Since an independent trace elemental analysis has been performed on the material, this granite sample is a known standard to compare both Monte Carlo simulations and elemental composition data derived from experimental measurements. When a solid extraterrestrial surface is bombarded with fast neutrons from a neutron generator, the nuclei in the planetary material are excited and emit gamma radiation characteristic of at least some of the isotopes of the elements present. While the conversion of a gamma-ray emission rate from a particular isotope of an element to the inferred elemental concentration is somewhat complicated, it is a direct procedure involving many microscopic and macroscopic nuclear parameters. The intensity of these characteristic gamma rays can then be converted to abundance of the specific elements that emitted them, assuming an isotopic distribution typical of other bodies in the solar system such as Earth. Incident neutrons will lose energy with successive interactions within the material at a rate that depends on material properties. By measuring count rates of the resulting slowed neutrons that reach the surface, we obtain information about density, hydrogen content, and subsurface layering. We thus gain valuable information from the instruments that detect both gamma rays and neutrons emerging from the object's surface. These gamma rays are emitted at specific energies characteristic of each element and thus the elemental abundances in the stimulated material can be readily determined from analysis of the emitted gamma rays. The probability of these processes occurring depends on neutron speed as they slow down from 14 MeV to epithermal (<500 keV) and thermal (≤0.4 eV) energies. Since the inelastic neutron scattering process requires the neutron to have significantly higher energy than the reaction threshold in the nucleus (~1–6 MeV), inelastic scattering events occur before the incident 14 MeV neutrons have had time to lose much energy. Thus, the emitted gamma rays from inelastic scattering processes are produced earlier in time than gamma rays produced by other processes. Characteristic gamma rays resulting from thermal or epithermal neutron capture interactions appear a short while later. Following the inelastic scattering or neutron capture processes, the interacting nuclei may be left in a radioactive ground state that lives for some time (fraction of a second to many days) and then usually decays by beta decay, often accompanied by gamma-ray emission from the daughter nucleus. Finally, there are naturally occurring radioactive elements such as K, Th, and U that are common in solid bodies in the solar system and emit characteristic gamma rays from radioactive decay.

No outside stimulation of these elements is needed for gamma-ray production. The intensity of the characteristic gamma-ray lines measured by the gamma-ray spectrometer can thus be used to infer absolute elemental abundances of the material. Accessible elements include C. H, O, P, S, Si, Na, Ca, Ti, Fe, Al, Cl, Mg, Mn, and V. Excellent reviews of the physics of neutron- induced gamma-ray techniques and how they are used for both remote and in situ geochemical analysis can be found in article by Grau et al. (1990), Boynton et al. (1993), and Evans et al. (1993).

The aim of the mission is determination of the absolute concentration of the chemical elements that make up the Martian surface. The mission will be done by the rover Curiosity (see Figure 7.72). The determination is performed by the analysis of gamma-ray line intensities, which in turn depend on the spectral density of the neutron flux interacting with the nuclei of the major rock-forming elements. The primary mission of neutron generator is to check for water-bearing minerals in the ground beneath the rover. The instrument, named Dynamic Albedo of Neutrons, or DAN, has two major components: the pulsed neutron generator on the starboard side of the rover (location indicated by dashed outline) and the detector and electronics module on the port side. The pulsed neutron generator will shoot high-energy neutrons into the ground. If there is hydrogen in the shallow subsurface, the injected neutrons will bounce

Figure 7.72 Curiosity is equipped with a neutron generator indicated with dashed box on the figure. (Image courtesy of NASA/JPL-Caltech.)

off the hydrogen atoms with a characteristic decrease in energy. Two detection devices in the detector and electronics module measure the rate and delay time of the reflected neutrons, yielding information about the amount and depth of any hydrogen. At the mission's near-equatorial landing area and in the oxidizing environment near the Martian surface, most hydrogen is expected to be in the form of water molecules or water-derived hydroxyl ions bound to minerals. DAN was developed by the Space Research Institute, Moscow, in close cooperation with the N. L. Dukhov All-Russia Research Institute of Automatics, Moscow, and the Joint Institute of Nuclear Research, Dubna.

In the course of the work of the DAN neutron detector, the scientists conducting the experiments have found out that the water content is different in the upper and lower layers of the soil. The boundary of these layers is located at a depth of about 20–30 cm. The soil of the top layer contains about 1.5% water by mass, and in the lower layer, the mass fraction of water is about 3%. The water content has been measured in 249 points and for some districts up to 6% of water has been detected in the lower layer.

Although the term "extreme conditions" covers many scenarios, we shall discuss in detail only the use of neutron generators for the inspection of objects on the sea floor. Environmental security of the coastal seafloor in the ports and waterways of the Mediterranean region is discussed by Obhodas et al. (2010), while the environmental security of the Adriatic coastal sea floor is presented in a paper by Valkovic et al. (2010).

After successful use of tagged neutron beams for localization and identification of threat materials in sea containers (EU FP6 project ERITRACK) and cars, it has been decided to investigate the possibility of using same technology underwater. An underwater system for explosive detection has been developed (Valkovic et al. 2007). Let us mention that Aleksakhin et al. (2013) have also described the use of the tagged neutron technique for detecting dangerous underwater substances.

For the detection of explosives in objects on the bottom of the sea, the following parameters were of importance:

1. Distance of the object, that is, explosive charge from the neutron sensor due to the effect of fast neutron attenuation in the sea water
2. Thickness of the object's wall (neutron attenuation in the iron shell, mainly; sometimes aluminum)
3. The choice of tagged neutron beam angle. This parameter can be satisfied either by
   i. The rotation of the neutron generator or by
   ii. The use of the multisectorial associated alpha particle detector inside the neutrongenerator so that all angles would be measured simultaneously

4. The proper selection of the gamma ray detector
5. The choice of the part of the object to be scanned for the explosive/threat material presence (this can introduce additional attenuation or complete miss of the explosive charge)

The three most important requirements on the remotely operated vehicles (ROV) functioning were as follows:

The requirement for the stability of ROV during the measurement time (>10 min) without touching the object. This was insured by landing the ROV above the inspected object on four hydraulic legs whose lengths were remotely controlled.

The requirement of fine (cm-scale) positioning of the ROV. This was performed by the use of eight brushless thrusters.

The requirement that the position and the size of the explosive charge (or some other threat material) could be determined by the neutron generator rotation (Sudac et al. 2009).

"Surveyor," an underwater system for threat material detection is described in detail by Valkovic et al. (2010). The essential hardware required in construction of such an inspector is composed of sealed tube neutron generator with α-particle detector, γ-ray detector with photomultiplier, and between them a cone-preventing detector to see directly the neutron source.

In order to satisfy the required performances, the system had to be housed in the water leak–proved container that would house all the required power supplies and electronics modules and variety of other sensors. These were ingredients needed for the realization of device called Surveyor (Valkovic et al. 2007).

Inspections of the objects on the sea floor by using 14 MeV tagged neutrons (Valkovic et al. 2012) related to the development of capabilities for the evaluation of environmental security of the coastal sea floor done at the test site in the base Kovcanje, island Lošinj, are presented next. The experimental setup included sealed-tube neutron generator, API-120, with the associated alpha particles detector and 3″ × 3″ Na(Tl) gamma-ray detector. The neutron generator was able to rotate by remote control allowing the tagged neutron beam to scan the investigated object. In addition to neutron generator and gamma detector, different power supplies and fast electronic modules for the measurement of gamma ray energy and TOF spectra were placed within the water proof container called "Surveyor." Control electronics and crane for handling the submarine is placed inside specially equipped ship container that also served as submarine storage facility; see Figure 7.73.

Measurements of 14 MeV neutron-induced gamma spectra require fixed geometry for the time interval up to 10 min. Two geometries were used during

Figure 7.73 Experimental container for housing control electronics and crane for submarine.

the measurements. In the first one, a special cage was constructed to keep submarine and the investigated object at the fixed relative positions; see Figure 7.74. Often divers are used to position the inspected object under the neutron exit, gamma entrance window; see Figure 7.75.

Different objects have been investigated during the series of measurements. In order to evaluate the importance of critical parameters, the measurements in "cage" geometry have been done with real ammunition. The following explosive devices (shown in Figure 7.76) were used in these measurements: (1) AT mine TMA-3 (total mass = 6.5 kg, TNT mass = 6.5 kg), (2) mine 120 mm for MB.LT.M62P3, w/o trigger (mass = 12.0 kg, TNT = 2.78 kg), (3) grenade 155 mm for H M1, TF (total mass 42.0 kg, TNT mass = 7.18 kg), (4) grenade 105 mm for H M1.Tf (total mass = 14.0 kg, TNT mass = 2.83 kg); (5) Grenade 122 mm for H D-30 TFG w/o trigger (total mass = 21.0 kg, TNT mass = 6.10 kg), and (6) rocket head 128 mm (total mass = 7.5 kg, TNT mass 4.76 kg).

Measurements with each individual device have been done in the seawater at the depth of 3 meters and performed for several rotation angles of the neutron generator. Individual measurement lasted up to 20 min with neutron beam intensity 1×10^7 n/s. Produced gamma rays were detected by 3″ × 3″ NaI (Tl) detector. The associated alpha particle detector had a collimator of 1.9 cm in diameter.

Figure 7.74 "Cage" geometry. Relative position of the inspected object and ROV is fixed before lowering the whole setup into the water.

Figure 7.75 Diver adjusting the position of the submarine "Surveyor."

Figure 7.76 Ammunition devices investigated for explosive content.

All measured spectra were fitted with the assumption that they contain only carbon, oxygen, and iron contributions. Contribution of the other elements like silicon or calcium could be ignored. The fitting procedure was done by using Equation 7.3 where the sum was done over the channel (ch) number. "Carbon," "Oxygen," and "Iron" are pure elemental spectra previously measured. Parameters "a," "b," and "c" are fitting parameters called carbon content, oxygen content, and iron content, respectively. The method of least squares states that the best values of "a," "b," and "c" are those for which chi-square, Equation 7.3, is a minimum:

$$\chi^2 = \Sigma\left[\text{a x Oxygen(ch)} + \text{b x Carbon(ch)} + \text{c x Iron(ch)} - \text{Target(ch)}\right]^2 / \text{Target(ch)} \tag{7.3}$$

TABLE 7.10 OXYGEN, CARBON, AND IRON "PARTS" IN INVESTIGATED AMMUNITION

Ammunition	O	C	Fe
155 mm grenade	0.50	0.06	0.45
Antitank mine TMA3	0.47	0.134	0.39
122 mm grenade	0.35	0.055	0.60
120 mm mortar shell	0.42	0.058	0.52
128 mm rocket head	0.42	0.078	0.50
105 mm grenade	0.40	0.067	0.53

where summation is done for ch values from ch_{min} to ch_{max}. The results of the fitting procedure are shown in Table 7.10. It should be noted that oxygen peak has contribution from the sea water also.

The measurements have been performed for all of the six explosive devices (shown in Figure 7.76) fixed under the ROV in the same geometry, lowered into the sea water and gamma ray spectra measurements for several neutron generator angles of rotation. In such a way, contribution of oxygen from the seawater was kept same for all devices; however, it did wary with the neutron beam angle. Maximum carbon contribution is taken as an indication of neutron beam aiming into the center of explosive charge.

Relative oxygen, carbon, and iron concentrations are shown in Table 7.10 and could be presented as two axes projection of triangle graph concentration values. All oxygen and iron concentration values have error ±0.01, while carbon values have errors from ±0.004 to ±0.006.

The same approach could be used in identifying shells and other objects containing chemical warfare agents. The results from the laboratory tests for the detection of the presence of chemical elements (P, S, Cl, F, As, or Br) characterizing the CW agents were also recently presented (Valkovic et al. 2010).

In the second set of measurements, the remotely controlled hydraulic legs were used for positioning the submarine above the investigated object and for stabilization during the measurements; see Figure 7.77. This is the first time that the concept of using 4 legs has been used for the ROV positioning above the object to be investigated. All the underwater photos are taken by using submarine "VideoRay" model "Scout."

In the tests with airplane bombshell (volume 2R = 18 × 88 cm with the housing of mass = 8.6 kg), the mixture of silicone oxide, melamine, and carbon was used to prepare a TNT stimulant having composition $Si_3C_7H_3N_3O_6$. The bomb was positioned at the seafloor as shown in Figure 7.78.

In order to improve the quality of measured gamma spectra resulting from the neutron interrogation of the airplane bomb, an air cushion has been

Figure 7.77 Surveyor positioned above the airplane bomb; slowly reducing the water layer between the object and the window.

Figure 7.78 The diver is releasing the bomb from the crane attachment and positioning it to the realistic environment.

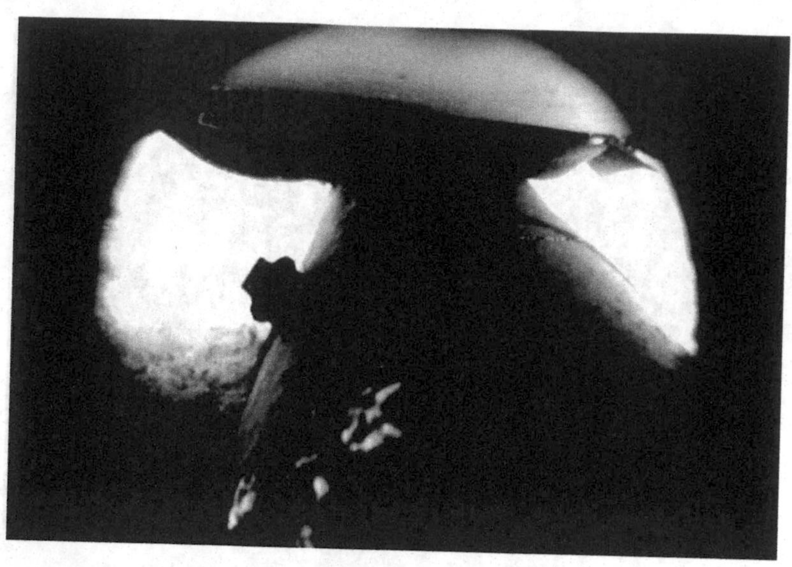

Figure 7.79 Ready for the measurement; view from Surveyor camera.

inserted on the neutron window eliminating water in-between neutron sensor and interrogated object; see Figure 7.79. The improvement in the quality of measured gamma spectrum is shown in Figure 7.80.

Matrix characterization is a key for the neutron prototype systems used for underwater threat materials inspections. Most of the background in the gamma spectra measured when investigating suspicious object on the bottom of the sea is generated by surrounding sea sediment and rocks. Therefore, the chemical composition of seafloor sediments and rocks has to be known. In addition, after years and decades of lying at the bottom of the sea, dumped ammunition could change to unrecognizing appearance, especially for untrained eye. By using only visual inspection, the military objects can be easily mistaken for rocks and vice versa.

In the paper by Obhodas et al. (2010), the results of underwater analysis (salinity of 38 g/l) of the main type of rocks found in the Adriatic Sea by using Fast Neutron Activation Analysis (FNAA) with associated alpha particle are presented. The chemical composition of two main types of rocks found in the Adriatic seafloor have been evaluated; prevailing limestones and dolomites (Ca, Mg–rocks) and magmatic rocks (Si-rocks) can be found in vicinity of the most distant islands in Eastern Adriatic (Jabuka, Svetac). Inspection of spectra obtained for limestones and dolomites, as shown in Figure 7.81, revealed that peaks attributable to oxygen, magnesium, and calcium were apparently strong while carbon, silicon, and iron were significantly weaker. In magmatic rocks, the peaks of oxygen, silicon, and iron were clearly visible while carbon and calcium were less distinguished. The nitrogen, which is a major component

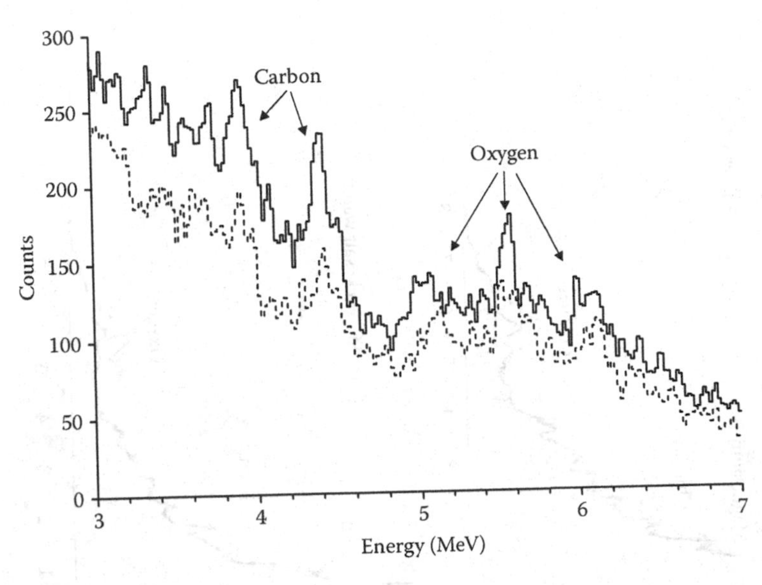

Figure 7.80 Gamma ray spectrum from airplane bomb with air cushion (upper) and without air cushion (lower) for neutron generator angle $\alpha = 26°$. Air cushion placed between ROV and object has a volume of $30 \times 30 \times 8$ cm^3.

of explosives, was not detected in any type of rocks. Compared to measurements in the air, underwater measurements suppress the peaks of carbon and calcium, but enhance peaks of oxygen and magnesium. In addition, the chlorine from the seawater interferes with some peaks of silicon and iron. The peaks corresponding to listed chemical elements in the obtained FNAA spectra have been detected approximately at 4.44 MeV for carbon; 6.13 MeV including escape peaks at 5.11 and 5.62 MeV for oxygen; 1.37 MeV for magnesium; 1.78 and 2.84 MeV for silicon; 1.22, 1.29, 1.73, 1.76, 2.13 MeV for chlorine; 0.755 MeV for calcium; and 0.85 and 1.24 MeV for iron.

Geochemical maps of coastal sea sediments are of importance if available for the area of interest. These geochemical maps can be used for evaluating background for neutron sensor applications. For example, mercury, lead, or other heavy metals are frequently found in explosive primers, thus, the presence of such atomic species may be indicative of the presence of military explosive. Similarly, phosphorous, sulfur, chlorine, fluorine, arsenic, bromine, etc., that are found in CW agents can indicate the presence of CW dump sites. Anyhow, the sediments can contain increased concentrations of aforementioned elements because of natural geological background or anthropogenic pollution of other source types that can interfere with the threat material inspection. Therefore, in addition to the use of chemical analysis of explosives, CW, and their "containers" by means

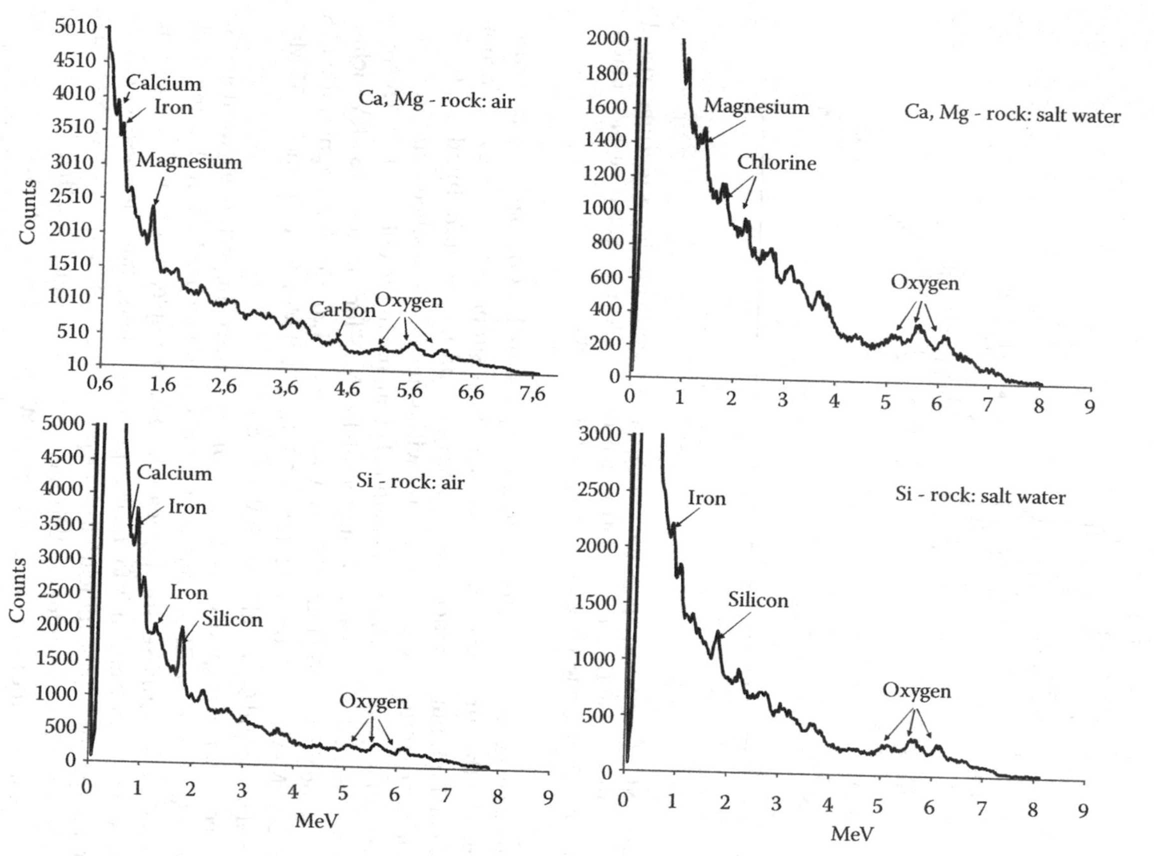

Figure 7.81 Gamma ray spectra from two types of stones in air and in water.

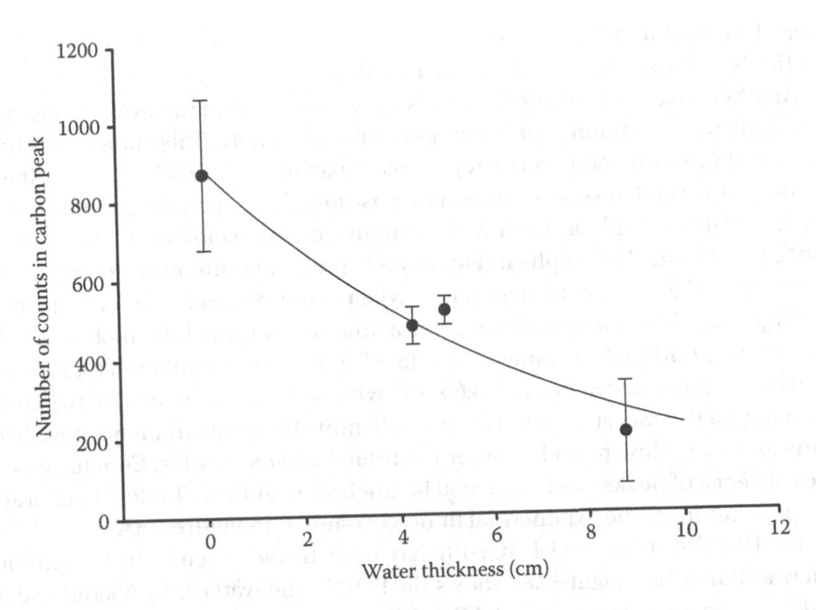

Figure 7.82 Number of events in the carbon peak as a function of water layer thickness between the explosive and the detection system (submarine window).

of the determination of presence or absence of particular elements of interest or the determination of their ratios, the matrix that produces background has to be well studied in order to not be misinterpreted as threat material.

Figure 7.82 shows the number of events in the carbon 4.44 MeV peak as a function of water layer thickness. Experimental data are fitted by

$$F = a \cdot Exp(-b\,x)$$

Parameters of the fit are

$$a = 883 \pm 68$$

$$b = (0.14 \pm 0.02)/cm$$

Theoretical value for attenuation can be obtained from the formula

$$b_{th} = \frac{\rho_{H_2O}}{M_{H_2O}}(2\sigma_H + \sigma_O)N_A$$

where

ρ is the water density
M is the water molar mass
σ is the total cross section for 14 MeV interaction with a given element
N_A is the Avogadro number

It can be calculated that $b_{th} = 0.09/cm$.

This is close enough to the measured value!

Another parameter of importance is the attenuation in the iron, the element from which the ammunition housings are usually made. This subject has been discussed in more detail in the report by Valkovic et al. (2007). In experiment reported, the thickness of the iron plate positioned between the submarine and explosive (mass 5 kg) has been varied. Submarine to explosive (mass 5 kg) distance was 11 cm. The graphs in Figure 7.83 show the number of counts in carbon 4.44 MeV peak (upper curve) and oxygen 5.62 MeV peak—first escape peak of oxygen 6.13 MeV line (lower curve) as a function of iron plate thickness. Solid lines correspond to the exponential fit ($a\ e^{-bx}$). The total number of tagged neutrons in each measurement was 3.6×10^8, with neutron beam of ~10^7 n/s corresponding to the measurement time of ~176 min. The measurements have been done for a long time period in order to obtain better statistics. Conclusions on the existence of peaks can be reached in much shorter time. Table 7.11 contains the parameters of the exponential fit ($a\ e^{-bx}$) shown in Figure 7.83.

The UNCOSS ROV model ACT-101 has been tested recently in the Adriatic Sea near Punat bay. Figure 7.84 shows the ROV in the water fully operational. In addition, the experimental and MCNP-simulated gamma-ray spectra for the

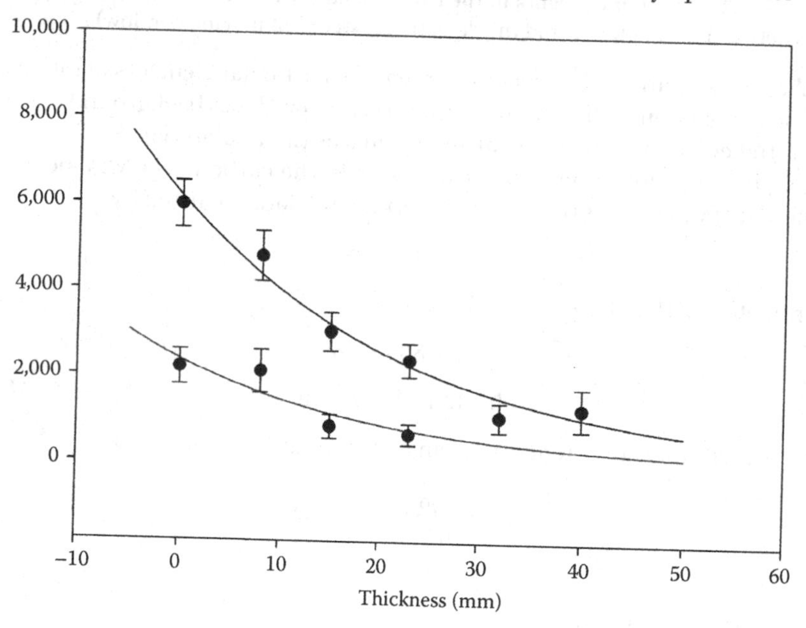

Figure 7.83 The number of counts in carbon 4.44 MeV peak (upper) and oxygen 5.62 MeV peak—first escape peak of oxygen 6.13 MeV line (lower) as a function of iron plate thickness.

TABLE 7.11 FITTING PARAMETERS

Element	a	b (1/mm)
Carbon	6169 ± 318	0.043 ± 0.004
Oxygen	2379 ± 411	0.05 ± 0.02

Figure 7.84 UNCOSS ROV model ACT-101 in the seawater fully operational.

UNCOSS neutron-based explosive detector have been discussed (Eleon et al. 2011). Acquisition of prompt gamma-ray spectra induced by 14 MeV neutrons and comparison with Monte Carlo simulations were discussed in detail by El Kanawati et al. (2011).

The field test on the use of alpha particle tagged neutrons for the inspection of objects on the seafloor for the presence of explosives is described in an article by Valkovic et al. (2013). The tests were performed with an airplane bomb (diameter of 20 cm and iron thickness of 2 mm) filled with ~30 kg of another TNT surrogate with the composition $Si_3C_7H_3N_3O_6$ (see Figure 7.85) and with a 120 mm grenade filled with ~3 kg of the $C_7H_6N_3O_6$ TNT surrogate. The TNT surrogate has been selected for the experiments since TNT is the most common explosive found in ammunition present on the seafloor. The results of the final field test (FFT) have shown that the system developed in the framework of the UNCOSS project can inspect the objects on the seafloor for the presence of threat materials by using alpha particle tagged neutrons from a sealed-tube d + t neutron generator to produce characteristic gamma rays within the interogated object. This was possible by construction and fabrication of a special

Figure 7.85 Underwater inspection of the 20 cm diameter airplane bomb at the depth of 10 m.

ROV able to position itself above the object to be inspected. In the report on the FFT, the maritime properties of ROV have been described and it has been shown that the measured gamma spectra for commonly found ammunition charged with TNT explosives are dominated by C, O, and Fe peaks enabling the determination of the presence of explosives inside an ammunition shell.

Figure 7.86 shows the time and energy spectra for the 16 cm diameter and 12 mm thick iron cylinder filled with TNT surrogate after the 10 min acquisition.

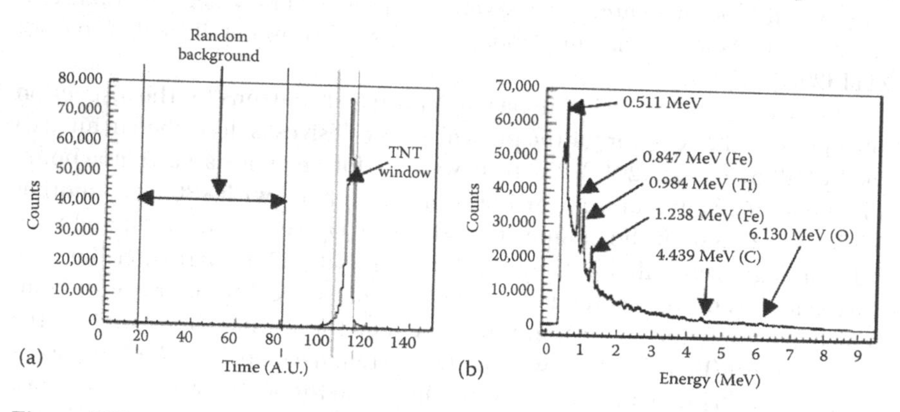

Figure 7.86 TOF of the 16 cm diameter and 12 mm thick iron cylinder filled with TNT surrogate ($C_7H_6N_3O_6$) (a) and energy spectra (b) measured in seawater for 10 min by the ROV. Average neutron emission was 2.6×10^7 n s^{-1}. (After Valkovic et al. 2013.)

The main characteristic peaks of iron (UXO shell), titanium (ROV window), carbon, and oxygen (TNT surrogate) can be observed. With respect to laboratory measurements of the sensor stability, underwater inspections were performed with a neutron emission rate of 2.6×10^7 n/s instead of the maximum 5.8×10^7 n/s. Indeed, neutron interactions with the surrounding environment strongly increase the gamma detector count rate and the associated random background, the last rising with the square of neutron emission.

The relative proportion of carbon, oxygen, and nitrogen is commonly used for explosive detection. In underwater inspection of thick iron shells, however, the characteristic peaks of nitrogen, mainly at 1.632, 2.313, and 5.106 MeV, are strongly attenuated and in addition, they are interfered by other lines such as the second escape of the 6.130 MeV gamma ray of oxygen, at 5.108 MeV, or the 1.555 and 2.375 MeV peaks of titanium. Therefore, for this application, we performed a simpler but robust data analysis based on the carbon 4.439 MeV and oxygen 6.130 MeV net areas, the C/O surface ratio being used to discriminate between metallic objects containing sediments or explosives. Figure 7.87 and Table 7.12 show that the C/O ratio allows to discriminate a TNT surrogate from the sea sediment inside a thick iron shell.

The importance of identification of the waste contained within barrels thrown into the sea is best illustrated by these two events. In the first one, more than 364 barrels full of different toxic substances were found on the Turkish Black Sea coast between July and December 1988. The first analysis of the waste was done in October 1988. The samples showed that the waste was comprised of paint or benzene and cellulose lacquer wastes. The dumping was clearly intentional as holes had been punched in the barrels to make them sink. It is presumed that the barrels that washed ashore represented but a small percentage of the entire shipment and that the vast majority of the barrels sank; see web page (http://www1.american.edu/TED/barrel.htm). It seems that this specific incident is not the only one, but just the tip of an iceberg. Hazardous waste pollution can contaminate a site very quickly. Pollution of this type can take years/decades and millions of dollars to cleanup. The opportunity to relatively quickly (in situ) evaluate the content of the barrels on an actual illegal dumpsite, allows Governmental officials to organize the removal of the waste safely with regard to the local population and environment.

Another example is when in the late 80s, local Nigerian officials discovered illegal toxic waste stored at the port of Koko. Over 100 workers from the Nigerian Port Authority were employed to remove the waste. Unfortunately, the waste was more toxic than realized at the time and many workers were hospitalized (http://www1.american.edu/TED/nigeria.htm). Obviously, prior to any action regarding cleanup/waste removal, the exact content of barrels must be determined.

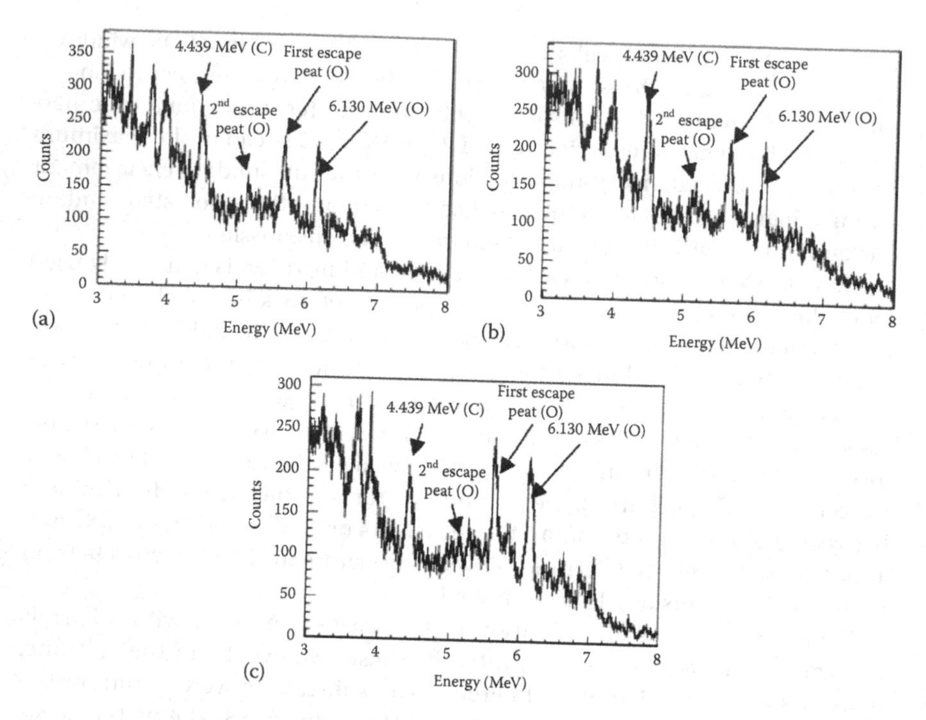

Figure 7.87 Energy spectra of the 16 cm diameter and 12 mm thick iron cylinders measured for 10 min in seawater with an average neutron emission of 2.6×10^7 n s^{-1} and by selecting the central alpha pixels #2, #5, and #8: the cylinder filled with ~13 kg of TNT surrogate (a) and an identical cylinder filled with local sand (b). (After Valkovic, V. et al., *Nucl. Instrum. Methods Phys. Res. Sect. A*, 703, 133, 2013.)

Barrel inspection utilizing a 14 MeV neutron beam and associate alpha particle method has been described by Sudac et al. (2012). Both, laboratory and field test have been described. The field test site was located in Bay Kovčanje on the island, Lošinj, in the Adriatic Sea. All the equipment components, including the Surveyor (shown in the Figure 7.88), were placed inside the standard 40 ft sea cargo container.

With a special crane, the submarine was moved out of the container and then down to the seafloor. The ROV used was equipped with the special hydraulic legs in order to be able to take the adequate position over the target (see Figure 7.89). A barrel with the dimension $\Phi = 35$ cm × 63 cm filled with 60 kg of graphite powder and blocks was used as a target. Two lead bricks with the mass of 10 kg were added to make the barrel heavy enough to sink. Neutron generator inside the ROV was rotated in order to determine the diameter of the graphite target inside the barrel.

Figure 7.90 shows the gamma ray spectrum of the graphite from the barrel for rotation angle of $\alpha - \alpha_0 = 26°$. Carbon lines were clearly visible together with

TABLE 7.12 C/O RATIOS FOR DIFFERENT SCENARIOS

Measurements	C(4.439 MeV) Counts	O(6.130 MeV) Counts	C/O Ratio
Iron cylinder filled with TNT surrogate (10 min) Run#1	252 ± 56	167 ± 52	1.51± = 0.58
Iron cylinder filled with TNT local sand (10 min) Run#1	91 ± 82	406 ± 66	0.22± = 0.21
Iron cylinder filled with TNT surrogate (10 min) Run#2	539 ± 76	365 ± 63	1.48± = 0.33
Iron cylinder filled with local sand (10 min) Run#2	118 ± 81	556 ± 60	0.21± = 0.15
Iron cylinder filled with TNT surrogate (10 min) Run#3	354 ± 65	259 ± 46	1.37± = 0.35
Iron cylinder filled with local sand (10 min) Run#3	234 ± 93	521 ± 66	0.45± = 0.19
Airplane bomb (10 min)	615 ± 66	433 ± 45	1.42± = 0.21
Grenade (10 min)	341 ± 95	355 ± 53	0.96± = 0.30
Grenade (20 min)	1158 ± 148	747 ± 104	1.55± = 0.29

Figure 7.88 Surveyor at its test site, ready to be lowered into the seawater.

Figure 7.89 Barrel content determination by using of the fast neutrons and associate alpha particle method. Barrel was filled with the graphite powder and blocks of density close to one gram per cubic centimeter.

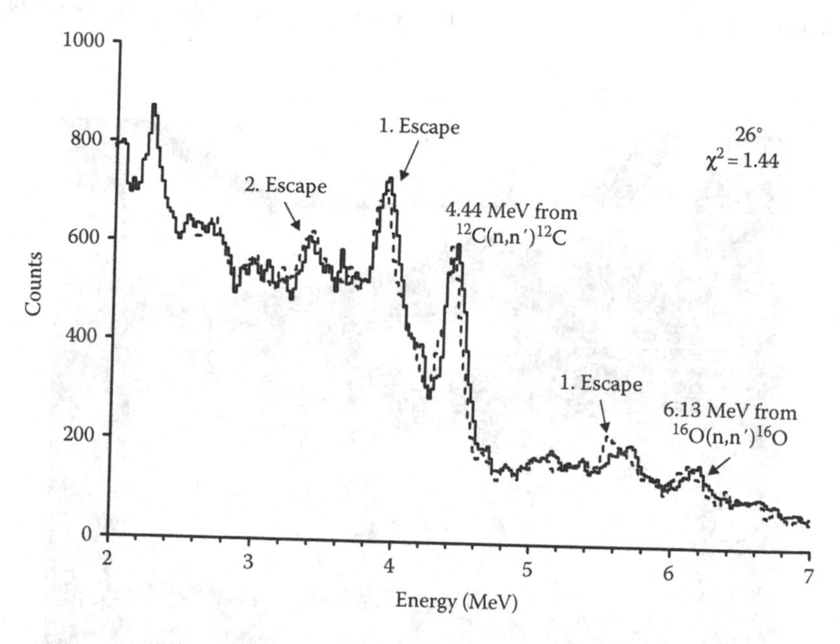

Figure 7.90 The gamma ray spectrum of the barrel filled with 60 kg of graphite powder and blocks (solid line) and the associated fitting curve (dashed line). Measurement time was 1366 s and neutron beam intensity ~10^7 n/s. Rotation angle was 26°.

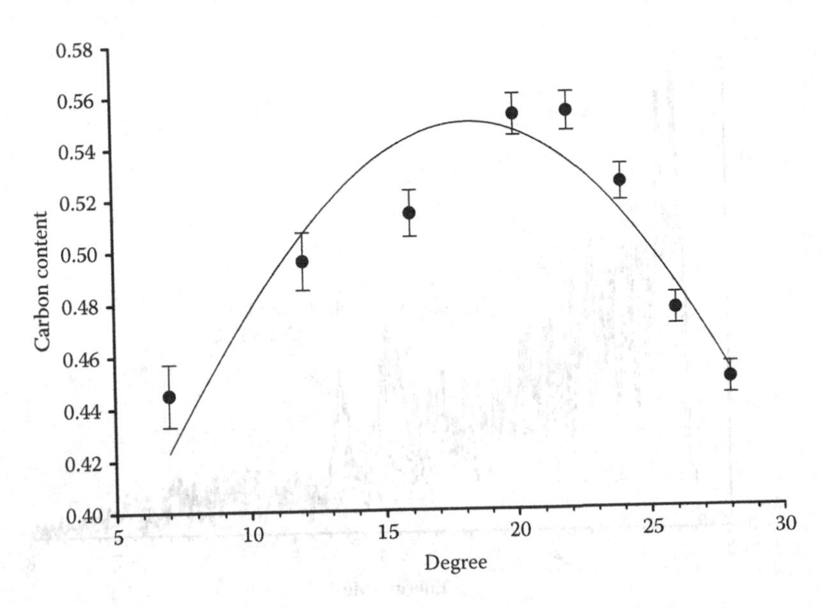

Figure 7.91 Carbon content in dependence on the neutron generator rotation angle. Fitting parameters are a = 0.55 ± 0.011, x 0 = (18.3 ± 0.6)°, and b = (15.6 ± 1.4)°.

the oxygen lines that came from the surrounding water. Spectrum was fitted with the assumption that it contains only carbon, oxygen, and iron contributions. Contribution of the other elements like chlorine or sodium was ignored. Fitting curve is shown in gray color.

Figure 7.91 shows the carbon content in dependence on the rotation angle. Associated errors were calculated from the error matrix. Experimental results were fitted with Gaussian distribution. Full width at half maximum was found to be FWHM = (36.7 ± 3.3)°, which on half a meter distance from the neutron source correspond to the (33.2 ± 3.2) cm barrel diameter. Figure 7.92 shows the gamma ray spectrum of the barrel (black) measured only for 35 s. For comparison, background was shown in gray color.

These measurements made show that the carbon content of the sunken barrel could be estimated by rotating the neutron generator or equivalently by using the adequately segmented alpha detector incorporated inside the neutron generator. For example, ING-27 produced by VNIIA Russia has 3 × 3 segmented alpha detector that allows simultaneous measurement in the nine different solid angles. Detection time depends on the carbon content inside the barrel and in the described case it was less than 35s. Barrel diameter was successfully measured to be (33.2 ± 3.2) cm in comparison to the actual diameter of 35 cm.

See also Bystritsky et al. (2013) describing the use of the tagged neutron technique for detecting dangerous underwater substances.

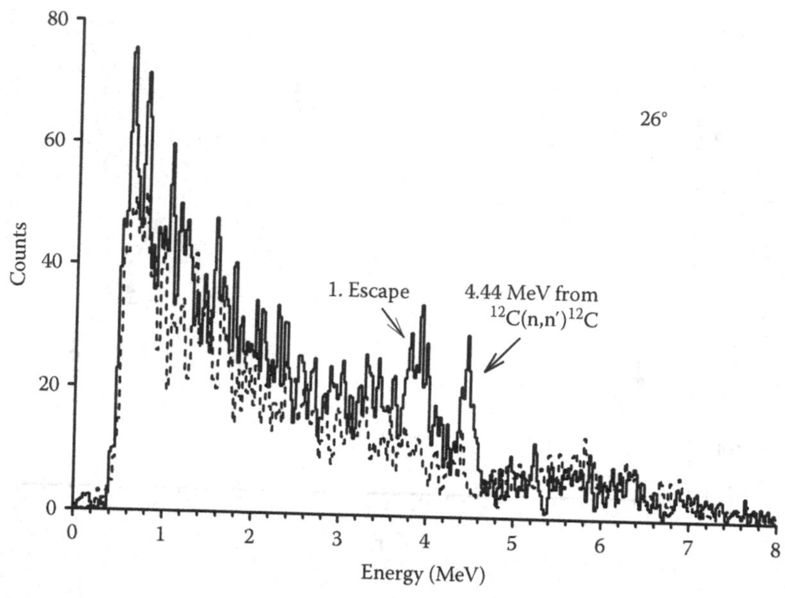

Figure 7.92 The gamma ray spectrum of the barrel filled with 60 kg of graphite powder and blocks (solid line) and the background (dashed line). Measurement time was 35 s.

7.6 MORE OBJECTS TO BE INSPECTED AND INSPECTION REQUIREMENTS

7.6.1 Inspecting Underwater Ship's Hull

It is often required to inspect ship hulls, either to detect potential anomalies attached to the hull or to determine the nature of materials within hull, especially of sunken ships. Older tanker walls have thickness up to 25 mm, while the modern ones are only 14–16 mm thick. In addition, the new tankers are required to have double hull construction with outer hull 14–19 mm thick (shaped to contour) and inner hull 12–14 mm thick (flat plate).

Inspecting the inside of the objects lying on the seafloor has been reported by Valkovic et al. (2010). In order to demonstrate the possibility of identifying the material within sunken ships and other objects on the seafloor, they have performed tests with the 14 MeV sealed tube neutron generator incorporated inside a small submarine submerged in the test basin filled with seawater.

Different targets were investigated: 10 L of diesel fuel, 5 kg of explosive, and variety of chemicals (expected components of chemical warfare agents) were placed behind 16 mm steel plate in the first measurement and behind sandwich 18 mm steel plate–10 cm air bag–16 mm steel plate (as shown in Figures 7.93, 7.94) in the second measurement, respectively.

Figure 7.93 Experimental setup for the double hull case. In double hull construction, the air bag is placed between two iron plates so that the whole setup can be immersed in the water.

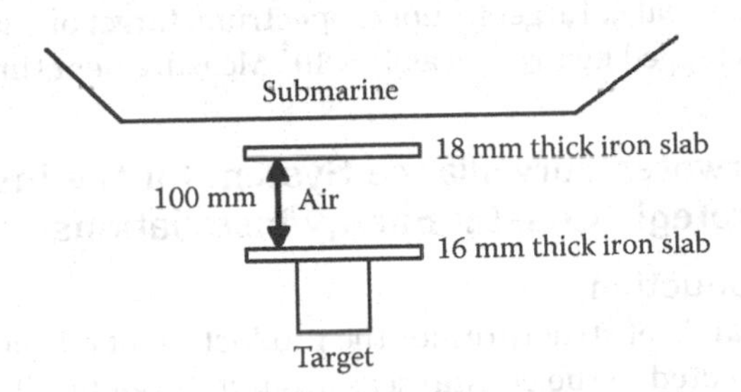

Figure 7.94 Targets were TNT explosive (5 kg) and diesel fuel (10 L). The submarine is positioned about 2 cm above the upper iron plate.

Using the window on the measured alpha—gamma time spectrum, gamma rays originating from the investigated volume were separated from the background radiation. From the inspection of the measured gamma spectra we were able to identify all of the investigated materials in both measurement geometries.

The measured gamma ray spectra from TNT (5 kg) target (upper spectrum) are shown in Figure 7.95 (linear scale). Target out: lower spectrum. The total number of the tagged neutrons is 24×10^7. Measurement time: ~6900 s.

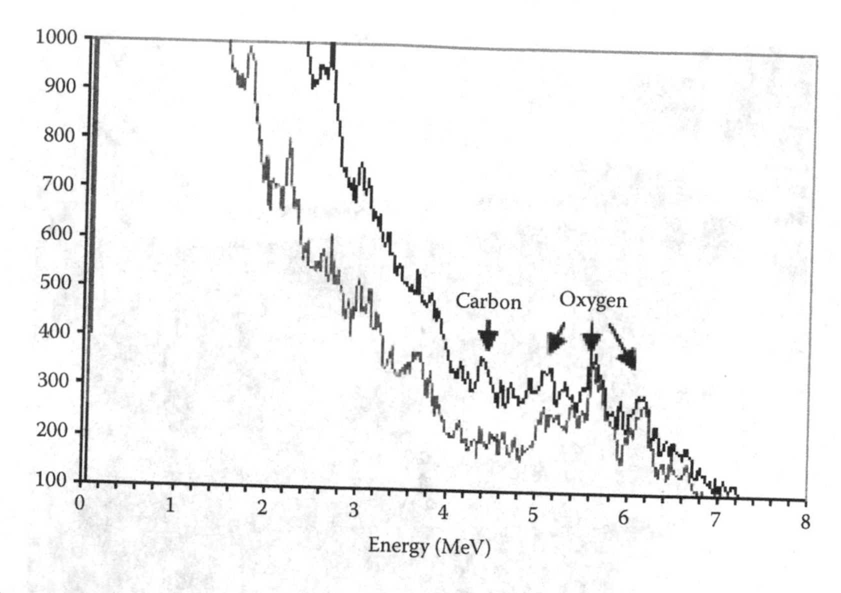

Figure 7.95 Measured spectra TNT (5 kg); linear scale. Target out: lower spectrum.

The experimentally measured gamma ray spectra from diesel fuel (10 L target), placed behind iron–air–iron sandwich in the sea water are shown in Figure 7.96 (linear scale). Target in: upper spectrum; target out: lower spectrum. Total number of tagged neutrons was 18×10^7. Measurement time: ~5240 s.

7.6.2 Underwater Surveillance System for the Inspection of Strategic Coastal Energy Installations

7.6.2.1 Introduction

Recently, thousands of structures for the production of oil, gas, and electricity have been erected in the coastal seas around the globe. Those are not the only structures in the sea; seafloor is intertwined by pipes providing water, electricity, gas, and communication to the islanders. In addition, infrastructure objects like bridges, power stations, ports, and dams have critical underwater components. They all need to be inspected for the service, repaired when they are malfunctioning, and kept in the environmentally accepted conditions.

Traditionally, divers have been the primary inspector. Visual inspection and photographic and TV documentation have been his primary tools. Inspection includes preliminary structure cleaning that can be a more arduous and time-consuming task than the inspection itself. The more demanding conditions and more complex structure have produced a need for structure inspection

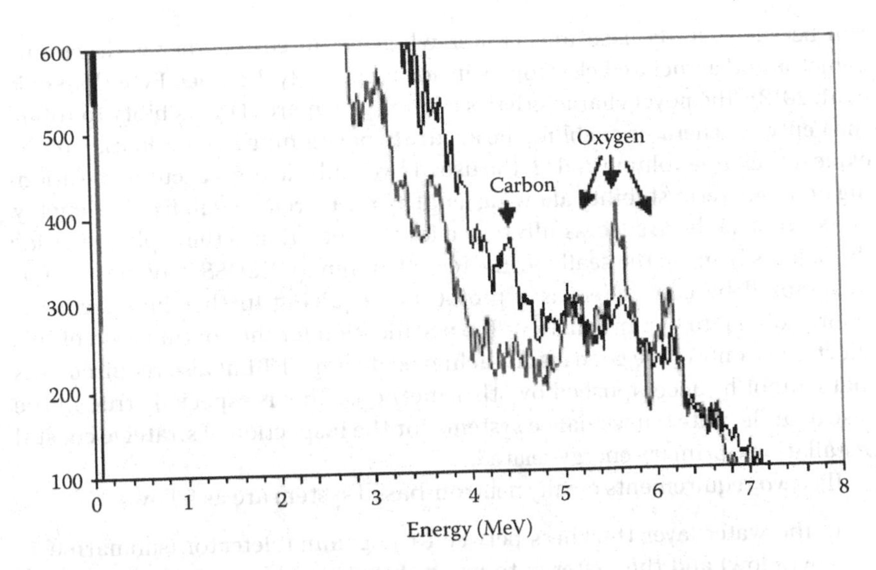

Figure 7.96 Gamma ray spectra from diesel fuel (10 L), upper spectrum, behind iron–air–iron sandwich in the seawater; linear scale. Target out: lower spectrum. Total number of tagged neutrons 18×10^7. Measurement time: ~5240 s.

that transcends simpler visual and photographic capabilities. As a result, nondestructive testing (NDT) methods have been introduced for underwater applications.

The NDT techniques used in the underwater structure surfaces inspection include

- Visual (surface crack detection)
- Magnetic particle (surface and shallow surface crack detection)
- Ultrasonic (thickness and surface–subsurface flaw and crack detection)
- Radiography (internal flaw and crack detection and thickness)
- Corrosion potential measurements

We should stress the role of ROVs often used to produce high quality visual and photographic inspections, and which bring some form of cleaning device (wire brush, chipping hammers, etc). In open waters they have been used quite successfully as pipeline inspection vehicles, but around and within steel structures they experienced difficulties with cable entanglement and location. ROVs offer a wide range of potential new developments for the underwater use of NDT.

A new addition to the underwater NDT family is FNAA induced by tagged 14 MeV neutrons. It has been shown that a sealed tube 14 MeV neutron generator can be successfully used underwater when mounted together with gamma detector and associated electronics inside a specially designed ROV (Valkovic et al. 2013). The novel characteristics of the system are (1) possibility to rotate the neutron generator enabling the accurate positioning and evaluation of the explosive charge volume and (2) the use of hydraulic legs for accurate positioning and long term stability allowing lengthy measurements in fixed geometry. The system has been successfully tested for the detection of the explosive inside the objects lying on the seafloor. The tested system (UNCOSS ROV model ACT-101) should be considered as a prototype requiring further improvements before going into test production. The justification for the continuation of this effort is not only in the good results achieved during FFT but also requirements that cannot be accomplished by other methods. This is especially true in the case of underwater surveillance systems for the inspection of strategic coastal installations, primary energy related.

The two requirements on any neutron-based system are as follows:

1. The water layer thickness between the gamma detector (submarine window) and the material to be analyzed should be <5 cm. In most cases this will be the distance between the submarine window and the investigated object.
2. This distance should be kept constant during the measurement (i.e., for at least 10 min).

Variations of the distance have influence on C/O ratio and on the measured spectrum interpretation.

The knowledge of iron shell thickness is mandatory for correct characterization of the investigated material inside the container for the C/O concentration ratio determination from the measured gamma spectrum. This information could be obtained from the (1) knowledge of construction details of investigated object and precise position of the neutron system and/or (2) from the iron peak in the measured gamma spectrum.

7.6.2.2 Objects to Be Inspected and Inspection Requirements

In the area of inspection of energy infrastructure objects there are several issues of importance:

- Links of coastal islands with the mainland for the provision of water, electricity, gas, and communication to islanders.
- Energy sources are often located in offshore sites and the produced gas and/or oil has to be brought to the mainland.
- Construction objects like bridges, platform, dams, etc.

Underwater pipelines have to be inspected for integrity, that is, possible leakage.

Platforms and other structures have to be inspected for

- Broken or bent elements
- Cracking and pitting
- Corrosion
- Marine fouling
- Debris accumulation
- Corrosion system effectiveness

All presently used testing techniques require that the structure to be inspected should be cleaned of marine organism. While brushing, chipping, and scraping will sometimes suffice, it is frequently required that a high pressure water jet be applied; the jet is cumbersome and potentially dangerous to the operator. Cleaning is not only arduous, but it can constitute the major expenditure of underwater time.

Locating the site to be inspected and positioning oneself to conduct the inspection test can be quite difficult, particularly on complex nodes or in the interior of a steel structure. If there are no markings to identify the work site, location is made much more difficult. If underwater visibility is near zero, location is virtually impossible and testing cannot be done with present techniques.

So far two techniques have been developed that can be used to monitor a fixed structure integrity: acoustic emission monitoring and vibration analysis.

Here we shall describe and establish priorities for specific tasks for technology development that should be undertaken to satisfy current and future requirements. This can be applied to the fixed offshore oil and gas structure, floating power platforms, offshore terminals, strategic coastal energy installation, and commercial ports and marinas. A variety of tasks concerning steel and concrete structures are proposed to look at all aspects in connection with safety, maintenance, and performance.

7.6.2.3 Concrete in the Seawater

There are problems related to integrity and stability of coastal and underwater constructions, namely, underwater parts of infrastructure objects like bridges, dams, etc.

The following tasks should be elaborated:

- Mechanism of corrosion of steel reinforcement in concrete immersed in the sea water.
- Influence of environment, stress, and materials on corrosion of reinforcement in concrete.
- Survey of existing reinforced concrete marine structures.

- Surface cleaning requirements. Cleaning tasks for steel and concrete structures are quite different and should be approached independently.
- Positioning should be developed to permit the ROV to rapidly and confidently locate the work site. Whereas acoustic techniques may be applicable to the exterior of the concrete structures, alternative techniques should be investigated for the steel structure. Accuracy of the positioning should be within ±1 m.

This type of work should be performed on energy transport objects, bridges caring pipelines, or dams making water reservoirs.

Sudac et al. (2013) describe the monitoring of concrete structures by using 14 MeV tagged neutron beam. Their experimental setup is shown in Figure 7.97 where iron plates and iron reinforcing bars (armature) were put in between two concrete blocks. The composite target was irradiated by 14 MeV tagged neutrons and resulting gamma rays measured by gamma detector placed inside the housing. In the first measurement, the 12 mm thick iron slabs were placed behind the 6 cm thick concrete block, as shown in Figure 7.98.

Surveyor was used to measure 6 cm thick concrete cover put over the iron slabs of different thicknesses on a platform placed below the submarine. In another experiment, iron slabs were replaced with the iron bars of 10 mm in

Figure 7.97 Experimental setup.

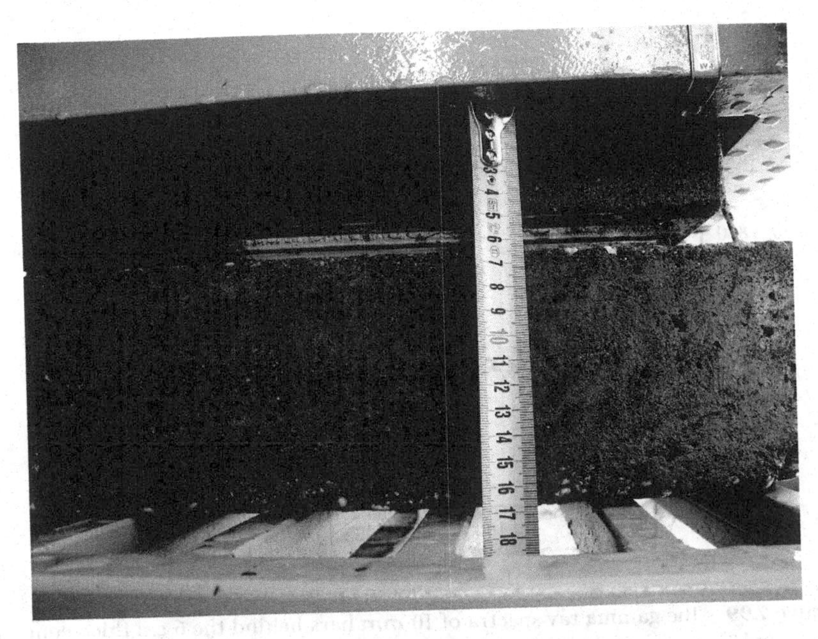

Figure 7.98 The 12 mm thick iron slabs behind the 6 cm thick concrete.

diameter. The submarine was submerged into the sea and gamma ray spectra were collected in each experiment for approximately half of hour depending on the neutron beam intensity that was around 2×10^6 n/s. Figures 7.99 and 7.100 show the gamma ray spectra of the 10 and 18 mm thick iron slabs, respectively, covered with the 6 cm thick concrete block and 2 cm sea layer together with the associated fitting curve. Fitting procedure was done as described in the last section, with the exception that here the oxygen contribution also was added. The normalized gamma ray spectra used for fitting procedure is shown in a Figure 7.101.

The fitting parameters for different targets are shown in Table 7.13. All iron targets were successfully detected. The 21° of the rotation angle correspond to the distance between the gamma ray detector and the iron bar. Two centimeters of the water layer was determined by the video camera. Here we present some preliminary results. Corrosion monitoring of reinforced concrete structure by using 14 MeV tagged neutron beams is discussed in detail in Sudac et al. (2012).

The experimental setup is shown in Figure 7.102. Two concrete blocks are placed in the water with the separation void being covered by green leafs while the $LaBr_3$ detector head is positioned above the void. The size of the void (simulating crack in the concrete wall) can be easily changed by increasing the separation between two blocks. Measured gamma spectra

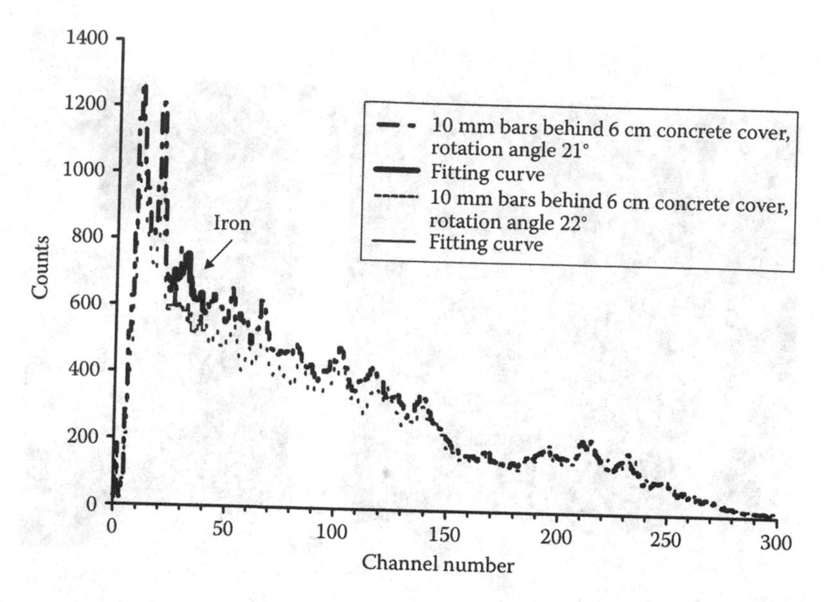

Figure 7.99 The gamma ray spectra of 10 mm bars behind the 6 cm thick concrete cover (and 2 cm of water layer) for different rotation angles of the neutron generator.

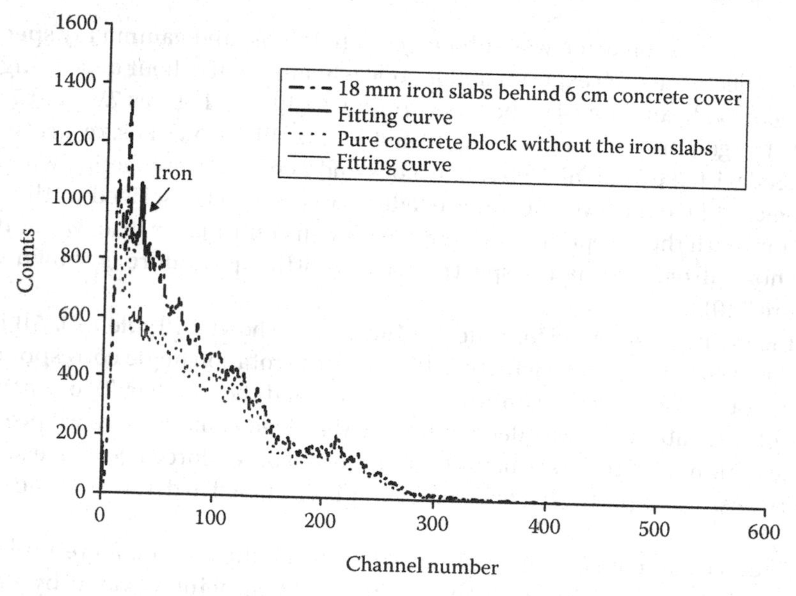

Figure 7.100 The gamma ray spectra of the concrete blocks covered with the 2 cm water layer with and without the 18 mm thick iron slabs placed under the blocks. Rotation angle of the neutron generator was 22°.

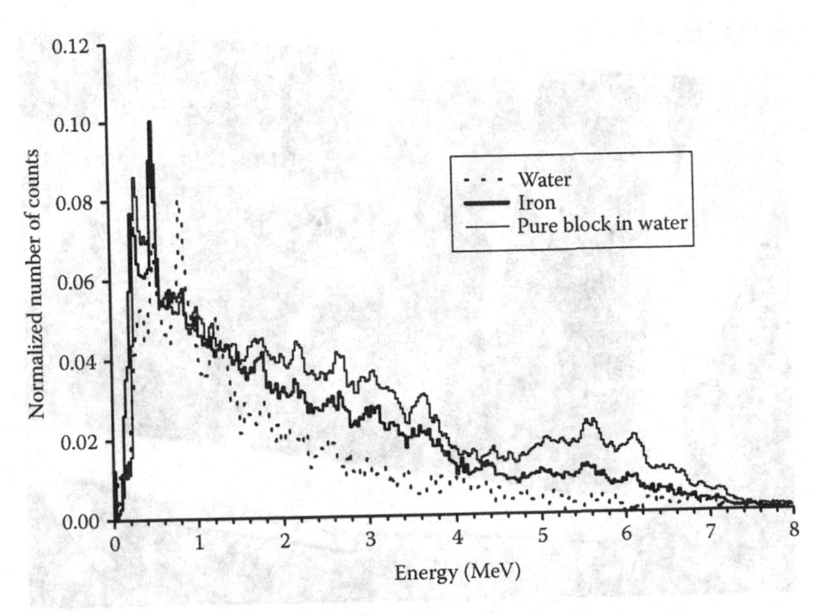

Figure 7.101 The normalized gamma ray spectra used for fitting procedure.

TABLE 7.13 FITTING PARAMETERS FOR DIFFERENT TARGETS

Target	Pure Block Content "a"	Iron Content "b"	Oxygen Content "c"
Iron slabs 18 mm thick	0.04 ± 0.18	0.33 ± 0.05	0.62 ± 0.16
Iron slabs 12 mm thick	0.18 ± 0.21	0.20 ± 0.05	0.61 ± 0.18
Iron slabs 9 mm thick	0.06 ± 0.18	0.27 ± 0.05	0.66 ± 0.16
Iron bars Φ10 mm, $\alpha = 21°$	0.71 ± 0.21	0.122 ± 0.055	0.17 ± 0.18
Iron bars Φ10 mm, $\alpha = 22°$	0.78 ± 0.23	-0.098 ± 0.058	0.32 ± 0.20
Pure concrete blocks in water	0.305 ± 0.23	0.06 ± 0.06	0.63 ± 0.20

are dominated by silicon peak (see Figure 7.103) whose depletion is an indication of voids in concrete structure.

Figure 7.104 shows the distribution of measured silicon content in concrete (upper) and distribution of silicon values in concrete with crack—silicon deficit (lower). The area of upper distribution left from threshold defines "false positive." The area of lower distribution left from threshold defines "detection probability." Note: detection probability + false negative = 1. Parameter "K" depends on value chosen for allowed false positive, for false positive = 10%, K = 1.282.

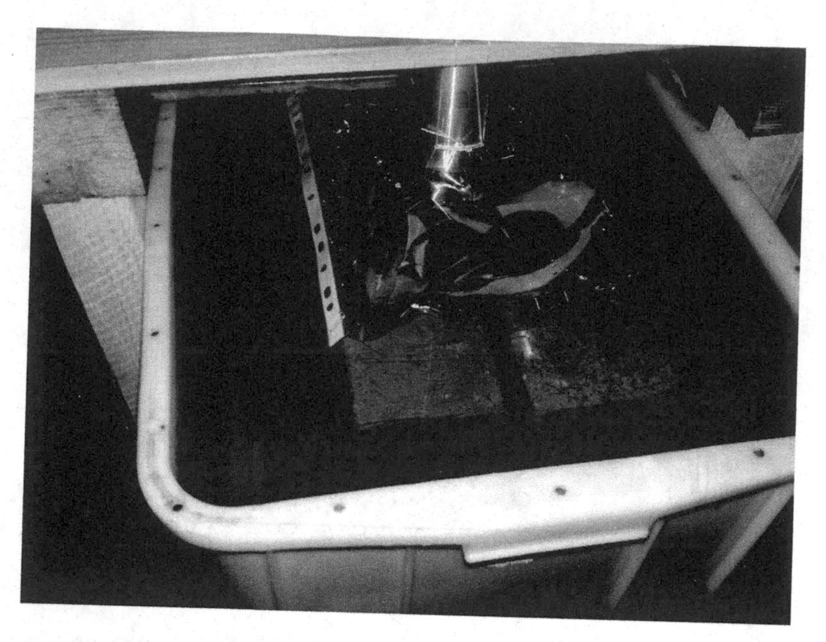

Figure 7.102 Experimental setup.

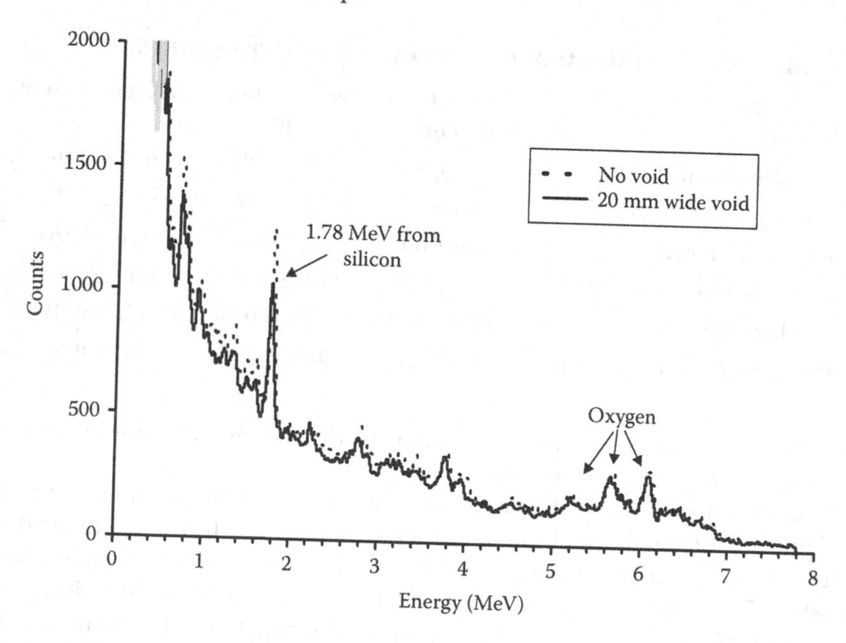

Figure 7.103 Measured gamma spectra are dominated by silicon peak (channel 490) whose depletion is an indication of voids in concrete structure.

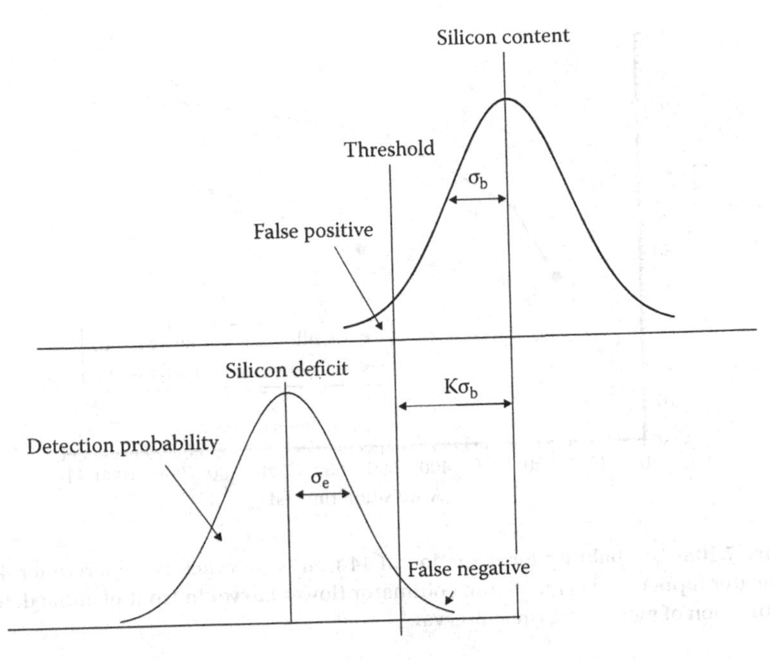

Figure 7.104 Distribution of measured silicon content in concrete (upper) and distribution of silicon values in concrete with crack—silicon deficit (lower).

Figure 7.105 shows the probability for detection of 14 mm wide crack in concrete for 4 mm collimator (upper curve) and 8 mm collimator (lower curve) in front of alpha detector as a function of measuring time interval. Silicon peak depletion in gamma ray spectrum is used as an indicator. False positive = 10%. Neutron beam intensity was 3×10^7 n/s in all measurements.

Figure 7.106 shows a probability for detection of 20 mm wide crack in concrete for 4 mm collimator (curve starts at lower acquisition time) and 8 mm collimator (curve starts at higher acquisition time) in front of alpha detector as a function of measuring time interval. Silicon peak depletion in gamma ray spectrum is used as an indicator. False positive = 10%. Neutron beam intensity was 3×10^7 n/s in all measurements.

There are four chief minerals present in a Portland cement grain: tricalcium silicate (Ca_3SiO_5), dicalcium silicate (Ca_2SiO_4), tricalcium aluminate ($Ca_3Al_2O_5$), and calcium aluminoferrite ($Ca_4Al_nFe_{2-n}O_7$). The formula of each of these minerals can be broken down into the basic calcium, silicon, aluminum, and iron oxides (Table 7.14). Traditional cement nomenclature abbreviates each oxide as shown in Table 7.14.

The composition of cement is varied depending on the application. A typical example of cement contains 50%–70% C3S, 15%–30% C2S, 5%–10%

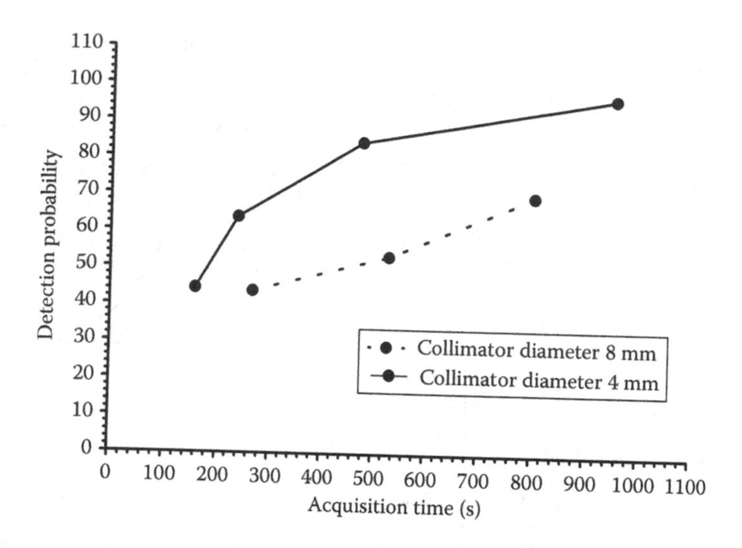

Figure 7.105 Probability for detection of 14 mm wide crack in concrete for 4 mm collimator (upper curve) and 8 mm collimator (lower curve) in front of alpha detector as a function of measuring time interval.

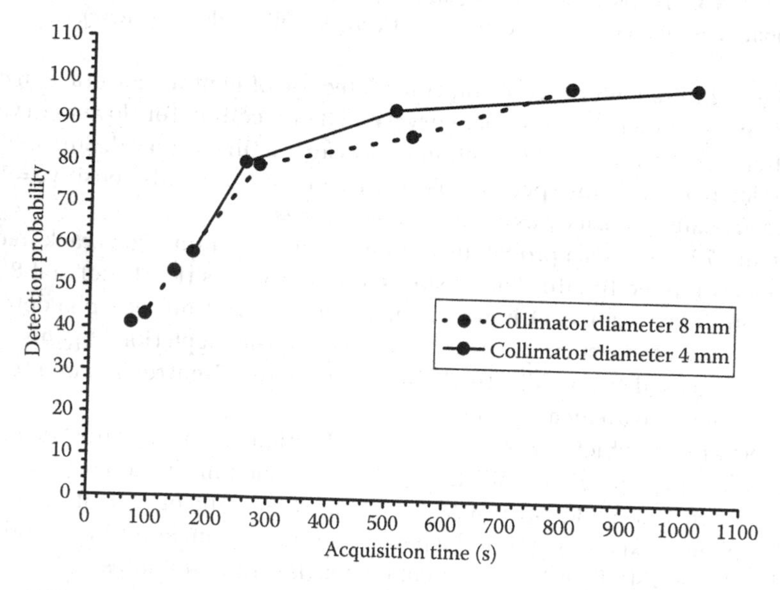

Figure 7.106 Probability for detection of 20 mm wide crack in concrete for 4 mm collimator (curve starts at lower acquisition time) and 8 mm collimator (curve starts at higher acquisition time) in front of alpha detector as a function of measuring time interval.

TABLE 7.14 CHEMICAL FORMULAE AND CEMENT NOMENCLATURE FOR MAJOR CONSTITUENTS OF PORTLAND CEMENT

Mineral	Chemical Formula	Oxide Composition	Abbreviation
Tricalcium silicate (alite)	Ca_3SiO_5	$3CaO.SiO_2$	C_3S
Dicalcium silicate (belite)	Ca_2SiO_4	$2CaO.SiO_2$	C_2S
Tricalcium aluminate	$Ca_3Al_2O_4$	$3CaO.Al_2O_3$	C_3A
Tetracalcium aluminoferrite	$Ca_4Al_nFe_{2-n}O_7$	$4CaO.Al_nFe_{2-n}O_3$	C_4AF

Abbreviation notations: C = CaO, S = SiO_2, A = Al_2O_3, F = Fe_2O_3.

C3A, 5%–15% C4AF, and 3%–8% other additives or minerals (such as oxides of calcium and magnesium). It is the hydration of the calcium silicate, aluminate, and aluminoferrite minerals that causes the hardening, or setting, of cement.

One needs to understand that the installation of the cement occurs without gaps, the cement cures properly, the cement is fully cured, and it has and maintains integrity over time. The conversion of calciumhydroxide in the concrete into calciumcarbonate from absorption of CO_2 over time period further strengthens the concrete and make it more resilient to damage. This reaction, called carbonation, lowers the pH of the cement pore solution.

By the use of tagged 14 MeV neutrons one could determine the depth variation of the stochiometric ratios for chemical elements of interest: Ca, Si, Al, Fe, O (cement constituents), and C introduced by additives and/or during curing process. Measured gamma energy spectrum resulting from 14 MeV irradiation of concrete block of mass 18.75 kg (3.6 kg Portland cement, 1.65 kg water, 4.5 kg gravel, and 9 kg quartz sand) is shown in Figure 7.107. The cement integrity could be assessed from the measurements of space and time variations of Ca, Si, Al, Fe, O, and C concentrations and concentration ratios.

7.6.2.4 Inspection/Testing of Unclean Structures

Devices that can perform without cleaning the structure could substantially reduce time and, ultimately, costs. Detecting a crack that is covered by a 4–6 cm-thick coating of organisms is a challenging task, with present techniques detecting an internal flow under these conditions is seemingly impossible. A capability of this type is very desirable not only to detect the crack but to quantify it as well. None of the present crack detection techniques, except acoustic emission, seem suitable for this task. A feasibility program to assess the possible means of addressing this problem should be undertaken. Active corrosion zones on the structure should be sought, as well as the cracks or failures.

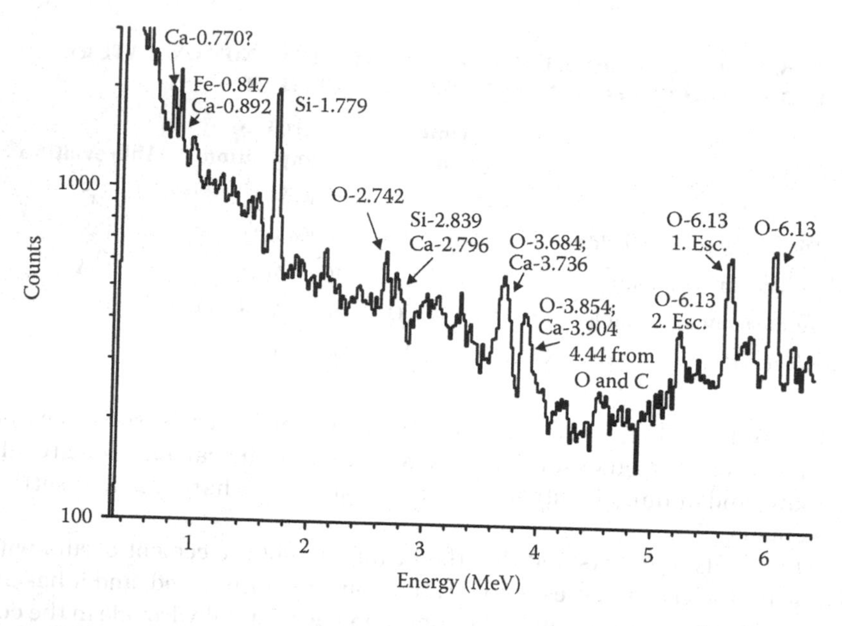

Figure 7.107 Gamma energy spectrum resulting from 14 MeV irradiation of concrete block of mass 18.75 kg (3.6 kg Portland cement, 1.65 kg water, 4.5 kg gravel, and 9 kg quartz sand).

Neutron-based system is capable to obtain information about the composition and state of the surface covered with 4–6 cm thick layer of marine organisms. The limitations of this approach should be established.

Monitoring of underwater concrete structures by using the 14 MeV tagged neutron beams was described by Sudac et al. (2013). The goal of their work was to study underwater corrosion processes in reinforced concrete. Concrete blocks of 40 × 20 × 10 cm^3 were made from 3.6 kg of Portland cement, 1.65 kg of water, 4.5 kg of coarse aggregate, and 9 kg of fine aggregate (SiO_2). After a month, its weight was 15.8 kg. In one block, 10% bwoc (weight of cement) of NaCl was put in order to check the possibility of the system to detect the chlorine. In another block 10% bwoc of graphite was put in order to simulate the fully carbonated concrete. The Portland cement contained ~75% of clinker and plaster and ~25% of slag and limestone. In one block, an iron bar 30 mm in diameter was put in order to check the possibility of the system to measure the concrete cover thickness and the bar diameter.

In comparison, Figure 7.108 shows the gamma ray spectra with and without the added graphite. Ten percent of graphite roughly represent fully carbonated concrete where all $Ca(OH)_2$ was converted to $CaCO_3$. Yield in 4.44 MeV counts shows that changes in carbon content can be measured.

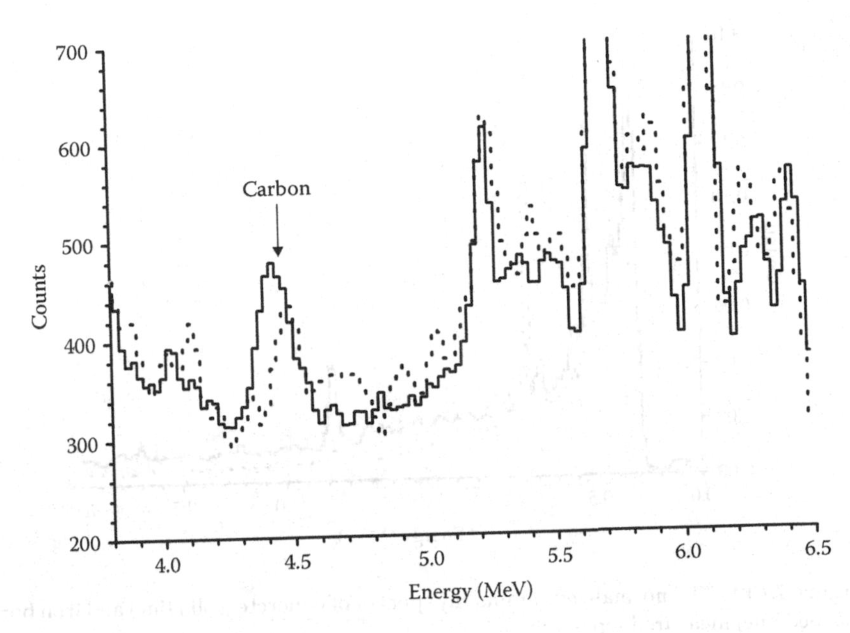

Figure 7.108 Part of the gamma ray spectrum of concrete (dashed line) and concrete with 10% bwoc of graphite (solid line).

In order to quantize the carbon content some kind of calibration is needed. Measurements lasted 9 min with the neutron beam intensity of 4.3 × 107 n/s. Block with the 10% bwoc of the added salt was also measured, but the chlorine lines were not detected.

Scanning of the concrete block with the iron bar was done by the rotation of the neutron generator in order to determine the bar diameter and the depth of the concrete cover. For each angle of rotation, the 9 min. gamma ray spectrum was collected. Spectra were fitted with the assumption that it does not contain impurities and just contribution from pure block and iron. Contribution of the other elements like chlorine or sodium was ignored. The fitting procedure was done using the following equations where the channel (ch) number was summed up as

$$\chi^2 = \text{summ}^2/(\text{ch}_{max} - \text{ch}_{min} - 2 + 1) \times \Sigma\,[a \times \text{Pure block(ch)} + b \times \text{Iron(ch)} - \text{Target1(ch)}]^2/\text{Target(ch)}$$

$$\Sigma\ \text{Pure block(ch)} = 1$$

$$\Sigma\ \text{Carbon(ch)} = 1$$

$$\Sigma\ \text{Target1(ch)} = 1$$

$$\Sigma\ \text{Target(ch)} = \text{summ}$$

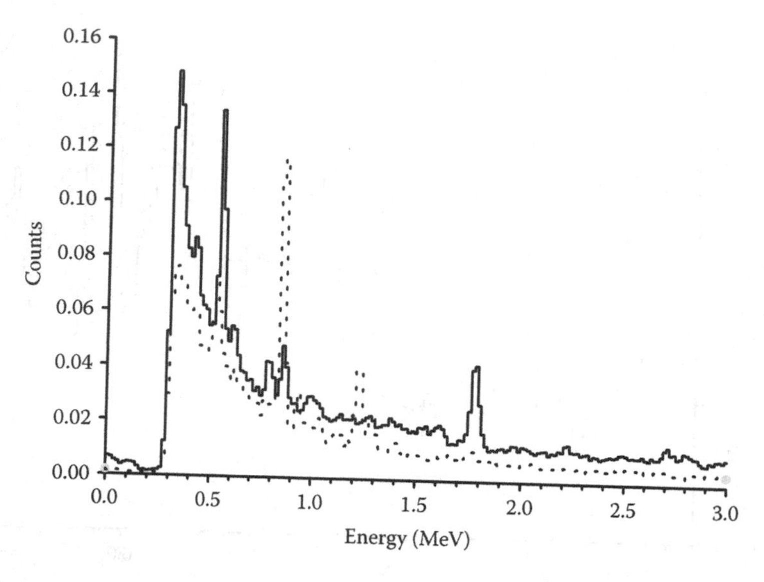

Figure 7.109 The normalized gamma ray spectra of concrete (solid line) and iron bar (dashed line) measured separately.

where summation, Σ, is done for ch values from ch_{min} to ch_{max}. "Pure block" and "Iron" normalized gamma ray spectra are shown in Figure 7.109. The method of least squares states that the best values of "a" and "b" are those for which chi-square is a minimum. Tables 7.15 and 7.16 show the results of fitting for concrete cover equal to 5 and 10 cm in depth, respectively. Figures 7.110 and 7.111 show the iron content in dependence of the rotation angle. Associated errors were calculated from the error matrix. Experimental results were fitted with the Gaussian distribution (Randall 1985).

TABLE 7.15 FITTING PARAMETERS IN DEPENDENCE ON THE NEUTRON GENERATOR ROTATION ANGLE IN A CASE OF 5 CM CONCRETE COVER THICKNESS

Angle of Rotation	Pure Block Content "a"	Iron Content "b"
41°	0.926 ± 0.011	0.096 ± 0.010
43°	0.9012 ± 0.0095	0.1218 ± 0.0087
45°	0.8523 ± 0.0089	0.1727 ± 0.0081
47°	0.8383 ± 0.0086	0.1798 ± 0.0079
49°	0.8582 ± 0.0087	0.1652 ± 0.008
51°	0.8771 ± 0.0088	0.1166 ± 0.008

TABLE 7.16 FITTING PARAMETERS IN DEPENDENCE ON THE NEUTRON GENERATOR ROTATION ANGLE IN A CASE OF 10 CM CONCRETE COVER THICKNESS

Angle of Rotation	Pure Block Content "a"	Iron Content "b"
41°	0.930 ± 0.011	0.067 ± 0.0095
43°	0.923 ± 0.011	0.0741± 0.0097
45°	0.9274 ± 0.0086	0.07 ± 0.0076
47°	0.9539 ± 0.0085	0.0443 ± 0.0075
49°	0.9553 ± 0.009	0.0429 ± 0.0079

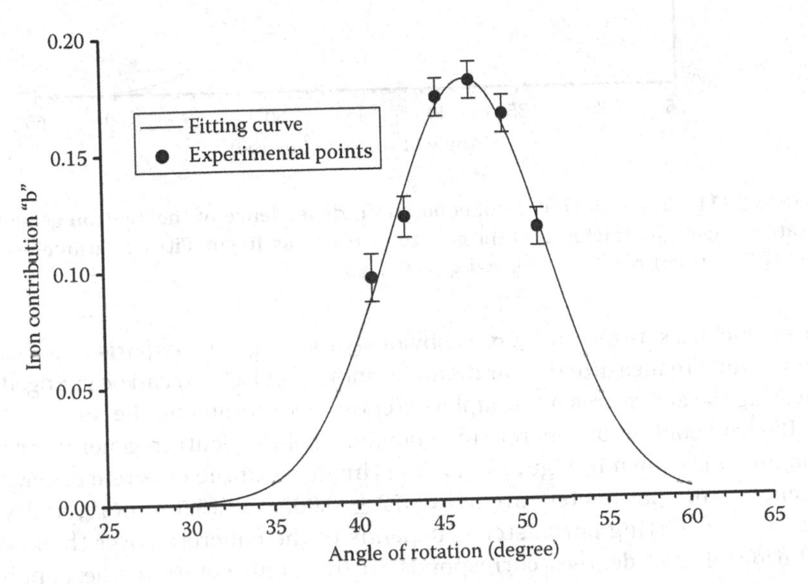

Figure 7.110 The iron content of concrete in dependence of the neutron generator rotation angle. Concrete cover was 5 cm. Fitting parameters were a = (0.1805 ± 0.006), b = (4.75 ± 0.32)°, and x_o = (46.7 ± 0.2)°.

$$F(x) = a \times \mathrm{Exp}[-(x - x_o)^{2}/2\sigma^2]$$

$$\sigma^2 = \sigma_1^{\ 2} + \sigma_2^{\ 2}$$

where "x" is the rotation angle and "a", "x_0," and "σ" are the fitting parameters. Fitting parameter σ^2 is sum of $\sigma_1^{\ 2}$ that measures the opening angle of the tagged neutron pyramid and $\sigma_2^{\ 2}$ that measures the opening angle of the iron bar. In our case $\sigma_1 \sim 4.55°$ and $\sigma_2 \sim 1.51°$, so $\sigma \sim 4.79°$. This number should be compared with (4.75 ± 0.32)° and with (5.5 ± 1.6)° for 5 and 10 cm concrete

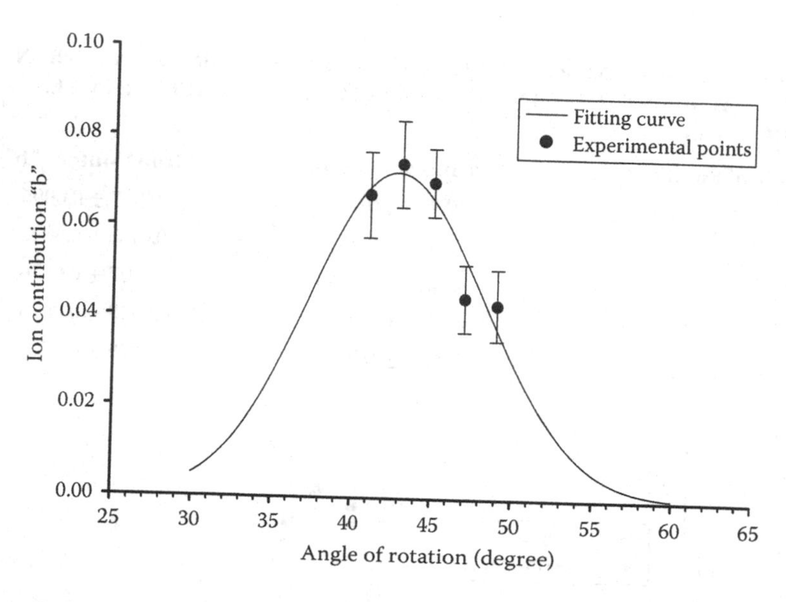

Figure 7.111 The iron content of concrete in dependence of the neutron generator rotation angle. The thickness of the concrete cover was 10 cm. Fitting parameters are a = (0.072 ± 0.005), σ = (5.5 ± 1.6), and x_0 = (42.7 ± 1.3).

cover thickness, respectively. σ_1 is obviously too large in comparison to the σ_2 if one wants to measure the bar diameter more precisely; σ_1 can be changed by lowering the active area of the alpha detector incorporated in the NG.

The iron content of concrete in dependence of the neutron generator rotation angle is shown in Figure 7.111. The thickness of the concrete cover was 10 cm. Fitting parameters are a = 0.072 ± 0.005, b = (5.5 ± 1.6)°, and x_0 = (42.7 ± 1.3)°. Fitting parameter x_0 depends of the concrete cover thickness. Rotation of four degrees corresponds to five centimeters of the concrete cover thickness.

7.6.2.5 Inspection/Testing of Clean Surfaces

Clean surface, or surface cleaned before the inspection process, can be investigated for the stochiometric ratios of chemical elements on the surface and within the interior of construction elements. This could be done by simultaneous use of XRF and tagged FNAA.

The XRF system can be built using separated x-ray source as shown in Figure 7.112. The XPIN detector has a 625 μm thick Si-PIN diode, an ultra-low-noise JFET, multilayer collimator, preamplifier, and is internally cooled with a two-stage thermoelectric cooler. Moxtek detectors use DuraBeryllium® windows uniquely coated with DuraCoat®, which is corrosion resistant and ideal

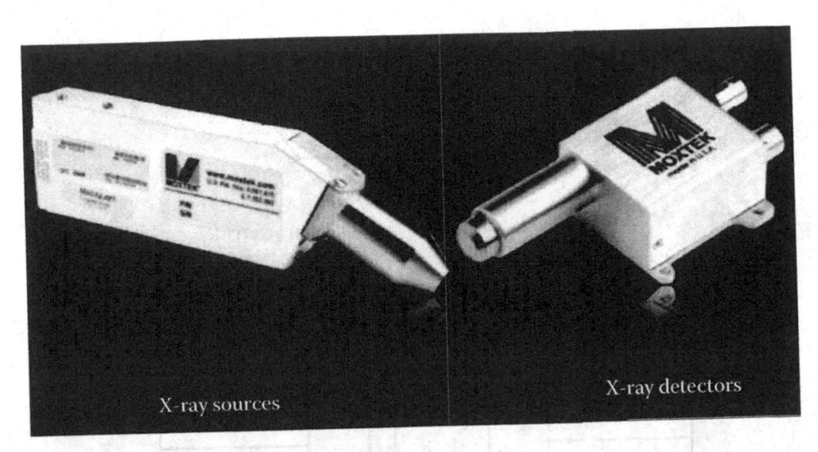

Figure 7.112 EDXRF system with MAGNUM x-Ray source and XPIN detector.

for harsh environments. All Moxtek detectors are sealed in vacuum enabling cool detector temperatures. The XPIN is available with 6 or 13 mm^2 active areas. The compact size allows for design flexibility. Another possibility is the development of a combined x-ray neutron source as schematically presented in Figure 7.113.

Work presented in the paper by Sudac et al. (2013) shows that neutron-based technique, especially in combination with the associated alpha particle method, can be used to measure the concrete cover thickness and iron bar diameter together with the changes in concrete moisture and carbon content. Detection of silicon can be used to quantify possible problems with alkali silicon reaction. In order to obtain required precision, active area of the alpha detector must be adequately adjusted. Some kind of calibration is needed if carbon and silicon content and concrete moisture need to be quantized. Standard concrete blocks with similar composition like the one to be inspected, with known quantities of carbon, silicon, and water in different proportion could be used for calibration. There are many concrete structures that have to be monitored. Previous experience shows that temperature and moisture content complicate data interpretation (Andrade and Martinez 2009). In the paper by Sudac et al. (2013) it was shown that, by using the neutron probe, sufficient quantitative data could be collected in order to predict future evolution of the concrete deterioration process.

7.6.3 Small-Scale Variations in the Carbon/Oxygen Ratio in the Environment

Often C/O ration needs to be determined for the identification of object investigated. Figure 7.114 shows determination of C/O ratio in the air. Figure 7.115

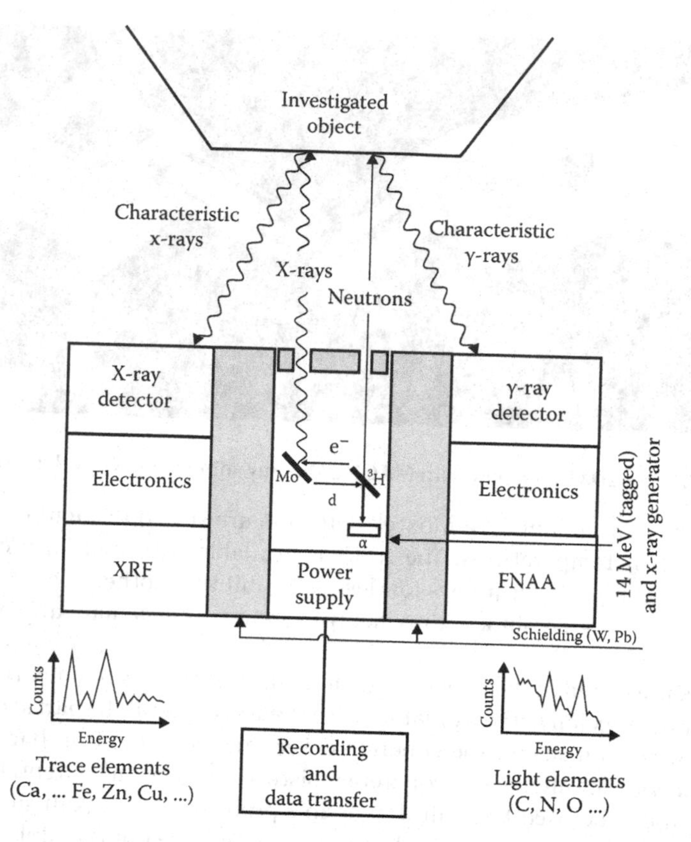

Figure 7.113 A combined x-ray and tagged neutron source.

shows the C/O ratio measured in water; it depends on the theoretical (real) C/O and on the distance that tagged neutrons must travel to the target through water. By knowing the functional dependence it is possible to find the real C/O if the distance to the object is known with the precision less than 1 cm. Functional dependence could be found by using the calibration procedure described earlier. C/O measured in water could also depend on the object shape and dimension and on the iron shirt thickness.

The C/O ratio is characteristic for many materials and it has been in the center of interest in many research areas. For example, the carbon-to-oxygen ratio in a planet provides critical information about its primordial origins and subsequent evolution. A primordial C/O greater than 0.8 causes a carbide-dominated interior as opposed to the silicate-dominated composition found on Earth; the atmosphere can also differ from those in the Solar System. The solar C/O is 0.54 (Asplund et al. 2005).

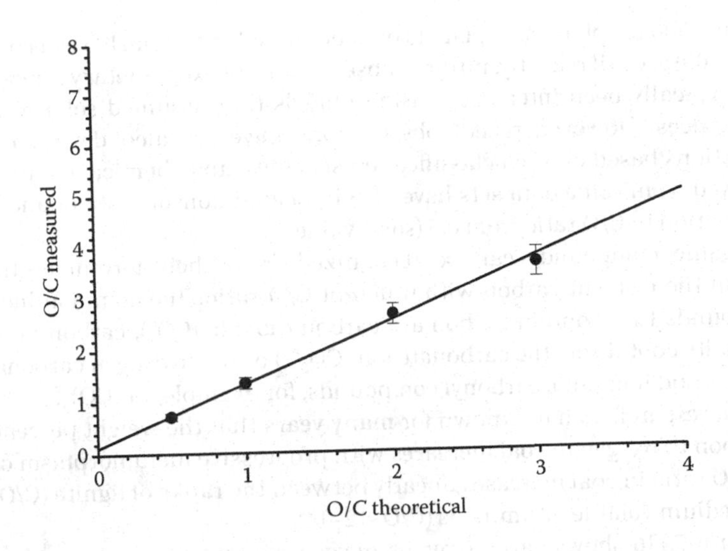

Figure 7.114 Measurement in the air: $O/C_{measured} = (0.123 \pm 0.05) + (1.215 \pm 0.05)^* O/C_{theoretical}$.

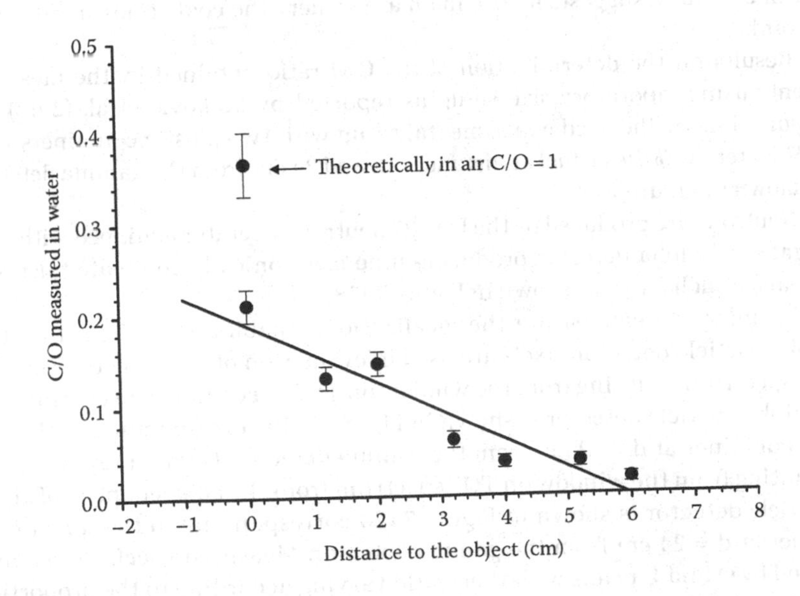

Figure 7.115 C/O ratio measured in water.

Some 500 exoplanets, or planets outside the solar system, have been discovered to date. Until recently, infrared observations of exoplanetary atmospheres have typically been interpreted using models that assumed solar elemental abundances. However, recent observations have revealed deviations from predictions based on such classification schemes, and chemical compositions retrieved from some data sets have also indicated nonsolar abundances, that is, deviation in C/O ratio from 0.5 (solar value).

Organic compounds can be recognized from their formulae—they all contain the element carbon with different C/O ratio. The common inorganic compounds that contain carbon are carbon dioxide (CO_2), carbon monoxide (CO), salts containing the carbonate ion (CO_3^{2-}) or the hydrogen carbonate ion (HCO_3^-), and inorganic carbonyl compounds, for example, $Co(CO)_6$.

Coal researchers have known for many years that the weight percent ratio of carbon to oxygen in coal increases with progressive metamorphism of coal. The C/O ratio in coal increases linearly between the ranks of lignite (C/O ~ 2.1) and medium volatile bituminous (C/O~12–14).

Figure 7.116 shows carbon versus oxygen content (dry, ash-free basis) of about 1200 specimens of coals ranging in rank from lignite to anthracite. Field of data points is enclosed by the dashed lines; approximate regression curve is shown by the solid line. Approximate equation for assumed linear regression curve between carbon values of 65% and 85% is $y = -0.92x + 91.5$; this relation was first introduced by Hickling (1931). The fact that the C/O ratio is an indicator of coal rank suggests its use in an area where the coal varies significantly in rank.

Results on the determination of the C/O ratio obtained by the measurements using laboratory size setup as reported by Valkovic et al. (2013) are reported next. The used experimental setup with two plastic containers with 50% water–50% diesel fuel at distances 12 and 24 cm from the gamma detector is shown in Figure 7.117.

Neutrons are produced by the ING-27 neutron generator equipped with 3 × 3 segmented alpha detector producing nine electronically collimated neutron beams in such a way as shown in Figure 7.118.

In order to demonstrate the localization capabilities of the setup, two alpha particle detector pixels are used in evaluation of TOF spectra. Gamma ray spectrum resulting from the window on TOF spectrum from central pixel of alpha particle detector is shown in Figure 7.119 corresponding to the target container at d = 12 cm from the gamma detector. Gamma ray spectrum resulting from the window on TOF spectrum from the adjacent pixel of alpha particle detector is shown in Figure 7.120 corresponding to the target container at d = 24 cm from the gamma detector. Measured spectra are dominated by O and C peaks with their ratio varying according to the proportion of diesel fuel in the target volume.

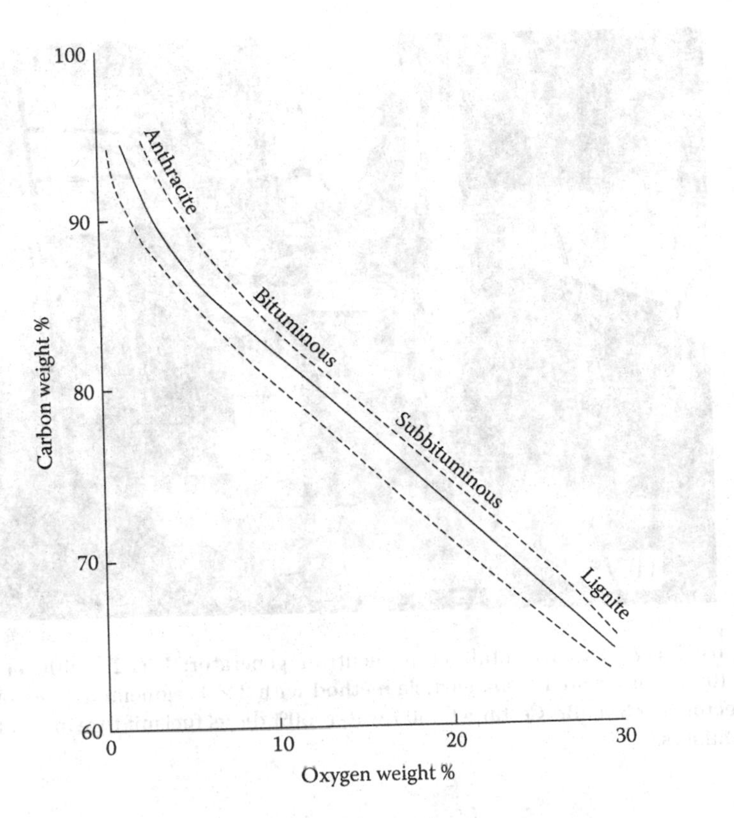

Figure 7.116 Carbon to oxygen ratio for different coal samples. (After Dyni, J.R. and Gaskill, D.L., *Contribut. Geochem. Geol. Surv. Bull.*, 1477, A1, 1980.)

7.6.4 Carbon in Soil

The forecast of massive and rapid global climate change due to the greenhouse gases appear every day more as a matter of fact. As a response, the relatively new concept of global carbon market has been introduced, requiring among others, the accurate estimation and mapping of soil carbon content on large scale. Avoiding devegetation and making relatively simple changes in cropping cycles and fertilizer use can help capture more carbon dioxide from the atmosphere and store it in the soil. Carbon credit that can be given for the abatement of carbon dioxide achieved by change in farming activities, usually expressed as a tone of the carbon dioxide removed from the atmosphere, is a property right that can be traded in carbon markets. It is a vivid market as it can be seen from the activity of ECX, the European platform for carbon emission trading. Looking at the

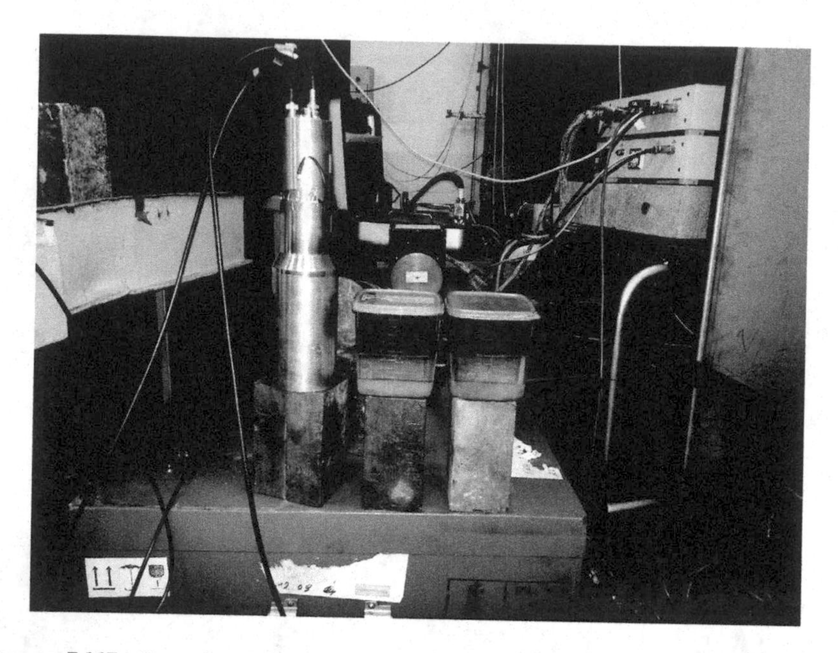

Figure 7.117 Experimental setup—neutron generator: ING-27 with 4π intensity 3×10^7 n/s; associated alpha particle method with 3 × 3 segmented detector; gamma detector: 3″ × 3″ $LaBr_3$:Ce; targets: 50% water–50% diesel fuel mixture in two 1 L plastic containers.

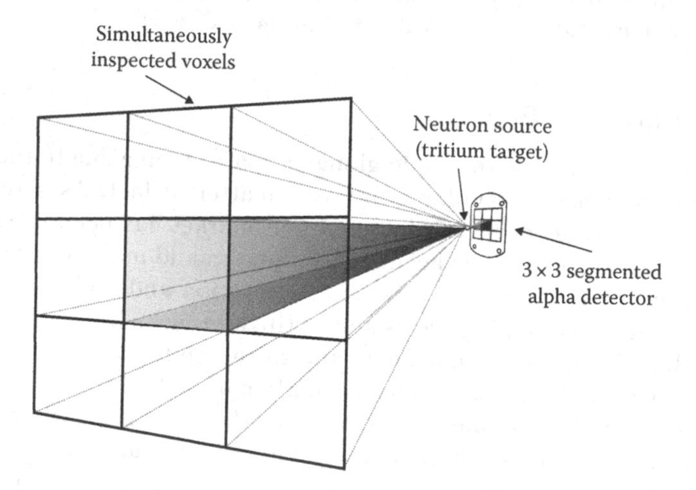

Figure 7.118 The nine electronically collimated neutron beams.

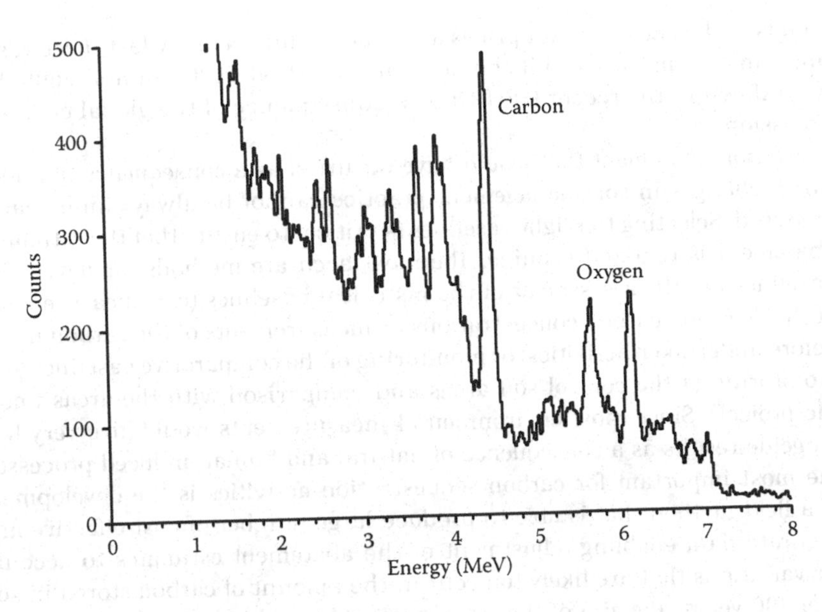

Figure 7.119 Gamma ray spectrum for 50% volume of diesel fuel in the 1 L target container at d = 12 cm from gamma detector.

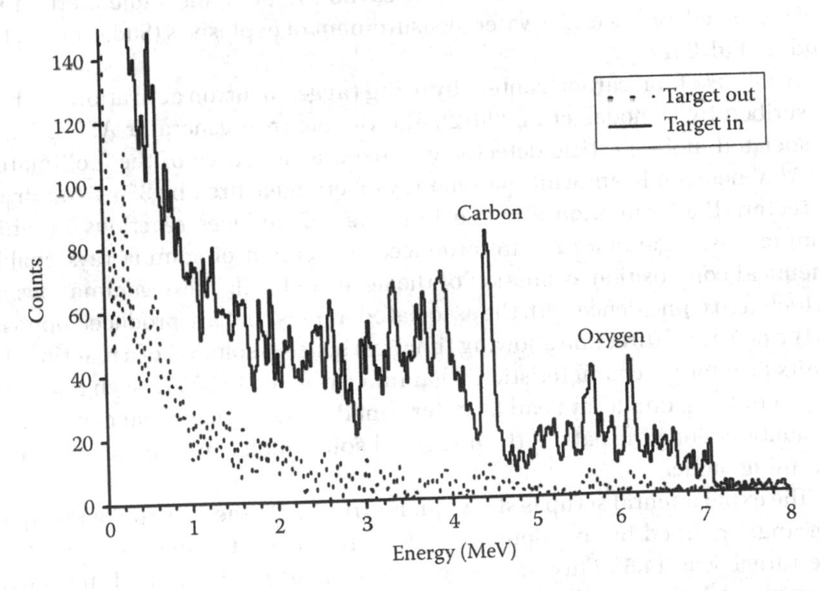

Figure 7.120 Gamma ray spectrum for 50% volume of diesel fuel in the 1 L target container at d = 24 cm from gamma detector. Lower spectrum is obtained with target out measurement.

trends of the carbon credit prices and trade volumes in the last three years, one can see that prices hit the peak of almost 30 EUR/t in mid-2008, but went down to the recent 6 EUR/t as a consequence of the global economy recession.

Carbon abatement that would have occurred as a consequence of undertaken changes in soil management practice cannot be always immediately observed. Selecting the right baselines is critical to ensure that the estimated abatement is real and genuine. The most accurate methods for developing baselines are the assessment of the historical baselines (e.g., measurements of the deep soil carbon concentrations or measurements of the carbon in soil before undertaken activities) or monitoring of the comparative baselines (e.g., monitoring of the control soil areas and comparison with the areas under the project). Since most environmental measurements would give very heterogenic results as a consequence of natural and human-induced processes, the most important for carbon sequestration activities is the development of a field method that is able to produce large number of cost-effective and accurate data enabling adjustment of the abatement estimates to account for variations that are likely to occur in the amount of carbon stored in soil over 100 years. The aim of the study presented in this paper was to develop methodology based on the FNAA with the associated alpha particle for fast, noninvasive, in situ measurements of carbon in soil. The same method has been utilized for the underwater measurement of explosives (Sudac et al. 2011; Sudac et al. 2012).

Analysis of soil carbon content by using tagged neutron activation has been described by Obhodas et al. (2012). The d-t neutron generator API-120 with associated alpha particle detector was used as a source of the "collimated" 14 MeV neutron beam while gamma rays were measured by 3″ × 3″ $LaBr_3$:Ce detector. The irradiation directed into the soil induces reactions in which characteristic gamma rays are produced. Detection of gamma rays enables chemical composition evaluation of the activated soil. These gamma rays are detected in coincidence with the associated alphas that are produced opposite to the neutrons' direction allowing time and space resolution of a reaction that emits promptly a characteristic C^{12} gamma ray at 4.43 MeV. In such a way, the origin of the gamma rays can be determined allowing chemical composition evaluation along the path of the inspected soil, which can be done in static or scanning mode.

The experimental setup is shown in Figure 7.121. The neutron beam opening angle defined by the alpha particle detector that is placed opposite to the target was 14.5°. Flux of neutrons was $2x10^6$ n/s by 4π and measuring time was 1 h. The irradiation can penetrate through approximately 10 cm soil layer.

Figure 7.121 Experimental setup.

TABLE 7.17 VOLUME AND MASS PROPORTIONS OF GRAPHITE IN QUARTZ SAND USED FOR SYSTEM CALIBRATION

No.	Graphite Mass (g)	Total Mass (g)	Total Volume (ml)	Carbon Mass Percentage (%)	Carbon Volume Percentage (%)
1	108.1	7038	4675	1.54	3.85
2	219.3	6700	4695	3.27	7.79
3	327.5	7000	4700	4.68	11.62
4	440.4	6300	4420	6.99	17.36
5	635	7000	4700	9.07	22.5
6	823	7002	4700	11.75	29.2
7	1171	6991	4700	16.75	41.59

Note: Graphite density = 0.6 g/cm^3.

Seven samples with different proportions of graphite in the quartz sand (SiO_2) were prepared for calibrating purposes as shown in Table 7.17.

Figure 7.122 shows the measured gamma ray spectra obtained by the neutron irradiation of the calibration sample No.1 and quartz sand with carbon and oxygen energy lines indicated. These two spectra can be clearly distinguished.

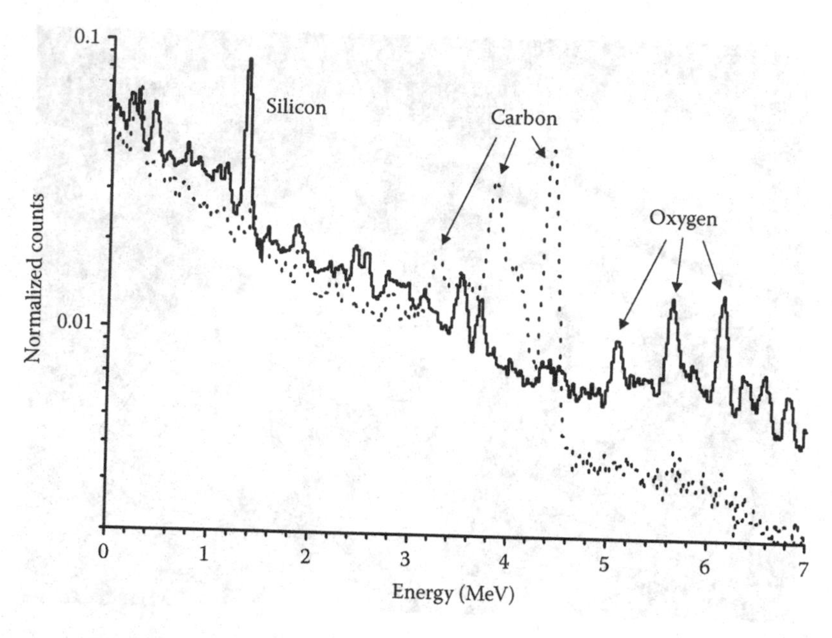

Figure 7.122 Quartz (solid line) sand and graphite (dashed line) gamma ray spectra.

The measured spectra were fitted with the assumption that they do not contain any other contributions apart from Si, C, and O. The following fitting equation was applied (least-square method):

$$\chi^2 = \frac{\text{summ}^2}{(\text{ch}_{\text{max}} - \text{ch}_{\text{min}} - 2 + 1)} \times \frac{\sum\left[\text{a x Sand}(\text{ch}) + \text{b x Carbon}(\text{ch}) - \text{Target1}(\text{ch})\right]^2}{\text{Target}(\text{ch})}$$

Channels with energies from 3.85–6.23 MeV were fitted and normalized in order to sum carbon and oxygen peaks together with the silicon contributions as follows:

$$\Sigma\ \text{Sand(ch)} = 1$$

$$\Sigma\ \text{Carbon(ch)} = 1$$

$$\Sigma\ \text{Target1(ch)} = 1$$

$$\Sigma\ \text{Target(ch)} = \text{summ}$$

Figure 7.123 shows fitted gamma ray spectra of the samples No.1–No.7. Parameters “a” and “b” obtained by Equation 7.1 are shown in Table 7.18.

In order to calibrate the system, the linear regression analysis was applied for fitting parameter “b” in relation to the mass and volume percentages of graphite in the measured samples as shown in Figures 7.124 and 7.125, respectively.

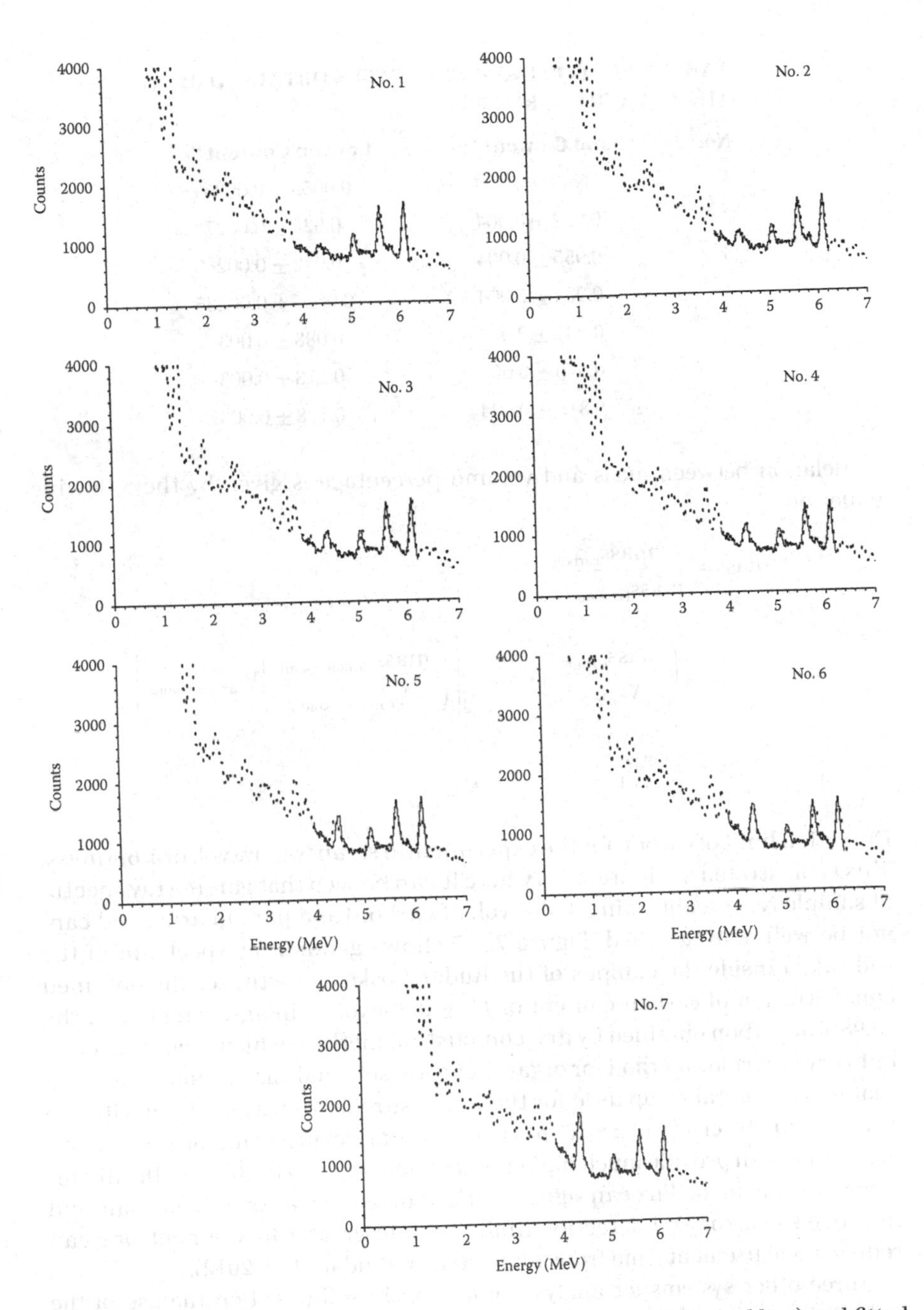

Figure 7.123 Gamma ray spectra of the samples No. 1–No. 7 (dashed line) and fitted spectrum (solid line).

TABLE 7.18 FITTING PARAMETERS OBTAINED BY THE EQUATION (SEE TEXT)

No.	Sand Content "a"	Carbon Content "b"
1	0.9903 ± 0.004	0.0055 ± 0.0026
2	0.972 ± 0.004	0.024 ± 0.0027
3	0.955 ± 0.004	0.042 ± 0.00265
4	0.934 ± 0.004	0.0615 ± 0.00285
5	0.941 ± 0.004	0.088 ± 0.003
6	0.876 ± 0.004	0.118 ± 0.003
7	0.815 ± 0.004	0.178 ± 0.003

Relation between mass and volume percentage is given by the following equation:

$$\%\text{mass} = \frac{\text{mass}_{\text{carbon}}}{\text{mass}_{(\text{carbon}+\text{sand})}}$$

$$= \left[\left(\frac{\text{mass}_{\text{carbon}}}{\text{V}_{\text{carbon}}}\right)\text{V}_{\text{carbon}}\right]\left[\left(\frac{\text{mass}_{\text{carbon}+\text{sand}}}{\text{V}_{\text{carbon}+\text{sand}}}\right)\text{V}_{\text{carbon}+\text{sand}}\right]$$

$$= \frac{\rho_{\text{carbon}}}{\rho\% \text{ vol.}}$$

Detection limit of carbon for the experimental setup was 4% vol. or 1.6% mass. This is illustrated in Figure 7.126 where it can be seen that gamma ray spectra of sample No. 1 (containing 3.85% vol. of carbon) and pure quartz sand cannot be well distinguished. Figure 7.127 shows gamma ray spectrum of the soil taken inside the campus of the Ruder Boskovic Institute. The obtained concentration of carbon content of 11 ± 0.8% vol. is in agreement with the 10.98% of carbon obtained by dry-combustion method, which is conventional but very laborious method for organic carbon soil analysis. It should be noted that experimental setup used for these measurements was working with the neutron flux intensity of 2×10^6 n/s while commercially available d-t neutron generators can provide much higher flux of neutrons up to 10^{10} n/s. The higher intensity of neutron flux can significantly shorten the measurement time and improve the error and detection limit. Tenfold greater flux of neutrons can reduce measurement time from 1 h to 10 min (Sudac et al. 2012).

Three other systems for analysis of carbon in soil based on the use of the pulsed d-t neutron generator have been described in the literature: one developed in Brookhaven National Laboratory (BNL), United States (Wielopolski

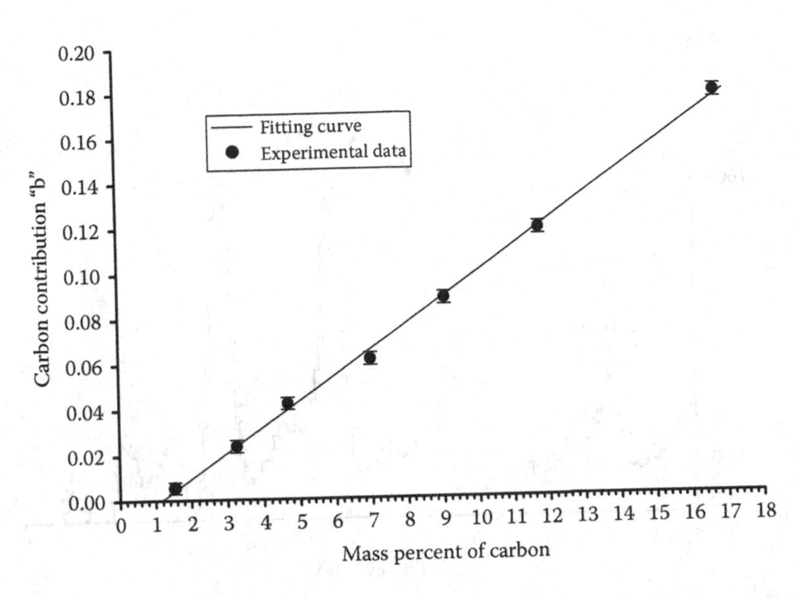

Figure 7.124 Linear regression analysis of the fitting parameter "b" in relation to the % mass of graphite (carbon) in the measured samples: Carbon content = (−0.013 ± 0.002) + (0.0113 ± 0.0002) x % mass (carbon).

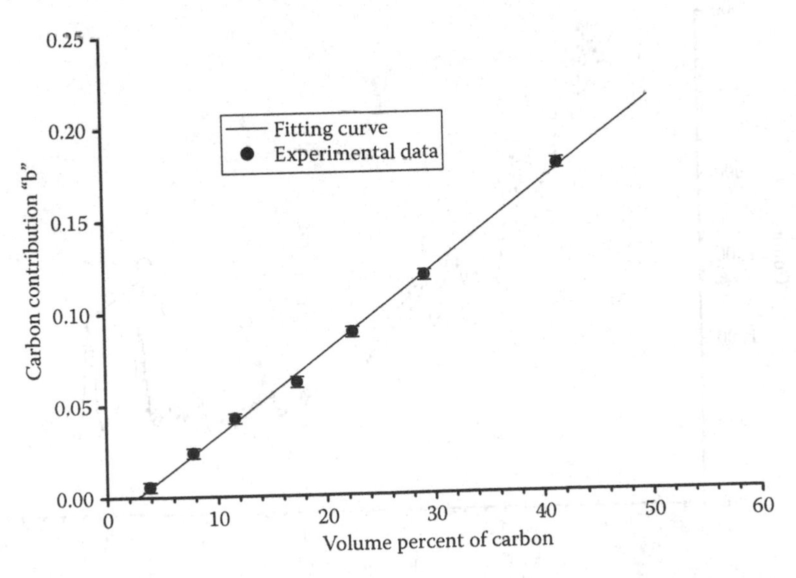

Figure 7.125 Linear regression analysis of the fitting parameter "b" in relation to the % volume of the graphite (carbon) in the measured samples: Carbon content = (−0.013 ± 0.002) + (0.0045 ± 0.00008) x % volume (carbon).

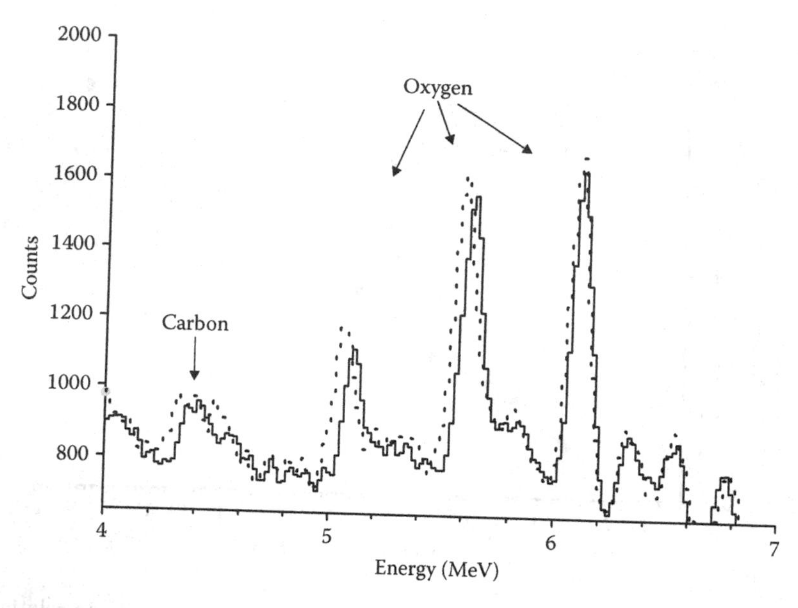

Figure 7.126 Gamma ray spectrum of sample No. 1 (solid line) and quartz sand (dashed line).

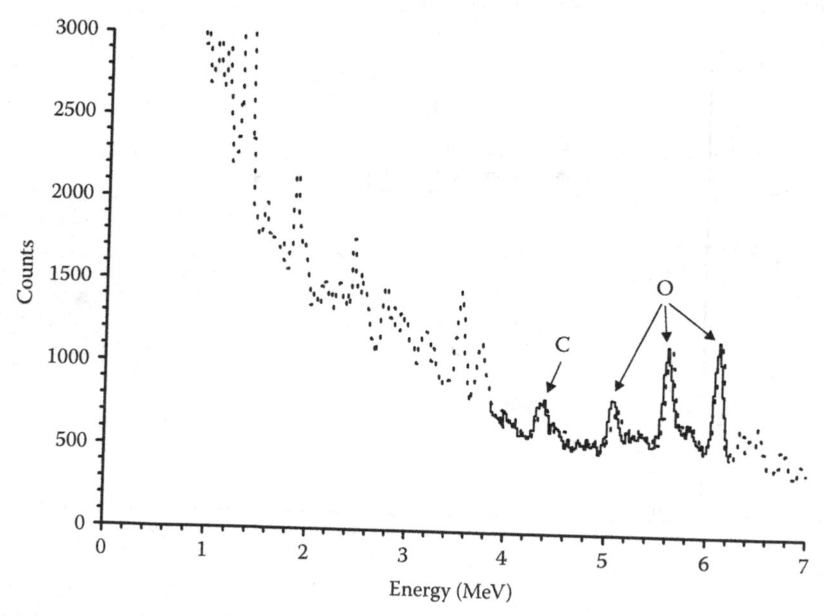

Figure 7.127 Gamma ray spectra of the soil taken from the Ruder Boskovic Institute campus (dashed line) and fitted spectrum (solid line).

et al. 2008, 2010), another one developed in Australian Nuclear Science and Technology Organization (Falahat et al. 2011), and the third developed at Goddard Geophysical and Astronomical Observatory in Greenbelt, MD, United States.

The BNL inelastic neutron scattering (INS) system was evaluated for use as a long-term, in-field monitor to detect cumulative changes in below-ground carbon resulting from the leakage of CO_2 stored in deep geological reservoirs. This system underwent tests at a facility constructed specifically for testing, under controlled conditions, various detection systems for monitoring near-surface transport and accumulations of CO_2 fluxes emanating from a shallow buried, slotted horizontal well. The INS System was assessed by comparing the results from placing it above the horizontal well at a spot with a known high CO_2 leak identified and quantified the previous years, with those obtained from background readings adjacent to the well. At two different "hot spots," a suppression of about 14% in 2008 and about 7% in 2009 in carbon content above the well in comparison to the background signal was observed (Wielopolski and Mitra 2010).

If we take 10 cm as the examined depth of soil, increase of 1 vol% of carbon content in soil equals soil carbon volume of 10 m^3 per 10,000 m^2 (1 ha). Carbon dioxide equivalent (CO_2-e) of emission reduction or removal for 10 m^3 of the soil carbon equals 18.3 kg:

$$pV = \left(\frac{m}{M}\right)RT,$$

where

p = 101,325 Pa
M = 44 g/mol
R = 8.314 J/mol K
T = 293.15 K (20°C)

Therefore, 1 vol% increase of soil carbon content should be obtained at the area of approximately 55 ha in order to achieve one carbon credit for CO_2–e.

We demonstrated the utility of a tagged neutron–based sensor to quantitatively analyze carbon in soil. The system has the potential to reduce the average measurement time to approximately 10 min per sampling point and to lower the maximum detection limit and error below 4% vol. and 10%, respectively. The additional work is required to test the system on soils with different texture and chemical composition. The future target is to analyze the ability of a system to work in a scanning mode.

7.6.5 Diamond Detection in Kimberlite

At present the kimberlite ore is processed in crushers or grinding rolls with subsequent grinding in wet mills down to a size of 0.2 mm and less. Ore is concentrated by x-ray fluorescent, gravitation, and flotation methods (Aleksakin et al. 2013). X-ray fluorescence separation uses diamond property to luminesce under x-ray exposure. An x-ray fluorescence separator automatically cuts out the bulk of the ore concentrate that contains detected diamonds. The basic disadvantage of the standard diamond processing technology is that crushing kimberlite ore can break the most valuable large diamonds of a few carats or larger.

The search for diamonds by the tagged neutron method is reduced to detection of excess carbon at a particular point of the kimberlite sample. Large penetrability of fast neutrons makes it possible to examine appreciably large samples of kimberlite. The ability of the tagged neutron method to determine the three-dimensional position of the explored object allows not only revealing the presence of a large-size diamond in a rock piece but also locating its position in the sample. Thus, rock pieces containing large-size diamonds can be identified before the crushing stage.

In the paper by Aleksakin et al. (2013) experimental diamond detection setup used to carry out experiments on estimation of detectable diamond size for various background conditions is described. The experimental setup for diamond detection in kimberlite consists of a portable neutron generator ING-27 with a built-in 64-pixel alpha detector, six gamma detectors based on BGO crystals, electronics of the data acquisition system for the alpha and gamma detectors, and power supply units for the neutron generator and the alpha and gamma detectors.

The six gamma detectors with their BGO crystals are protected with steel shielding against direct neutrons emitted by ING-27. The area of the tagged neutron beams irradiating the investigated object is outlined with six laser line generators fixed on the neutron generator housing. The diamond simulant is attached to a set of suspension wires, which can also hold several kimberlite samples. The diamond simulant is typically located behind the kimberlite sample. In some cases the diamond simulant is placed between two samples of natural kimberlite.

Since diamonds entirely consist of carbon, the procedure of their detection consists in detecting excess carbon in any area of the kimberlite sample. Measurements with six kimberlite and three core samples do not show any local excess of the number of events above the average in the carbon line region, whereas measurements with a combination of diamond simulants and kimberlite samples of different size and weight show the local excess of the number of events around 4.44 MeV. The minimal detectable mass of a diamond simulant is 1.15 g with a 1-cm-thick screening kimberlite layer in front of the simulant.

The subsequent check of the identification algorithm on 33 kimberlite samples with a total weight of 55 kg confirmed the potential of the tagged neutron method: false alarms were not detected and the sample with a local carbon content excess from 3.3 to 5.8 % was found to have two inhomogeneous diamond inclusions up to 7 mm in diameter consisting of small particles with a size of 1–2 mm (Aleksakin et al. 2013).

REFERENCES

Aleksakhin, V. Yu., Bystritsky, V. M., Zamyatin, N. I. et al. 2013a. Detection of diamonds in kimberlite by tagged neutron method. Preprint of the JINR, P18-2013-132.

Aleksakhin, V. Yu., Bystritskii, V. M., Zamyatin, N. I. et al. 2013b. Use of the tagged neutron technique for detecting dangerous underwater substances. *Phys. Particles Nuclei Lett.* 10(7): 860–867.

American University, Washington, DC, 2012a. TED Case Studies, Black Sea Pollution. http://www1.american.edu/TED/barrel.htm. Accessed September 20, 2014.

American University, Washington, DC, 2012b. Nigeria Waste Imports from Italy. http://www1.american.edu/TED/nigeria.htm. Accessed September 20, 2014.

Andrade, C. and Martinez, I. 2009. Embedded sensors for the monitoring of corrosion parameters in concrete structures, in *NDTCE'09, Non-Destructive Testing in Civil Engineering*, Nantes, France, June 30–July 3, 2009.

Asplund, M., Grevesse, N., and Sauval, A. 2005. Cosmic abundances as records of stellar evolution and nucleosynthesis, in Barnes, T. G. III and Bash, F.N. (Eds.), *Proceedings of a Symposium*, Austin, TX, June 17–19, 2004,pp. 25–38, ASP Conference Series 336.

Barman, I. V. and Shevchenko, A. A. 1983. Soil scooping mechanism for the venera 13 and venera 14 unmaned interplanetary spacecraft. *Cosm. Res.* 21(2): 118–122.

Batyaev, V. F., Belichenko, S. G., and Bestaev, R. R. 2013. Ultimate levels of explosives detection via tagged neutrons, Presented at 2013 *International Conference on Applications of Nuclear Techniques*, Crete, Greece, June 23–30, 2013.

Batyaev, V. F., Belichenko, S. G., Bestaev, R. R., and Gavryuchenkov, A. V. 2014. Ultimate levels of explosives detection via tagged neutrons. *Appl. Nucl. Techn. (CRETE13), Int. J. Modern Phys. Conf. Ser.* 27: 1460131 (8pa).

Belichenko, S. G., Anan'ev, A. A., Bogolyubov, E. P. et al. 2009. Tagged neutrons fram portable neutron generator for detection of high explosives and fissile materials in cargo containers. Reported at *Neutron Based Techniques for the Detection of Illicit Materials and Explosives*, Vienna, Austria, May 4–8, 2009, SM/EN-P11.

Beyerle, A., Hurley, J., and Tunnell, L. 1990. Design of an associated particle imaging system. *Nucl. Instrum. Methods Phys. Res. A* 229: 458–462.

Blackston, M. A., Hausladen, P., Bingham, P. R. et al. 2009. Using fast neutrons to image induced fissions, in *IEEE Nuclear Science Symposium*, Orlando, FL.

Blagus, S., Sudac, D., and Valković, V. 2004. Hidden substances identification by detection of fast neutron induced γ rays using associated α particle technique. *Nucl. Instrum. Methods B* 213: 434–438.

Boynton, W. V., Evans, L. G., Reedy, R. C., and Trombka, J. I. 1993. The Composition of Mars and Comets by Remote and In-Situ Gamma-Ray Spectrometry. In Pieters, C., Englert, P. (Eds.), *Remote Geochemical Analysis: Elemental and Mineralogical Composition*, Cambridge University Press, Cambridge, U.K., pp. 395–411.

Boynton, W. V. et al. 2007. Concentration of H, Si, Cl, K, Fe, and Th in the low and mid latitude regions of Mars. *J. Geophys. Res. Planets* 112: E12S99. doi:10.1029/2007JE002887.

Bradley, J. G., Schweitzer, J. S., Truax, J. A., Rice, A., and Tombrello, T. A. 1994. A neutron activation gamma ray spectrometer for planetary surface analysis. Brown Bag Preprint BB-127 (Cal Tech), Report IAA-L-0202P.

Briesmeister, J. F. 2000. MCNP–A general Monte Carlo n-particle transport code. Los Alamos NAtional Laboratory, Los Alamos, NM, LA-13079-M.
Bruchini, C. et al. 1998. Ground penetrating radar and imaging metal detector for antipersonnel mine detection. *J. Appl. Geophys.* 40: 59–71.
Bystritsky, V.M., Zamyatin, N. I., Zubarev, E. V. et al. 2013. Stationary setup for identifying explosives using the tagged neutron method. *Phys. Particles Nuclei Lett.* 10(5): 442–446.
Bystritsky, V. M. et al. 2013. Use of the tagged neutron technique for detecting dangerous underwater substances. *Phys. Particles Nuclei Lett.* 10(7): 860–867.
Carasco, C., Deyglun, C., Pérot, B. et al. 2012. Detection of special nuclear materials with the associate particle technique. *AIP Conf. Proc.* 1525: 220. http://dx.doi.org/10.1063/1.4802323.
Carasco, C., Perot, B., Mariani, A. et al. 2010. Material characterization in cemented radioactive waste with associated particle technique. *Nucl. Instrum. Methods Phys. Res. A* 619: 419–426.
Das, B. S., Hendrickx, J .M. H., and Brochers, B. 2001. Modeling transient water distributions around landmines in bare soils. *Soil Sci.* 166(3): 163–173.
Donzella, A., Boghen, G., Bonomi, G. et al. 2006. Simulation of a tagged neutron inspection system prototype. *J. Phys. Conf. Ser.* 41: 233–240.
Dyni, J. R. and Gaskill, D. L. 1980. Relation of the carbon/oxygen ratio in coal to igneous intrusions in the somersewt coal field, Colorado. *Contribut. Geochem. Geol. Surv. Bull.* 1477: A1–A20.
Eleon, C., Perot, B., Carasco, C., Sudac, D., Obhodas, J., and Valkovic, V. 2011. Experimental and MCNP simulated gamma-ray spectra for the UNCOSS neutron-based explosive detector. *Nucl. Instrum. Methods Phys. Res. A* 629: 220–229. doi:10.1016/j.nima.2010.11.055.
El Kanawati, W., Perot, B., Carasco, C. et al. 2011. Acquisition of prompt gamma-ray spectra induced by 14 MeV neutrons and comparison with Monte Carlo simulations. *Appl. Radiat. Isot.* 69: 732–743.
Evans, L. G., Reedy, R. C., and Trombka, J. I. 1993. Introduction to Planetary Remote Sensing Gamma-Ray Spectroscopy. In Pieters, C., Englert, P. (Eds.), *Remote Geochemical Analysis: Elemental and Mineralogical Composition*, Cambridge University Press, Cambridge, U.K., pp. 167–198.
Evsenin, A. V., Gorshkov, I., Kuznetsov, Yu., Osetrov, A. V., and Vakhtin, O. I. 2009. Detection of explosives and other illicit materials by nanosecond neutron analysis. Presented in *Neutron Based Techniques for the Detection of Illicit Materials and Explosives*, Vienna, Austria, May 4–8, 2009, Paper SM/EN-11.
Falahat, S., Koble T., Schumann, O., Waring, C., and Watt, G. 2012. Development of a surface scanning soil analysis instrument. *Appl. Radiat. Isot.* 70: 1107–1109.
Faust, A. A., Chesney, R. H., Das, Y., McFee, J. E., and Russell, K. L. 2005. Canadian teleoperated landmine detection systems. Part I: The improved landmine detection project. *Int. J. Syst. Sci.* 36 (9): 529–543.
FLIR, 2011, http://gs.flir.com/detection/radiation. The identiFINDER. Handhold Radiation Detectors. Accessed September 21, 2014.
Gordon, C. M. and Peters, C. W. 1990. A fast-neutron probe for tomography and bulk analysis, *Int. J. Radiat. Appl. Instrum. Part A* 41: 1111–1116.
Grau, J. A., Schweitzer, J. S., and Hertzog, R. C. 1990. Statistical uncertainties of elemental concentrations extracted from neutron induced gamma-ray measurements. *IEEE Trans. Nucl. Sci.* 37: 2175–2178.

Grogan, B. and Mihalczo, J. T. 2012. Identification of Lithium Isotopes using time tagged neutron scattering, in *INMM 53rd Annual Meeting on Institute of Nuclear Materials Management*, , Orlando, FL, July 2012.

Hickling, H. G. A. 1931. The geological history of coal. *Fuel Sci. Pract.* 10: 212–232.

Kanawati, W. El., Carasco, C., Perot, B. Mariani, A., Valkovic, V., and Sudac, D. 2010. Gamma-ray signatures improvement of the euritrack tagged neutron inspection, system database. *IEEE Trans. Nucl. Sci.* 57(5): 2879–2885.

Kanawati, W. El., Perot, B, Carasco, C., Eleon, C., Valkovic, V., Sudac, D., and Obhodas, J. 2011. Conversion factors from counts to chemical ratios for the EURITRACK tagged neutron inspection system. *Nucl. Instrum. Methods Phys. Res. A* 654: 621–629.

Kazunori, T., Preetz, H., and Igel, J. 2011. Soil properties and performance of landmine detection by metal detector and ground-penetrating radar-Soil characterization and its verification by field test. *J. Appl. Geophys.* 73: 368–377.

Kuznetsov, A., Evsenin, A., Osetrov, O., Vakhtin, D., and Gorshkov, I. 2006. SENNA—Device for explosives' detection based on nanosecond neutron analysis. *Proc. SPIE* 6213(621306): 1–12.

Le Tourneur, P., Dumont, J. L., Groiselle, C. et al. 2009. ULIS : A portable device for chemical and explosive detection, in *Proceedings of the International Topical Meeting on Nuclear Research Applications and Utilization of Accelerators, IAEA*, Vienna, Austria, Paper SM/EN-17.

Lopera, O. and Milisavljevic, N. 2007. Prediction of the effects of soil and target properties on the antipersonnel landmine detection performance of ground-penetrating radar: A Columbian case study. *J. Appl. Geophys.* 63: 13–23.

Megahid, R. M., Osman, A. M., and Kansouh, W. A. 2009. Low cost combined systems for detection of contraband hidden in cargo containers. Reported at *Neutron Based Techniques for the Detection of Illicit Materials and Explosives*, Vienna, Austria, May 4–8, 2009, SM/EN-P03.

Mihalczo, J. T. 1973. Prompt neutron decay in plutonium metal using 252Cf as a randomly pulsed neutron source. Y/DR 111, Union Carbide Corporation Nuclear Division, Oak Ridge, TN, Y12 Plant.

Mihalczo, J. T., Bingham, P. R., Blackston, M. A. et al. 2013. Fast-neutron imaging with API d-t neutron generators, in Barmakov, Yu. N. (Ed.), *Proceedings of the 2012 International Scientific and Technical Conference, All-Russia Research Institute of Automatics-VNIIA*, Moscow, Russia, pp.198–212.

Mihalczo, J. T., Mullens, J. A., Mattingly, J. K., and Valentine, T. E. 2000. Physical description of nuclear materials identification system (NMIS) signatures. *Nucl. Instrum. Methods Phys. Res. A* 450: 531–555.

Miller, T. W., Hendrickx, J. M. H., and Brochers, B. 2004. Radar detection of buried landmines in field soils. *Vadose Zone J.* 3(4): 1116–1127.

Mullens, J. A., McConchie, S., Hausladen, P., Mihalczo, J., Grogan, B., and Sword, E. 2011. Neutron radiography and fission mapping measurements of nuclear materials with varying composition and shielding, in *INMM 52nd Annual Meeting on Institute of Nuclear Materials Management*, Palm Desert, CA, July 2011.

Nebbia, G., Pesente, S., Lunardon, M. et al. 2005. Detection of hidden explosives in different scenarios with the use of nuclear probes. *Nucl. Phys. A* 752: 649C–658C.

Obhodas, J., Sudac, D., Matjacic, L., and Valkovic, V. 2012. Analysis of carbon soil content by using tagged neutron activation. *Proc. SPIE* 8371: 83711B. doi: 10.1117/12.918316.

Obhodas, J., Sudac, D., and Valkovic, V. 2009. Matrix characterization in threat material detection processes, in *AIP Conference Proceedings of the 12th International Conference on Application of Accelerators in Research and Industry*: March 10, 2009, Vol. 1099, pp. 578–582. doi:10.1063/1.3120103. American Institute of Physics, College Park, Maryland.

Obhodas, J., Sudac, D., Valkovic, V. et al. 2010a. Analysis of containerized cargo in the ship container terminal. *Nucl. Instrum. Methods Phys. Res. A* 619: 460–466.

Obhodas, J., Sudac, D., Valkovic, V. 2010b. Matrix characterization of the sea floor in the treat material detection processes. *IEEE Trans. Nucl Sci.* 57(5), Part: 2: 2762–2767.

Obhodas, J., Valkovic, V., Sudac, D., Matika, D., Pavic, I., and Kollar, R. 2010c. Environmental security of the coastal seafloor in the ports and waterways of the Mediterranean region. *Nucl. Instrum. Methods Phys. Res. A* 619: 419–426.

Parsons, A., Bodnarik, J., Evans, L. et al. 2011. Active neutron and gamma-ray instrumentation for in situ planetary science applications. *Nucl. Instrum. Methods Phys. Res. A* 652: 674–679.

Perot, B., Carasco, C., Valkovic, V., Sudac, D., and Franulovic, A. 2009. Detection of illicit drugs with the EURITRACK system, in *AIP Conference Proceedings of the 12th International Conference on Application of Accelerators in Research and Industry*, March 10, 2009, Vol. 1099, pp. 565–569. doi:10.1063/1.3120100. American Institute of Physics, College Park, Maryland.

Perot, B., Perret, G., Mariani, A. et al. 2006. The EURITRACK project: Development of a tagged neutron inspection system for cargo containers, in Vorvopoulos, G. and Doty, F.P. (Eds.), *Proceedings of the SPIE on Non-Intrusive Inspection Technologies*, Vol. 6213, p. 621305. SPIE, Bellingham, Washington.

Perret, G. and Perot, B. 2007. Dose calculations for the EURITRACK project. Unpublished Report, http://www.euritrack.org. EUropean Riposte against Illicit TR@fiCking project Website.

Pesente, S., Lunardon, M., Nebbia, G., Viesti, G., Sudac, D., and Valkovic, V. 2007. Monte Carlo analysis of tagged neutron beams for cargo container inspection. *Appl. Radiat. Isot.* 65: 1303–1396.

Pesente, S., Nebbia, G., Lunardon, M. et al. 2004. Detection of hidden explosives by using tagged neutron beams with sub-nanosecond time resolution. *Nucl. Instrum. Methods A* 531: 657–667.

Pesente, S., Nebbia, G., Lunardon, M. et al. 2005. Tagged neutron inspection system (TNIS) based on portable sealed generators. *Nucl. Instrum. Methods Phys. Res. B* 241: 743–747.

Radle, J. E., Archer, D. E., Carter, R. J. et al. 2010. Fieldable nuclear material identification system, in *51st INMM Annual Meeting*, Institute of Nuclear Materials Management, Baltimore, MD, July 11–15, 2010.

Radle, J. E., Archer, D. E., Mullens, J. A. et al. 2009. Fieldable nuclear material identification system (FNMIS), in *50th INMM Annual Meeting*, Institute of Nuclear Materials Management, Tucson, AZ, July 12–16, 2009.

Randall, R. R. 1985. Application of accelerator sources for pulsed neutron logging of oil and gas wells. *Nucl. Instrum. Methods Phys. Res.* B10/11: 1028–1032.

Rhodes, E. and Dickerman, C. E. 1997. Associated-particle sealed-tube neutron probe for nonintrusive inspection, in Duggan, J.L. and Morgan, I.L. (Eds.), *CP392, Applications of Accelerators in research and Industry*, AIP Press, New York, pp. 833–836.

Smith, H. 1997. Conversion coefficients for use in radiological protection against external radiation. International Commission on radiological Protection, ICRP. ICRP Publication 74. Ann. ICRP 26 (3-4), 1996. SAGE Publications, Thousand Oaks, California.

Sudac, D., Blagus, S., and Valkovic, V. 2004. Chemical composition identification using fast neutrons. *Appl. Radiat. Isot.* 61/1: 73–79

Sudac, D., Blagus, S., and Valković, V. 2005. Inspections for contraband in a shipping container using fast neutrons and the associated alpha particle techniques: Proof of principle. *Nucl. Instrum. Methods Phys. Res. B* 241: 798–803.

Sudac, D., Blagus, S., and Valkovic, V. 2007. The limitations of associated alpha particle technique for contraband container inspections. *Nucl. Instrum. Methods*, 263(1): 123–126.

Sudac, D., Majetić, S., Kollar, R., Nad, K., Obhodaš, J., and Valković, V. 2012. Inspecting the minefield and residual explosives by fast neutron activation method. *IEEE Trans. Nucl. Sci.* 59: 1421–1425.

Sudac, D., Matika, D., and Valković, V. 2008. Identification of materials hidden inside a sea-going cargo container filled with an organic cargo by using the tagged neutron inspection system. *Nucl. Instrum. Methods Phys. Res. A* 589(1): 47–56.

Sudac, D., Matika, D., and Valkovic, V. 2009. Detection of explosives in objects on the bottom of the sea, in *AIP Conference Proceedings of the 12th International Conference on Application of Accelerators in Research and Industry*: March 10, 2009, Vol. 1099, pp. 574–577. doi:10.1063/1.3120102.

Sudac, D., Matika, D., Nad, K., Obhodas, J., and Valkovic, V. 2012. Barrel inspection utilizing a 14 MeV neutron beam and associate alpha particle method. *Appl. Radiat. Isot.* 70: 1070–1074.

Sudac, D., Nad, K., Obhodas, J., and Valkovic, V. 2012. Corrosion monitoring of reinforced concrete structures by using the 14 MeV tagged neutron beams. *SPIE Defense, Security,Sensing* 2012: 8366–8410.

Sudac, D., Nad, K., Obhodas, J., and Valkovic, V. 2013. Monitoring of concrete structures by using 14 MeV tagged neutron beams. *Radiat. Meas.* 59: 193–200.

Sudac, D., Pesente, S., Nebbia, G., Viesti, G., and Valkovic, V. 2007. Identification of materials hidden inside a container by using the 14 MeV tagged neutron beam. *Nucl. Instrum. Methods Phys. Res. B* 261: 321–325.

Sudac, D. and Valkovic, V. 2006. Vehicle and container control for the presence of a «dirty bomb». *Proc. SPIE* 6204: 620402, 620410.

Sudac, D., Valkovic, V., Nad, K., and Obhodas, J. 2011. The underwater detection of TNT explosive. *IEEE Trans. Nucl. Sci.* 58 (2): 547–551.

Ussery, L. E. and Hollas, C. L. 1994. Design and development of the associated-particle three-dimensional imaging technique. Los Alamos National Laboratory, Los Alamos, NM, Report LA-12847-MS.

Valkovic, V. 2006. Applications of nuclear techniques relevant for civil security. *J. Phys. Conf. Ser.* 41: 81–100.

Valkovic, V. 2009. New/future approaches to explosive/chemicals detection, in *AIP Conference Proceedings of the 12th International Conference on Application of Accelerators in Research and Industry*, March 10, 2009, Vol. 1099, pp. 574–577, doi:10.1063/1.3120102. American Institute of Physics, College Park, Maryland.

Valkovic, V., Blagus, S., Sudac, D., Nad, K., and Matika, D. 2004. Inspection of shipping containers for threat materials. *Radiat. Phys. Chem.* 71: 897–898.

Vakhtin, D. N., Evsenin, A. V., Gorshkov, I. Yu., Kuznetsov, A. V., Osetrov, O. I., and Rodionova, E. E. 2009. Detection of explosives and other illicit materials by nanosecond neutron analysis, in Presentation at *International Topical Meeting Neutron Based Techniques for the Detection of Illicit Materials and Explosives*, Vienna, Austria, May 4–8, 2009.

Vakhtin, D. N., Gorshkov, I. Yu., Evsenin, A. V., Kuznetsov, A. V., and Osetrov, O. I. 2006. SENNA—Portable sensor for explosives detection based on nanosecond neutron analysis, in Schubert, H. and Kuznetsov, A. (Eds.), *Detection and Disposal of Improvised Explosives*, NATO Security through Science Series, Springer, Dordercht, the Netherlands, pp. 87–96.

Valkovic, V., Kollar, R., Sudac, D., Nad, K., and Obhodas, J. 2010. Inspecting the inside of the objects lying on the sea floor, in Harmon, R.S., Holloway, J.H., and Broach, J.T. (Eds.), *Proceedings of the SPIE Detection and Sensing of Mines, Explosive Objects and Obscured Targets XV*, Vol. 7664, pp. 78840U. doi:10.1117/12.849507. SPIE, Bellingham, Washington.

Valkovic, V., Matika, D., and Kollar, R. 2007. An underwater system for explosive detection. *Proc. SPIE* 6540: 654013 (On line publication date: May, 4. 2007).

Valkovic, V., Matika, D., Kollar, R., Obhodas, J., and Sudac, D. 2010. environmental security of the adriatic coastal sea floor. *IEEE Trans. Nucl. Sci.* 57(5): 2724–2731.

Valković, V., Sudac, D., Blagus, S., Nad, K., Obhodaš, J., and Vekić, B. 2007. Fast neutron inspection of sea containers for the presence of "dirtybomb". *Nucl. Instrum. Methods Phys. Res. B Beam Interact. Mater. Atoms* 263(1): 119–122.

Valkovic, V., Sudac, D., Kollar, R. et al. 2012. Inspection of the objects on the sea floor for the presence of explosives. *Proc. SPIE* 8357: 83571P. doi: 10.1117/12.918125.

Valković, V., Sudac, D., and Matika, D. 2006. Port security: Container cargo control. *Promet* 18(3): 235–244.

Valkovic, V., Sudac, D., and Matika, D. 2010. Fast neutron sensor for detection of explosives and chemical warfare agents. *Appl. Radiat. Isot.* 68: 888–892.

Valkovic, V., Sudac, D., Nad, K., Obhodas, J., Matika, D., and Kollar, R. 2010. "Surveyor": An underwater system for threat material detection, in *IAEA Proceeding Series CD*, STI/PUB/1441, reference number ISBN 978-92-0-152010-4. ISSN 1991-2374. Vienna, Austria.

Valkovic, V., Sudac, D., Obhodas, J., Matika, D., Kollar, R., Nad, K., and Orlic, Z. 2012. Inspection of the objects on the sea floor by using 14 MeV tagged neutrons. *IEEE Trans. Nucl. Sci.* 59: 1237–1244.

Valkovic, V., Sudac, D., Obhodas, J. et al. 2013. The use of alpha particle tagged neutrons for the inspection of objects on the sea floor for the presence of explosives. *Nucl. Instrum. Methods Phys. Res. Sect. A* 703: 133–137.

Viesti, G., Pesente, S., Nebbia, G., Lunardon, M., Sudac, D., and Nad, K. 2005. Detection of hidden explosives by using tagged neutron beams: Status and perspectives. *Nucl. Instrum. Methods Phys. Res. B* 241: 748–752.

Wielopolski, L., Hendrey, G., Johnsen, K. H. et al. 2008. Nondestructive system for analyzing carbon in the soil. *Soil Sci. Soc. Am. J.* 72: 1269–1277.

Wielopolski, L. and Mitra, S. 2010. Near-surface soil carbon detection for monitoring CO_2 seepage from a geological reservoir. *Environ. Earth Sci.* 60: 307–312.

Wielopolski, L Yanai, R. D., Levine, C. R., Mitra, S., and Vadeboncoeur, M. A. 2010. Rapid, non-destructive carbon analysis of forest soils using neutron-induced gamma-rays pectroscopy. *Forest Ecol. Manage.* 260(7): 1132–1137.

Zaitseva, L. and Steinhauser, F. 2004. Illicit trafficking of weapons-usable nuclear material: Facts and uncertainties. *Phys. Soc. Newslett.* 33(1): 5–8.

Chapter 8

Nuclear Reactions Logging

8.1 NUCLEAR WELL LOGGING

Well logging is the practice of measuring the properties of the geologic strata through which a well has been or is being drilled. A well log is the trace or record of the data from a downhole sensor tool plotted versus well depth. Its most common application is by the oil and gas industries that seek out recoverable hydrocarbon zones. For oil and gas production, companies would like to have several kinds of information about a geologic layer, such as the hydrocarbon content. To measure these properties, sources and sensors loaded into housings called sondes can be lowered into an existing borehole (a technique called wireline logging) or can be mounted on a collar behind the drilling head for taking measurements while the well is being drilled (called logging while drilling [LWD]). LWD records bottomhole data acquired incrementally from sensors located in the drill string near the bit in a drilling well. Recording LWD is done during drilling. Data can be transmitted to the surface in real time by pressure pulses through the mud inside the drill pipe. Large data files (e.g., waveforms) are temporarily stored in the memory of the tool for later recovery.

Nuclear techniques in underground mining are discussed by Przewlocki (1984) and many others (Edwards et al. 1967, IAEA 1974, Caldwell et al. 1974, Borsaru et al. 2004, 2006). Nuclear and nonnuclear well logging tools are used in concert with each other to obtain information about the geologic media through which a borehole has been drilled. There are five main nuclear well logging tools: the density porosity tool, using ^{137}Cs; the neutron porosity and elemental analysis tools, typically using americium–beryllium (Am-Be) radioisotope sources; and the neutron absorption and carbon/oxygen (C/O) tools, which use 14.1 MeV neutrons from d-t accelerators. The ^{137}Cs source in a density log is a vitrified Category 3 source. The Am-Be sources range from Category 3 to Category 2. It would not be difficult to replace the Am-Be source in elemental analysis logs with a d-t accelerator. Replacing the Am-Be porosity tools is more difficult, although Schlumberger does now market d-t accelerator nuclear porosity logging tools for LWD and logging drilled holes (wireline). A major reason why such logs have not been adopted widely is that well log analysis relies on a large body of data that has been accumulated for the porosity logs using Am-Be sources. These data would be less useful in analyzing the results from the 14.1 MeV neutrons from d-t accelerators. ^{252}Cf sources might also be used to replace Am-Be sources, but they also suffer (somewhat less) from a similar lack of supporting data. In addition, ^{252}Cf sources have a half-life of only about 2.45 years and would have to be replaced in about two half-lives. These replacement source approaches are presently being studied by Monte Carlo simulation and in-hole experiments and demonstrations.

Nuclear well logging is a method of studying the materials surrounding exploratory boreholes. A tool consisting of a neutron or gamma-ray source and one or more detectors is lowered into the borehole. The response of the detectors to radiation returning from outside the borehole depends in part on the lithology, porosity, and fluid characteristics of the material. In principle, the characteristics of the materials outside the borehole can be inferred from the response of the detectors (Hearst and Carlson 1970, Dupree 1988). A typical well logging configuration is shown in Figure 8.1, the sonde contains the radiation source and two gamma detectors, "near" and "far" detector. Well logging is widely used in the oil industry. The interpretation of the data is based on benchmark measurements in a known environment, comparison with data from other types of measurements, and computer simulation of the measurements with neutron and gamma-ray transport codes. For technology developments see (Lebreton et al. 1963, Caldwell 1969, Flanagan et al. 1991, Borsaru et al. 2006).

In wireline logging, sondes and supporting electronic cartridges are strung together and lowered into an uncased borehole on a cable that has an electronic signal wire. As the string is raised, the sensors measure some or all of the following properties as functions of the depth: electrical resistivity, electron density, sound velocity, neutron moderation, thermal-neutron absorption, natural and artificial (induced) radioactivity, gamma-ray spectra, Compton

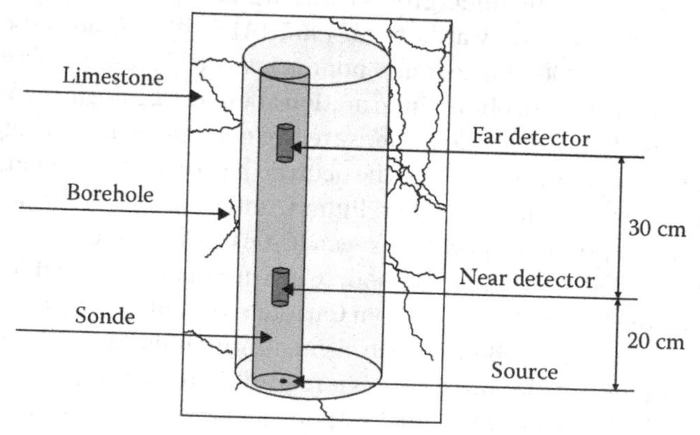

Figure 8.1 A typical well logging configuration.

scattering, borehole dimension, and, occasionally, nuclear magnetic resonance. The data are transmitted through the wire to computers at the surface where the data are logged. Similar measurements can be made in a cased borehole, although it is much more challenging to carry out the measurements through the steel casing. The effect of the bore hole on nuclear logging measurement is discussed by McGregor and Eisler 1983. Even with that difficulty, however, there is an increasing demand for logging of previously utilized production wells.

Recent developments in technology of neutron generator manufacturing have enabled the application of fast neutron activation analysis to a variety of problems. A major breakthrough was accomplished by using the detection of associated alpha particles in the neutron-producing reaction $d + t \rightarrow \alpha + n$. The cone of detected α-particles defines the cone of neutrons, "neutron beam," and if the neutron-produced γ-rays are detected in coincidence with α-particles, the volume where they are produced is defined by the overlap of "neutron beam" cone and γ-detector solid angle. Measurement of time between these two events gives the information on distance of investigated voxel from the neutron source.

Computer simulation of nuclear logging devices technology uses parallel deterministic numerical methods to solve the three-dimensional multigroup, neutron/radiation transport equation on massively parallel architectures, enabling high resolution simulations of nuclear well logging tools. One is performing high resolution computer simulations of nuclear well logging tools to aid in determining the lithology, porosity, and fluid characteristics of the surrounding materials near a borehole. The computer simulation is done by calculating the response of a tool to a given material using the linear Boltzmann equation and libraries of neutron and photon cross-section data widely available. The ability to calculate responses greatly enhances the ability to design optimum tools and interpret their response, because different tool designs or material combinations can be investigated with calculations rather than with expensive experiments, see (Jacobson et al. 1990, Tittle 1992, Patchett and Wiley 1994).

In practice the computational approach has been proven difficult for the following reasons:

- The problem is intrinsically three dimensional, which means that all six dimensions of phase space appear as independent variables.
- Some tools depend on a time-dependent response, which adds another variable.
- The problem is inherently "thick," because the particles (neutrons or gamma rays) must travel into the material, interact, and travel back to one of the detectors.
- The geometry is awkward, because the critical dimensions of the tool are small, but the interaction with the material takes place over quite a large volume.

The current practice is to use the Monte Carlo method for these calculations. In Monte Carlo, one takes a random sample of the phase space. This method often makes it possible to get answers to problems that are too complex to solve by more conventional methods. Because a random process is used, all results come with some statistical uncertainty. Different answers are obtained for the same problem simply by changing the sequence of random numbers. This may be particularly troublesome in problems where the desired answer is the ratio of responses of two detectors, each with its own statistical error. The additional statistical error makes it difficult to compare responses from two similar materials, because the statistical error may be comparable in magnitude to the material effects. This type of error decreases only as one divided by the square root of H, where H is the number of histories. Thus, the answer converges slowly as more computer power is applied.

Deterministic methods for solving complex transport problems were used at Lawrence Livermore National Laboratory (LLNL) for a number of years. Some of the codes they used in applied physics studies are among the fastest and most efficient in the world. In particular, they have used deterministic codes with great success in solving three-dimensional physics problems previously considered tractable only by Monte Carlo methods. Also, they have developed methods for using massively parallel computers to solve transport algorithms quickly and efficiently. Problem Solution is done by the use a finite element solution of the transport equation. The method developed at LLNL allows us to represent complicated geometry with a grid optimized for transport purposes. The density profiles of the different materials in the problem can be integrated directly with the functions used to represent the neutron or gamma fluxes. The acquired experience indicates that accurate solutions for complex geometry are obtained in this manner, even in three dimensions, using a reasonable number of zones. Brown (1996) has developed methods for solving this class of equations efficiently on modern computers. He also uses special methods for treating the relatively small sources and detectors in these problems. These methods include a special calculation of the first flights of the particles, as well as techniques for mitigating the persistent "ray" effects typically present in deterministic solutions of the Boltzmann equation due to finite spatial and angular zoning (further information in Brown 1996).

In the paper by Bliven and Nikitin (2010), challenges of nuclear well logging related to the lack of the gamma ray and neutron detection technologies capable of operating reliably in the downhole hostile environment are discussed. Different nuclear well logging techniques are described, the parameters of nuclear detectors that would satisfy the requirements imposed by these techniques are considered, and the results of the analysis in the well logging industry's demand for different types of nuclear detectors is presented. The impact of several technological innovations introduced recently in nuclear well logging, including detectors based on $LaBr_3$:Ce and GYSO scintillation materials, are discussed. The possibility of replacing ^3He ionization detectors with ^6Li glass scintillation detectors was considered. Also some of the reliability problems of well logging nuclear detectors were analyzed and their possible solutions are discussed (Bliven and Nikitin 2010).

Whereas, in borehole logging, variations in the geometrical configuration (the source-rock-detector system) mainly relate to the diameter of the borehole and the probe and the position of the probe within the borehole, in online measurements, large variations in source-detector separation and in the mass and profile of the mineral are possible. An experimental approach to parameterize this situation is extremely difficult and tedious, so that resort to calculation is strongly preferred, if this is possible.

Examples of spring-loaded inline centralizers used for centralization of the logging string in the well are shown in Figure 8.2.

From what has been stated above, it is apparent that very significant amount of data are required for a very large number of elements. The neutron cross-sections relevant to deterministic and Monte Carlo calculations in nuclear geophysics are as follows:

1. Total crosssection: σ (E)
2. Crosssections for elastic scattering:
   a. Total crosssections: $\sigma_n(E)$
   b. Angular distributions of elastically scattered neutrons: $\sigma_n(E,\theta)$
3. Crosssections for inelastic scattering:
   a. Total crosssections: $\sigma_n'(E)$
   b. Energy distributions of the scattered neutrons: $\sigma_n'(E,E',\theta)$
4. Crosssections for neutron multiplicative processes:
   a. Fission crosssection: $\sigma_f(E)$
   b. Crosssections for the reaction (n,2n): $\sigma_{2n}(E)$, $\sigma_{2n}(E,E')$
5. Crosssections for processes in which the neutron disappears:
   a. Radiative capture: σ_γ
   b. Charged particle reactions, for example, (n,p), (n,d), (n,α)

Apart from microscopic cross-section data, there is a variety of macroscopic data that must be evaluated in order to make progress with the more effective use of nuclear methods in the geosciences. In this category we can include the use of neutron migration lengths as link parameters in the design of porosity probes in oilwell logging and the use of λ_0 values (λ_0 is the total epithermal neutron flux per unit lethargy interval per unit thermal flux) to describe the shape of the neutron spectrum in different types of rock. The initial presentation of evaluated microscopic nuclear data for application in nuclear geophysics is presented in IAEA (1993), see also (Oleson 1990, Schneider et al. 1994).

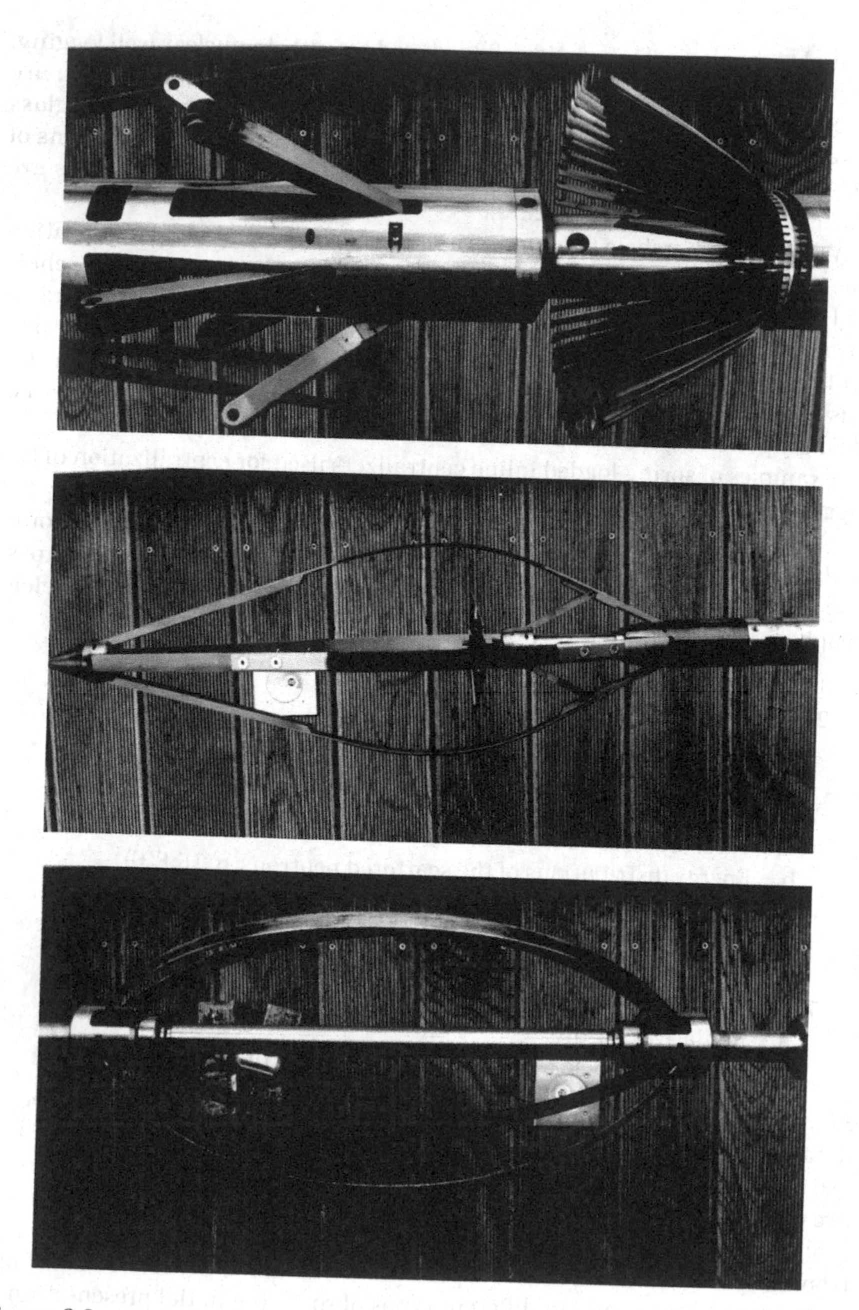

Figure 8.2 Examples of spring-loaded inline centralizers used for centralization of the logging string in the well.

Well integrity measurements will be the most required measurements in the future. These measurements will be done not only by nuclear techniques but also by using magnetic thickness measurements (ODT, EMDS), cement evaluation by RBT tool with the help of a weel scatch manager software. There are test wells in several locations since the calibration of the tools is an important factor difficult to evaluate.

8.1.1 Nuclear Well Logging in Hydrology

The optimum development of groundwater resources requires a quantitative evaluation of local aquifers and of the physical and chemical properties relevant to the recharge and to the withdrawal of water from them. Measurements of many hydrological and geological parameters can be made in situ by nuclear geophysical well-logging methods (IAEA 1971). Nuclear well-logging techniques have for some decades been capable of solving a number of subsurface measurements problems faced by hydrogeologists. However, the methods of drilling commonly used for hydrogeological investigations restrict the accuracy and reliability of the subsurface information from the strata penetrated. The purpose of well logging is to obtain information about the three broad categories of hydrogeological parameters: (1) characteristics of the strata, (2) characteristics and movements of fluids, and (3) engineering conditions of the well.

Information about the characteristics of the strata includes the following:

- Estimation of total porosity, bulk density, specific yield, effective porosity, and moisture content
- Identification of lithology and stratigraphic correlation
- Estimation of relative permeability, recognition of aqulfer type

Information on the characteristics and movements of fluids includes the following:

- Location of water levels
- Recognition of levels and rates of entry or exit of water into or out of the well
- Measurement of the resultant rate and direction of movement of the water column
- Identifying the physicochemical characteristics of the water in the well and observing changes in quality over time

Information on the engineering conditions of the well includes the following:

- Measurement of well diameter
- Location of casing and casing failures
- Determination of the optimum construction

For more details see IAEA (1971), Morin and Hess (1991).

In most hydrological investigations, the value of nuclear logging can be increased considerably by its use in combination with other nonnuclear logging methods. As an example, let us mention the work by Paillet (1988) describing full suite of geophysical logs including nuclear, electric, acoustic transit-time, acoustic waveform, and acoustic televiewer logs, and high-resolution flowmeter measurements have been used to investigate the lithologic and hydrologic properties of three igneous plutons located on the southern margin of the Canadian shield. Geophysical logs were used to identify lithologic boundaries, determine the properties of unfractured granitic or gabbroic rocks, interpret and calibrate the results of surface geophysical surveys, and characterize permeable fracture zones that could serve as conduits for fluid migration. Nuclear and acoustic transit-time logs provided good quantitative correlation with changes in lithology. Electric logs yielded consistent qualitative correlations, with lower resistivities associated with more mafic lithologies. Lithologic contacts identified on logs generally confirmed the results of surface electromagnetic, seismic, and gravity surveys. All major fracture zones intersected by boreholes were clearly indicated by the geophysical logs. Electric, epithermal-neutron, and acoustic transit-time logs gave the most consistent indications of fracturing, but the lithologic responses associated with some thin mafic intrusions were difficult to distinguish from possible fractures, and some steeplydipping fractures were not indicated by conventional acoustic transit-time logs. Electric and neutron log response is attributed to the effect of day mineral alteration products in the vicinity of fractures. This alteration may be indirectly related to permeability, but no direct relationship between resistivity or neutron attenuation and permeability appears to exist. Tube-wave attenuation determined from acoustic waveform logs was related to the transmissivity of equivalent infinite, plane fractures; these results agree qualitatively and possibly quantitatively with packer isolation and injection tests if the combined effects of differing scales of investigation and borehole enlargements in fracture zones are taken into account. Tube-wave attenuation in waveform logs also compares well with the permeability distributions determined from tube-wave generation in vertical seismic profiles.

Comparison of conventional geophysical logs, acoustic televiewer images of the borehole wall, and fracture frequency distributions measured on core samples indicates that many fractures are completely sealed and have no effect on log response, whereas many more apparently sealed fractures have been slightly opened during drilling and do provide some log response. High resolution flowmeter measurements of natural flow in boreholes and comparison of packer isolation tests with log data indicate that a relatively few individual fractures often provide a large proportion of fracture zone transmissivity in the immediate vicinity of the borehole, and that the orientation of these fractures may not coincide with fracture zone orientation. These results indicate that the scale problem in relating borehole logs to regional configuration of fracture flow systems may be the most important consideration in the application of geophysical well logging to the characterization of ground water flow in crystalline rock bodies (Paillet 1988).

Most logs provide information related to lithology and can be used for stratigraphic correlation; however, the response of many logs is different above and below the water table. Certain resistivity and fluid logs may be used to locate and identify contaminants in the materials surrounding monitoring wells. Fluid logs can be used to obtain data on borehole flow and the distribution of permeability. Fractures and other secondary permeability features may be identified and characterized by certain acoustic and borehole imaging logs. Another important use of logs is to provide baseline data so that changes with time can be measured.

Nuclear logs provide information through steel casing and annular materials. Induction logs can add useful information to this suite in nonconductive casing. A wide variety of logs can be run in uncased boreholes but the usefulness of each may be limited by borehole diameter, type of rock or soil penetrated, and salinity of borehole and interstitial fluids (Keys et al. 1989).

Borehole geophysics is frequently applied to environmental investigations for such objectives as geohydrology to aid site selection, monitoring, determining well construction, and planning remediation. In planning a logging program for environmental applications, one of the most difficult questions to answer is what geophysical logs will provide the most information for the funds available. *A Practical Guide to Borehole Geophysics in Environmental Investigations* (Keys 1996) explains the basic principles of the many tools and techniques used in borehole logging projects. In this book, applications are presented in terms of broad project objectives, providing a hands-on guide to geophysical logging programs, including specific examples of how to obtain and interpret data that meet particular hydrogeologic objectives. For more on this subject, see the report produced by the United States Department of Transportation (Wightman et al. 2003).

Neutron-activation logging has potential for application to groundwater quality problems, because this technique permits the remote identification of elements present in the borehole and adjacent rocks under a wide variety of borehole conditions. Neutron activation produces radioisotopes from stable isotopes; the parent or stable isotope may be identified by the energy of the gamma radiation emitted and its half-life, using a gamma spectral probe (Keys and Boulogne 1969).

8.1.2 Nuclear Mineral Logging

A compendium of nuclear data to be used for neutron borehole logging and neutron activation analysis of mineral samples, meeting the major requirements of the nuclear geophysics community for microscopic crosssection and decay data has been published by IAEA (1993). The geological interpretation of well logs has been discussed, among others, by Rider (1996).

In the report by Harvey and Lovell (1992) the inversion of nuclear-derived chemistry into quantitative mineralogy is considered. The techniques, problems, and philosophy of the inversion process are discussed. The roles of mineral choice and individual mineral composition are shown to be important in estimating an accurate modal mineralogy. Compositional colinearity in which three or more of the phases sought lie on, or close to, the same compositional plane and can produce unstable mathematical solutions. The absence of water and carbon dioxide data together with the inability to measure Na and Mg individually contribute additional difficulties to the inversion process. Despite these concerns, a geologically meaningful mineralogy may be produced, though adequate validation of a particular solution still remains.

There are many reports describing applications of prompt gamma neutron activation analysis (PGNAA) to in situ characterization of ore locations. For example, Borsaru et al. (2002) describe the determination of Fe, Si, and Al content in Iron Ore Blast Holes, in situ determination of sulfur in coal seams and overburden rock (2004), and the automated lithology prediction from PGNAA and other geophysical logs (2006).

Comparison of spectrometric neutron-gamma and gamma-gamma techniques for in situ assaying for iron grade in large diameter production holes has been discussed by Charbucinski (1993). The same group has reported prompt gamma neutron activation analysis method and instrumentation for copper grade estimation in large diameter blast holes (Charbucinski et al. 2003, 2004) and for reserves estimation and mine planning at open-cut coal mines (Charbucinski and Nichols 2003). In the series of reports Smith and Barry (2005a–c) have described the applications of spectrometric borehole logging for exploration and mining in various South African mines.

In the paper by Trofimczyk et al. (2009), the PGNAA probe, called SIROLOG, a spectrometric borehole logging tool developed by the CSIRO in Australia is described. The principle of the technique is the measurement of the characteristic intensity and energy of gamma rays that result from nuclear interactions when fast neutrons bombard the formation. The intensity and energy of the emitted γ rays is related to the elemental composition of the rock, thus the volumetric occurrence of the elements in a bulk rock mass intersected by a borehole, can be directly estimated.

Downhole nuclear logging has certain benefits over conventional core and drill-chip assaying. A much larger volume of the material surrounding the borehole is sampled than the sample taken for laboratory analysis, thus providing better sampling statistics. Provided a suitable calibration exists, data processing can be automated for quick turnaround of results (near real-time processing). Since a continuous log is measured in situ, the data is not limited by zones of core loss or breakage or potential sample contamination.

The SIROLOG PGNAA probe was demonstrated on four Anglo-American mines in 2004. At Namakwa sands heavy mineral sands mine in South Africa, the technique indicated the potential to deliver quantitative estimates of whole rock geochemistry, including heavy mineral concentrations. At Sishen iron ore mine in South Africa, quantitative in situ determination of Fe (iron) grade was demonstrated. This is in line with published results from studies done at other iron ore mines. Encouraging results were obtained for phosphorous, one of the important penalty elements, although low concentrations and a limited range of grade values in the small data sample affected the outcome. At Skorpion mine in Namibia, a good estimate of Zn grade, the primary ore mined, was obtained in blast boreholes; however, the penalty element, calcium, could not be accurately predicted due to the low abundance of this element intersected by the test boreholes. Similarly, low abundance of Cu returned poor calibrations in a study on waste dumps at Mantos Blancos mine in Chile. The results demonstrate that the PGNAA technique can quantitatively predict the abundance of certain elements in situ, although site-specific calibrations are required for best results. However, calibrations are difficult to establish where the abundance and the range of grade values of the target element are low, or the typical signature of the target element overlaps that of another element. The development of higher resolution $LaBr_3$ γ-ray detectors can potentially alleviate these limitations. Another drawback of downhole nuclear logging is the hazard associated with the use of chemical radioactive sources. Alternative, safe sources, such as neutron activation systems are preferred, after Trofimczyk et al. (2009).

Application of gamma-ray spectral analysis to subsurface mineral exploration and sensitivity of the method for in situ borehole neutron activation for the noble elements has been discussed by Senftle (1980) and Senftle et al. (1984).

8.1.3 Nuclear Well Logging in Oil Industry

It was Conrad Schlumberger who in 1912 gave the idea of using electrical measurements to map subsurface rock bodies. The gamma ray and neutron log were first applied in 1941.

In 1959 Mott and Ediger (1959) wrote that during the past 20 years nuclear well logging has grown to become one of the most important peaceful applications of the principles of nuclear physics. Nuclear logs are used to resolve a variety of exploration and production problems in both cased and uncased wells and in many areas have provided information not obtainable by other means. Nevertheless, developments have been slow, and as a consequence nuclear logging must still be regarded as being in its infancy. Many of the most promising methods have yet to be tried out experimentally in a borehole, and methods that have been in use for many years are continuously undergoing improvements. The paper by Mott and Ediger (1959) reviews the development and principal applications of nuclear well logs at that time. An aim is to evaluate critically the status of nuclear well logging in general and because of its fundamental aspects, gamma-ray logging in particular (see also Hilchie 1977).

Later, an overview is provided (Kerr and Worthington 1988) of the role of nuclear-based well-logging techniques, principally in the petroleum industry. The nuclear processes that provide the physical basis for logging tools were summarized together with information on the fluid/rock system that these interactions provide. Log interpretation strategy was briefly traced from its beginnings in the late 1920s in order to demonstrate the motivation behind these measurements.

Nuclear well logging is a well-established method of studying the materials surrounding exploratory boreholes. A tool consisting of a neutron or gamma-ray source and one or more detectors is lowered into the borehole. The response of the detectors to radiation returning from outside the borehole depends in part on the lithology, porosity, and fluid characteristics of the material. In principle, the characteristics of the materials outside the borehole can be inferred from the response of the detectors. Combination formation density and neutron porosity measurements while drilling is reported by Wraight et al. (1989).

The elements of interest for inelastic gate of the inelastic mode are carbon, oxygen, silicon, calcium, sulfur, and iron. H, Si, Ca, Cl, Fe, and S yields are obtained in the capture mode and from capture gate of the inelastic mode. The most fundamental reservoir parameters—oil, gas, and water content—are critical factors in determining how each oilfield should be developed. The carbon-to-oxygen ratio contains valuable information concerning the location of oil and/or gas bearing rock.

Nuclear well logging is done also by using radioactive sources, the most common being RaBe and ^{252}Cf. The Am-Be source has 480 years half-life and produces cca 2.5×10^6 n/s per Ci of 4.18MeV average neutron energy. The ^{252}Cf source has 2.7 years half-life and produces cca 4.1×10^9 n/s per Ci of 2.2 MeV average neutron energy. The characteristics of these two sources are as follows:

1. Am-Be—Plus: long half-life (flux stability); most of the tools are characterized for Am-Be. Minus: relatively expensive; uncertainty of the supply
2. ^{252}Cf—Plus: lower price than Am-Be; predictable supply. Minus: short Lifetime (only 2.7 years), most of the tools are not characterized for ^{252}Cf

The obvious alternative to the use of radioactive sources is the use of neutron generators, either d + d or d + t. There are at least eight reasons why to replace radioactive sources with neutron generators (Knapp 2013), which includes required paperwork for obtaining the source, high price and long lead time for the chemical sources, transportation safety (lost source, accidents, terrorism), transportation paperwork, personnel and exposure safety, lost in hole safety, new upcoming legal regulation, and added value.

The arguments for the replacement are following:

- Neutron generators are not neutron emitters while not active.
- d + t generators are β radioactive, but the activity is below the safety threshold. They fall into category UN2910/2911 (exempt activity) and there are no special measures for transportation required.
- As a potential dual-use product, licenses from the Ministry of Commerce are required for international transportation.
- Transportation paperwork for the chemical sources is required for each trip to the well.
- Law regulations are getting stricter every day.
- New regulation will require special vehicles with rotating light for the sources above 15Ci.
- This 15Ci limit will probably go lower in the near future.
- There are about 9000 registered neutron sources in Texas only. About 480 sources were lost in hole in the period from 1956–2006 (Knapp 2013). Based on this, the probability to lose the source in hole is about 0.1% per year. The costs connected to this and the environmental impact of this risk has to be calculated into the price of the logging jobs with neutron sources.

The radiation risks to nuclear well loggers are discussed by Fujimoto et al. (1985). There are several well-known manufacturers of oil logging equipment. We shall here mention only some less known to us and newcomers to the game.

GOWell Petroleum Equipment Co. Ltd. has a product called the Neutron Lifetime Logging Tool (SMJ-E) that contains a pulsed neutron generator producing bursts of high-energy neutrons, two radiation detectors to detect radiation resulting from nuclear interactions between the neutrons emitted from the generator and nuclei comprising the materials in the borehole region and the formation, circuitry to record the detection times relative to a time reference related to the neutron production bursts, downhole and uphole circuitry to control operation of the tool, and surface system process the spectrum of detection times (die-away or decay spectrum) to estimate earth formation properties useful in the evaluation of oil and gas reservoirs. The logging tool is passed through a borehole (uncased or cased and with or without tubing depending on the application), measurements are made as a function of depth, and a log of the results of the data processing is recorded as a function of depth. The detector system is designed to detect gamma rays resulting from the capture of thermal neutrons produced as a result of slowing down and thermalization of the high-energy source neutrons or the thermal neutrons themselves.

Since fresh water and crude oil have the same capture ability for neutrons, for fresh water oil field, sigma (Σ) logging is not able to identify oil or water layer. Fortunately, there is sodium chloride in the formation water. The chlorine atom has stronger ability to capture neutron than carbon, hydrogen and oxygen atoms in the fluid (water, oil) of the formation. Then the total cross-sectional area of water layer is bigger than the cross-sectional area of the reservoir.

TABLE 8.1 SPECIFICATION OF NEUTRON LIFETIME LOGGING TOOL

Dimensions and ratings	
Max. temperature	155°C
Max. pressure	100 Mpa
Tool OD	45 mm
Tool length	4.86 m/5160 mm
Measurement characteristics	
Neutron beam intensity	1.5×10^8 n/s
Output	Σ_F(FSIG), Σ_N(FSIG), N, F, Σ_F/Σ_N, τ_F, τ_N, $N_{1\Sigma}$, $F_{1\Sigma}$, RNFC, GR, and CCL
	Main parameters: Σ_F, N, F, RNFC, GR, and CCL
Σ Measurement range	7.6–91 c.u.
Measurement accuracy	For Σ_F and Σ_N in three standard water wells noncontinuously logging: Relative standard deviation: ≤3% Relative system error: ≤±3%
Vertical resolution	50 cm
Logging speed	360 m/h

If the salinity is higher and porosity is bigger, the difference will more obvious. For high salinity oilfield, there are four main applications:

- As conventional logging, according to recorded Σ curve, it can directly identify oil, water, and gas. With other method and date, it is able to calculate out the saturation of water (Sw).
- With "logging–water injection–logging" method, it can accurately achieve current oil saturation (So).
- With "time decay" logging, it can identify the boundary of the water, oil, and gas displacement and estimate fluctuations of Sw.
- With "water injection-logging" residual oil saturation (Sor) can be calculated. This method is not limited by the salinity of oilfields.

The tool specifications are shown in Table 8.1.

8.2 LOGGING WITH PULSED NEUTRON BEAM

The development of a pulsed neutron method of borehole testing has been already discussed in 1960s and 1970s; see, for example, Bespalov et al. (1971). Pulsed neutron–neutron logging is a method that can determine water saturation by means of the formation macroscopic absorption cross section according to thermal neutron time spectra by using ^{3}He-detector. In the paper by Zhang et al. (2007), the thermal neutron time spectra under the conditions of different formation water salinity, porosity, saturation, and borehole were simulated by using the Monte Carlo method. The relationship of formation macroscopic cross section and water salinity was studied. It is concluded that the suitable formation water salinity of the pulsed neutron neutron (PNN) logging is about 10–100 g/L, and the suitable porosity minimum is about 10% when the formation water salinity measures up to 50 g/L in theory. The formation macroscopic absorption cross section was less affected by borehole fluid although thermal neutron count rate is different. The porosity can be determined by using the thermal neutron count ratio of two different spacing detectors. The evaluation method of matrix and water saturation is put forward according to the formation macroscopic absorption cross section versus porosity under the condition of different lithology and saturation, and then the oil, water and gas reservoir can be identified by the PNN logging. As a whole the PNN logging method is preferable to the low salinity and porosity over the other ways to determine the remaining oil saturation (Zhang et al. 2007).

We shall here describe a PNN system as designed by Hotwell being a new approach to saturation evaluation as a part of reservoir monitoring in oil and gas wells. The principle is relatively old, well known, and utilized by many different companies and in many different versions of downhole tools and surface data acquisition systems. Each of the earlier developed systems has its own characteristics with some strong and some weak parts of the system. There is some degree of inconsistency in the logging data processing, analysis, and final presentation that all are not compatible to the internationally recognized standards used in oil and gas producing industry.

Another company and its activities we shall describe in some details is Hotwell, Austria. Their goal was to develop a new approach to saturation measurement that will utilize most of the advantages of thermal neutron recording instead of gamma ray as a product of thermal neutron capturing, together with a sophisticated and powerful data processing and data analysis. This new PNN system has been developed as portable and easy for field application on any kind of well logging unit with the downhole tool based on well-known VNIIA d-t fast neutron generator ING-101, which is fairly simple and extremely easy for field use and maintenance and with new data acquisition, processing, and analysis software. The pulsed neutron tool is a thermal neutron decay type of tool that is detecting thermal neutrons. Its components are communication section with CCL, natural gamma ray section, neutron detectors section, and neutron generator section (see Figure 8.3). When using this tool, the formation is bombarded with pulses of high energy neutrons (14 MeV) generated by the tool. Neutrons interact with the surrounding atoms and, during the lapse of time between high-energy neutron pulses, the thermal neutron population that reaches the neutron detectors is sampled by two detectors with 60 time channels each. Per channel neutrons counted are used to compute the rate of decay—it is equivalent to measuring the rate at which thermal neutrons are absorbed into the formation. The greater fluids and minerals capture thermal neutrons, the higher the value of sigma.

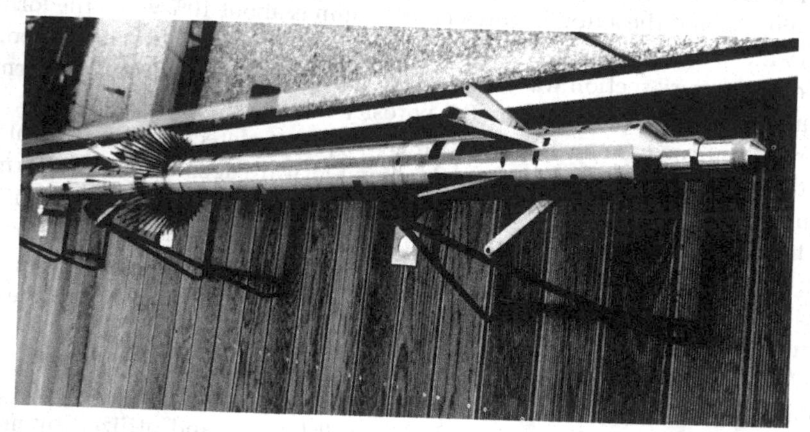

Figure 8.3 Hotwell PNN probe.

The key part of the system is data processing software package that is human driven.

This means that the processing software parameters should be set by the log analyst. The software feature, which at the first look appears as disadvantage and the whole software package weak point, in fact is one of the key strengths of the system. Each particular case can be processed with the most suitable parameters and with care about the details because the data processing is controlled and navigated by log analyst and just the calculations executed by computer. Initially, the PNN system has been designed to be affordable to smaller independent logging companies, to be portable, and to be used on any logging unit regardless of the existing data acquisition system built into the unit.

The PNN system consists of three main parts:

- Downhole tool string.
- Surface data acquisition system.
- Data processing and data interpretation software package.

Downhole tool and surface data acquisition system are used to run the log and to collect all raw data. The system is designed to send all raw data to the surface where the raw data are stored in the file format compatible to data processing software package written by Hotwell. The concept of sending all raw data to the surface allows data processing with different parameters to be repeated as many times as necessary. This concept has been proven as one of big advantages of PNN system introduced by Hotwell over the similar competitive systems which don't have this ability due to some calculations done downhole. Besides the possibility of using raw data unlimited number of times for unlimited number of processing parameters set up, this concept provides an extra tool to Hotwell geologists and this is a possibility to treat each field, or even each well with the maximum care taken to the geological specifics of the area.

Most of the well-known saturation systems take their origin in big worldwide well-known logging companies that have database portfolio containing thousands of well data from different geological conditions that gives them the ability to set up the system for particular area. Hotwell did not have it at the very beginning and individual approach to each particular well was necessary to achieve the best possible result. The data processing and data analysis system as it has been introduced by Hotwell, and still exists as a powerful PNN data handling tool, allowed Hotwell geologists to cope with the most difficult cases of saturation logging in the extreme conditions as low salinity, low porosity, or complicated lithology.

The other big PNN system advantage confirmed in practice is recording thermal neutrons instead of recording gamma rays as the result of the thermal neutron capturing what most of the competitive systems do. Detailed PNN tool and the whole system description can be found in literature released by

Hotwell through different conference papers or in technical description of the system; see http://www.hotwell.at. When they decided to develop the system, a few problems were in front of them. Following the idea as to how to collect and analyze the data from the tool it was obvious that low frequency neutron tube, Minitron, is more appropriate than the high frequency tube that is used in most of the competitive systems. Neutron generator manufactured by VNIIA was the only neutron generator available to Hotwell at that time. The first tests confirmed the expectations and the data processing principles, but still a lot of practical problems were on the way to build a reliable tool for practical field use in oil and gas well logging.

One of the key problems was temperature rating of only 75° (167°F), which was far from enough for serious appearance in real well logging field work. The first tool designed was successfully tested in a shallow well with a downhole temperature of 60°C (140°F), and this was the final confirmation of the principle used for both measurement and data processing but not enough to appear in the market.

The whole neutron generator consists of the two units: neutron emission unit, the so-called Monoblock, and power supply and control unit. Further work showed that more critical was the power supply and control unit, which had two problems: too low operating temperature rate and too short life time if powered at or close to maximum operating temperature than neutron emission unit. Power supply basically is a dc/dc converter that converts 150 V from logging line to 3 kV needed to supply neutron emission unit, and this part was not critical for design. The most critical part of the power supply and control unit is the switch that has to discharge 150 nF capacitor charged to 3 kV in less than 1 μs across primary coils of high voltage transformer inside the Monoblock. As gas-filled tube used in original power supply unit has been confirmed as not reliable for well logging application, a new solution with IGBT was tested and found more reliable. Soon after the first tests were performed, new power supply and control unit has been successfully tested at 125°C, which was the temperature limit of Monoblock at that time.

In parallel to the work on new power supply unit, completion work on the Monoblock temperature upgrade has been accomplished by VNIIA. In the beginning of 2002, the 150°C (300°F) downhole tool has been released for field use, and since that time it, with minor changes, is successfully used worldwide. One of the Monoblock limitations is the neutron tube's (Minitron's) relatively short life time. To minimize this problem the whole PNN system is designed to fire the generator at lower firing rate to save the tritium target in the Minitron tube and extend the life time of the tube. The firing frequency of 13 firings per second has been found as an optimum between the neutron output and the neutron generator life time.

Intensive work on the new software package for the data processing and interpretation has been started in parallel to the work on hardware. This was an ambitious project that finally resulted with several independent software modules of unique approach to saturation log interpretation and the start of Hotwell Log Analysis Center (LAC) that specialized in all kinds of well logging data processing and analysis.

Soon after the PNN system has been for the first time presented for real field work in 1998, the system started to be regularly used as new reliable saturation service. There were some difficulties in explaining the new approach to saturation logging and convincing the customers that the new system is capable to provide good result in formation water salinity environment as low as 2000–4000 ppm and porosity as low as 6%–7%. In addition, more important was an ongoing intensive work on the system operating temperature upgrade to 150°C (300°F). Still, since the first field introduction until the downhole tool has been accomplished as planned, many international references from China, Croatia, Egypt, Romania, and Serbia have been collected and the PNN system proved as a new reliable saturation system in low-salinity, low-porosity formation environment.

Shortly after the 150°C (300°F) tool was successfully tested, the PNN services in Gulf of Mexico and Venezuela have been started. Maracaibo lake region in Venezuela was not critical with respect to temperature because the downhole temperature is below 125°C (257°F), but it was an excellent reference to prove the system reliability in low salinity environment of 2000–4000 ppm and porosity of 10%–20%. At the same time, the well logging environment in Gulf of Mexico was different; downhole temperature was on regular base higher than 130°C (266°F), very often above 150°C (300°F) and close to 160°C (320°F) and sometimes with extremely high salinity borehole fluid, multiple casing well completion, relatively high formation salinity, and moderate porosity.

The PNN service has been expanded to East Venezuela where the downhole temperature is at the edge or, very often, slightly above the tool limit, and to the United States (Texas, Louisiana, Oklahoma, Montana, Colorado, and North Dakota), Canada (Alberta and Saskatchewan), China, Azerbaijan, Egypt, Syria, Iraq, Turkey, and many other locations worldwide. Hotwell LAC already has a database with more than 4000 wells analyzed from more than 20 countries. Besides, data processing and data analysis software packages are sold to many customers who are providing complete service including data processing and interpretation and the total number of wells logged with PNN system is not precisely known. More than 300 PNN tools have been produced until now with the production rate of 30–40 tools per year, which is stabilized during the past 5 years.

More than 4000 wells have been logged for more than 480 operating companies worldwide since the system has been put in field use. The service is provided by 65 well logging service companies that are either the customers of Hotwell who have their own PNN system or are providing service together with Hotwell where the service company is responsible for the field work and Hotwell for the data processing and the data analysis. The number of wells logged as given earlier is based on Hotwell LAC database, which contains all wells analyzed by LAC. There is an unknown number that can be estimated as 2000 additional wells logged by the tools owned by the companies that are customers of Hotwell that have their own data processing and data analysis software purchased from Hotwell.

All of this makes PNN system offered by Hotwell not only very competitive but also highly reliable well logging service that is already recognized by the long-term solid relationship with high number of satisfied customers worldwide. The PNN system is proven as reliable well logging system that is very simple for field use and easy for maintenance. There were some problems with 3 kV power supply unit at the very beginning of the field use of the tool and at temperatures close to the tool rating of 150°C (300°F). The key problem was high-voltage high-current fast switch. Some time was needed to find the appropriate design of IGBT firing circuit to make the switch stable a televated temperature. This still can be classified as a technology problem, and the main limitation is the need to upgrade the tool string to targeted 175°C (350°F). An intensive work is still going on to upgrade the switch and improve the tool temperature rating to 175°C (350°F), which is very often the real field demand.

Lifetime of the neutron emission unit, Monoblock, often appears as a field problem too. Fortunately the Monoblock neutron output decays with time and at the same time the data processing routine is not, up to a certain point, sensitive to countrate; so this problem is easier to cope with. But still, if logging is performed at the temperature close to the tool limit, then the Monoblock lifetime of less than 100 working hours is a matter of the customers' complains. There were some logging situations where the tool has been exposed to temperatures above 170°C (338°F). In many cases, the tool was working properly even at the temperature much above the rated temperature limit. In some cases it happened that the temperature compensation of the silicon oil dilatation into the Monoblock was overloaded, which caused the Monoplock unit oil leak. The problem was treated seriously by VNIIA, the Monoblock manufacturer, and Hotwell together in order to provide fully reliable tool.

The rest of the tool string encounter even less problems in everyday field use and can be classified as fully reliable with a failure rate below the average for this kind of equipment.

A lot of different experiences have been collected during the 15 years of ING-101 neutron generator usage in PNN system. Most of the experience accumulated confirms that well logging systems based on electronically controlled neutron generators and neutron measurement has huge potential for the future development. The accumulated experience should be transformed into new ideas in the future generation of the tool design. In general it can be said that the future tool design should tend to two different directions:

1. To build simple, reliable, and, for the independent service companies, affordable tool for saturation measurement in general and in low salinity and low porosity environment in particular
2. To build complex tool that will utilize energy spectrum analysis of natural gamma ray, inelastic gamma ray, gamma ray as result of thermal neutron capturing, thermal neutron, and epithermal neutron at different distances from the fast neutron source

Accomplishment of the first task is relatively simple and should be a logical continuation in PNN development and maintaining the system at the competitive level to the similar systems. After many years of experience and thousands of wells logged and analyzed, Hotwell is able to offer a serious upgrade of the PNN system that should result in more reliable and much shorter tool string and in the acquisition software, which will offer real time data processing in most of the standard conditions; just special cases that require the system fine tuning or detailed data analysis with complete reports will be still kept as the LAC office work.

The more complex tool as proposed requires a new approach to the tool design from the aspect of hardware and software design. The intention is to work on the new generation of neutron generator that should result in long life, high output stability, and small dimensions neutron generator. The future tool should utilize several new features that are not a part of any known competitive tool. Besides just saturation and fluid contact determination, the new system should be able to do an accurate element analysis to make the system applicable in all cases of oil and gas exploration but also in the different cases of mining application and especially in underwater and deep water mining, which is becoming more and more a matter of daily practice.

Of course, to build such a tool is not an easy task and is a matter of compromising in many details due to physical dimensions available for downhole tool, technology available at the time of the new tool development, and many other practical details. Starting with the new tool based on the ING-101 neutron generator, Hotwell engineers had a key goal—to provide well logging world with simple cheap and reliable saturation system that will be affordable to small and medium size independent well logging service companies. In parallel to the tool development, the system for data processing and data analysis has also been developed to fully support the customers with the complete saturation measurement and interpretation. The system has been successfully introduced into the field operation 15 years ago and is still in use without significant changes, proven as reliable and competitive (Novak 2013).

Low frequency neutron generator ING-101 manufactured by VNIIA is confirmed and proven as fully reliable and appropriate for the application. At the same time, many advantages of the neutron generator used are confirmed in daily practice, especially in difficult logging environment as low formation water salinity, low porosity, extremely high borehole fluid salinity, complicated well completion, and some other tough well logging conditions. The results achieved through PNN system experience are encouraging for Hotwell designing team to go forward with the new ideas and some new approach in the well logging systems design.

PNN measurement offers a good set of curves that helps to cover the lack of data at these wells and to offer full petrophysical interpretation. Standard PNN interpretation is based on capture cross section (Sigma) curve, which combined with independent petrophysical interpretation (porosity and lithology) gives a good present quantitative saturation evaluation. However, even in the cases when the good openhole interpretation is unavailable, PNN measurements in most of cases are giving other curves that enable performing the full petrophysical interpretation (porosity and lithology as well as saturation).The new interpretation technique as well as algorithm and program PNNQI (PNN quick interpretation) is developed in order to enable us full petrophysical interpretation solely based on PNN data (Markovic 2013).

Let us mention the reported specific investigation of well integrity using PNN with other methods (Köhler et al. 2013). The optimal production and injection capacity of natural gas storage sites in aquifers depends predominantly on the properties of the reservoir and the technical properties of the installed production wells. In order to maintain the performance of a well one must evaluate the well conditions continuously. Considering safety aspects and the long period of service of operation of wells at natural gas storage sites, a continuous control of the well integrity is required. Therefore, appropriate geophysical measurement methods have to be used.

The example described by Köhler et al. (2013) refers to the operation wells of the aquifer storage site Buchholz, Germany (near Berlin), which are producing a lot of water during gas withdrawal. The water influx also occurs at the beginning of the withdrawal phase (high reservoir pressure). The main objectives of the geophysical measurement performed were (1) to identify the source of the produced water and to check if the water comes from the next sandstone formation above the reservoir and (2) to get an indirect evidence for the tightness of the cementation above the reservoir. In the measurements performed, a combination of the three measurement systems PNN, SIPLOS, and AFL was applied: PNN (flowing and shut-in conditions) to evaluate the present contacts of gas and water in the borehole and in the reservoir with saturation interpretation;

SIPLOS measurement during shut-in and flowing conditions to evaluate production profile; AFL runs to check water inflow from shallower strata and to confirm the water flow from deeper water-saturated zones.

The results were the following. The AFL runs for the flow down between casing and tubing or possibly between casing and formation, confirmed that there is no flow in this zone. This result confirmed the tightness of the cementation indirectly. Based on the combination of the three measurement systems used, the water inflow was detected. By analyzing all of the logs measured it was clearly confirmed that the water was coming from the water-saturated zone of the reservoir itself. Additionally, the main gas production zones were also clearly identified and quantified with a full production profile. With this information, it is possible to adjust the well operation scheme in order to get an optimized performance of the well (Köhler et al. 2013).

8.3 CARBON–OXYGEN LOG

The inelastic elemental yields have been studied for years to measure oil saturation by determining a carbon-to-oxygen ratio (COR); see Schultz and Smith (1974), Lock and Hoyer (1974), Roscoe and Grau (1988), and Roscoe et al. (1991). Carbon–oxygen logging is used primarily to estimate oil saturation in cased-hole conditions when the formation water is fresh or of unknown salinity. The C/O ratio has proved to be a reliable indicator of hydrocarbons; it proved to be a reliable indicator of hydrocarbons in sandstone formations, independent of formation water salinity and, in most instances, independent of shaliness.

Carbon–oxygen tools typically use gamma ray spectroscopy measurements to directly sense the presence of carbon atoms in oil and oxygen atoms associated with water. The ratio of the carbon to the oxygen measurement, or C/O, allows for the evaluation of differences in water and oil saturations independent of formation water salinity. Historically, the C/O log has experienced limited use due to the poor accuracy and precision of the measurement available to provide valid answers. To achieve the sophisticated answers required by operators, service companies built C/O tools that used large diameter detectors and required excessively slow logging speeds. These large diameter tool designs could only be conveyed into wells where tubing was removed prior to logging.

The C/O ratio, effectively, can indicate the presence or absence of hydrocarbons. Since the formation rock also may contain carbon (e.g., calcium carbonate), a comparison of the relative amounts of silicon and calcium is used to differentiate between carbon in the fluid and in lithology. The silicon and calcium measurements are performed using two different methods. A Si/Ca ratio is measured using capture gamma rays, while a Ca/Si ratio is obtained using inelastic (prompt) gamma rays. Each mode has certain advantages; the carbon and oxygen measurements are done only in the inelastic mode. The ratios, as measured by the C/O instrument, are proportional to the actual atomic ratios but should be considered as indices for comparative purposes rather than as absolute atomic ratio values; see, for example, Lawrence (1981).

Let us mention one of the C/O tools available in the market. The (Halliburton 2014) RMT™i tool is a slimhole pulsed neutron logging system for monitoring and managing the production of hydrocarbon reserves. This unique through-tubing carbon/oxygen (C/O) system has two to three times higher measurement resolution than other systems. Its high-density bismuth germanium oxide (BGO) detectors let the RMTi tool achieve resolutions previously available only with larger-diameter C/O systems. The RMTi tool can even be conveyed into a well with tubing completions unlike larger diameter C/O systems that can only log through casing.

8.4 SMALL-SCALE VARIATIONS IN C/O RATIOS

The aim of the described research (Valkovic et al. 2013) is the development of a methodology for the measurement of small-scale variations in carbon–oxygen ratio Log. The rationale for this is the following.

The most fundamental reservoir parameters—oil, gas, and water content—are critical factors in determining how each oilfield should be developed. It is well-established that carbon to oxygen ratio log yields accurate and repeatable data that can be used to identify and monitor reserves depletion. Typical injection well schematic view from above, following Rowe (2012) is shown in Figure 8.4.

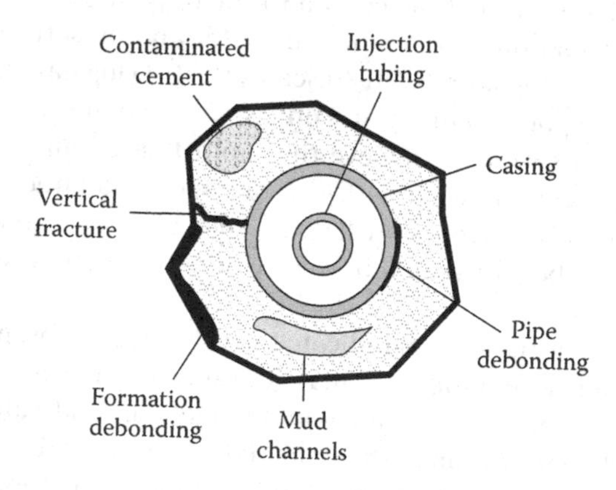

Figure 8.4 Typical injection well schematic view from above.

Different methods and tools are used to obtain information from the well lodging activities. The parameters of interest are depth of investigation and resolution of logging tools, where the depth is a distance from the inner tubing. The resistivity logging tool has a depth of investigation of 160 cm (laterolog) to 260 cm (induction log) and resolution of 80 cm in both cases. Within nuclear logging, one should differentiate between neutron log, gamma-ray log, and density log, the depth of investigation and resolution of logging tools being 70 and 40 cm, 55 and 30 cm, 40 and 20 cm, respectively. Sonic digital tool has a depth of only 10 cm and resolution of ~60 cm. Microresistivity logging tool can have a depth of 10 cm and a resolution of 5 cm, while dipmeter depth and resolution are of the order of 2 cm.

The evaluation of cement integrity is done with an ultrasonic transducer (0.2–0.7 MHz) using resonance technique to locate and give information on casing weld, gas microannulus, mud channels, perforations, well-centered casing washouts, and eccentered casing.

Nuclear well logging is a well-established method of studying the materials surrounding exploratory boreholes. A tool consisting of a neutron or gamma-ray source and one or more detectors is lowered into the borehole. The response of the detectors to radiation returning from outside the borehole depends in part on the lithology, porosity, and fluid characteristics of the material. In principle, the characteristics of the materials outside the borehole can be inferred from the response of the detectors. The elements of interest for inelastic gate of the inelastic mode are C, O, Si, Ca, S, and Fe. The yields from elements H, Si, Ca, Cl, Fe, and S are obtained in the capture mode and from capture gate of the inelastic mode.

Carbon–oxygen log is a log that presents a measure of the relative abundance of C–O derived from the detection of the gamma rays produced from both elements by the inelastic scattering of 14 MeV neutrons. The gamma rays are measured within energy spectrum windows representing the gamma-ray escape peaks of C and O. The ratio of counting rates provides a means of predicting the relative amounts of hydrocarbons and water. The log is an alternate means for detecting hydrocarbons (particularly oil) behind casing in formations not subject to flushing or reinvasion by borehole fluids. The C/O ratio is relatively independent of formation water salinity and shaliness. In order to differentiate the carbon in hydrocarbon molecules from that in the rock framework (i.e., carbonate solid matter), a Si/Ca ratio is also determined.

The COR contains valuable information concerning the location of oil and/or gas-bearing rock. These inelastic elemental yields have been studied for years to measure oil saturation by determining a COR. Carbon–oxygen logging is used primarily to estimate oil saturation in cased-hole conditions when the formation water is fresh or of unknown salinity.

The COR model is based on the fact that the ratio of C and O yields is a measure of the amount of oil around the tool. This ratio depends on many variables including oil saturation, borehole oil fraction, porosity, lithology, borehole diameter, casing diameter, and casing weight.

Although the ratio of carbon and oxygen yields is a measure of the amount of oil around the tool, it should be realized that a carbon signal can originate from several sources including the borehole, the cement behind the casing, the formation rock, and the formation fluid. In order to evaluate these contributions individually, we are proposing the modification of the neutron generator by the insertion of segmented associated alpha particle detector. From the measurement of time of flight spectra (alpha particle detector—start signal; gamma ray detector—stop signal), it will be possible to determine the location of gamma ray producing voxel and that way to determine radial variations in COR ratio.

The neutron sensor needs to operate inside the oil well casing pipes. The wall thickness of oil well casing pipes depends on their size, weight, and outside diameter (168–508 cm) and it is in the 5.9–16.1 mm range. The casing tube walls are of thicknesses that should not present an obstacle to such approach. The knowledge of iron shell thickness is mandatory for correct investigated material inside the container for the C/O concentration ratio determination from the measured gamma spectrum. This information could be obtained (1) from the knowledge of construction details of investigated object and precise position of the neutron system, and/or (2) from the iron peak in the measured gamma spectrum.

Recent improvements in neutron generator and gamma detector technologies resulted in small devices that allowed through-tubing measurements. Although the ratio of carbon and oxygen yields is a measure of the amount of oil around the tool, it should be realized that a carbon signal can originate from several sources including the borehole, the cement behind the casing, the formation rock, and the formation fluid.

$$C/O = A\,(C_M + C_{PS} + C_{BH})/(O_M + O_{PS} + O_{BH})$$

In order to evaluate these contributions individually, we are proposing the modification of the neutron generator by the insertion of segmented associated alpha particle detector. From the measurement of the time of flight spectra (alpha particle detector—start signal; gamma ray detector—stop signal), it will be possible to determine the location of gamma ray producing voxel and in such a way to determine radial variations in COR ratio.

The use of tagged 14 MeV neutrons allows the determination of the radial variation of stoichiometric ratios for the following chemical elements: carbon, oxygen, silicon, calcium, sulfur, and iron. The casing tube walls are of thickness that should not present an obstacle to such approach.

Schematic presentation of the production and use of the neutron beam "collimated" by the detection of the associated alpha particles is shown in Figure 7.1 presenting the physical principle of material inspection by tagged neutron beam; see also Perot et al. (2006) and Valkovic et al. (2013).

We have proposed the new tube housed in the 4.3 cm diameter housing, as shown in Figure 8.5. An associated alpha particle detector will be composed of 4 YAP crystal detectors with light guides and PM tubes. Accelerator-produced 100 keV deuteron beam will pass through the central tube and hit the tritium target. The shielding material behind the tritium target will prevent the produced neutrons to hit gamma detector and produce noise. Gamma detector, most probably BGO or NaI(Tl) crystal (to be determined during the test period),

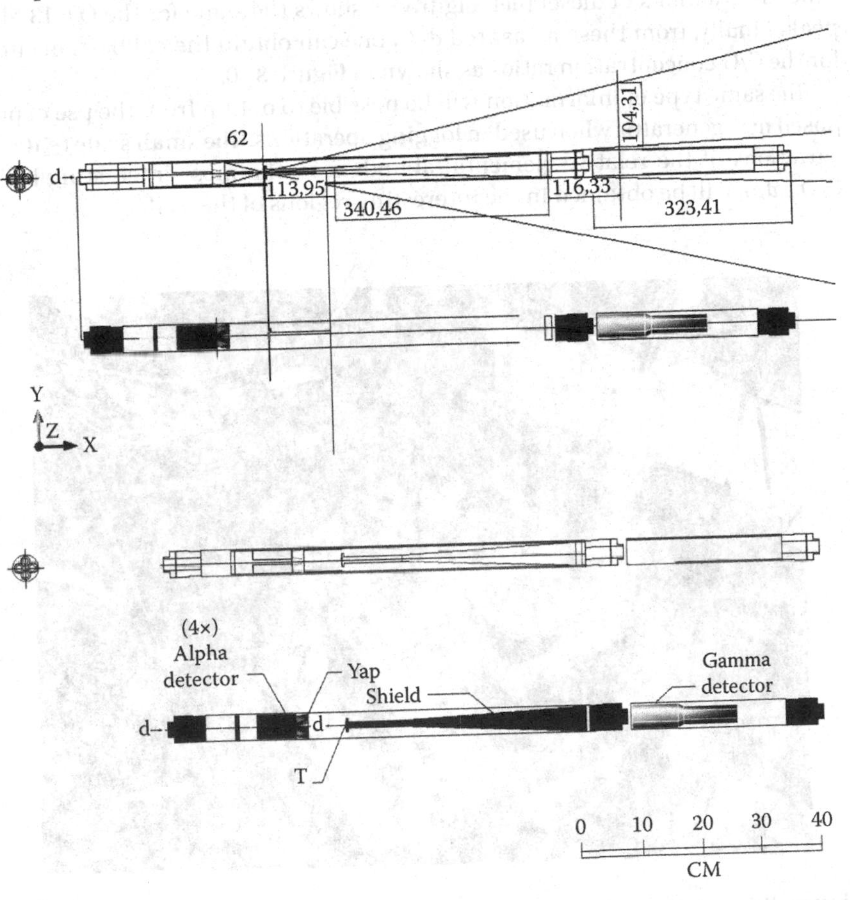

Figure 8.5 Proposed new tube for the investigation of small-scale element concentration variations by using the detection of associated alpha particles.

will be followed by a photomultiplier tube. Two parameters will be measured: alpha–gamma t-o-f spectrum and gamma energy spectra. With t-o-f resolution of 2 ns, approximately 10 cm voxels could be measured for relative elemental abundances, most interesting one being C/O ratio. Such a tube can be tested using 100 keV deuteron beam from the laboratory size Cockcroft–Walton accelerator.

The expected results from the use of the construction of such a neutron generator are indicated by the preliminary results obtained by laboratory size setup. The experimental setup used is shown in Figure 8.6, the measured gamma spectrum for 50%–50% mixture of water and diesel fuel is shown in Figure 8.7, while Figure 8.8 shows the variation of C 4.44 MeV peak intensity for different fractions of diesel fuel. Figure 8.9 shows the same for the O 6.13 MeV peak. Finally, from these measured data one can obtain the calibration curve for the C/O concentration ratios as shown in Figure 8.10.

The same type of information will be possible to obtain from the use of proposed new generator when used in logging operations. The small scale (≤10cm) variations of the relative elemental abundances, the most interesting being C/O ratio, will be obtained in the interesting regions of the well.

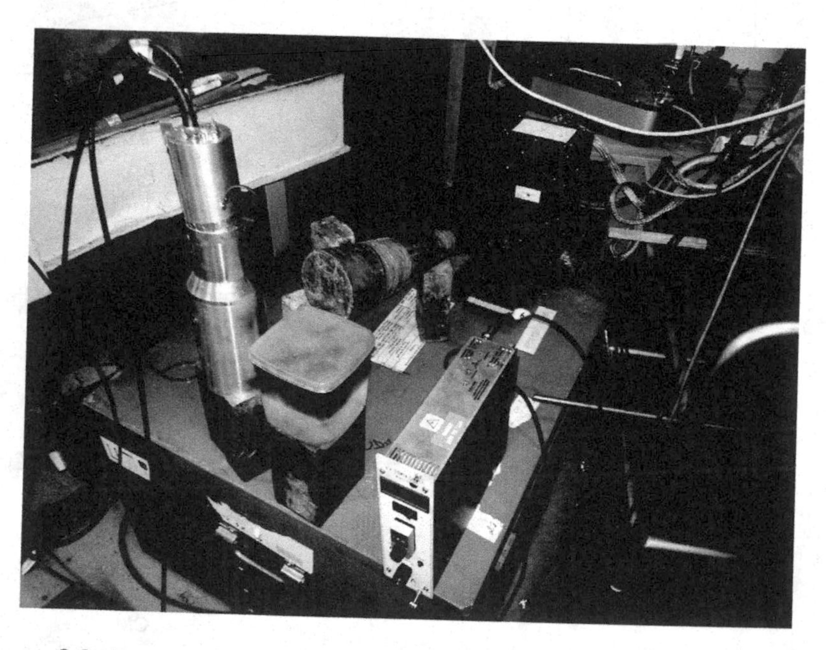

Figure 8.6 Experimental setup—neutron generator: ING-27 with 4π intensity 3×10^7 n/s, associated alpha particle method; gamma detector: 3″ × 3″ LaBr3:Ce; target: water–diesel fuel mixture in 1 L plastic container.

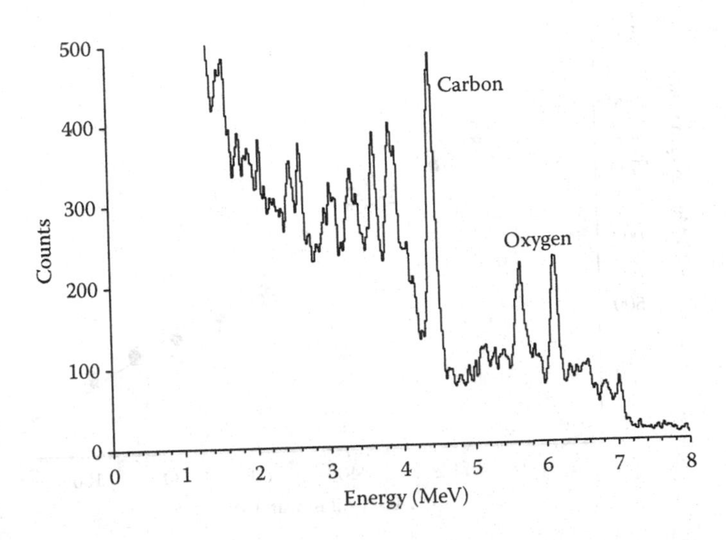

Figure 8.7 Gamma ray spectrum for 50% volume of diesel fuel and 50% water in target container.

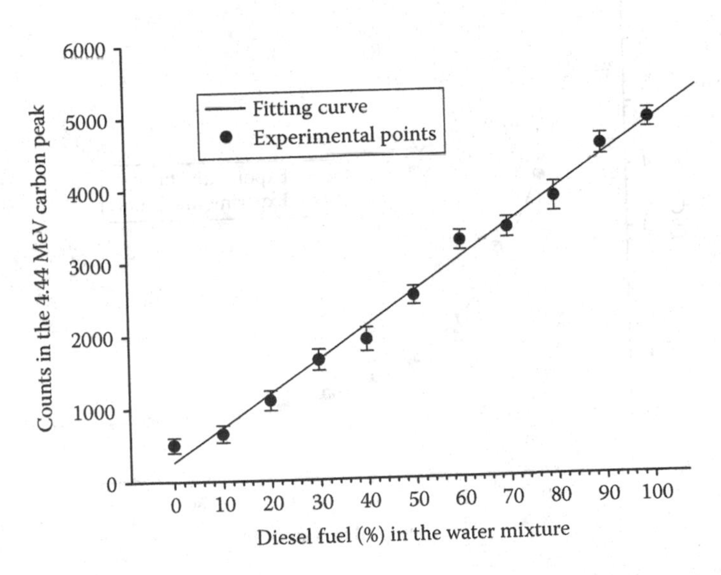

Figure 8.8 Number of counts in carbon peak, experimental points, as a function of diesel fuel volume percentage in the water mixture, linear fit.

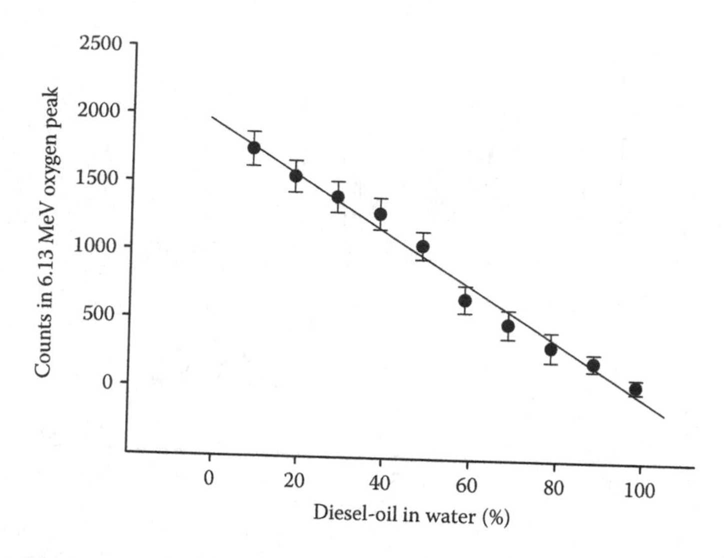

Figure 8.9 Number of counts in oxygen peak, experimental points, as a function of diesel fuel volume percentage in the water mixture, linear fit.

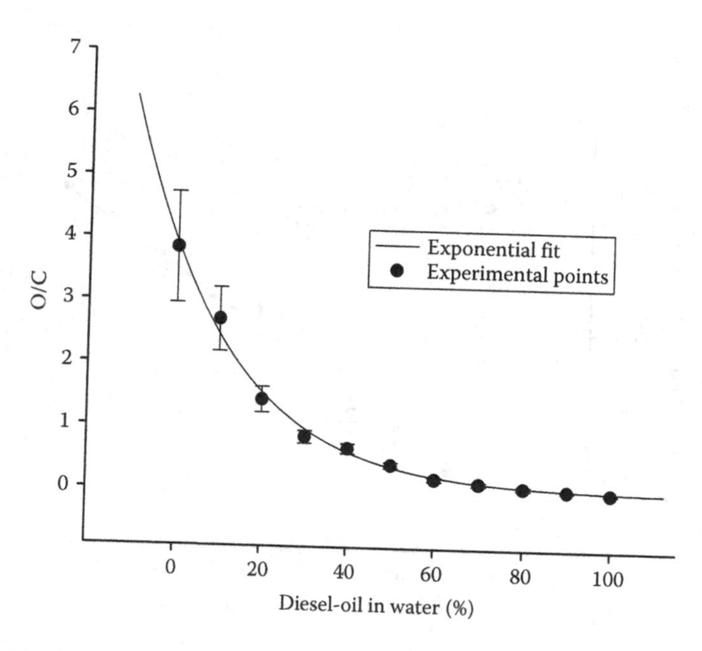

Figure 8.10 Experimentally determined O/C ratio as a function of diesel fuel volume percentage in the water mixture with exponential fit to data points.

By acquiring the inelastic gamma ray and time of flight spectra, the C and O concentration ratios could be obtained to determine small-scale variations in oil saturation and borehole oil fraction. Information on small-scale variations in lithology, porosity, and salinity can also be obtained. Logging is proposed to be performed as station measurements in this mode, but for higher porosities, low speed (≤20 m/h) continuous passes will be tested. A carbon signal can originate from the borehole, the cement behind the casing, the formation rock, and the formation fluid. In the proposed approach, the fraction of the ratio of C and O yields being a measure of the amount of oil around the tool will be spatially characterized.

Measurements of C/O Log are not performed at present on many oil fields around the world. They are must if one wants to avoid incomplete exploration of oil reserves in the particular field. Financial benefit for the countries and companies involved could not only be in the confirmation of the increase in oil reserves, but also in benefits resulting from the introduction of new technique for well characterization.

By the use of tagged 14 MeV neutrons, one could determine the depth variation of the stoichiometric ratios for chemical elements of interest: Ca, Si, Al, Fe, O (cement constituents), and C introduced during curing process. The casing tube walls are thick, which should not represent an obstacle to such an approach.

We shall next describe the activation flow log tool. This unique tool uses two d-t generators that emit 14 MeV neutrons in short pulses in order to activate ^{16}O into ^{16}N (Hotwell 2009). Neutron flux is on the order of 10^8 n/s. After approximately 7 s, these ^{16}N nuclei revert to ^{16}O emitting primary gamma rays at 6.13 MeV with the half-life of 7.1 s. There are four gamma-ray detectors that are spaced at known intervals from the two generators, such that any transport of the activated isotopes due to water flow may be detected by the difference in count rates between the four detectors. Up or down flow may be detected by using either the upper or lower generator to create activated isotopes of oxygen. It is also sometimes possible to measure the flow rates of water behind the casing.

In addition, some activation of hydrocarbons will also occur causing a release of gamma rays with energy levels between 3 and 6 MeV. This secondary effect can be useful for measuring the flow of oil or gas. Both generators are fired, one at a time, under program control for periods of as much time as is required to accumulate a reasonable amount of gamma ray count data. Flow rates may then be calculated using Hotwell analysis software (Dasic and Pletikosic 2014).

A case study on using this tool (activation flow tool [AFT]) in Columbia is presented by Markovic and Popovic (2014). The tool they used was 43 mm in diameter and 3.7 m long.

Different measurement parameters have been chosen depending on weather the water flow is slow or fast. The software used allowed the choice of activation time (fire time), accumulation time, and cycles. In addition, the logging up or down is possible in order to compare and/or confirm the data when some uncertainties are present.

REFERENCES

Bespalov, D. F., Erozolimski, B. G., Ivankin, V. P. et al. 1971. Present status and future prospects of the development of a pulsed neutron method of borehole testing, in *International Conference on the Peaceful Uses of Atomic Energy. Proceedings of the Conference in Geneva*, IAEA, Vienna, Austria, September 6–16, 1971, Vol. 7, pp. 63.

Bliven, S. and Nikitin, A. 2010. Needs of well logging industry in new nuclear detectors, in *Nuclear Science Symposium Conference Record (NSS/MIC), 2010 IEEE*, Knoxville, TN, October 30–November 6, pp. 1214–1219.

Borsaru, M., Rojc, A., Ronaszeki, J., and Smith, C. 2002. The determination of Fe, Si, and Al content in iron ore blast holes by nuclear logging, in *Proceedings of the Iron Ore 2002 Conference*, Perth, Australia. September 9–11, 2002.

Borsaru, M., Berry, M., Bigg, M. and Rojc, A. 2004. In situ determination of sulphur in coal seams and overburden rock by PGNAA. *Nucl. Instrum. Methods Phys. Res. B* 213: 530–534.

Borsaru, M., Zhou, B., Aizawa, T., Karashima, H., and Hashimoto, T. 2006, Automated lithololgy prediction from PGNAA and other geophysical logs. *Appl. Radiat. Isot.* 64: 272–282.

Brown, P. N. 1996. Nuclear Well Logging, http://computation.llnl.gov/casc/well_log/.

Caldwell, R. L. 1969. Isotope technology development—Nuclear logging methods. *Isot. Radiat. Technol.* 6 (3): 257–273.

Caldwell, R. L., Desai, K. P., and Mills, W.R., Jr. 1974. Geophysical well logging using nuclear techniques, in *Proceedings of International Atomic Energy Agency Panel Nuclear Techniques in Geochemistry and Geophysics*, Vienna, Austria, November 25–29, 1974, pp. 3–19.

Charbucinski, J. 1993. Comparison of spectrometric neutron-gamma and gamma-gamma techniques for in situ assaying for iron grade in large diameter production holes. *Nucl. Geophys.* 7: 133–141.0

Charbucinski, J., Duran, O., Freraut, R., Heresi, N., and Pineyro, I. 2004. The application of PGNAA borehole logging for copper grade estimation at Chuquicamata mine. *Appl. Radiat. Isot.* 60: 771–777.

Charbucinski, J., Malos, J., Rojc, A., and Smith, C. 2003. Prompt gamma neutron activation analysis method and instrumentation for copper grade estimation in large diameter blast holes. *Appl. Radiat. Isot.* 59: 197–203.

Charbucinski, J. and Nichols, W. 2003. Application of spectrometric nuclear borehole logging for reserves estimation and mine planning at Callide coalfields open-cut mine. *Appl. Energy* 74: 313–322.

Dasic, M. and Pletikosic, N. 2014. 1–3/8 Activation Flow Log Tool general description and communication, in *Contribution to HOTWELL's 9th PNN/Well Logging Conference*, Klingenbach, Austria, September 18–20, 2014.

Dupree, J. H. 1988. Cased-hole nuclear logging interpretation, in *Society of Professional Well Log Analysts Logging Symposium, 29th, San Antonio, Tex., Transactions*, Prudhoe Bay, North Slope, Alaska, September 29–October 2, 1988, pp. TT1–TT23.

Edwards, J. M., Ottinger, N. H., and Haskell, R. E. 1967. Nuclear log evaluation of potash deposits, in *Society of Professional Well Log Analysts Annual Logging Symposium, 8th, Transactions*, Denver, CO, 12–14 June, pp. L1–L12.

Flanagan, W. D., Bramblett, R. L., Galford, J. E., Hertzog, R. C., Plasek, R. E., and Olesen, J. R. 1991. A new generation nuclear logging system, in *Society of Professional Well Log Analysts Logging Symposium, 32nd, Transactions*, Midland, TX, June 16–19, 1991, pp. Y1–Y25.

Fujimoto, K., Wilson, J. A., and Ashmore, J. P. 1985. Radiation exposure risks to nuclear well loggers. *Health Phys.* 48 (4): 437–445.

Halliburton. 2014, http://www.halliburton.com/public/lp/contents/data_sheets/web/h/h02621.pdf.

Harvey, P. K. and Lovell, M. A. 1992. Nuclear logging and mineral inversion in sedimentary consequences. *IEEE Trans. Nucl. Sci.* 39: 1007–1013.

Hearst, J. R. and Carlson, R. C. 1970. Well-logging research for geonuclear technology. *Nucl. Appl. Technol.* 8: 276–282.

Hilchie, D. W. 1977. Nuclear well logging for petroleum. *Dev. Econ. Geol.* (7): 201–213.

Hotwell 2009. 1-3/8 Activation Flow Log Tool Manual P/N:1250.0000. Hotwell, Kligenbach, Austria.

IAEA. 1971. Nuclear well logging in hydrology. International Atomic Energy Agency, Vienna, Austria, Technical report series no. 126, 88pp.

IAEA. 1974. Nuclear techniques in geochemistry and geophysics, in *Panel on Nuclear Techniques in Geochemistry and Geophysics, Proceedings*, Vienna, Austria, 25–29 November, 271pp.

IAEA. 1993. Handbook on nuclear data for borehole logging and mineral analysis. Technical reports series No. 357, STI/DOC/10/357, International Atomic Energy Agency, Vienna, Austria.

Jacobson, L. A., Wyatt, D. F., Jr., Gadeken, L. L., and Merchant, G. A. 1990. Resolution enhancement of nuclear measurements through deconvolution, in *Society of Professional Well Log Analysts Logging Symposium, 31st, Transactions*, Lafayette, LA. June 24–27,1990, pp. TT1–TT15.

Keys, W. S. 1996. *A Practical Guide to Borehole Geophysics in Environmental Investigations.* CRC Press, Boca Raton, FL.

Keys, W. S. and Boulogne, A. 1969. Well logging with Californium-252, in *Transactions, Society of Professional Well Log Analysts 10th Annual Logging Symposium, Society of Professional Well Log Analysts*, Houston, TX, May 25–28 1969, pp. 1–25.

Keys, W. S., Crowder, R. E., and Henrich, W. J. 1989. Selecting geophysical logs for environmental applications, http://info.ngwa.org/GWOL/pdf/930158326.PDF.

Kerr, S. A. and Worthington, P. F. 1988. Nuclear logging techniques for hydrocarbon, mineral, and geological applications. *IEEE Trans. Nucl. Sci.* 35: 794–799.

Knapp, K. 2013. Replacement of neutron sources used in well logging by neutron generators, in *8th PNN & Well Logging Conference*, Eisenstadt, Austria, 26–27, September.

Köhler, S., Wolf, M., and Becker, W. 2013. Specific investigation of well integrity with PNN, SIPLOS and AFL at the natural gas storage site Buchholz, Germany, in *8th PNN/Well Logging Conference*, Eisenstadt, Austria, 26–27 September.

Lawrence, T. D. 1981.Continuous carbon/oxygen log interpretation techniques. *J. Petrol. Technol.* 33 (8): 1394–1402.

Lebreton, F., Youmans, A. H., Oshry, H. I., Wilson, B. F. 1963. Formation evaluation with nuclear and acoustic logs, in *Society of Professional Well Log Analysts Annual Logging Symposium, 4th, Transactions*, Oklahoma City, OK. May 23–24, 1963, pp. IX1–IX18.

Lock, G. A. and Hoyer, W. A. 1974. Carbon–Oxygen (C/O) log: Use and interpretation. *J. Petrol. Technol.* 26: 1044–1054.

Markovic, Z. 2013. Petrophysical interpretation using only PNN data, in *Contribution to HOTWELL's 8th PNN/Well Logging Conference*, Eisenstadt, Austria, September 26–27.

Markovic, J. and Popovic-Gluhak, B. 2014. AFT Examples (case study), in *Contribution to HOTWELL's 9th PNN/Well Logging Conference*, Klingenbach, Austria, September 18–20.

McGregor, B. J. and Eisler, P. 1983. The effect of the bore hole on nuclear logging measurements. Australian Atomic Energy Commission Research Establishment, AAEC/E581, 21 p.

Morin, R. H. and Hess, K. M. 1991. Preliminary determination of hydraulic conductivity in a sand and gravel aquifer, Cape Cod, Massachusetts, from analysis of nuclear logs, in *U.S. Geological Survey Toxic Substance Hydrology Conference*, Monterey, CA [Proceedings] March 11–15, 1991. U.S. Geological Survey Water-Resources Investigations report 91-4034: 23–28.

Mott, W. E. and Ediger, N. M. 1959. Nuclear well logging in petroleum exploration and production, in *5th World Petroleum Congress*, May 30– June 5, NY.

Novak, M. 2013. Pulse Neutron Neutron system—15 Years' experience and future development possibilities, in *Presentation at 8th PNN & Well Logging Conference*, Eisenstadt, Austria, September 26–27. See, http://www.hotwell.at.

Oleson, J.-R. 1990. A new calibration, wellsite verification and log quality-control system for nuclear tools, in *Society of Professional Well Log Analysts Logging Symposium, 31st, Transactions*, Lafayette, LA, June 24–27, 1990, pp. PP1–PP25.

Paillet, F. L. 1988. Geophysical well-log analysis in characterizing the hydrology of crystalline rocks of the Canadian shield. MIT Earth Resources Laboratory, Lab Reports Archive, http://erl.mit.edu/lab-reports-archive.php.

Patchett, J. G. and Wiley, R. 1994. Inverse modeling using full nuclear response functions including invasion effects plus resistivity, in *Society of Professional Well Log Analysts Annual Logging Symposium, 35th, Transactions*, Tulsa, OK, June 19–22, 1994, pp. H1–H23.

Perot, B., Perret, G., Mariani, A. et al. 2006. The EURITRACK project: Development of a tagged neutron inspection system for cargo containers, in *Proc. SPIE*, Vol. 6213, Orlando, FL, 621302–1,6.

Przewlocki, K. 1984. Nuclear techniques in underground mining, in *Proceedings of IAEA Consultants' Meeting on nuclear Data for Bore-Hole and Bulk-Media Assay Using nuclear Techniques. IDC(DS)-ISI/L.* IAEA, Vienna, Austria.

Rider, M. 1996. *The Geological Interpretation of Welllogs*, 2nd edn. Whittles Publishing, Dunbeath, Scotland.

Roscoe, B. A. and Grau, J. A. 1988. Response of the carbon/oxygen measurement for an inelastic gamma ray spectroscopy tool, in *SPE Formation Evaluation*, March, pp. 76–80.

Roscoe, B. A., Lenn, C., Jones, T. G. J., and Whittaker, A. C. 1991. A new through tubing oil saturation measurement system, in *SPE Middle East Oil Show*, Bahrain, November 16–19.

Rowe, W. 2012. Well monitoring and logging. April 18, 2012, http://www.slb.com/carbon services.

Schneider, D., Hutchinson, M., and Deady, R. 1994. Processing and quality assurance of unevenly sampled nuclear data recorded while drilling, in *Society of Professional Well Log Analysts Annual Logging Symposium, 35th, Transactions*, Tulsa, OK, June 19–22, 1994, pp. RR1–RR25.

Schultz, W. E. and Smith, H. D. 1974. Laboratory and field evaluation of carbon/oxygen well logging system. *J. Petrol. Technol.* 26: 1103–1110.

Senftle, F. E. 1980. Application of gamma-ray spectral analysis to subsurface mineral exploration, in Muecke. G. K. (Ed.) *Short Course in Neutron Activation Analysis in the Geosciences.* John Wiley and Sons, New York.

Senftle, F. E., Mikesell, J. L., Tanner, A. B., and Lloyd, T. A.1984. Sensitivity of in situ Borehole neutron activation for the noble elements, in *Proceedings of IAEA Consultants' Meeting on Nuclear Data for Bore-Holeand Bulk-Media Assay Using Nuclear Techniques, IAEA. I DC(DS)-ISI/L.* IAEA, Vienna, Austria, pp. 159.

Smith, C. P. and Berry, M. 2005a. The application of spectrometric borehole logging for exploration and mining at Namakwa Sands Mine, South Africa. Final report, February 2005, Exploration and mining report no. P2004/69.

Smith, C. P. and Berry, M. 2005b. The application of spectrometric borehole logging for mining at Sishen Iron Mine, South Africa. Final report, January 2005, Exploration and mining report no. P2004/68.

Smith, C. and Berry, M. 2005c. The application of spectrometric borehole logging for mining at Skorpion Mine, Namibia. Final report, February 2005, Exploration and mining report no. P2004170.

Tittle, C. W. 1992. Diffusion theory models of invasion for nuclear porosity tools, in *Society of Professional Well Log Analysts Annual Logging Symposium, 33rd, Transactions*, Oklahoma City, OK, June 14–17, 1992, pp. N1–N12.

Trofimczyk, K., Saraswatibhatla, S., and Smith, C. 2009. Spectrometric nuclear logging as a tool for real-time, downhole assay—Case studies using SIROLOG PGNAA, in *Proceedings of 11th SAGA Biennial Technical Meeting and Exhibition*, Swaziland, September 16–18, pp. 161–171.

Valkovic, V., Sudac, D., Obhodas, J. et al. 2013. The use of alpha particle tagged neutrons for the inspection of objects on the sea floor for the presence of explosives. *Nucl. Instrum. Methods Phys. Res. A* 703: 133–137.

Wightman, W. E., Jalinoos, F., Sirles, P., and Hanna, K. 2003. Application of geophysical methods to highway related problems. Federal Highway Administration, Central Federal Lands Highway Division, Lakewood, CO, Publication No. FHWA-IF-04-021, http://www.cflhd.gov/resources/agm/.

Wraight, P., Evans, M., Marienbach, E., Rhein-Knudsen, E., and Best, D. 1989. Combination formation density and neutron porosity measurements while drilling, in *Society of Professional Well Log Analysts Logging Symposium, 30th, Transactions*, Denver, CO, June 11–14, 1989, pp. B1–B23.

Zhang, F., Xu, J.-P., Hu, L.-M., Xiu, C.-H., and Sun, J.-M. 2007. Monte Carlo simulation for the pulsed neutron-neutron logging method. *Chin. J. Geophys.* 50 (6): 1654–1661.

Chapter 9

Medical Applications of 14 MeV Neutrons

9.1 INTRODUCTION

Attenuation of neutron beam, as shown in Figure 9.1, by the specimen depends on thickness and the attenuation coefficient similar to gamma ray as follows:

$$I_t = I_0 \exp(-\Sigma t)$$

where

I_t is the transmitted neutron intensity (n/cm²/s)
I_0 is the incident neutron intensity (n/cm²/s)
t is the specimen thickness (cm)
Σ is the macroscopic cross section (cm^{-1})

Σ is equivalent to the linear attenuation coefficient of gamma ray (μ) and is the characteristic of elements in the specimen. Σ and μ are the products of atom density of elements contained in the specimen (in atoms/cm³) and their effective microscopic cross sections (σ) to the reactions of interest (in cm²). σ is the effective cross section, not the actual physical cross section of the nucleus. It indicates probability of occurrence for each neutron interaction. For example, $\sigma_{(n,\gamma)}$ indicates the probability of (n, γ) reaction and σ_s indicates the probability of scattering reaction that combines elastic (n, n) and inelastic (n, n′) scattering cross sections. Σ and σ of pure elements and common compounds or mixtures (such as water, heavy water, and concrete) can be found in literatures.

9.2 BIOLOGICAL AND MEDICAL APPLICATIONS

Large numbers of 14 MeV neutrons can be produced using the $^3H(d,n)^4He$ reaction. The neutron energy, fluence, and dose rate are nearly independent of angle so that planning irradiations and designing fixtures to hold samples are relatively easy. A significant fraction of the energy deposited in tissue by 14 MeV

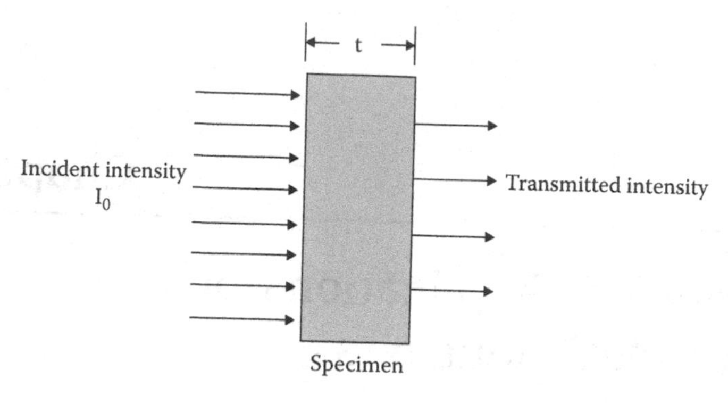

Figure 9.1 Attenuation of neutrons by a specimen.

neutrons is from alpha particles and heavy-ion recoils. Approximately 70% of the energy deposition is from proton recoils.

Let us describe in some detail the Radiological Research Accelerator Facility, Nevis Laboratories, Irvington, NY (RARAF 2014). Neutron irradiations are performed in two caves. Normally, a horizontal charged-particle beam is used but a vertical beam is available for irradiating biological systems for which it may be better suited. Neutron irradiation fixtures for radiobiology and physics are designed in consultation with the experimenter to meet the needs of both the researcher and the dosimetrist.

Although the target assemblies were designed to minimize absorbing material, the neutron dose at large angles from the beam direction is not azimuthally uniform. To irradiate large numbers of samples uniformly, most fixtures provide a means of rotation about the beam axis. They have a Ferris wheel–like fixture used to irradiate rats and mice, flasks, test tube, and dishes. Above, the wheel has been modified for irradiation of cell monolayers growing in commercial cell culture flasks. This arrangement has been used in studies of transformation induction in mouse cells. Higher dose rates to cells are delivered using a fixture mounted on the vertical beam line. This apparatus was originally used for irradiating hamster cells in small vials made from plastic pipettes.

Dosimetry for neutron irradiations is performed using tissue-equivalent (TE) ionization chambers for total dose measurements and neutron-insensitive dosimeters to measure gamma-ray dose. The dosimetry measurements are relative to the response of a TE ionization chamber in a fixed location, which is used as a monitor. Radiation doses are then delivered based on the response of the fixed monitor chamber. The gamma-ray dose and the incident beam current on the target are also monitored.

The total dose ionization chambers have walls made of A-150 muscle TE plastic and have methane-based TE gas sealed inside or flowing through.

Insulators are made of tissue-like materials such as styrene or Lucite. The chamber is placed at the same position as the sample would be during the irradiation. A chamber is selected so that the chamber volume is similar to the sample volume, and the wall thickness is adjusted to match the amount of material between the center of the sample and the neutron-producing target. Chambers of various sizes and geometries from 1/4 in. diameter spherical to 1 in. diameter by 3/16 in. deep parallel plate arrangements are available so that measurements can be made even for small volumes or large areas. If the sample is rotated about the beam axis or if there are several samples being irradiated simultaneously, measurements are made at various positions around the target.

Gamma-ray dosimetry is performed using either a compensated Geiger–Mueller dosimeter or an aluminum-walled, argon-filled (Al–Ar) ionization chamber. The large response of the Al–Ar chamber to 14 MeV neutrons makes it unfeasible to use for that energy. Measurements of gamma-ray dose often cannot be made at the sample position with the G-M-type dosimeter but are made as close to it as possible. The gamma-ray dose rate is essentially isotropic about the target so that only inverse square law corrections are necessary.

The dosimeters are calibrated using either a 50 mg radium-226 or 7 Ci cesium-137 gamma-ray source, both of which have been calibrated by the National Bureau of Standards. Corrections to the dosimetry measurements are made for any positional differences, for dose buildup or attenuation factors due to differences in material thickness, for variations of W_n with neutron energy, and for neutron kerma difference due to composition differences between the samples and the chamber. Computer calculations of the mean neutron energy, energy spread, and the relative dose rate at various points on the sample are provided (RARAF 2014).

For its high sensitivity to a great number of elements of biological concern, NAA, together with high-resolution Ge(Li) spectrometry, has been used for the determination of elements such as Na, K, Cl, Mn, Fe, Rb, and Co and in some cases Ca, Zn, Cs, Sc, and Cr (see Csikai 1987). Many research groups have applied the NAA in a wide range of samples in life and agricultural sciences, see, for example, Crambes et al. (1967), Meloni et al. (1993) and references in Chapter 6.

9.3 IN VIVO NEUTRON ACTIVATION ANALYSIS

The use of NAA techniques for medical applications was first reported in 1964 for the measurement of sodium in the body. Anderson et al. (1964) introduced the in vivo activation analysis method to estimate the whole body content of sodium by whole-body exposure to 14 MeV neutrons using nuclear reaction

$$^{23}Na + n \rightarrow {}^{24}Na + \gamma \; (E_\gamma = 1.37 \text{ and } 2.75 \text{ MeV})$$

This new technique allows the elements in the living human body to be determined more easily than by other methods. There are, however, two difficulties associated with in vivo activation analysis: the calibration and the radiation hazards. The effect of flux in homogeneity can be eliminated by the use of a phantom containing known concentrations of the various elements. The radiation hazard can be decreased to the permissible level by pulsed activation analysis, producing short-lived isotopes. The investigations show that the reproducibility and accuracy of in vivo measurements make possible the determination of the absolute quantity of elements in man, using whole-body counters.

Let us mention some early work in this field. For example, Chamberlain et al. (1968) reported the measurement of body calcium and sodium and described techniques for whole-body NAA and pulsed NAA.

$$^{48}Ca + n \rightarrow ^{49}Ca + \gamma \, (E_\gamma = 3.1 \text{ MeV})$$

Both the total body and partial body in vivo activation analyses are applied. Cohn and Dombrowski (1971) have measured the absolute levels of Ca, Na, Cl, N, and P in the whole body for a number of patients with various diseases by in vivo activation analysis. They found that the precision of the measurements repeated on the same patient is better than ±4%. A γ-ray spectrum after 5 min irradiation of a patient with 15 min counting time, choosing 6 min for cooling, is sufficient for decision making. For the determination of total body Ca in humans, a very sensitive method was developed by Palmer (1973) based on the detection of fluorescent yield of the 2.63 keV chlorine x-rays emitted during ^{37}Ar decay, using a gas proportional counter. This technique enables the measurement of total body Ca with a very small radiation dose lower by a factor of 50–200 than that used for present in vivo neutron activation methods. Mitra and his coworkers used in vivo NAA to provide accurate measurement of relative changes in whole body Ca and in regional Ca and P contents (see Mitra et al. 1990, 1993, 1995a,b, 1998; Garrett and Mitra 1991).

Kehayias and Zhuang (1993) used a small sealed d+t neutron generator for the pulsed (4–8 kHz) production of fast neutrons. Carbon and oxygen were detected in vivo by counting the 4.44 and 6.13 MeV gamma rays resulting from the inelastic scattering of the fast neutrons.

Hydrogen was detected by thermal neutron capture. BGO detectors (127 × 76 mm) were found more tolerant to neutron exposure and improved the signal to background ratio for the carbon detection by a factor of 6, compared to 152 × 152 mm NaI(Tl). The elemental analysis of the body is used to study the changes of body composition with aging. They investigated the causes of depletion of lean body mass and the development of ways of maintaining functional capacity and the quality of life of the elderly.

Blood flow in an organ can be measured in vivo by irradiating with short pulses of 14 MeV neutrons a cylinder of thoracic tissues, which includes the heart. After triggering and contraction of the heart, the blood invades the organ under consideration a few seconds later, and this can be observed by counting the gammas from ion activity. Some general and particular aspects of in vivo NAA are discussed in IAEA (1973). A number of simultaneous 14 MeV NAA of the body elements in man-like models have been reported. In diseases of bone, the Ca and P content can change significantly. Hyvonen-Dabek et al. (1979) have described a method based on 14 MeV NAA for in vitro measurement of the P content of bones. The Ca/P ratio for compact bone was also determined and the suitability of the $^{31}P(n,\alpha)^{28}Al$ reaction induced by 14-MeV neutrons for studying bone mineral composition in vitro has been discussed.

FNAA was used to study the chemical composition of fossil bones by systematic measurements on a large number of samples. Concentrations of N, F, Si, and Fe were determined by 14 MeV NAA. An analysis of the data led to the conclusion that a correlation exists between the concentration of nitrogen and the climate at the burial time of bones (Badone and Farquhar 1982). This correlation enabled the temperature for different eras to be estimated for the last 104 years. Correlation was also found between the Al, Si, and Fe infiltered into the organic part of bones. It was found that the content of Mg, Al, Si, Mn, Fe, Sr, and Ba is higher in fossil bones than in recent ones, while the concentration of N and Na is lower, see also Holmberg et al. (1978).

Cohn and Dombrowski (1971) reported the measurement of calcium, sodium chlorine, nitrogen, and phosphorus in the human body through in vivo NAA. Since then, NAA and PGNAA have been used for a variety of applications, such as the measurement of nitrogen, carbon and oxygen, cadmium, and manganese in the body and in trace element research to identify cancerous tissue. Inelastic neutron scatter analysis (INSA) using fast neutrons use 14 MeV neutrons from a (d,t) sealed-tube neutron generator to determine whole body carbon content as a measure of energy expenditure in the body; this was reported by Kyere et al. (1982).

The prompt- and delayed-gamma neutron activation techniques have been used for the noninvasive measurement of human body composition. In recent years, neutron irradiators have used only transuranic isotopic sources (^{238}PuBe, ^{241}AmBe, ^{252}Cf). However, in today's security-minded environment, the use of alternate neutron sources may provide some advantages. Shypailo and Ellis (2005) have examined several designs for an irradiator that would use a high-output, miniature d–t neutron generator (MF Physics). The use of this type of neutron source will lessen the storage, security, and transport issues associated with continuous-output isotopic neutron sources. To determine the scientific impact of this decision, they have performed Monte Carlo simulations (MCNP-4B2; Los Alamos National Laboratory) to aid in the design of the irradiator system, evaluating shielding materials, collimation, and source-to-subject distance, for the measurement of total body nitrogen (TBN). Based on internal flux distributions within the simulated body region of a subject, several design options were identified. The final design will be selected based on the optimization of precision, dose, and exposure time.

Computation of activities induced by 14 MeV neutrons for the elemental analysis of human tissues, skin, and tooth enamel has been performed by Khanchi et al. (1994). For planning experiments for neutron activation analysis investigations at 14 MeV, suitable induced beta activities in Bq/g/neutron flux have been computed via appropriate reactions for the estimation of ^{19}F, ^{23}Na, ^{24}Mg, ^{31}P, ^{32}S, $^{35,37}Cl$, ^{39}K, ^{40}Ca, ^{55}Mn, ^{56}Fe, ^{66}Zn, $^{63,65}Cu$, $^{107,109}Ag$, and ^{208}Pb. The computational work for (n,p), (n,α), (n,n′), and (n,2n) reactions induced with 14 MeV neutrons is based on the preequilibrium emission mechanism and also the compound nucleus theory with optical model potential and pairing energy corrections.

Neutron stimulated emission computed tomography (NSECT) is being developed as a noninvasive spectroscopic imaging technique to determine element concentrations in the human body. NSECT uses a beam of fast neutrons that scatter inelastically from atomic nuclei in tissue, causing them to emit characteristic gamma photons that are detected and identified using an energy-sensitive gamma detector. By measuring the energy and number of emitted gamma photons, the system can determine the elemental composition of the target tissue. Such determination is useful in detecting several disorders in the human body that are characterized by changes in element concentration, such as breast cancer (Kapadia et al. 2008).

In the paper by Kapadia et al. (2008) experimental implementation of a prototype NSECT system for the diagnosis of breast cancer is described and experimental results from sensitivity studies using this prototype are presented. Results are shown from three sets of samples: (1) excised breast tissue samples with unknown element concentrations, (2) a multielement calibration sample used for sensitivity studies, and (3) a small-animal specimen to demonstrate detection ability from in vivo tissue. Preliminary results show that NSECT has the potential to detect elements in breast tissue. Several elements were identified common to both benign and malignant samples, which were confirmed through neutron activation analysis. Statistically significant differences were seen for peaks at energies corresponding to ^{37}Cl, ^{56}Fe, ^{58}Ni, ^{59}Co, ^{79}Br, and ^{87}Rb. The spectrum from the small animal specimen showed the presence of ^{12}C from tissue, bone, and elements ^{39}K, ^{27}Al, ^{37}Cl, ^{56}Fe, ^{68}Zn, and ^{25}Mg. Threshold sensitivity for the four elements analyzed was found to range from 0.3 to 1 g, which is higher than the microgram sensitivity required for cancer detection. Patient dose levels from NSECT were found to be comparable to those of screening mammography (Kapadia et al. 2008).

In the work by Garrett and Mitra (1991) the feasibility of using the time correlated associated particle technique for in vivo 14-MeV neutron activation analysis has been investigated. Gamma rays following neutron inelastic scattering with nitrogen, carbon, and oxygen have been measured with a 12.5 × 10 cm NaI (T1) detector. The results have been scaled to a proposed facility comprising four such detectors past which the subject would be scanned. Based on counting statistics, the precision of estimation of these elements has been determined to be 2.1%, 1.0%, and 1.1%, respectively, for experimental measurements on a sample containing physiological concentrations of the major body elements. The average body dose level would be restricted to 0.3 mSv.

For plant samples, nondestructive 14 MeV NAA is suitable mainly for primary constituents, C, O, K, Cl, Ca, and Fe; to observe microelements (Cu, Zn, Y, Mn, Mo, and Co), the neutron generator can more favorably be used as a thermal neutron source. The 14 MeV neutron generators are widely used for the determination of protein content in plants and cornmeal via the $^{14}N(n,2n)^{13}N$ reaction. For example, the soya, one of the oldest domesticated plant in mankind, plays an ever-increasing role in world economics. The soya bean contains 30%–40% crude proteins, up to 20% oils, and 30% carbohydrates. Although its chemical structure and composition are genetically determined, the constituents can be influenced strongly by cultivation circumstances; the oil and protein content, in particular, depend considerably on producing conditions.

To establish a correlation between the protein output and the elemental composition of the soil, it is necessary to investigate separately the roots, stalks, leaves, blooms, and crops of the plant in different phases of its growth, together with the concentration of elements in the samples taken from the soil in these phases. Investigations show that different chemical fertilizers can influence the nitrogen and crude protein content of plants. Generally, the cornmeal and plants contain K and P in addition to N, which produce spectral interferences. Both the P and K content can be determined by an independent method, for example, using the x-ray fluorescence technique.

9.4 NEUTRON RADIOGRAPHY

Neutron radiography with slow neutrons has been well described in the literature; see, for example, Chankow (2012). The overview of slow neutron radiography is presented in publications by Barton (1976), Garrett and Berger (1977), Lamarsh (1983), Harms (1977), and Von der Hardt and Röttger (1981).

Although different in many aspects from the neutron radiography with 14 MeV neutrons, some of the fundamentals are the same. Neutron radiography requires parallel beam or divergent beam of low energy neutrons having intensity in the range of only 10^4–10^6 neutrons/cm^2/s to avoid the formation of significant amount of long-lived radioactive isotope from neutron absorption within the specimen. The transmitted neutrons will then interact with neutron converter screen to generate particles or light photons that can be recorded by film or any other recording media. Free neutrons emitted from all sources are fast neutrons while neutron radiography prefers low energy neutrons. To reduce neutron energy, neutron sources are normally surrounded by a large volume of hydrogenous material such as water, polyethylene, transformer oil, and paraffin. Neutron collimator is designed to bring low energy neutron beam to the test specimen.

Numerous are efforts to thermalize fast neutrons in order to achieve improvements in the performance for radiography (Cluzeau et al. 1997). In a small neutron radiography system such as DIANE (made by Sodern), which is based on a sealed-tube neutron generator, the maximum possible efficiency of fast neutron thermalization must be achieved, consistent with realistic industrial and manufacturing practices. To this end, MCNP simulations and experiments have been performed for further enhancing the performance of the moderator/collimator assembly in the DIANE. These calculations and experiments have shown that a fast neutron reflector/multiplier can be useful in increasing the number of fast neutrons entering the moderator. Various heavy metals with high inelastic cross sections for fast neutrons and low capture cross sections have been tested. Results indicate that the best moderating materials for this application are beryllium, zirconium hydride, and high-density polyethylene, and the best reflector/multiplier material for this use is tungsten. The MCNP calculations indicate that for a fast neutron output of 4×10^{11} n/s in 4π steradians, a DIANE can be fabricated that produces a thermal neutron beam for radiography having a flux of approximately 2.1×10^5 n/cm^2/s at an effective collimator ratio of 30, or about 7.5×10^4 n/cm^2/s at a collimator ratio of 50.

The paper by Brzosko et al. (1992) evaluates the potentials of fast-neutron radiography (FNR) for the inspection of bulky, solid objects. Data for both a fast (E_n = 14.7 MeV) and a slow (E_n = 0.1 eV) neutron source are compared. The reproduction of images consists of Monte Carlo simulations of (1) the neutron random walk in a slab (iron, SiC ceramic, and polyethylene $(CH_2)_n$ plastic) with a void, (2) the process of neutron recording in a detector, and (3) a printout of images. For a general analysis, 3D-MCSC-RWR software operates without simplification of either the FNR design or the nuclear data files. The results first show the feasibility of the use of 14 MeV neutron radiography, then the superiority of FNR over slow-neutron radiography in the field when the thickness of the full slab is over 1 cm and requires a resolution better than 0.1 mm. Examples of some numerically simulated images as well as FNR scaling functions are shown. A review of the available fast-neutron sources reveals that only plasma-focus machines would simultaneously meet all FNR requirements: $Y_n \geq 10^{13}$ n/pulse, small-source dimensions, and mobility.

It should be pointed out that the attenuation coefficient of gamma ray increases with increasing atomic number of element while the attenuation coefficients of neutron are high for light elements like hydrogen, lithium, and boron, as well as some heavy elements such as gadolinium, cadmium, and dysprosium. In contrast, lead has very high attenuation coefficient for gamma ray but very low for neutron. Neutron radiography, therefore, can make parts containing light elements such as polymer, plastic, rubber, and chemical visible even when they are covered or enveloped by heavy elements.

After neutrons pass through the specimen they interact with the converter screen to produce radioisotopes, alpha particles, or light, which can be recorded by film, imaging plate, optical camera, or video camera. Image recording medium must be selected to match with the particles or light emitted from the neutron converter screen so as to obtain the maximum efficiency. The neutron converter screen/image recording device assemblies commonly used in neutron radiography are described in the following.

1. *Metallic foil screen/film*: Metallic foil with high neutron cross section is employed to convert slow neutrons to beta particles, gamma rays, and/or conversion electron while industrial x-ray film is normally used as the image recorder. Gadolinium, Gd, foil is the best metallic screen for neutron radiography in terms of having extremely high neutron absorption cross section, giving the best image resolution and not becoming radioisotope after neutron absorption. ^{155}Gd and ^{157}Gd are found to be 14.9% and 15.7% of natural Gd isotopes with neutron absorption cross sections of 61,000 and 254,000 barns, respectively. ^{155}Gd and ^{157}Gd absorb neutrons and then become ^{156}Gd and ^{158}Gd correspondingly, which are not radioactive. Prompt captured gamma rays emitted during neutron absorption can cause film blackening. More importantly, prompt gamma rays may hit atomic electrons resulting in ejection of electrons from the atoms (so-called conversion electron), which are more effective to cause film blackening. It should be noted that less than a few percentage of gamma-ray photons cause film blackening. Electrons and beta particles are preferred because they interact with film much more than gamma rays.
2. *Light emitting screen/film*: Light-emitting screen is a mixture of scintillator or phosphor with ^{6}Li and/or ^{10}B. Neutrons interact with ^{6}Li or ^{10}B to produce alpha particles via (n,α) reaction. Light is then emitted from energy loss of alpha particles in scintillator or phosphor. Light sensitive film, digital camera, or video camera can be used to record image. This makes real-time and near real-time radiography possible. The most common light-emitting screen is NE426 available from NE Technology, which is composed of ZnS(Ag) scintillator and boron

compound. Gadolinium oxysulfide (terbium) [Gd_2O_2S (Tb), GOS] and lithium-loaded glass scintillator are also common in neutron radiography. GOS itself is a scintillator. Conversion electrons as well as low energy prompt gamma rays emitted from the interaction of neutrons with Gd cause light emission. Glass scintillator is sensitive to charged particles such as alpha and beta particles. Lithium is added into the glass scintillator so that alpha particle will be emitted from $^{6}Li(n,\alpha)^{3}H$ reaction resulting in the emission of light.

3. *Alpha-emitting screen/track-etch film*: Alpha-emitting screen is made of lithium and/or boron compound. Particles emitted from $^{6}Li(n,\alpha)^{3}H$ and $^{10}B(n,\alpha)^{7}Li$ reactions interact with track-etch film (or so-called solid state track detector [SSTD]) to produce damage tracks along their trajectories. The detector is later put into hot chemical solution to enlarge or "etch" the damage tracks. After etching, the damage tracks can be made visible under an optical microscope with a magnification of 100X and up. Radiation dose and/or neutron intensity can be evaluated by counting the number of tracks per unit area. The area where track density is so large becomes translucent while the area with low track density is more transparent. The degree of translucence depends on track density resulting in the formation of visible image on the film. However, contrast of the image is poor while sharpness is comparable to the Gd foil/x-ray film assembly. Methods for viewing the image is needed to improve image contrast such as reprinting the image on a high contrast film.

Let us mention some interesting work in this field. Neutrons with the initial energy of 14 MeV produced by the neutron generator of the Department of Reactor Physics, Chalmers University of Technology, Göteborg, Sweden, were detected by a camera system in connection with scintillator screens (Lehmann et al. 2005). Images were produced for 14 MeV and moderated neutrons with suited scintillator screens. Qualitative and quantitative evaluations were made to characterize the performance of the experimental setup.

A presentation of new results for the transport of 14 MeV neutrons in the atmosphere is described by Marcum (1963). Two corrections have been made to the RAND neutron Monte Carlo transport code. Data from previous reports are presented so that comparisons can be made with the new results and the previous incorrect results can be changed. Also, for the first time, results are given for the transport of 14 MeV neutrons near an air–ground interface and in an infinite homogeneous air medium.

The U.S. Army had plans to use the neutron radiography system to take images of artillery shells and other critical subcomponents of military equipment. Neutron radiography is similar to taking an x-ray, but neutron radiographs give additional information not available from traditional x-rays. In particular, neutrons have the ability to pass through thick layers of metal, relatively unaffected, and give information about low density, low atomic number materials inside an object.

Fast neutron imaging offers the potential to be a powerful nondestructive inspection tool for evaluating the integrity of thick sealed targets (Dietrich et al. 1997) and for the interrogation of air cargo containers (Sowerby et al. 2009). This is particularly true in cases where one is interested in detecting voids, cracks, or other defects in low-Z materials (e.g., plastics, ceramics, salts, etc.) that are shielded by thick, high-Z parts. The authors presented the conceptual design for a neutron imaging system for use in the 10–15 MeV energy range and discussed potential applications in the area of nuclear stockpile stewardship; see also Watterson (1997).

A system has been designed and a neutron generator installed to perform fast neutron radiography (Klann 1997). With this system, objects as small as a coin and as large as a 19 L container have been radiographed. The neutron source was an MF Physics A-711 neutron generator that produced 3×10^{10} neutrons/s with an average energy of 14.5 MeV. The radiography system used x-ray scintillation screens and film in commercially available light-tight cassettes. The cassettes have been modified to include a thin sheet of plastic to produce protons from the neutron beam through elastic scattering from hydrogen and other low Z materials in the plastic. For film densities from 1.8 to 3.0, exposures range from 1.9×10^7 n/cm^2 to 3.8×10^8 n/cm^2 depending on the type of screen and film. The optimum source-to-film distance was found to be 150 cm. At this distance, the geometric unsharpness was determined to be approximately 2.2–2.3 mm (or 8.7%–9.1% of the object size) and the smallest hole that could be resolved in a 1.25 cm thick sample had a diameter of 0.079 cm.

The initial use of the generator was for inspecting explosive materials within munitions in order to detect defects. During the fabrication process, cracks or bubbles can form within explosive substances. If these defects go undetected, they can cause an artillery shell to prematurely detonate, potentially killing members of the firing squad. Many such defects cannot be detected with x-rays, so the Army is seeking another nondestructive evaluation method to detect them. PNL's neutron radiography system could be the solution. The widespread use of neutron radiography for this application has been previously limited by the weak source strength of existing electronic neutron generators. This results in extremely long imaging times, making implementation of neutron-based nondestructive testing in factory settings impractical. PNL's neutron source is approximately 1000 times stronger than existing technologies and thus allows for the use of neutron radiography in a production environment (Phoenix Nuclear Labs 2013).

9.5 FAST NEUTRON RADIOTHERAPY

A beam of fast neutrons of several MeV is sufficiently penetrating for the therapy of deep-seated cancer. A monoenergetic beam of 14 MeV neutrons has depth-dose characteristics similar to those of megavolt x-rays (Bewley 1966). The neutron beam at Hammersmith (M. R. C. Cyclotron Unit, Hammersmith Hospital, Ducane Road, London, W. 12, England) is produced by directing a beam of 16 MeV deuterons onto a thick beryllium target. The spectrum extends up to 19 MeV with a modal value at about 6 MeV. The depth-dose distributions are similar to those of 250 kV x-rays, making therapy possible at many sites. The LET spectrum in tissue is complex, extending from low values characteristic of gamma radiation up to nearly 1000 keV/μ. It is not meaningful to describe the distribution in terms of a single effective LET since different parts assume varying degrees of importance depending on circumstances, for example, test material, size of dose, presence of oxygen or chemical protectors, etc.

The rationale for using fast neutrons or other high LET radiations in radiotherapy lies in their reduced oxygen enhancement ratio. The maximum dose of radiation that can be delivered depends on the response of normal tissues that are mostly well oxygenated; the oxygen enhancement ratio indicates the reduction in effective dose received by anoxic regions in a tumor. Consequently, the ratio of the two oxygen enhancement ratios, that of x-rays to that of neutrons, gives the increase in the effective dose received by these regions when neutrons rather than x-rays are used. This ratio has been called the gain factor by Alper (1963).

Neutrons are of interest for cancer therapy because they produce protons and other densely ionizing particles in tissue, which are much more effective in dealing with anoxic tumor cells than electrons or x-rays. Anoxic tumor cells are less sensitive to the ionizing effect of conventional x-ray irradiation than well-oxygenated normal cells. The protection of anoxic tumor cells diminishes as the ionization density is increased.

Current radiation therapy practice requires from the neutron source a high dose rate, good depth dose, well-defined horizontal and vertical beams movable about the patient, and a variable beam size up to about 20 × 20 cm. A minimum average yield of 4×10^2 n/s as a source strength, 125 cm as the specification on the source-skin distance, and 2 cm as the maximum allowable source diameter is required for cancer therapy. A typical collimator arrangement to produce a 10 × 10 cm^2 14 MeV neutron beam has been described by Greene and Major (1965). The isodose curves for this neutron field in air and in a water phantom were determined along the lines perpendicular to the central axis of the neutron beam. The collimator system ensures that only 15% γ-ray component contributes to the total dose around the central axis. The authors discussed measurements on 14 MeV neutron beams defined by a number of different collimator designs. It is shown that the differences produced in the beam definition, under scattering conditions, are fairly small, and limited in their effect by the presence of a large component of scattered radiation outside the main beam, in a phantom. A high proportion of the scattered radiation is gamma rays.

Only neutrons above 5 MeV have the necessary penetration in tissue; this factor, too, stresses the usefulness of 14 MeV neutrons in radiotherapy. The radiobiological effect of fast neutrons has been discussed in Elkind (1971) and Berry (1969). Tissue is predominantly composed by H, C, N, and O. Taking into account their cross sections and abundances at 14 MeV, about 65% of the absorbed dose is due to neutron–proton elastic scattering. Inelastic scattering and reactions may also occur in tissues and contribute to the dose. According to this, the absorbed dose is proportional to the H content of the organs, for example, it is about 15% higher in fatty tissues than in muscle tissues, while it is much lower in bones. The reverse is true with conventional x-ray therapy. A comparison of the measured isodose curves for 14 MeV neutrons at an SSD of 125 cm and for ^{60}Co at the usual treatment distance of 80 cm shows that 14 MeV neutrons have a satisfactory depth of penetration and in this respect are as good as ^{60}Co gamma rays. The dose buildup reaches a saturation value in the first few millimeters of the irradiated surface, which is related to the skin-sparing effect.

The 14 MeV neutrons have almost the same skin-sparing effect as the neutron beam from a cyclotron that accelerates deuterons to 30 MeV. A consequence of central axis depth dose curves is that the dose required for tumor cells located at a depth of 10 cm within the body is unacceptably high to healthy tissues near the surface, even if opposing radiation fields are used. According to experimental results, the relative biological effectiveness (RBE) for fast neutrons is higher than for x-rays and increases when an extended series of small dose fractions is applied.

Some practical advantages of sealed tube 14 MeV neutron generators for total body in vivo activation analysis have been discussed by Boddy et al. (1974), which summarizes the practical advantages of using two sealed tube 14 MeV neutron generators with a scanning geometry in conjunction with a shadow-shield whole body counter for this purpose. The requirements of neutron sources include uniformity of fast and thermal neutron fluence through the body; the simultaneous measurement of as many body elements as possible, especially Ca, P, N, Na, and Cl; an acceptable neutron dose; reproducibility; an irradiation procedure causing as little inconvenience and trauma as possible to the patient; and acceptable costs.

The clinical experience on patients with advanced malignant diseases is promising. The treatment schedule usually involves two fractions per week, modal tumor doses of 1.2–1.3 Gy of neutrons per fraction, and total doses in the range of 14–21 Gy. With this schedule, no severe normal tissue reactions were observed. The fractionating scheme employed has succeeded in establishing that regression of tumors could be achieved with fast neutrons in some cases.

To give statistical significance to the results, a large number of patients with specific types of cancer should be involved in such trials, using the relatively simple and inexpensive neutron generators or sealed-off tubes. The present status of fast neutron radiotherapy has been reviewed by Fowler (1966) who concluded that clinical results to date are controversial. It should be noted, however, that until then, very limited data had been available on clinical experiments.

Several authors have measured the dependence of RBE, the O enhancement ratio (OER), the linear energy transfer (LET), and the hypoxic gain factor (HGF) on neutron energy. It was clearly proved, for example, that the RBE value is higher for hypoxic cells than for well-oxygenated tumor cells. The OER value is between 2.5 and 3.0, while for neutrons for law LET radiation, it was found to be 1.5–1.6 in the energy range 1–15 MeV. The results are contradictory both for the neutron energy and dose fractionation dependence of OER. Measurements and calculations were carried out for the determination of a 14 MeV neutron dose delivered to the body via the reaction products of the major and trace elements.

9.5.1 Boron–Neutron Capture Therapy

Boron–neutron capture therapy (BNCT), originally proposed by Locher (1936), is an effective cancer treatment (see Haque et al. 1995; Elshahat et al. 2007). Because a high neutron flux is needed, a neutron reactor or a d–t neutron generator can be used as the neutron source in BNCT. Compared with the reactor neutron source, the d–t neutron generator is an optimal neutron source for this technique because of advantages such as removability, low cost, and no radioactivity after being turned off. In BNCT, the ratio of the fast neutron flux to the neutron flux in the tumor (RFNT) must be less than 3% to reduce neutron-induced damage and the neutron flux in the tumor (NFT) should exceed $10^9/cm^2/s$ to increase therapy efficiency. When a d–t neutron generator with a yield of 10^{13} n/s is used as a thermal neutron source in BNCT, the key point is that the 14 MeV neutron moderator must be designed appropriately to reduce the RFNT and increase the NFT. Hydrogen-rich materials are usually used as the neutron moderator, but the thermal neutron absorption cross section (TNACS) of hydrogen is too large to make the RFNT less than 3%. This means that the d–t neutron generator cannot be directly used in BNCT, no matter how high the yield of the neutron generator is. In order for it to be used in BNCT, heavy water, lead or graphite is usually employed to moderate the 14 MeV neutrons. Although their TNACSs are lower, the thermalization efficiencies for 14 MeV neutrons are so low that the yield of the d–t neutron generator must exceed 10^{13} n/s in BNCT (Agosteo and Curzio 2002). In order to increase the thermalization efficiency, tungsten, lead, and diamond were used to moderate the 14 MeV neutrons by the Monte Carlo neutron–photon transport code (MCNP) in paper by Cheng et al. (2012).

The boron capture therapy is a promising method in the treatment of malignant tumors. The incorporation of B can release more radiation energy in the tumor than in normal tissues via the reaction $^{10}B(n,\alpha)^{7}Li$ induced by thermal and epithermal neutrons. The extent of the dose enhancement can be measured by small BF_3 counters, by the track-etch method, or by activation foil techniques. The magnitude of the dose enhancement depends upon the depth, field size, and neutron beam energy. The dose enhancement at a depth of 8 cm was found to be 0.32% for each microgram of ^{10}B uptake per gram of tissue at En = 2.4 MeV incident neutron energy. The RBE value is about twice that of the fast neutron dose in the absence of boron because of the high LET of the reaction products. Results from Tokyo have reported longer survival of patients treated by boron-capture therapy at the time of brain surgery than by ^{60}Co and surgery. The main difficulty is to ensure a uniform and known high concentration of B in all parts of the tumor. Recently, results obtained by three dosimetric methods in a TE-phantom irradiated by fast neutrons were compared, and the neutron gamma ray doses at different depths were determined. For the determination of the fast neutron fluence and energy spectrum in the body produced by 14 MeV incident neutrons, unfolding methods should be improved. Cross and Ing (1985) have compiled and derived from the recent data the values of conversion factors between neutron fluence and various dosimetric quantities, including the kerma, dose equivalent (DE), and the quality factor as a function of neutron energy. They have given empirical analytical expressions for the energy variation of some of these conversion factors.

The energy of the neutron is usually reduced by the collision, especially elastic collision with the atomic nucleus. After an elastic collision, the average energy loss of a neutron is

$$\Delta E_{av.} = \frac{\left[2A/(A+1)^2\right]}{E_0}$$

where

E_0 is the initial energy of the neutron
A is the mass number of the nucleus

In principle, the smaller the A, the higher the moderating ability. In the paper by Cheng et al. (2012), diamond, heavy water, water, and polythene were used to moderate the 14 MeV neutrons. Although graphite is also often used as a neutron moderator, it was not studied in their paper because it has a lower density than diamond. Because the large diamond is too expensive to be used as a

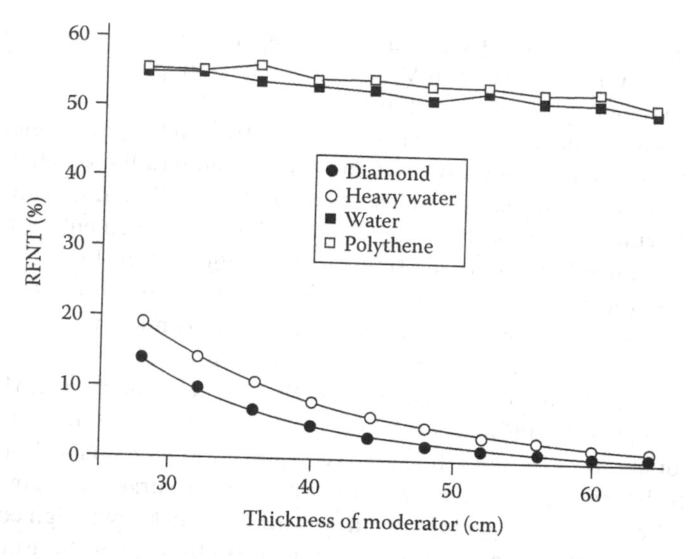

Figure 9.2 RFNT changed with the thickness of the moderator. (After Cheng et. al. (2012). Drosophila G-protein-coupled receptor kinase 2 regulates cAMP-dependent Hedgehog signaling. http://dx.doi.org/10.1242/dev.068817, *Development*, 139(1): 85–94.)

moderator, small artificial diamond particles (abbreviated as diamond) whose density is supposed to be 3.2 g/cm^3 were studied instead. The change of the ratio of the fast neutron flux to the neutron flux in the tumor, RFNT, with the thickness of the moderator is shown in Figure 9.2. The results from Figure 9.2 are as follows:

1. Neither polythene nor water, composed of carbon, hydrogen, and oxygen, can be used as a neutron moderator in BNCT. The TNACSs of carbon, hydrogen, and oxygen are about 3 mb, 0.19 b, and 0.1 mb, respectively. In water and polythene, the number densities of hydrogen atoms are high, which can reduce the fast neutron flux. But the TNACS of hydrogen is large, which will reduce the thermal neutron flux. So, it is difficult to estimate the relationship between the RFNT and the number density of hydrogen atoms. Based on the simulations, they found that when the thickness of polythene or water was 60 cm, the RFNT is still more than 50%. If they are used as the moderators in BNCT, the RFNT is hardly less than 3%.
2. Heavy water and diamond can be used as a moderator in BNCT. They are made of deuterium, oxygen, and carbon. The TNACS of deuterium is about 0.0005 b. Although the moderator abilities of heavy water and diamond are low, their TNACSs are small, which can increase the thermal neutron flux and reduce the RFNT. If the thickness of heavy water or diamond is more than 60 cm, the RFNT is less than 3%.

3. The RFNT of diamond is the smallest if the moderators have the same thicknesses. So, diamond was used to moderate the 14 MeV neutrons in the paper by Cheng et al. (2012). If the neutron flux is normalized to one source neutron in the simulation, the maximum NFT (abbreviated as MNFT), moderated by diamond, was 1.007×10^{-4} cm^{-2} s^{-1} at the RFNT, which is less than 3% (Cheng et al. 2012).

These authors also considered double- and three-layers moderators. Namely, if the 14 MeV neutrons are inelastically scattered by the heavy metal, the energies of the neutrons are often lower than 3 MeV and the peak value of the neutron spectrum is about 2 MeV. If a double-layer moderator composed of heavy metal and diamond is used to moderate the 14 MeV neutrons, the NFT can be greatly increased under the condition of RFNT, which is less than 3%. Uranium and lead are often used to first moderate the 14 MeV neutrons. In this work, uranium was not studied because of its high price and large TNACS. Tungsten is a very heavy metal, whose ability to moderate the 14 MeV neutrons is very strong. The (n,2n) cross sections of lead and tungsten are both about 2 b, which can increase the neutron flux. In this study, lead or tungsten was first used to moderate the 14 MeV neutrons, and then diamond. Taking the double-layer moderator composed of tungsten and diamond as an example, the simulation process is as follows:

1. The radius of tungsten (RW) was increased from 1 to 24 cm in steps of 1 cm.
2. The radius of diamond (RD) was increased from (RW + 1) cm to 63 cm at each RW, and then, this model was simulated by the MCNP code
3. Fixing the RW, RD was changed to get the MNFT (maximum value of the NFT at each RW) under the condition of RFNT, which is less than 3%. The MNFT could be considered as the NFT of tungsten with RW (cm) thickness.

Following conclusions could be drawn from the measurements:

1. When the thickness of the heavy metal is less than 7 cm, the NFT with tungsten is slightly higher than that of lead. Because the density of tungsten is much higher than lead, the moderating ability of tungsten is stronger than that of lead.
2. When the thickness of the heavy metal is more than 7 cm, the NFT with lead is much higher than tungsten. The moderating ability of tungsten is stronger than lead, but its TNACS (11 b) is much larger than lead (0.1 b). With the increase of the thermal neutron flux, the NFT moderated by tungsten and diamond becomes lower than that moderated by lead and diamond.

Under the condition of RFNT, which is less than 3%, the MNFT, moderated by the double-layer moderator composed of lead and diamond, is 2.590×10^{-4} cm^2/s.

Considering the stronger moderating ability of tungsten and the smaller TNACS of lead, a moderator composed of three layers was designed, and the 14 MeV neutrons were first moderated by tungsten, then by lead, and at last by diamond. The simulation process is as follows (Cheng et al. 2012):

1. The RW was increased from 1 to 7 cm in steps of 1 cm.
2. The radius of lead (RPb) was increased from (RW + 1) cm to 20 cm at each RW.
3. The RD was increased from (RPb + 1) cm to 63 cm at each RPb, and then this model was simulated by the MCNP code.
4. Fixing the RW, RPb and RD were changed to get the MNFT at the RFNT, which is less than 3%. Keeping the RFNT less than 3%, the MNFT, moderated by the three-layer moderator composed of a 3 cm–thick tungsten layer, a 14 cm–thick lead layer, and a 21 cm–thick diamond layer, was 2.882×10^{-4} cm^2/s.

In conclusions, under the RFNT conditions of less than 3%, the MNFTs moderated by a single-layer moderator, a double-layer moderator, and a three-layer moderator were 1.007×10^{-4}, 2.590×10^{-4}, and 2.882×10^{-4} cm^2/s, respectively. A three-layer moderator was composed of a 3 cm–thick tungsten layer, a 14 cm–thick lead layer, and a 21 cm–thick diamond layer. If it is used to moderate the 14 MeV neutrons in BNCT, the yield of the d–t neutron generator only needs to be 3.470×10^{12} n/s (Cheng et al. 2012).

Radio frequency (RF)–driven ion sources are being developed in Lawrence Berkeley National Laboratory (LBNL) for sealed-accelerator-tube neutron generator application. By using a 5 cm diameter RF-driven multicusp source H^+, yields over 95% have been achieved. These experimental findings will enable one to develop compact neutron generators based on the d–d or d–t fusion reactions. In this new neutron generator, the ion source, the accelerator, and the target are all housed in a sealed metal container without external pumping.

Recent moderator design simulation studies have shown that 14 MeV neutrons could be moderated to therapeutically useful energy ranges for BNCT. The dose near the center of the brain with optimized moderators is about 65% higher than the dose obtained from a typical neutron spectrum produced by the Brookhaven Medical Research Reactor (BMRR) and is comparable to the dose obtained by other accelerator-based neutron sources. With a 120 keV and 1 A deuteron beam, a treatment time of ~35 min is estimated for BNCT (Verbeke et al. 1999).

9.6 COLLIMATED NEUTRON BEAMS

9.6.1 Slow Neutron Collimators

Neutrons in moderator are scattered in all directions, which is not suitable for radiography. Neutron collimator is a structure designed to extract slow neutron beam from the moderator to the specimen. Ideally, parallel neutron beam is preferred because it gives best image sharpness. If this is the case, Soller or multitube collimator is used. However, divergent collimator is easier to construct and gives good image sharpness depending on the geometrical parameters as will be discussed later.

1. Soller or multitube collimator: This collimator is constructed with neutron absorbing materials such as boron, cadmium, and gadolinium as illustrated in Figure 9.3 so as to bring parallel neutron beam to the test specimen. Neutrons can only get into the collimator from one end which is in the moderator then get out to the other end. Neutrons that are not traveling in parallel with the collimator axis will hit the side of the tube or plate and then will be absorbed allowing only neutrons traveling in parallel with the tube axis to reach the test specimen. This type of collimator

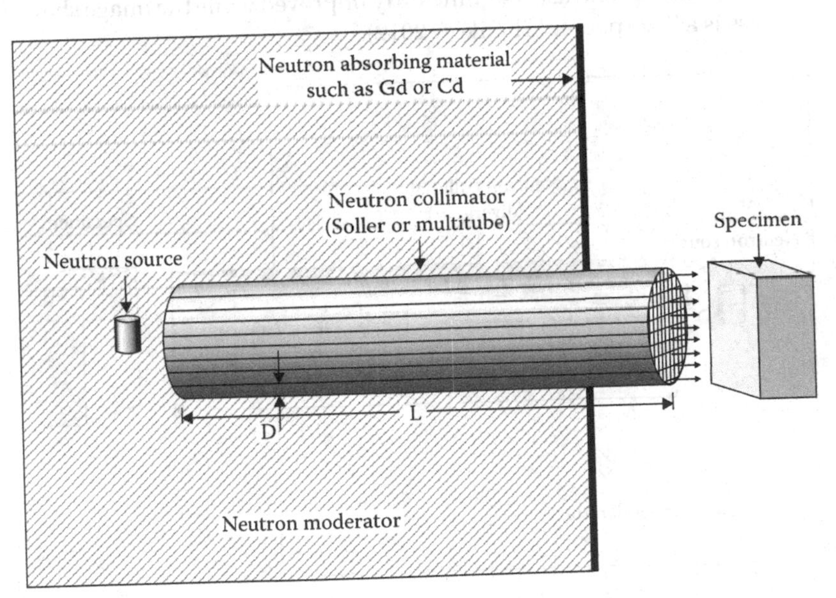

Figure 9.3 Multitube neutron collimator following Chankow (2012). (From Chankow, N., Neutron radiography, in Omar, M., Ed., *Nondestructive Testing Methods and New Applications*, InTech, Rijeka, Croatia, 264pp., 2012.)

is applicable to nuclear reactor where input neutron intensity to the collimator is high. The drawbacks are that the pattern of parallel plates or tubes may be seen on the image and it is more costly to be constructed in comparison to the divergent collimator (Chankow 2012).

2. Divergent collimator: Divergent collimator is designed in the way that neutrons are allowed to get into the collimator only through a small hole from one end and then diverge at the other end; see Figure 9.4. The collimator is lined with neutron absorber to absorb unwanted scattered neutrons. It is easy to be constructed and can be used with nonreactor neutron source like radioisotope and accelerator where slow neutron input is low. The drawback is that image sharpness may not be as good as the Soller collimator. For low neutron intensity as in radioisotope system, neutron output at the specimen position can still be increased by making part of the collimator on the input or source side free from neutron absorber as shown in Figure 9.5. Neutrons can thus enter the collimator through this part resulting in increasing of neutron intensity. From experience with ^{241}Am/Be and ^{252}Cf sources, neutron intensity can be increased approximately by 10%–60% and the cadmium ratio can also be increased from about 5 to 20. In doing so, the image contrast is significantly improved while the image sharpness is a little poorer (Chankow 2012).

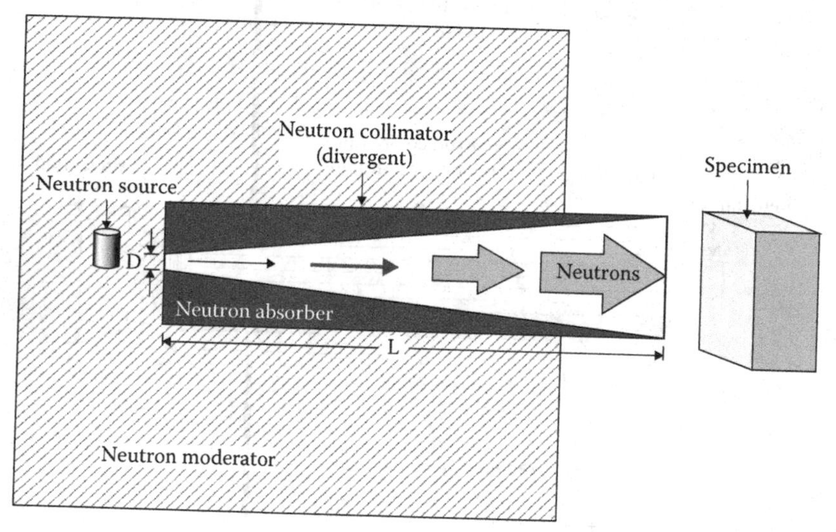

Figure 9.4 Divergent neutron collimator allowing neutrons to get into the collimator only through the hole of diameter "D." (After Chankow, N., Neutron radiography, in Omar, M., Ed., *Nondestructive Testing Methods and New Applications*, InTech, Rijeka, Croatia, 264pp., 2012.)

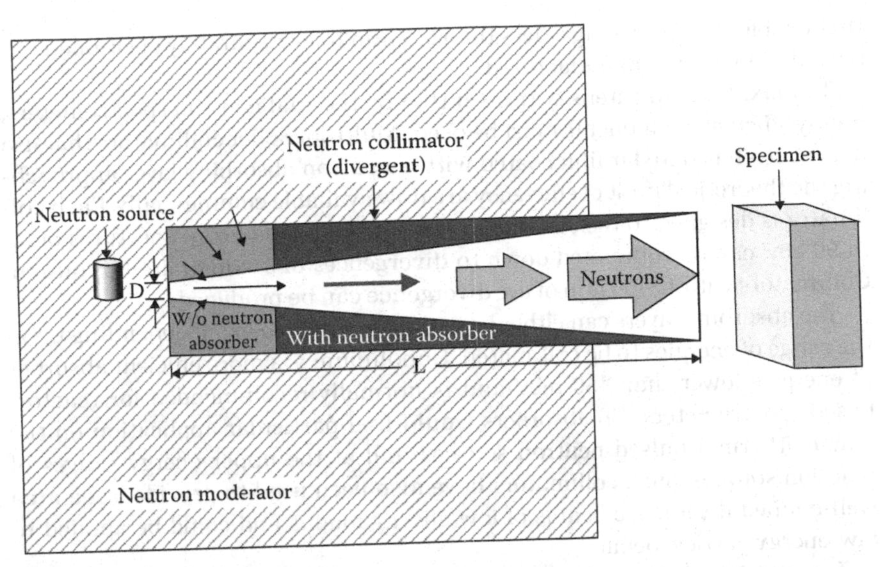

Figure 9.5 Divergent neutron collimator with part of the source side contains no neutron absorber allowing more neutrons to get into the collimator. (After Chankow, N., Neutron radiography, in Omar, M., Ed., *Nondestructive Testing Methods and New Applications*, InTech, Rijeka, Croatia, 264pp., 2012.)

The company (Euro Collimators 2012) has successfully designed, developed, and produced Soller, diverging, double diverging, honeycomb, radial and oscillating radial collimators utilizing a variety of materials and methods, which include foils made of Mylar, high performance thermal plastic (PEEK), aluminum, and stainless steel coated with gadolinium oxide or ^{10}B absorbers. Euro Collimators' most frequently produced collimator is the highly successful Soller slit collimator with stretched Mylar foils coated with gadolinium oxide or ^{10}B absorber and optimized to produce transmissions greater than 95% of the theoretical maximum.

This type of collimator has been further developed to include collimators with either aluminum or PEEK foils. Each has excellent resistance to gamma radiation, capable of absorbing over 1000 rads of irradiation as well as over 10 rads of alpha or beta irradiation—the latter are more reliable and equally as efficient. All types of foil can be coated with gadolinium oxide or ^{10}B and are suitable for in-pile collimation. The efficiency of Soller slit collimators for limiting the divergence neutron beams depends upon the uniformity of the slit spacing throughout the collimator length, the thickness of the foil, and the straightness of the foil edges, as well as the degree of absorption of the foil. These points are particularly important when fine collimation (less than 10 min of arc) is required. A system for the tensioning of small divergence collimators that enables blade spacing of less than 0.5 mm to be accurately positioned has been developed and manufactured.

To maximize the intensity transmitted by thermal neutron spectrometers, highly efficient components must be used. Multislot Soller collimators that use foils of stretched Mylar film coated with a neutron absorbing layer approaching the theoretical limit of efficiency are now available commercially. Each collimator is designed to the customer's requirements and beam apertures of up to 50 cm^2 can be collimated down to divergences of 5 min of arc if required. Collimators with 10–60 min of arc divergence can be produced.

The absorbing layers can either be gadolinium based or ^{10}B depending upon the range of energies to be collimated. Gadolinium is a more efficient absorber of energies lower than 150 meV and is normally recommended for reactor-based spectrometers. ^{10}B absorbers would find particular application on the new epithermal pulsed neutron sources and at the upper energy ranges of reaction sources. Such collimators have also been used to provide large-area well-defined divergence beams of low energy muons and could be applied to low-energy particle beams.

For special applications, converging Soller collimators have been built using the aforementioned technique and crossed collimator arrangements find applications in small-angle scattering diffractometers; see http://www.eurocollimators.com/Products.html. From their production line we present here the double diverging rotating radial collimator (see Figure 9.6) and oscillating radial collimator (see Figure 9.7).

Technical characteristics of double diverging rotating radial collimator are

Angle between centers of foils: 0.19°
Effective foil length: 450 mm
Total collimator angle of both planes: 7.79°
Foil material: Mylar

Figure 9.6 Double diverging rotating radial collimator made by Euro collimators. (From http://www.eurocollimators.com/Products.html.)

Figure 9.7 Oscillating radial collimator, made by Euro collimators. (From http://www.eurocollimators.com/Products.html.)

Absorber: Gadolinium oxide
Number of foils: 40
Coated foil thickness: 0.076 mm
All materials are nonmagnetic

The collimator is held in an adjustable ring so that it can be moved along its longitudinal axis without the need to remove from its support assembly.

Technical characteristics of oscillating radial collimator are the following:

Angular aperture: 170°
Eighty-four vanes 0.20 mm thick coated with 0.05 mm of gadolinium oxide absorber
Height of vanes: 250 mm
Inside radius: 300 mm
Outside radius: 400 mm

Each vane is individually tensioned and capable of being replaced without the instrument being dismantled.

Collimated neutron beam for neutron radiography is also discussed by Dinca et al. (2005); the obtaining of a collimated neutron beam on the tangential channel of the ACPR reactor from INR Pitesti, Romania, aimed to satisfy the requests of a neutron radiography facility is presented. The collimation of neutrons means the elimination from the neutron beam those neutrons that have trajectories that are not inside the space defined by walls or successive apertures that are made of neutron absorbent materials. The assembly that assures the collimation of neutrons, named collimator, is optimized using MCNP 4B code based on Monte Carlo method for neutrons and gamma radiation.

An ideal neutron beam should be parallel, monoenergetic, with big intensity, free of other contaminant radiation and uniform on its cross section. In practice it is intended to have experimental arrangements to accomplish neutron beam parameters as closely as possible to ideal ones. For this purpose a collimator is used. The neutrons pass through a collimator from the entrance aperture placed near the neutron source to the exit window used for neutron radiography investigations. The inner space of a collimator is evacuated or filled with air, or better filled with helium. A characteristic parameter of a collimator that defines the degree of divergence of the neutron beam is the L/D ratio, where L is the length of the collimator and D is the diameter (or generally the opening) of the entrance aperture:

- The thermal neutron beam intensity at least 5×10^5 n/cm^2/s
- The collimation ratio, L/D, at least 90
- The exit window, 250 mm in diameter
- The n/γ ratio at least 1×10^6 n/cm^2/mrem (that determines used investigation methods)
- The divergent angle under 4°
- The cadmium ratio above 17

In order to obtain a thermal neutron beam with such parameters, the following experimental setup was used as reported by Dinca et al. (2005): (1) a graphite illuminator placed on channel near to the reactor core to scatter neutrons toward the exit of the channel, (2) a mobile monocrystalline bismuth filter for the attenuation of the gamma radiation and scattering of fast neutrons that will allow performing direct neutron radiography investigations and also γ radiography investigations, and (3) a set of successive apertures from boral, indium, and lead for the formation of the divergent collimator. The position and dimensions of these components were optimized by calculations made with MCNP 4B code based on Monte Carlo method both for thermal neutrons and gamma radiation.

9.6.2 Use of Collimated Fast Neutron Beam

The most important mechanism of interaction between neutrons and the inspected object is scattering. However, the desired attenuated signal one would like to measure in the detector is the result of primary neutrons. The result of scattered neutrons that are measured in the detector is image contrast degradation in the radiography images and artifacts such as streaks and cupping in tomography images. In conventional x-ray CT, this problem is mostly dealt with hardware solutions such as high Z material antiscatter grids that are part of the detecting system, preventing the scattered photons from reaching the detector. There are also numerous attempts to solve the issue by using software corrections (Sabo-Napadensky and Amir 2005; Star-Lack et al. 2009; Grogan 2010). Both are complicated, do not solve the problem completely, and may eventually induce new artifacts. In the case of neutrons, the hardware solution that exists for x-ray is impossible.

In the work described by Sabo-Napadensky et al. (2012), the main effort was to lower the amount of scattered neutrons that arrive at the detector by either limiting the amount of scattered neutrons in the object under inspection or by lowering the amount of scattered neutrons from the inspected object or the surrounding to get into the detector. The experimental setup consists of an off-the-shelf d–t generator (VNIIA) producing a monochromatic 14.2 MeV fast neutron beam of 2×10^9 n/s in 4π. As a consequence, it is limited on one hand by the flux to short distances, which means more scattered neutrons arrive at the detectors and on the other hand to smaller scanned objects, which means less scattered neutrons are scattered in the object itself. The generator is followed by an upstream beam collimator, the scanned object, a downstream collimator in the fan collimation configuration, and at the end the detection system (see Figure 9.8). The detection system consists of a polystyrene scintillating fiber array screen, with an active area of 200 × 200 mm, 30 (or 10) mm depth,

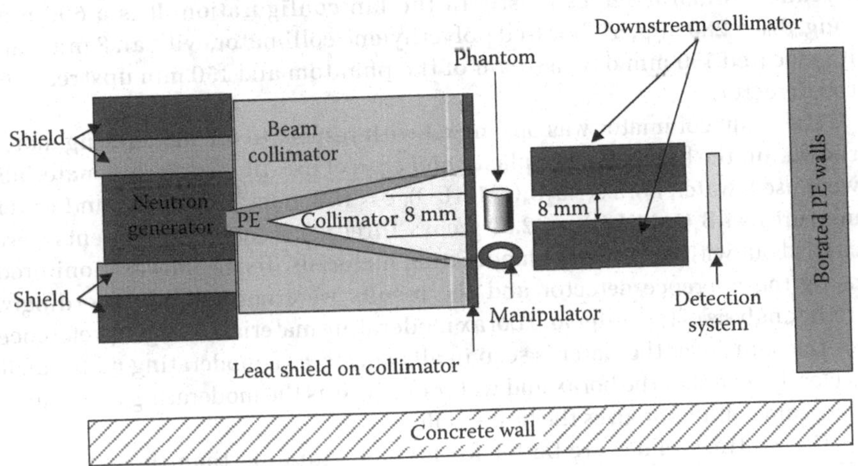

Figure 9.8 The scheme of collimator configuration in the experimental layout in the work described by Sabo-Napadensky et al. (2012). (From Sabo-Napadensky, I. et al., Research and development of a dedicated collimator for 14.2 MeV fast neutrons for imaging using a D-T generator, in *2nd International Workshop on Fast Neutron Detectors and Applications*, Ein Gedi, Israel, November 6–11, 2011, IOP Publishing Ltd and Sissa Medialab srl, doi:10.1088/1748-0221/7/06/C06005, 2012.)

and 0.5 (or 1.0) mm pixel size. The neutrons arrive at the detector, interact in the fiber scintillating screen, and produces light image. Optics transfer the image to the image-intensifier and a CCD camera views the image created at the phosphor of the image intensifier.

Two phantoms were used in their experiment:

1. A point spread function (PSF) phantom: a stainless steel cylinder shaped with a concentric cylinder hole in it
2. A modulation transfer phantom (MTF): a box shaped stainless steel with slits in different sizes

Next to the generator there was a plastic scintillator neutron detector that served as a reference detector for beam monitoring. The upstream collimator was a stainless steel "barrel-shaped" container. It is 1000 mm long, with a radius of 275 mm, and a side wall of 2 mm wide material. The barrel is filled with one of the three different moderating materials examined. The results for the three different moderating materials are compared as part of the optimization studies (Sabo-Napadensky et al. 2012).

In order to define the fan beam configuration, a fan beam definer made of 5% borated polyethylene is put into the cone hole. Once it is in the hole it defines a horizontal fan beam of 8 mm gap. The width of the fan beam gap was optimized in order to get maximum SNR and contrast. The downstream collimator serves mostly in the fan configuration. It is a 600 mm long, 320 × 280 mm, 5% borated polyethylene collimator, with an 8 mm gap. It is located 150 mm downstream of the phantom and 150 mm upstream of the detector.

The beam collimator was optimized with respect to the moderating material within the barrel. For simplicity and ease of use, three different materials were used: water, borax ($Na_2B_4O_7$-$10H_2O$), $\rho = 1.73$ g/cm^3, and borax and water mixture ($Na_2B_4O_7$-$14H_2O$), $\rho = 2.09$ g/cm^3. Three consecutive experiments were carried out with the different moderating materials. The beam was monitored using the reference detector and the results were normalized accordingly. In the analysis of the water and borax moderating materials, a slight preference for the borax over the water is seen in all aspects, as a moderating material. It is clearly seen that the borax and water mixture as the moderating material in the beam collimator gives the best results.

The contrast values are better for fan configuration than those for cone configuration for all the moderating materials. Comparing the results of the borax and water moderating materials, one can see that there is a slight advantage to the borax on water. The scatter of the surroundings is reduced meaningfully using a dedicated massive generator shield, which was built based on experimental and simulation studies and tested successfully for proven efficiency.

As a consequence a dedicated collimator for a d–t generator 14.2 MeV fast neutrons is developed for imaging purposes, using the massive generator shield and the borax and water mixture as the moderating material by Sabo-Napadensky et al. (2012).

A new collimator system was constructed to produce a new collimated d–t neutron beam for new integral benchmark experiments at the first target room of the Fusion Neutronics Source (FNS) facility in the Japan Atomic Energy Agency (JAEA). The collimator system had been designed and optimized with a neutron transport calculation code and the performance of the collimated d–t neutron beam was tested with an imaging plate, activation foil, and a scintillation counter. The d–t neutron flux at the exit of the collimator hole was 2.2×10^6 cm^2/s, which was 239 times as large as that at the 20 cm off-centered position. It was confirmed that the new d–t neutron beam had a good performance as expected (Ohnishi et al. 2011a). They have designed a new collimator system with a neutron transport code to produce a d–t neutron beam in the target room. In this study, the new collimator system was constructed based on this design and the characteristics of the d–t neutron beam were measured. Integral benchmark experiments for nuclear data benchmarking, in situ experiments (Maekawa et al. 1998), and TOF (time-of-flight) experiments (Oyama et al. 1990) have been carried out with d–t neutrons at the FNS facility in JAEA. Neutron spectra, reaction rates, gamma-heating rates, etc., were measured inside a thick experimental assembly in the in situ experiments, where uncollided and strongly forward scattered neutrons are mainly detected in the whole energy region. Angular leakage neutron spectra above 0.1 MeV from a thick experimental assembly are measured in the TOF experiments. These experimental data are very useful for nuclear data validation, but only a small angle part in the angular distribution of nuclear data could be mainly targeted.

The collimator system was constructed with the assemblies used in the past ITER shielding experiments (Konno et al. 1994) in JAEA/FNS. The assemblies consisted of the source reactor and hollow plates made of type 316 stainless steel (SS316) as shown in Figure 9.9. While the void regions inside of the hollow plates in Figure 9.9 had been fully filled with 5 × 5 × 5 cm blocks to simulate ITER blankets and the vacuum vessel in the past experiments, in this work, a 1 mm thick aluminum rectangular pipe whose width, length, and height were 5, 5, and 45 cm, respectively, was installed on the center of the hollow plates to make the rectangular collimator hole of 5 × 5 cm^2. The other void regions were filled with various blocks as described later. The source reflector was also reused as a source shield in order to reduce leakage neutrons, which became a large background source by scattering in the room wall. The source shield had a 4 × 8 cm rectangular hole to insert a beam line duct with a tritium target. The tritium target was inserted into the reflector so that the distance from the entrance of the collimator hole was 40 cm. The hollow plates were filled with four layers. The first layer was made with

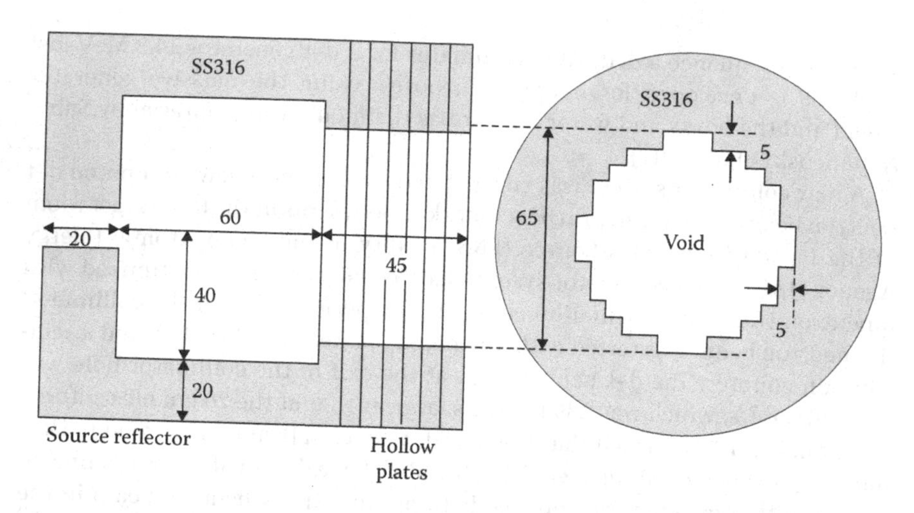

Figure 9.9 A collimator system at ITER assemblies, as described in Konno et al. (1994). All dimensions shown are in cm; various 5 × 5 × 5 blocks can be installed inside the hollow plates. (From Konno, C. et al., *Bulk Shielding Experiments on Large SS316 Assemblies Bombarded by D-T Neutrons, Volume I: Experiment*, JAERI-Research 94-043, Japan Atomic Energy Research Institute, Tokai, Japan, 1994.)

heavy materials to attenuate d–t neutrons. As the second layer, light materials were used to moderate neutrons. Neutron absorbers were used in the third layer to absorb moderated neutrons. Finally, lead blocks were employed as the forth layer to shield secondary gamma rays. Considering materials that are stored in FNS, the four layers had been optimized with a neutron transportation calculation and authors adopted 15 cm thick tungsten and 10 cm thick SS316 as the first layer, 10 cm thick polyethylene as the second layer, 5 cm thick lithium oxide as the third layer, and 5 cm thick lead as the fourth layer as shown in Figure 9.10.

Four measurement points were set as shown in Figure 9.10: (1) EP1, exit of the collimator hole; (2) EP2, 20 cm apart from EP1 along the collimator hole center; (3) OP1, 20 cm offset from EP1; and (4) OP2, 20 cm offset from EP2.

They used the Monte Carlo radiation transport code (MCNP5.14, 2003) and the cross section library, FENDL/MC-2.1 (Aldama and Trkov 2004), in order to analyze the experiment in detail. The profiles of the neutron beam were measured with imaging plates (IPs) at EP1 and EP2. They chose $^{27}Al(n,\alpha)^{24}Na$ reaction to convert neutrons to gamma-emitting radionuclides because of the interest in d–t neutrons. Then, aluminum plates of 0.1 cm in thickness, 20 cm in width, and 15 cm in height were placed at EP1 and EP2. After d–t neutron irradiation for an hour, they were closely attached to gamma-sensitive imaging plates, which were irradiated by photon from ^{24}Na, for 2 days. The output images of the IPs were shown in Figure 9.11. The d–t

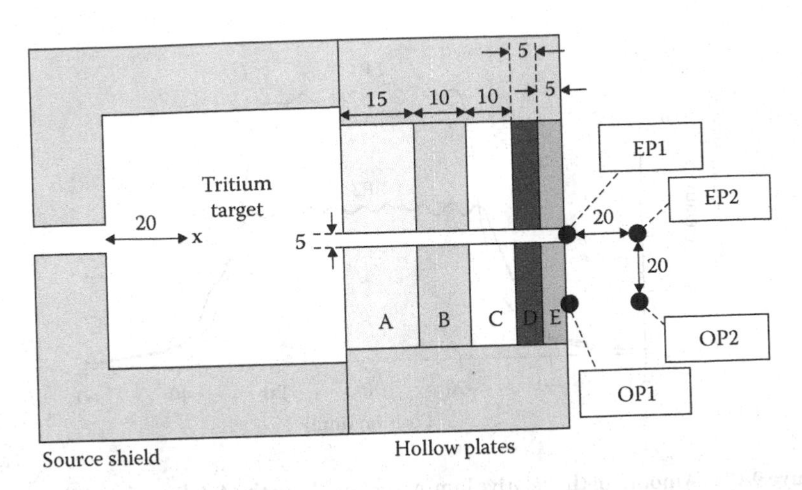

Figure 9.10 Multilayer collimator system (stainless steel, polyethylene, tungsten, lithium oxide, and lead) and four measuring points (EP1, EP2, OP1, and OP2); see text. All dimensions are in cm. The entrance into collimators is made of 0.1 cm thick aluminum rectangular frame. (After Konno, C. et al., *Bulk Shielding Experiments on Large SS316 Assemblies Bombarded by D-T Neutrons, Volume I: Experiment*, JAERI-Research 94-043, Japan Atomic Energy Research Institute, Tokai, Japan, 1994.)

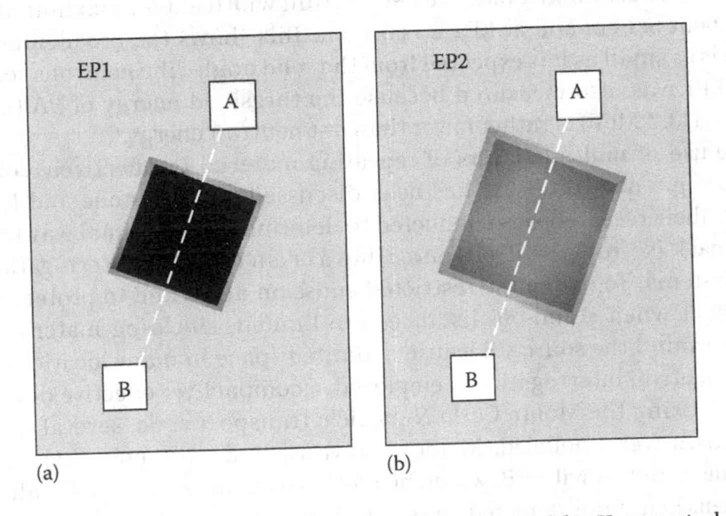

Figure 9.11 The d–t neutron 2D distributions as reported by Konno et al. (1994). Intensity distributions shown in Figure 9.12 are taken along A-B lines. (From Konno, C. et al., *Bulk Shielding Experiments on Large SS316 Assemblies Bombarded by D-T Neutrons, Volume I: Experiment*, JAERI-Research 94-043, Japan Atomic Energy Research Institute, Tokai, Japan, 1994.)

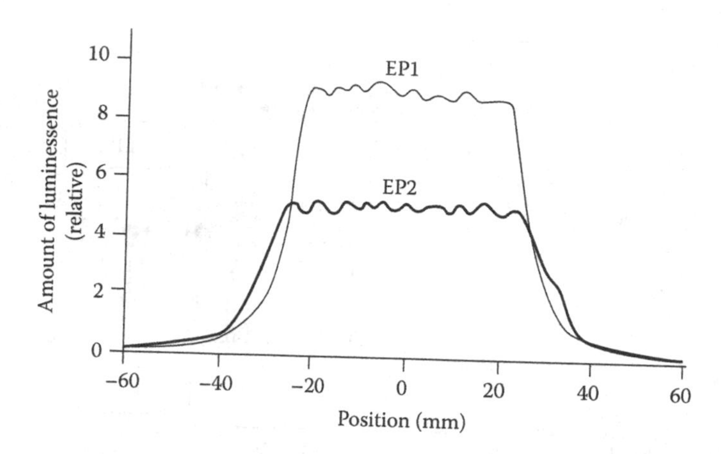

Figure 9.12 Amount of the relative luminescence along the A-B lines from Figure 9.11 as reported by (Konno et al. 1994).(From Konno, C. et al., *Bulk Shielding Experiments on Large SS316 Assemblies Bombarded by D-T Neutrons, Volume I: Experiment,* JAERI-Research 94-043, Japan Atomic Energy Research Institute, Tokai, Japan, 1994.)

neutron one-dimensional distributions along the line A–B in Figure 9.11 was read and their background counts were subtracted. They were shown in Figure 9.12. These figures show that the neutron distributions over the collimator hole are almost flat. The FWHM (full width at half maximum) at EP1 was about 5.3 cm and at EP2 was 6.4 cm. This shows the broadening of the beam is as small as it is expected from the solid angle. The neutrons scattered near EP1 was also measured because the threshold energy of $^{27}Al(n,\alpha)^{24}Na$ reaction (3.2 MeV) is rather lower than d–t neutron energy.

The use of multiple layers of repeating material to effectively collimate an isotropic neutron beam has been discussed in Whetstone and Kearfott (2011). Their research was conducted to determine the optimal way to shield a compact, isotropic neutron source into a beam for active interrogation neutron systems. To define the restricted emission angle and to protect nearby personnel when stand-off distances are limited, shielding materials were added around the source. Because of limited space in many locations where active neutron interrogation is employed, a compact yet effective design was desired. Using the Monte Carlo N-particle transport code, several shielding geometries were modeled. Materials investigated were polyethylene, polyethylene enriched with ^{10}B, water, bismuth, steel, nickel, INCONEL® alloy 600, tungsten, lead, and depleted uranium. Various simulations were run testing the individual materials and combinations of them. It was found that at a stand-off distance of 1.5 m from the source, the most effective shielding configuration is a combination of several layers of polyethylene and steel.

Without any shielding, the dose is 3.71×10^{-15} Sv/source particle. With a shielding consisting of multiple layers of steel totaling 30 cm thickness interspersed with several layers of polyethylene totaling 20 cm thickness, the dose drops to 3.68×10^{-17} Sv/emitted neutron at π radians opposite the shield opening. The layered shielding approach is more effective at reducing dose equivalent and neutron fluence than shields made out of single continuous layers of the same material and thicknesses. Adding boron to the polyethylene and substituting tungsten for steel would make the shielding more effective but would add mass and cost.

In another paper Ohnishi et al. (2011b) have reported a new collimator that is designed and optimized to produce a new d–t neutron beam with a small target in the first target room of the Fusion Neutronics Source facility in Japan Atomic Energy Agency. The characteristics of the collimator are calculated with the DORT code and the FENDL/MG-2.1 multigroup cross section library. It is concluded that the calculated neutron fluence above 14 MeV at the exit of the collimator is 180 times as large as that at the 10 cm offset position from the beam axis.

The multilayer collimators shown in Figure 9.13 were examined. In order to make use of neutron absorption reaction of SS, it is desirable to install SS layers behind PE ones. On the contrary, it is desirable to install PE behind SS, where neutrons below 1 MeV are significant, in order to make full use of moderating power of PE (both parts are made of type 316 stainless steel, SS316); Thus a multilayer structure is adopted; polyethylene (PE) layers to moderate neutrons in MeV region, SS layers to attenuate 14 MeV

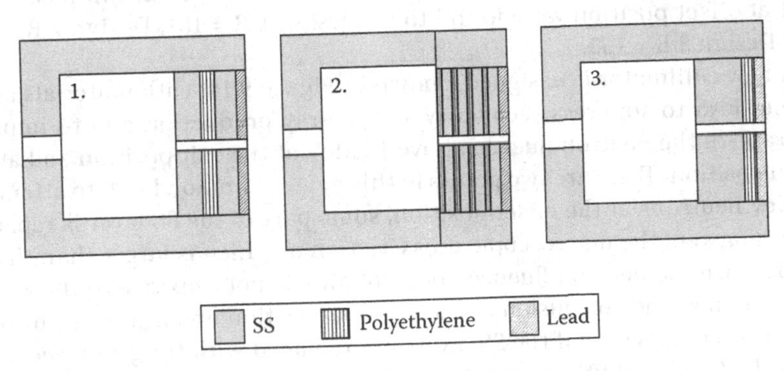

Figure 9.13 Multilayer collimators examined by Ohnishi et al. (2011b). Design 1: Middle positioning moderator—SS 15 cm + PE (polyethylene) 10 cm + SS 15 cm + Pb 5 cm. Design 2: Distributed moderator—(PE 2.5 cm + SS 7.5 cm) × 4 + Pb 5 cm. Design 3: Post-positioning moderator—SS 30 cm + PE 10 cm + Pb 5 cm. (From Ohnishi, S. et al., *Prog. Nucl. Sci. Technol.*, 1, 73, 2011.)

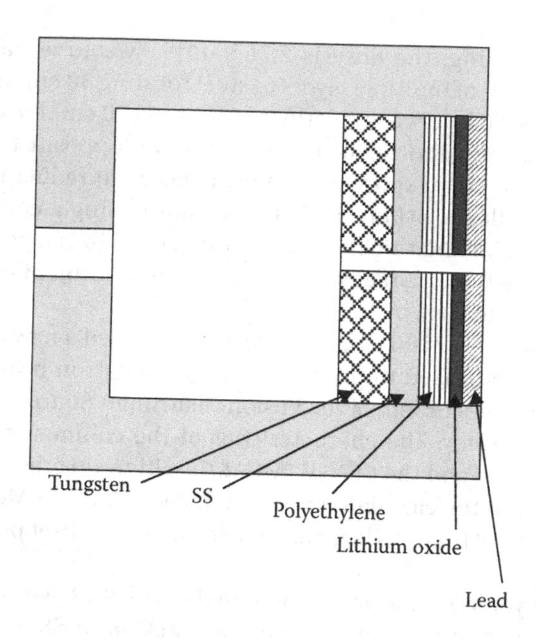

Figure 9.14 FNS collimator designed by Ohnishi et al. (2011b). Materials used are tungsten (15 cm); stainless steel (10 cm); polyethylene (10 cm); lithium oxide, Li_2O (5 cm); and lead (5 cm). (From Ohnishi, S. et al., *Prog. Nucl. Sci. Technol.*, 1, 73, 2011.)

neutrons and to absorb thermal neutrons, and a lead layer to shield secondary gamma rays. Ratios of neutron fluence above 14 MeV at exit position to that at offset position were found to be Design 1 R = 151; Design 2 R = 152; and Design 3 R = 152.

A new collimator is designed (shown in Figure 9.14) with materials available at FNS to suppress secondary gamma-ray production and to improve the ratio of the neutron fluence above 14 MeV at the exit position and at the offset position. There are two points in this new collimator. First, to attenuate 14 MeV neutrons at the offset position, some part of the SS layer is replaced with tungsten, the macroscopic cross section of which is larger than that of SS. Because the neutron fluence above 14 MeV is not sensitive to the shielding structure, the conclusion of the previous section about moderator position still remains even if the SS layers are replaced with tungsten. Secondly, 5 cm thick natural lithium oxide (Li_2O) and lead layers are added into the collimator to absorb thermal neutrons and to attenuate secondary gamma ray respectively. The collimator (FNS collimator) of five layers is illustrated in Figure 9.14.

A new collimator is designed for providing a new d–t neutron beam at JAEA/FNS TR1. To make the neutron spectra monoenergetic and to concentrate the neutron beam on the axis of the collimator, multilayer collimators are designed and optimized. Among those multilayer collimators, the FNS collimator shows especially good performance in the whole energy region and the neutron fluence above 10 MeV at the exit of the collimator is 180 times as large as that at the offset point. Scattered neutrons in the wall of TR1 are very few in MeV region. An experiment of neutron measurement with this FNS collimator geometry is planned.

In another work Maglich and Nalcioglu (2010) have reported the use of so-called smart collimator for in vivo cancer diagnostics. The first remark that should be made concerning the design of the smart collimator is that the device one seeks to construct resembles a reflector much more than a collimator, since the principle aim of this device is to enhance the flux at the target rather than decreasing the flux outside the target as traditional collimator would. Based on the experiments on differential femto oximetry, Maglich and Nalcioglu (2010) conducted a computer simulated study of the feasibility of conceptual design for their noninvasive malignancy probe, Oncosensor, to diagnose hypoxia of malignancy $M = -0.90$, measured by pO2—which correspond to volume averaged hypoxia $M' = -0.09$— in 1, 3, and 5 cm DIA tumors embedded in the middle of a 10 cm DIA breast. M' is further masked by background γ's from the in vivo tissue by factor $x = 4.4$–7 for subcutaneous and central tumor, respectively, to apparent $M'' = M'/X$ which, in turn, renders hypoxia nondiagnosable for 1 cm tumors; marginally so for 3 cm ones with specificity $S = 75\%$; and fully diagnosable with $S = 95\%$ in 5 cm ones. To diagnose 1–3 cm and smaller tumors, the authors propose to enhance M'' by a factor of ≈ 3 by replacing air breathing with that of carbogen (O_2 95%, CO_2 5%). With carbogen breathing, simulations predict hypoxia detection in 1 cm subcutaneous tumor with $S = 68\%$ and in 3 cm ones with $S = 95\%$–99.9%. Carbogen renders possibly two additional diagnostic tests for redundancy. Significant improvements of the aforementioned measurement accuracies are projected. Oncosensor will be tested in vivo with R3230 tumors in Fischer rats at UCI's Center for Functional Onco-Imaging. Oncosensor requires imaging guidance.

The instrument called Nanosniper, as developed by Bioatom (see: http://www.calseco.com) and its subsidiary Centurion Inc. is shown in Figure 9.15.

In an earlier work May et al. (2007) have designed a transportable collimated neutron source to perform the neutron interrogation of improvised explosive devices (IEDs). Their most used design tool was MCNP—Monte Carlo N-particle code. They used it to simulate the neutron generator and collimator systems. The code allowed the calculation of the dose around the surface of the collimator very quickly and efficiently. Using the same code they also determined the flux of neutrons at the target.

The material for the collimator was to be a polyethylene due to its high hydrogen density and the ability to shield neutrons. The design limit for weight was set

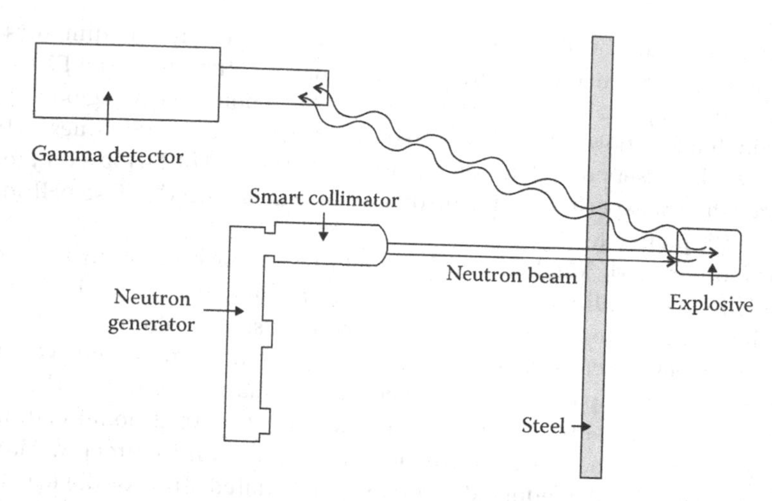

Figure 9.15 Nanosniper emits directed neutrons; it is directional since it analyzes gamma rays from target only.

at 1000 kg, which is set to standardize between concepts and keep the size down. Borated polyethylene is a polyethylene with boron, a neutron absorber, mixed in, and it provides the best results per mass of any material they tested. One problem that they have encountered in the design of the collimator is one of manufacturability. The collimator was very large, totaling 1.6 m in length and 1.15 m in diameter.

Polyethylene cannot be bonded together using adhesives, so it must be welded together or molded as one unit. Both techniques have their advantages and disadvantages. One possible manufacturing technique would be to mold the collimator into the proper shape. This involves fabricating a mold of the collimator geometry, and then pouring or injecting molten plastic into it. This prevents having to bond sections together but does present the problem of locating accompany with the capability of molding a part of the large size needed. The largest capability that they found was of the order of 800 mm length and 100 mm maximum wall thickness. This was not large enough to construct the collimator. This would make it pertinent to mold the collimator into several smaller parts that would be within the manufacturing capabilities of most plastic molding companies, and then assemble these together. This technique reduces waste by reducing the machining losses, but also requires finding a source of borated polyethylene in a form that could be molded. It is also possible that in the molding process the composition of the material could be adversely affected, such as all the boron settling out of the polyethylene in the molten state, giving poorer shielding results than expected.

The borated polyethylene is sold in sheets, which would be bonded together to make the final collimator shape. This type of bonding involves either heat welding or ultrasonic welding for the bonding process, which actually melts the plastic together. Manufacturing in this method would increase machining losses due to the form that the material is available in. The best form for this would be to layer it from the top to the bottom as it would be when in use. This would prevent the weight force of the collimator itself from generating shearing forces to break the layers apart as it would if the layers were vertical. This method will have increased cost due to machining losses, but would be feasible without finding a large scale molding facility.

The end solution for the manufacturing of such collimator will probably be a combination of the two methods, using available industrial molding facilities to construct the collimator in sections that would then be welded together. Most likely construction would be of horizontal sections to increase the structural integrity of the collimator.

Reflecting collimator design is described by many authors (see, for example, Cussen 1998). The device they seek to construct resembles a reflector much more than a collimator, since the principle aim of this device would be to enhance the flux at the target rather than decreasing the flux outside the target as traditional collimator would. Some authors have chosen to refer to such device as a reflecting collimator. An overview of the cross sections of known materials demonstrates the absence of an ideal high energy neutron reflecting material. In the articles some advanced reflector concepts are presented together with the theory behind their development. The order of magnitude calculations discussed reveal the necessity of a unique, highly efficient reflecting collimator design. In order to realize such a unique design, it is necessary to develop a conceptual understanding of high energy neutron transport through reflecting media. A simple analogy provides an instructive mental picture. Neutron transport in high Z materials is analogous to visible light transport in a diffuse scattering medium. In both cases scattering is nearly isotropic in the laboratory coordinate system and the fraction of scattering interactions that lead to absorption can be quite small so that the number of scatters is quite large. Hence, "creating a collimator to reflect fast neutrons into a beam is like making a flashlight reflector with a cloud." From this analogy one can readily identify the nature of the problem. There are no "surface effects" and there is no specular reflection for high energy neutrons.

The first baseline case is that of a point source and detector located 10 m from the target. The fraction of source particles that lead to a photon passing through the detector is calculated. From this simple analysis one can see that the collimator efficiency will be vital to the success of a remote detection system even with a source strength of 10^{10} n/s. The planar reflector is certainly not a concept that should be considered for implementation; however, the reflection factor resulting from this arrangement helps to provide an understanding for the affects of modifying other collimator geometries.

MCNP should be used to determine the relationship between thickness of the reflecting plane and the flux at the target. From these values, the reflection factor F_R should be determined by

$$F_R = \text{Flux with collimator/flux without collimator (at target)}$$

This definition permits quick comparison of a new geometry to already existing geometries. Ideal reflector would have no leakage in any direction but the target direction. According to the reflecting light with cloud analogy, one would not see the collimator, unless looked down its axis. A trend line has been fit to the data points for each curve. As expected, the field diverges nearly as $1/r^2$. The ratio of flux for the 40 cm reflector to that with 0 cm reflector is $F_R = 2.23$. The most useful information gathered from this simulation is the thickness of reflector material that approximates an infinite medium. The mean free path (mfp) for 14 MeV neutrons in graphite is about 6.4 cm. It seems that the flux at 6 mfp approaches the limit for an infinite medium. From this analysis, it is clear that other more efficient reflector geometries must be developed.

The about 6.4 cm mfp for neutrons in graphite necessitates the use of a very large collimator. Materials with 1/10 of this mfp might enable the construction of a portable collimator. It is prudent to search for such a material. The ideal reflector material would have the following characteristics:

1. High scattering cross section at 14 MeV ($\sigma_S > 50$ barns, actual 4)
2. High ratio of elastic scattering to other scatterings ratio ($R = \sigma_{Elastic}/\sigma_S$ >100, actual 3.5)
3. High atomic number (to limit thermalization, actual realizable)
4. Inelastic cutoff above 6.5 MeV (does not exist)
5. Inexpensive (realizable)

A quick look at neutron cross sections for some common materials reveals that an ideal material does not exist. A neutron interaction cross section from JENDL-3.3 (JAERI 2002) illustrates the shortcomings of iron as a typical material. Even materials like cadmium, effectively "black" to thermal neutrons, have a small 14 MeV cross section. Lead and tungsten are similar.

The concept generation process following questions should be addressed toward developing a conceptual understanding of reflector design. These include the following:

1. Is it possible to create an ideal reflector—redirecting 100% of source particles to target?
2. If an ideal reflector exists, does it exist as a result of the dimensions being stretched to infinity, or is there a finite version that reasonably approximates the ideal case?

3. Is there a way to determine whether a small change to a given reflector shape will enhance or decrease efficiency?
4. Can the above method of determining $\Delta F_R = f(\Delta_{geom})$ be proceduralized and converted to yield a useful computer program that can serve as a design aid?
5. Is it possible to create a "white box with a hole?" Can geometry serve to contain neutrons, constantly randomizing their directions until these directions align with the target, in which case escape probability becomes high and neutrons leave toward the target?

The ultimate goal is of course to develop the best reflector geometry. If an ideal geometry exists, then its basic concept should be applied to known concepts to maximize the efficiency of the design.

In such a geometry more than 50% of emitted neutrons are directed through the target. In order to achieve this there must be no leakage in two dimensions in the limit as dimensions become infinite. The guiding idea that emerged was that an ideal collimating reflector would redirect neutron current by minimizing the macroscopic cross section in the direction of the target, while maximizing cross section in other direction. The first geometry contrived to meet this criterion was a two-dimensional stack of scattering slabs (Figure 9.16). The slabs would alternate between materials of high and low cross sections.

Farr et al. (2001) have designed a multileaf collimator (MLC) for installation on the super-conducting cyclotron at the Gershenson Radiation Oncology Center, Karmanos Cancer Institute, Detroit, Michigan. This MLC was produced with

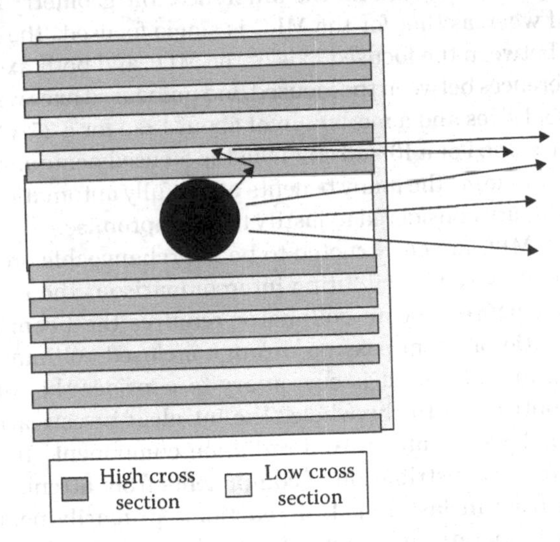

Figure 9.16 A schematic of an ideal collimating reflector.

the aim to replace the existing multirod collimator and the increased efficiency thus achieved should allow for a 50% increase in the number of patients treated. A study of the penumbra region of the neutron beam with focused and unfocused collimator leaves has been completed, together with activation measurements in steel and tungsten. Results of these studies were used to finalize the collimator leaf design. A steel collimator leaf with a 5 mm projection at the isocenter and a wedge-shaped section has been chosen, to provide beam divergence in the direction perpendicular to the leaf motion. The leaf profile is "stepped" to prevent neutron leakage. The rationale for this leaf design is discussed. The overall design of the collimator system and the incorporation of a remote wedge-changing device are presented. Each leaf is positioned using a stepping motor; the leaf position is independently confirmed using an optical system incorporating a coherent fiber optic and a CCD camera. The control system is being designed to allow for the implementation of intensity modulated neutron radiation therapy (IMNRT).

The radiological validation of this MLC was reported in the literature; see Farr et al. (2007). Prior to clinical application, the basic radiological properties of the fast neutron MLC were studied. Complicating the evaluation was the mixed neutron and gamma radiation field environment encountered with fast neutron beams. As a reference the MLC performance is compared to an existing multirod collimator (MRC) used at the facility for more than 10 years. The MLC aggregate transmission is found to be about 4%, slightly outperforming the MRC. The measured gamma component for a closed collimator is 1.5 times higher for the MLC, compared with the MRC. The different materials used for attenuation, steel and tungsten, respectively, account for the difference. The geometry for the MRC is double focused whereas that for the MLC is single focused. The resulting penumbrae agree between the focused axis of the MLC and both axes of the MRC. Penumbra differences between the focused and unfocused axes were not observable at small field sizes and a maximum of about 1 cm for a 25×25 cm^2 field at 2.5 cm depth in water. For a 10×10 cm^2 field the focused penumbra is 9 mm and the unfocused is 12 mm. The many benefits of the fully automatic MLC over the semimanual MRC are considered to justify this compromise.

The MLC and MRC are constructed to be interchangeable on the cyclotron yolk (primary collimation) facilitating intercomparison. The radiologic characterization of a neutron beam collimator requires the beam transmission defined as the ratio of beam passing through a closed collimator relative to the open beam. In addition, it is customary to consider the interleaf transmission as it contributes to the whole. The intraleaf investigational method was complicated by a variable mixed radiation component. This was due to changes in the relative distribution of components from attenuation in 30 cm of steel in part from inelastic nuclear reactions, primarily neutron-gamma. An experimental apparatus was fabricated to allow the interleaf spacing to be varied. Two milled steel stock pieces with blocking steps were used to provide

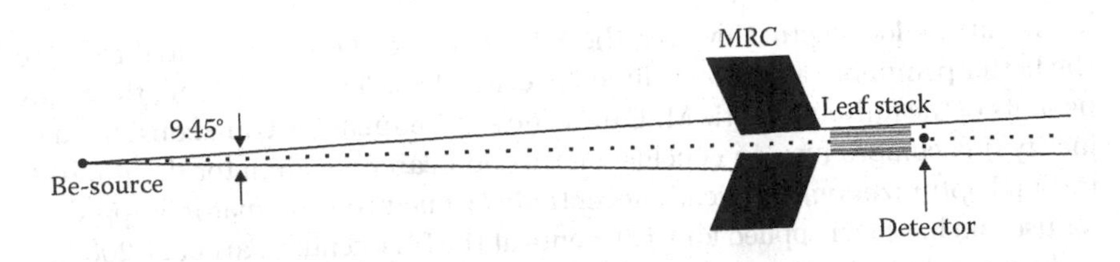

Figure 9.17 Experimental setup for leaf gap characterization. Eleven central leaves from the MLC are mounted in a precision holder. Precise alignment is provided by rotating the leaf stack by small angles until the transmission is at a maximum (after Farr et al. 2007).

constraint and divergence for a stack of 11 MLC leaves. A series of precision brass shim stock was used to provide leaf spacing that could be varied over the range 0.025–0.41 mm. Figure 9.17 shows the experimental setup with the leaf stack aligned with the beam axis shown approximately as a dotted line. Field sizes of 5×5 cm^2 and 10×10 cm^2 were set on the MRC immediately proximal to the experimental apparatus. Because the leaf stack was positioned slightly distal to the MRC, the physical leaf divergence was not precisely aligned with the beam divergence. However, the source to isocenter distance for the machine is 185.4 cm so the slight geometric mismatch is considered to be acceptable for transmission testing and for the small fields considered. A precision turntable was also included to align the apparatus with the radiation beam. The affect on transmission from a series of different gap widths (0.10–0.41 mm) was evaluated using an ionization chamber and film. A 1 cm^3 ionization chamber with A150 buildup cap was used (Farr et al. 2007). Because in this case the measurements were relative between the different leaf spacings, they were made in ambient air.

In conclusion, the basic radiological properties of the MLC have been validated against the existing MRC. Design decisions such as material choice, leaf geometry, and spacing contribute to the overall measured collimator performance. Interleaf spacing was shown to have a measurable effect on MLC transmission. If interleaf radiation blocking steps are included the interleaf spacing must be carefully considered. In this case the interleaf step width is adequate but in hindsight could have been made slightly more generous. At about 4% aggregate transmission, the results from the MLC indicate a reduction in average transmission in comparison to the MRC. Interestingly, when closed, the steel MLC contributes a higher gamma component than does the tungsten MRC. The result is understood based on the physical processes involved in the two materials. The MLC penumbra results indicate the importance of geometry choice and indicate a similar performance to the MRC in the focused direction. The penumbra in the unfocused direction was measured to vary with respect to the doubly focused MRC results from no observable difference at the smallest field sizes to about 1 cm for the largest at shallow depth. However, the MLC can and should be rotated to place the better penumbra axis along clinically critical borders. In addition, the many benefits of the fully automatic MLC over the semimanual MRC are considered to justify this compromise. In conclusion, through careful collimator design, testing, and optimization, a clinically acceptable fast neutron collimator is approved for use and has been applied for treatments at the FNT center (Farr et al. 2007).

In general, the most promising reflecting materials should be selected based on the value of 14 MeV elastic cross section. When lead is considered as material one should keep in mind that ^{208}Pb, which satisfies this condition, is only one of the four stable isotopes. The others have much higher capture cross sections and the inelastic cutoff is at lower energy. Otherwise, lead, with its high density and reasonable cross section, might be used to construct a reasonably small collimator.

As an example, we present here the 14 MeV mfp calculation for iron:

$$\Sigma_S = \frac{\sigma_S \rho N_A}{M} = \frac{[2\times10^{-24}\,\text{cm}^2/\text{atom}(7.87\,\text{g/cm}^3)6.022\times10^{23}\,\text{atom/mol}]}{55.8\,\text{g/mol}}$$

$$= 0.169\,\text{cm}^{-1}$$

$$\text{mfp} = \frac{1}{\Sigma_S} = 5.89\,\text{cm}$$

9.7 QUALITY ASSURANCE PROGRAMS FOR DIAGNOSTIC FACILITIES

A quality assurance program is a system of plans, actions, reviews, reports, and records whose purpose is to ensure that diagnostic facilities achieve consistent high quality imaging and other diagnostic results, while maintaining radiation output and personnel doses within limits prescribed by the department.

Each radiation facility conducting diagnostic x-ray and/or radioactive materials procedures, except dental, pediatric, and veterinary facilities, should implement a quality assurance program including the following at a minimum:

1. The adoption of a manual containing written policies and procedures for radiation protection and describing the facility's quality assurance program. Policies and procedures must be consistent with the types of equipment and services provided, including but not limited to, use of gonad or scoliosis shielding, personnel monitoring, protection of pregnant workers and patients, and holding of patients. The quality assurance manual must describe the various processing, generator, and

systems quality control tests appropriate for the types of equipment and services provided in sufficient detail to ensure that they will be performed properly.
2. The performance of quality control tests and the correction of deficiencies as specified in the quality assurance manual.
3. The maintenance of equipment records for each diagnostic imaging system, containing test results, records of equipment repairs, and other pertinent information.
4. The provision of a formalized in-service training program for employees, including, but not limited to, quality assurance and radiation safety procedures.
5. The measurement of radiation output at the point of skin entry for common x-ray examinations.
6. The measurement of the amount of activity of each dose of a radiopharmaceutical administered to a patient.
7. The calculated absorbed dose for diagnostic procedures involving radioactive materials.
8. The provision of the information described in (5), (6), and (7) to any patient upon request.
9. The conduct of an ongoing analysis of repeated, rejected, or misadministered diagnostic studies that are designed to identify and correct problems and to optimize quality.

A quality assurance program for external beam therapy and brachytherapy is a system of plans, actions, reviews, reports, and records whose purpose is to ensure a consistent and safe fulfillment of the dose prescription to the target volume, with minimal dose to normal tissues. Each licensee or registrant who uses external beam therapy and/or brachytherapy in humans should implement a quality assurance program that includes at a minimum the adoption of a quality assurance manual containing written policies and procedures designed to assure effective supervision, safety, proper performance of equipment, effective communication, and quality control. These must include policies and procedures to assure the following:

1. Each patient's evaluation and intended treatment is documented in the patient's record.
2. A written, signed, and dated order for medical use of radiation or radioactive material is made for each patient in accordance.
3. All orders and other treatment records are clear and legible.
4. Staff will be instructed to obtain clarification before treating a patient if any element of the order or other record is confusing, ambiguous, or suspected of being erroneous.

5. Each patient's response to treatment is assessed by a physician knowledgeable in external beam therapy and/or brachytherapy and that unusual responses are evaluated as possible indications of treatment errors.
6. Complete treatment records containing data recorded at the time of each treatment are maintained.
7. The treatment charts of patients undergoing fractionated treatment are checked for completeness and accuracy at weekly intervals.
8. Final plans of treatment and related calculations are checked for accuracy before 25% of the prescribed dose for external beam therapy or 50% of the prescribed dose for brachytherapy is administered. If a treatment plan and related calculations were originally prepared by a radiation therapy it may be checked by the same person using a different calculation method. Treatment plans and related calculations prepared by other personnel must be checked by a second person using procedures specified in the treatment planning procedures manual.

There is quality control for all physical components of radiation therapy such as equipment function and safety (including treatment planning equipment), treatment planning procedures and computer codes, treatment application procedures, dosimetry, and personnel radiation safety. The quality control tests to be performed are documented, including the following:

1. Detailed procedures for performing each test
2. The frequency of each test
3. Acceptable results for each test
4. Corrective actions to be taken
5. Record keeping and reporting procedures for test results

Each licensee or registrant shall ensure that a radiation therapy physicist possessing the specified qualifications prepares a procedures manual describing how radiation therapy treatment planning is to be performed at the licensee's or registrant's facility. The treatment planning manual may be part of the quality assurance manual required and shall include the calculation methods and formulae to be used at the facility (including the methods for performing the checks of treatment plans and related calculations as required. The treatment planning manual should be reviewed annually by a radiation therapy physicist and should be included in training given to facility staff that plan to participate in treatment planning.

Each licensee or registrant shall ensure that all equipment used in planning and administering radiation therapy is properly functioning and designed for the intended purpose, is properly calibrated, and is maintained in accordance with the manufacturer's instructions and the quality assurance program described in the licensee or registrant's quality assurance manual.

Each licensee or registrant shall implement procedures for auditing the effectiveness of the radiation therapy quality assurance program. Audit procedures must specify either of the following:

1. External audits will be conducted at intervals not to exceed 12 months by radiation therapy physicists possessing the qualifications specified, and physicians who are in the active practice of the type of radiation therapy conducted by the licensee or registrant. These must be individuals who are not involved in the conduct of the therapy program being audited; the individuals who conduct the audit will prepare and deliver to the licensee or registrant a report that contains an assessment of the effectiveness of the quality assurance program and makes recommendations for any needed modifications or improvements, and the licensee or registrant shall promptly review the audit findings, address the need for modifications or improvements, and document actions taken. If recommendations are not acted on, the reasons for this will also be documented.
2. Internal audits will be conducted at intervals not to exceed 12 months by program staff who will prepare and deliver to the licensee or registrant a report as specified in (1), and external audits will be conducted at intervals not exceeding 5 years by an organized review program supervised by the American College of Radiology (in the United States), or a program found equivalent by the department based on the scope of the audit and the experience of the sponsoring organization in performing such audits; and the licensee or registrant shall promptly review the audit findings, address the need for modifications or improvements, and document actions taken. If recommendations are not acted on, the reasons for this will also be documented.

Each licensee or registrant must maintain written records for review by the department that document quality assurance and audit activities.

A quality assurance program for radiopharmaceutical therapy is a system of plans, actions, reviews, reports, and records whose purpose is to ensure a consistent and safe fulfillment of the dose prescription.

9.8 PRODUCTION OF RADIOISOTOPES USING FAST NEUTRONS

A radioisotope with a half-life ($T_{1/2}$) of less than several days plays an important role in nuclear medicine. The daughter nuclide of ^{99}Mo ($T_{1/2}$ = 66 h), that is, ^{99m}Tc ($T_{1/2}$ = 6 h), is used for diagnostics, and ^{90}Y ($T_{1/2}$ = 64 h) is used for cancer therapy.

Most medical radioisotopes, including ^{99}Mo and ^{90}Y, are imported in many countries. An unscheduled shutdown of aging research reactors in which most of the ^{99}Mo nuclide was produced has resulted in a critical shortage of ^{99}Mo worldwide. Therefore, the establishment of a reliable production method for ^{99}Mo is very important to ensure the continued medical applications of ^{99m}Tc.

Minato and Nagai (2010) have proposed a new route for producing medical radioisotopes using accelerator neutrons. Neutrons (about 10^{15} n/s) with an E_n of about 14 MeV are produced in the ^{nat}C(d,n) reaction using 40 MeV 5 mA deuteron beams provided by an accelerator. Such accelerators are currently under construction in several countries. In fact, a variety of medical radioisotopes can be produced using accelerator neutrons as the production cross section of a sample nucleus is large at E_n = 10–15 MeV. For example, the ^{99}Mo production cross section of the ^{100}Mo(n,2n)^{99}Mo reaction is 1.5 b at $E_n \approx$ 10–20 MeV, while the ^{97}Zr production cross section of the ^{100}Mo(n,α)^{97}Zr reaction is 0.002 b. Note that ^{97}Zr, a radioactive impurity nucleus, is produced in small amounts. Minato and Nagai (2010) evaluated the angular and depth distributions of ^{99}Mo that was produced by using the accelerator neutrons to study the effective use of the neutrons. It is shown that the ^{99}Mo yield is restricted to a narrow region at an extremely forward angle with respect to the deuteron beam direction; this observation assisted us in obtaining high-specific-activity ^{99}Mo. The ^{90}Y production cross section of the ^{90}Zr(n,p)^{90}Y reaction is also large. High-quality ^{99m}Tc and ^{90}Y can be separated from the irradiated Mo and Zr samples by sublimation and ion exchange, respectively. Quasimono energetic, high intensity accelerator neutrons (E_n = 10–15 MeV), therefore, are very useful for the production of medical radioisotopes.

In order to achieve a neutron yield of 10^{14} neutrons/s, a large multicusp source together with a multiaperture extraction system to produce an ion beam current of 1 A, accelerated to 120 kV, and impinging on a well-cooled target is required. The main components of the sealed d–t neutron tube are the ion source, the 120 kV accelerator column, the water-cooled target, and the vacuum system. Such is a sealed d–t neutron generator, a scale up version of the compact neutron tube that was developed at Lawrence Berkeley National Laboratory (LBNL). The multicusp generator is a 30 cm diameter cylindrical stainless-steel chamber surrounded with columns of samarium–cobalt magnets. The plasma is produced by RF induction discharge. In order to deliver RF power to the plasma, a coupler in the form of a multiturn induction coil is used. The RF power supply is a broadband power amplifier driven at 13.56 MHz by a signal generator. To maximize the neutron output at the target, it is necessary to produce high d^+ and t^+ concentrations in the extracted beam. Experiments have been carried out with a 5 cm diameter ion source to determine the fractions of hydrogen ion species (Verbeke et al. 1999).

9.9 DOSIMETRY

The document "Protocol for Neutron Beam Dosimetry" is one in the series of American Association of Physicists in Medicine (AAPM) reports. It contains recommendations for the dosimetry of high energy neutron beams. These recommendations are intended to serve the immediate needs of particular neutron therapy centers and to be used as a common basis for clinical neutron basic calibration. This report was prepared by Task Group 18 under the direction of Peter Wootton (Wootton 1980). The Group has drawn up a number of recommendations concerning dose calculations, dose-measurements procedures, the nature of the phantoms to be used in depth-dose measurements, and the determination of displacement correction factors. It has generated studies of such items as the elemental composition of the tissue-equivalent (TE) plastic used in ion chambers, effects of phantom size on depth dose, and gamma-ray contamination, and the accuracy attainable in neutron beam dosimetry. The U.S. protocol has been developed in parallel with the European neutron dosimetry protocol, in a cooperative effort with the European Clinical Neutron Dosimetry Group (ECNEU). (Broerse and Mijnheer 1976, 1979). The ECNEU was established in Europe by the Fast Particle Therapy Project Group under the sponsorship of the European Organization for Research on Treatment of Cancer (EORTC). The main items of the two protocols were summarized by Broerse et al. (1979).

Neutron fields are always accompanied by gamma rays originating from the neutron-producing target, the primary shielding, the field-limiting or collimating system, the biological object or phantom being irradiated, and from the surroundings. The proportion of the total absorbed dose due to the photon component of the mixed neutron–photon field increases markedly with increasing depth of penetration of the incident beam in a phantom and with the field size at a fixed depth. Because of the differences in biological effectiveness (the magnitude of which can depend on the specific biologic end point) of these two radiation components, it is necessary to determine the separate neutron absorbed dose, D_n, and the gamma-ray absorbed dose, D_γ, of the radiation field at all points in tissue. The general dosimetric methods have received extensive discussion in the literature (ICRU 1977).

List of reactions in the JENDL Dosimetry File 99 (JENDL/D-99) is available (Kobayashi et al. 2002). The JENDL Dosimetry File 99 (JENDL/D-99), which is a revised version of the JENDL Dosimetry File 91 (JENDL/D-91), has been compiled and released for the determination of neutron flux and energy spectra. This work was undertaken to remove the inconsistency between the cross sections and their covariances in JENDL/D-91 since the covariances were mainly taken from IRDF-85 although the cross sections were based on JENDL-3. Dosimetry cross sections have been evaluated for 67 reactions on 47 nuclides together with covariances. The cross sections for 34 major reactions and their covariances were simultaneously generated and the remaining 33 reaction data were mainly taken from JENDL/D-91. Latest measurements were taken into account in the evaluation. The resultant evaluated data are given in the neutron energy region below 20 MeV in both point-wise and group-wise files in the ENDF-6 format. In order to confirm the reliability of the evaluated data, several integral tests have been carried out: comparisons with average cross sections measured in fission neutron fields, fast thermal reactor spectra, d–t neutron fields, and Li(d,n) neutron fields. It was found from the comparisons that the cross sections calculated from JENDL/D-99 are generally in good agreement with the measured data. The contents of JENDL/D-99 and the results of the integral tests are described in this report. All of the dosimetry cross sections are shown in a graphical form in the Appendix of the Report.

The development of dosimeters to determine neutron exposure has not been as successful as the development of gamma ray dosimeters. For many years, exposure to gamma and x-rays were determined from film badges and, more recently, glass dosimeters and TLD's. With these dosimeters, personnel exposure can be determined with acceptable accuracy, the largest uncertainty being the variation in readings caused by the orientation of the wearer to the source during exposure. Neutron dosimeter advances have been much slower and less successful. For many years, NTA films was the only fast neutron dosimeter available, but now there are additional practical dosimetry systems: albedo-neutron dosimeters and fission fragment track registration. Albedo-neutron dosimeters have recently been placed into service by several organizations, but probably more widespread interest and study has been given to fission fragment track registration.

In the report by Ruddy et al. (2006) silicon carbide semiconductor neutron detectors have been irradiated with ^{252}Cf fission neutrons, thermal-neutron-induced fission neutrons from ^{235}U, and 14 MeV neutrons from a d–t neutron generator. In the latter case, reaction peaks corresponding to $^{12}C(n,\alpha)^{9}Be$ and $^{28}Si(n,\alpha)^{25}Mg$ reactions have been observed. Multiple reaction branches to the ^{25}Mg ground state and several excited states are observed for the $^{28}Si(n,\alpha)^{25}Mg$ reaction, and the detector energy calibration for these reactions can be derived from the peak positions. Although only the ground state branch is observed for the $^{12}C(n,\alpha)^{9}Be$ reaction, the shift of the reaction energy with angle relative to the 14 MeV source can be used to derive an energy scale based on this reaction. These energy response measurements will form the basis for neutron spectrum unfolding methods for inferring incident neutron energy spectra from the resulting recoil-ion energy spectra in silicon carbide detectors (Ruddy et al. 2006).

The 14 MeV neutrons inevitably bring forth the secondary radiations of scattered neutrons and γ-rays, whose intensity depends on the geometry and construction materials of the target room. The samples in the target room are irradiated by the primary 14 MeV neutrons and by the secondary radiations. For the study of irradiation effects of 14 MeV neutrons, it is essentially important to measure the total absorbed dose and to separate the total absorbed dose into the dose due to 14 MeV neutrons and the secondary radiations. The alanine dosimeter (Regulla and Deffner 1982) is one of the free radical dosimeters and has been used as a convenient dosimeter for γ-rays and fast neutrons in a nuclear reactor.

In the work by Ikezoe et al. (1984), the absorbed dose due to the mixed radiations in the FNS target room was measured both by the alanine dosimeter and by the Au activation method. Intercomparison of the results obtained by the two methods made it possible to separate the total dose into the dose due to 14 MeV neutrons and that due to the secondary radiations. They heated the reagent alanine to about 100°C and impregnated it with paraffin. At room temperature, the 80% alanine 20% paraffin mixture was pressed in the shape of rods (sensors) of 3.5 mm diameter and approximately 10 mm length, or hollow cylinders (sheaths) of 3.7 mm ID, 9.4 mm OD, and approximately 10 mm length. For irradiation in the FNS target room, the sensor was usually inserted into the sheath. In some cases, the sensor was covered by polyethylene film of 0.2 mm thickness or by a quartz tube of 1 mm thickness instead of the sheath. The samples along with small pieces of Au foil were set and irradiated for 50.33 h in the target room of the FNS, where 14 MeV neutrons were generated at an average rate of 2.2×10^{12} n/s. When alanine is irradiated by ionizing radiations, a stable radical $CH_3CHCOOH$ is formed. The concentration of the radical formed in the sensor was measured by an electron spin resonance (ESR) spectrometer, after cutting off each end of the sensor rod about 2 mm. The measurements were made at 100 kHz modulation and 10 G modulation width of the magnetic field and 1 mW input of the x-band microwave. The ESR signal intensity is proportional to the absorbed γ dose in a wide range between 10^2 and 10^6 rad. The flux of 14 MeV neutrons was obtained by measuring the induced activity of ^{196}Au by the nuclear reaction ^{197}Au(n,2n)^{196}Au.

REFERENCES

Agosteo, S. and Curzio, G. 2002. Characterization of an accelerator-based neutron source for BNCT versus beam energy. *Nucl. Instrum. Methods Phys. Res. A* 476: 106–112.

Aldama, D. L. and Trkov, A. 2004. The cross section library: INDC(NDS)-467, IAEA, Vienna, Austria.

Alper, T. 1963. Pre-therapeutic experiments with the fast neutron beam from the Medical Research Council cyclotron. V. Comparison between the oxygen enhancement ratios for neutrons and x-rays, as observed with *Escherichia coli* B. *Br. J. Radiol.* 36: 97–101.

Anderson, J., Osborn, S. B., Tomlinson, R. W. et al. 1964. Neutron activation analysis in man in vivo. A new technique in medical investigation. *Lancet* 2: 1201–1205.

Badone, E. and Farquhar, R. M. 1982. Application of neutron activation analysis to the study of element concentration and exchange in fossil bones. *J. Radioanal. Chem.* 69: 291–311.

Barton, J. P. 1976. Neutron radiography—An overview, in Berger, H. (Ed.) *Practical Applications of Neutron Radiography and Gauging*, ASTM STP 586. American Society of Testing and Materials, West Conshohocken, PA, pp. 5–19.

Berry, R. J. 1969. The role of factors modifying neutron damage in irradiation of mammalian cells in vitro and in vivo, in *Proceedings of the Symposium on Neutrons in Radiobiology*, Oak Ridge, TN, November 11–14, 1969, p. 483.

Bewley, K. K. 1966. Radiobiological research with fast neutrons and the implications for radiotherapy. *Radiology* 86: 251–257.

Boddy, K., Holloway, I., and Elliott, A. 1974. Some practical advantages of sealed tube 14 MeV neutron generators for total body in vivo activation analysis. *Phys. Med. Biol.* 19: 379–381.

Broerse, J. J. and Mijnheer, B. J. 1976. Second draft protocol for neutron dosimetry for radiobiological and medical applications, in Broerse, J. J. (Ed.) *Workshop on Basic Physical Data for Neutron Dosimetry*, Rijswijk, the Netherlands, May 19–21, 1976, p. 311.

Broerse, J. J. and Mijnheer, B. J. 1979. Fourth draft of a European protocol for neutron dosimetry for external beam therapy. European Clinical Neutron Dosimetry Group (ECNEU), Rijswijk, the Netherlands.

Broerse, J. J., Mijnheer, B. J., Eenmaa, J., and Wootton, P. 1979. Dosimetry intercomparison and protocols for therapeutic applications of fast neutron beams, in Barendsen, G. W., Broerse, J. J., and Breur, K. (Eds.) *High-LET Radiations in Clinical Radiotherapy*. Pergamon Press, Oxford, U.K., p. 117, Suppl. *Eur. J. Cancer*.

Brzosko, J. S., Robouch, B. V., Ingrosso, L., Bortolotti, A., and Nardi, V. 1992. Advantages and limits of 14-MeV neutron radiography. *Nucl. Instrum. Methods Phys. Res. B* 72(1): 119–131.

Chamberlain, M. J., Fremlin, J. H., Peters, D. K., and Philip, H. 1968. Total body calcium by whole body neutron activation: New technique for study of bone disease. *Br. Med. J.* 2: 581–583.

Chankow, N. 2012. Neutron radiography, in Omar, M. (Ed.) *Nondestructive Testing Methods and New Applications*. InTech, Rijeka, Croatia, 264pp.

Cheng, D.-W., Lu, J.-B., Yang, D., Liu, Y.-M., Wang, H.-D., and Ma, K.-Y. 2012. Designing of the 14 MeV neutron moderator for BNCT. *Chinese Physics C – CPC* 36(9): 905–908.

Cluzeau, S., Le Tourneur, P., and Dance, W. E. 1997. Upgrade of the DIANE: Performance improvement in thermalization of fast neutrons for radiography, in Duggan, J. L. and Morgan, I. (Eds.) *Application of Accelerators in Research and Industry*, CP392. AIP Press, New York, pp. 887–890.

Cohn, S. H. and Dombrowski, C. S. 1971. Measurement of total-body calcium, sodium, chlorine, nitrogen and phosphorus in man by in vivo neutron activation analysis. *J. Nucl. Med.* 12(7): 499–505.

Crambes, M., Nargolwalla, S. S., and May, L. 1967. Elemental analysis of proteins by 14 MeV neutron activation. *IEEE Trans. Nucl. Sci.* 10: 63.

Cross, W. G. and Ing, H. 1985. Conversion and quality factors relating neutron fluence and dosimetric quantities. *Radiat. Prot. Dosimetry* 10(1–4): 29–42.

Csikai, J. 1987. *CRC Handbook of Fast Neutron Generators.* CRC Press, Boca Raton, FL.

Cussen, L. D. 1998. A design for improved neutron Collimators. *Nucl. Instr. Meth. Phys. Res.* A414: 365–371.

Dietrich, F., Hall, J., and Logan, C. 1997. Conceptual design for a neutron imaging system for thick target analysis operating in the 10–15 MeV energy range, in Duggan, J. L. and Morgan, I. (Eds.) *Application of Accelerators in Research and Industry*, CP392. AIP Press, New York, pp. 837–840.

Dinca, M., Pavelescu, M., and Iorgulis, C. 2005. Collimated neutron beam for neutron radiography. *Rom. J. Phys.* 51(3–4): 435–441.

Elkind, M. M. 1971. Summary of general discussion on radiobiological aspects of fast neutrons in radiotherapy. *Eur. J. Cancer* 7: 249–257.

Elshahat, B. A., Naqvi, A. A., Khalid, A. et al. 2007. Design calculations of an accelerator based BSA for BNCT of brain cancer. *J. Radioanal. Nucl. Chem.* 274(3): 539–543.

Euro Collimators. 2012. Euro Collimators Ltd., Gloucestershire, U.K. http://www.eurocollimators.com. Accessed September 15, 2014.

Farr, J. B., Maughan, R. L., Yudelev, M., Forman, J. D., Blosser, E. J., and Horste, T. 2001. A multileaf collimator for neutron radiation therapy, in Marti, F. (Ed.) *Sixteenth International Conference on Cyclotrons and their Applications 2001*, CP600. American Institute of Physics, College Park, Maryland, pp. 154–156.

Farr, J. B., Maughan, R. L., Yudelev, M., Blosser, E. J., Brandon, J., Horste, T., and Forman, J. D. 2007. Radiological validation of a fast neutron multileaf collimator. *Med. Phys.* 34(9): 3475–3484.

Fowler, J. F. 1966. Radiation biology as applied to radiotherapy, in *Current Topics in Radiation Research*, Vol. II, 2nd edn. North Holland, Amsterdam, the Netherlands, pp. 303–364.

Garrett, D. A. and Berger, H. 1977. The technological development of neutron radiography. *At. Energy Rev.* 15(2): 123–142.

Garrett, R. and Mitra, S. 1991. A feasibility study of in vivo 14 MeV neutron activation analysis using the associated particle technique. *Med. Phys.* 18: 916–920.

Greene, D. and Major, D. 1965. Collimation of 14 MeV neutron beams. *Eur. J. Cancer* 7: 121–127.

Grogan, B. R. 2010. The development of a parameterized scatter removal algorithm for nuclear materials identification system imaging. PhD dissertation, University of Tennessee, Knoxville, TN.

Haque, A. M., Moschini, G., Valkovic, V., and Zafiropoulos, D. 1995. Boron–neutron capture therapy. *Proc. SPIE* 2339: 514, doi:10.1117/12.204194.

Harms, A. A. 1977. Physical processes and mathematical methods in neutron radiography. *At. Energy Rev.* 15(2): 143–168.

Holmberg, P., Hyvonen, M., and Tarvainen, M. 1978. In vitro activation of bone with 14 MeV neutrons. *J. Radioanal. Chem.* 42: 169–175.

Hyvonen-Dabek, M., Tarvainen, M., Holmberg, P., and Dabek, J. T. 1979. In vitro measurement of phosphorus in small bone biopsies using 14 MeV neutron activation analysis. *Ann. Chim. Res.* 11: 179–183.

IAEA. 1973. *In Vivo Neutron Activation Analysis. Proceedings of a Panel, Vienna, Austria, April 17–21, 1972.* IAEA, Vienna, Austria, 1973.

ICRU. 1977. Neutron dosimetry in biology and medicine. ICRU Report 26. International Commission on Radiation Units and Measurements, Washington, DC.

Ikezoe, Y., Sato, S., Onuki, K., Morishita, N., Nakamura, T., Katsumura, Y., and Tabata, Y. 1984. Alanine dosimetry of 14 MeV neutrons in FNS target room. *J. Nucl. Sci. Technol.* 21(9): 722–724.

Kapadia, A. J., Sharma, A. C., Tourassi, G. D. et al. 2008. Neutron stimulated emission computed tomography for diagnosis of breast cancer. *IEEE Trans. Nucl. Sci.* 55(1): 501–509.

Kehayias, J. J. and Zhuang, H. 1993. Use of the zetatron D-T neutron generator for the simultaneous measurement of carbon, oxygen, and hydrogen in vivo in humans. *Nucl. Instrum. Methods Phys. Res. B* 79: 555–559.

Khanchi, R. S., Bansal, S. L., Aggarwal, S., Khurana, S., and Mohindra, R. K. 1994. Computation of activities induced by 14 MeV neutrons for the elemental analysis of human tissues, skin and tooth enamel. *Appl. Radiat. Isot.* 45(1): 129–132.

Klann, R. T. 1997. A system for fast neutron radiography, in Duggan, J. L. and Morgan, I. (Eds.) *Application of Accelerators in Research and Industry*, CP392. AIP Press, New York, pp. 883–886.

Kobayashi, K., Iguchi, T., Iwasaki, S. et al. 2002. JENDL dosimetry file 99 (JENDL/D-99), JAERI 1344. Japan Atomic Energy Research Institute, Tokai, Japan.

Konno, C., Maekawa, F., Oyama, Y., Ikeda, Y., Kosako, K., and Maekawa, H. 1994. *Bulk Shielding Experiments on Large SS316 Assemblies Bombarded by D-T Neutrons, Volume I: Experiment*, JAERI-Research 94-043. Japan Atomic Energy Research Institute, Tokai, Japan.

Kyere, O. B., Oxby, C. B., Burkinshaw, L., Ellis, R. E., and Hill, G. L. 1982. The feasibility of measuring total body carbon by counting neutron inelastic scatter gamma rays. *Phys. Med. Biol.* 27: 805–817.

Lamarsh, J. R. 1983. *Introduction to Nuclear Engineering*, 2nd edn. Addison-Wesley, New York.

Lehmann, E., Frei, G., Nordlund, A., and Dahl, B. 2005. Neutron radiography with 14 MeV neutrons from a neutron generator. *IEEE Trans. Nucl. Sci.* 52(1): 389–393.

Locher, G. L. 1936. Biological effects and therapeutic possibilities of neutrons. *Am. J. Roentgenol.* 36(1): 1–13.

Maekawa, F., Konno, C., Kasugai, Y., Oyama, Y., and Ikeda, Y. 1998. JAERI-Data/Code 98-021. Japan Atomic Energy Research Institute, Tokai, Japan.

Maglich, B. C. and Nalcioglu, O. 2010. ONCOSENSOR for noninvasive high-specificity breast cancer diagnosis by carbogen-enhanced neutron femto-oximetry, in *Proceedings of the ASME 2010 First Global Congress on Nano Engineering for Medicine and Biology*, Houston, TX, February 7–10, 2010, pp. 57–58.

Marcum, J. I. 1963. Monte Carlo calculations of the transport of 14 MeV neutrons in the atmosphere. RAND Corporation research memorandum series.

May, E. W., Franklin, A., Schuh, N., Sudlow, R., and Egley, A. 2007. Transportable neutron source for explosives detection. Final report, Kansas State University, Manhattan, KS.

Meloni, S., Oddone, M., Bottazzi, P., Ottolini, L., and Vannucci, R. 1993. Neutron activation analysis investigations in updating REE composition of gabbro GOG-1 reference sample: A comparison with SIMS and ICP-AES data. *J. Radioanal. Nucl. Chem.* 168(1): 115–123.

Minato, F. and Nagai, Y. 2010. Estimation of production yield of 99Mo for medical use using neutrons from natC(d,n) at Ed = 40 MeV. *J. Phys. Soc. Jpn* 79(9): 093201-1–093201-3.

Mitra, S., Sutcliffe, J. F., and Hill, G. L. 1990. A proposed three phase-counting system for the in vivo measurement of the major elements using pulsed 14 MeV neutrons. *Biol. Trace Elem. Res.* 26: 423–428.

Mitra, S., Plank, L. D., and Hill, G. L. 1993. Calibration of a prompt gamma in vivo neutron activation facility for direct measurement of total body protein in intensive care patients. *Phys. Med. Biol.* 38: 1971–1975.

Mitra, S., Wolff, J. E., Garrett, R., and Peters, C. W. 1995a. Application of the associated particle technique for the whole-body measurement of protein, fat and water by 14 MeV neutron activation analysis – A feasibility study. *Phys. Med. Biol.* 40: 1045–1055.

Mitra, S., Wolff, J. E., Garrett, R., and Peters, C. W. 1995b. Whole body measurement of C, N and O using 14 MeV neutrons and the associated particle time-of-flight technique. *Asia Pacific J. Clin. Nutr.* 4: 187–189.

Mitra, S., Wolff, J. E., and Garrett, R. 1998. Calibration of a prototype in vivo total body composition analyzer using 14 MeV neutron activation and the associated particle technique. *Appl. Radiat. Isot.* 49: 537–539.

Ohnishi, S., Kondo, K., Sato, S. et al. 2011a. Implementation of a collimated DT neutron beam at the 1st target room of JAEA/FNS for new integral benchmark experiments. *J. Korean Phys. Soc.* 59(2): 1949–1952.

Ohnishi, S., Sato, S., Ochiai, K. et al. 2011b. Collimator system design for a DT neutron beam at the first target room of JAEA/FNS. *Prog. Nucl. Sci. Technol.* 1: 73–76.

Oyama, Y., Yamaguchi, S., and Maekawa, H. 1990. Experimental Results of Angular Neutron Flux Spectra Leaking from Slabs of Fusion Reactor Candidate Materials (I). JAERI-M 90-092. Japan Atomic Energy Research Institute, Tokai, Japan.

Palmer, H. E. 1973. Feasibility of determining total-body calcium by measuring 37Ar in expired air after low-level irradiation with 14 MeV neutrons, PL-493121, 203. IAEA, Vienna, Austria.

Phoenix Nuclear Labs. 2013. Press release February 1, 2013: Phoenix Nuclear Labs neutron radiography system installed at U.S. Army Picatinny Arsenal.

RARAF. 2014. Radiological Research Accelerator Facility Nevis Laboratories. http://www.raraf.org/neutronsfast.html. Accessed September 20, 2014.

Regulla, D. F. and Deffner, U. 1982. Dosimetry by ESR spectroscopy of alanine. *Int. J. Appl. Radiat. Isot.* 33: 1101–1114.

Ruddy, F. H., Seidel, J. G., and Dulloo, A. R. 2006. Fast neutron dosimetry and spectrometry using silicon carbide semiconductor detectors. *J. ASTM Int.* 3: 8pp.

Sabo-Napadensky, I. and Amir, O. 2005. Reduction of scattering artifact in multislice CT. *Proc. SPIE* 5745: 983–991.

Sabo-Napadensky, I., Weiss-Babai, R., Gayer, A. et al. 2012. Research and development of a dedicated collimator for 14.2 MeV fast neutrons for imaging using a D-T generator, in *2nd International Workshop on Fast Neutron Detectors and Applications*, Ein Gedi, Israel, pp. 1–18. November 6–11, 2011, IOP Publishing Ltd and Sissa Medialab srl, doi:10.1088/1748-0221/7/06/C06005.

Shypailo, R. J. and Ellis, K. J. 2005. Design considerations for a neutron generator-based total-body irradiator. *J. Radioanal. Nucl. Chem.* 263: 759–765.

Sowerby, B. D., Cutmore, N. G., Liu, Y., Peng, H., Tickner, J. R., Xie, Y., and Zong, C. 2009. Recent developments in fast neutron radiography for the interrogation of air cargo containers, in *IAEA Conference*, Vienna, Austria, May 4–8, 2009, Paper SM/EN-01.

Star-Lack, J., Sun, M., Kaestner, A., Hassanein, R., Virshup, G., Berkus, T., and Oelhafen, M. 2009. Efficient scatter correction using asymmetric kernels. *Proc. SPIE* 7258: 72581Z.

Verbeke, J. M., Lee, Y., Leung, K. N. et al. 1999. Neutron tube design study for boron neutron capture therapy application, in *Proceedings of the 1999 Particle Accelerator Conference*, New York, pp. 2540–2542. March 29–April 2.

Von der Hardt, P. and Röttger, H. (Eds.). 1981. *Neutron Radiography Handbook*. D. Reidel Publishing Company, Dordrecht, the Netherlands.

Watterson, J. I. W. 1997. Computational modeling and the assessment of fast neutron radiography, in Duggan, J. L. and Morgan, I. (Eds.) *Application of Accelerators in Research and Industry*. AIP Press, New York, pp. 909–912.

Whetstone, Z. D. and Kearfott, K. J. 2011. Use of multiple layers of repeating material to effectively collimate an isotropic neutron source. *Nucl. Technol.* 176: 395–413.

Wootton, P. 1980. Chairman Task Group No. 18 Fast Neutron Beam Dosimetry Physics, Radiation Therapy Committee. Protocol for Neutron Beam Dosimetry, American Association of Physicists in Medicine, AAPM Report No. 7.

Index

D

G

H

I

R

S

T

Printed in the United States
by Baker & Taylor Publisher Services